International Symposium of the Society of Core Analysts (SCA 2019)

Core Analysis in a Digital World

E3S Web of Conferences Volume 146 (2020)

Pau, France
26 - 30 August 2019

Editors:

Josephina Schembre-McCabe
Holger Ott

ISBN: 978-1-7138-0685-1

Printed from e-media with permission by:

Curran Associates, Inc.
57 Morehouse Lane
Red Hook, NY 12571

Some format issues inherent in the e-media version may also appear in this print version.

For additional information, please contact EDP Sciences – Web of Conferences at the address below.

EDP Sciences – Web of Conferences
17, Avenue du Hoggar
Parc d'Activité de Courtabœuf
BP 112
F-91944 Les Ulis Cedex A
France

Phone: +33 (0) 1 69 18 75 75
Fax: +33 (0) 1 69 28 84 91

contact-edps@webofconferences.org

Additional copies of this publication are available from:

Curran Associates, Inc.
57 Morehouse Lane
Red Hook, NY 12571 USA
Phone: 845-758-0400
Fax: 845-758-2633
Email: curran@proceedings.com
Web: www.proceedings.com

Pau, France

Aug 26 – 30, 2019

Final Program

PAUFRANCE
2019
SCA SOCIETY OF CORE ANALYSTS

Core Analysis in a Digital World

From the 2019 President

Dear Colleagues,

On behalf of the Society and the Board of Directors, I am delighted to welcome you to the 2019 International Symposium of the Society of Core Analysts (SCA). This year's symposium theme is "Core Analysis in a Digital World". The symposium is being held in the beautiful French city of Pau, situated just north of the Pyrenees Mountains and home to Total's research group.

Special thanks to this year's VP Technical (I told you it was a lot of work), Holger Ott, VP Arrangements, Benjamin Nicot, and (of course) Melanie Young. A symposium such at this does not happen on its own. I am confident you will find this year's symposium as rewarding and fun as previous ones.

With the ups and down of our industry it is becoming increasingly important to rely on science and technology in our decision-making process. I firmly believe that integration of technologies, core analysis, well logging, digital rock physics, etc., is the answer to full understanding of our reservoirs. I have enjoyed coming to the SCA symposium (my first one was in Toronto 2005) over the years and developing long lasting friendships and colleagues.

I would like to thank the board and all members for my year as president. It has been a pleasure to contribute back into a society that I have gotten so much from.

Safe travels,

Derrick Green

SCA President 2019

From the 2019 VP Arrangements

Bienvenue à Pau!

On behalf of the Organizing Committee, I would like to welcome the attendees of the 2019 SCA Annual Symposium in the Palais Beaumont in Pau. We hope that this city full of history and scenic views over the Pyrenees Mountains will stimulate both young and seasoned colleagues from all around the world under our Symposium theme "Core Analysis in a digital world".

During our conference in the heart of Pau, you will hopefully discover and enjoy a city with many different aspects and lots of history. From the Palais Beaumont and the very close Boulevard des Pyrenees you will be able to enjoy the scenic views of the Pyrenees Mountains.

The history of Pau is marked by the birth of Henry of Bourbon in 1553 in the castle that you can visit in the medieval quarter. He gained access to the throne of France in 1589 under the title of Henry IV.

The city also developed during the late 19th early 20th century with a massive influx of foreign tourists. From this period dated the first golf course on the continent, this is where our golf event will be held on Sunday 25th. This period also saw Pau becoming one of the capitals of the nascent aerospace industry under the influence of the Wright brothers. In the city you will see many villas and mansions dating from this period and generally called "les villas anglaises". Our Gala dinner on Wednesday 28th will be held in one of the beautiful villa anglaise called Villa Navarre.

During the late 20th century Pau turned into oil and gas with the development of a big gas field in the nearby city of Lacq. Since then, Pau has been tightly linked to the oil and gas sector with Elf and now Total, and we will visit on Tuesday 27th the latest log calibration center installed in the vicinity of pau, and I would like to thank SPWLA for their help in organizing this visit.

Pau is also a city very much turned into sports with the proximity of the mountains for hiking and skiing, the proximity of the ocean for surfing, and the wild river "gave de pau" flowing in the very city. On Tuesday 27th the young professionals will have a chance to experience wild water rafting at Pau wild water stadium where the 2017 canoe world championship was held! Hold on tight young professionals!

We will discover more of the mountains around and the geology with the traditional Friday field trip.

With a large variety of vendors, technical veterans, industry leaders, and rising young professionals, the SCA meeting is a great opportunity for professional interaction, to share ideas, innovations, knowledge, best practices, products and services. In the exhibition halls the vendors showcase their latest technical innovations and offerings in equipment and services.

I would like to thank the sponsors and vendors for their support, the authors and technical committee of the SCA for their dedication to make the SCA a passionate conference, and a special thank you for all the team from Total who helped in the organization.

Welcome to Pau!

Benjamin NICOT,
VP Arrangements

From the 2019 VP Technology

Dear Colleagues and Friends

Welcome to the 2019 SCA International Symposium in Pau, France!

This year's symposium is titled "Core Analysis in a Digital World". Even though we have long operated digitally, everyone understands something else under digitization and there is much need for discussion. This year's symposium will contribute to it; next to the various classical sessions, we have contributions to the symposium theme starting on Monday with the short courses on numerical interpretation of SCAL measurements and the numerical interpretation of pore scale experiments by topological means. In addition, there will be many occasions such as coffee breaks and social events to discuss digitization.

As you may have noticed, there were some changes in the selection and review processes of the submitted abstracts and manuscripts. With the goal of increasing the quality of the papers, the board decided to ask for an extended number of full manuscripts associated to oral and alternate oral presentations, but no longer for poster presentations. The reactions from the community were mixed; some poster presenters welcomed the idea where others would have liked to publish a manuscript. In addition, the already tough review process has been extended to include a final assessment and ranking of the already reviewed manuscripts. From 150 received abstracts, 44 manuscripts and 47 abstracts were accepted populating the 2019 technical program with 36 orals and 55 poster presentations.

Although I have gone through the SCA review process several times as author and reviewer myself, I was surprised at the number and quality of technical discussions during the review process and the commitment and dedication of everyone! It was a very good personal experience. For this I would like to thank the authors and the Technical Committee who form the backbone of this symposium and who are responsible for the high technical standard of the SCA. Last but not least, my special thanks go to Melanie Young, our SCA Executive Director, for her invaluable support in all matters. With these contributions we have again a solid technical program, and I am sure that we will have excellent presentations and fruitful technical discussions that will help us progressing as a community. I am looking forward to seeing you in Pau!

Holger Ott

SCA VP of Technology

Current SCA Board

EXECUTIVE DIRECTOR	Melanie Young Society of Core Analysts
PRESIDENT	Derrick Green Green Imaging Technologies, Inc
VP TECHNOLOGY	Holger Ott Montanuniversität Leoben
VP MEMBERSHIP	Matthias Appel Shell
VP PUBLICATIONS	Josephina Schembre-McCabe Chevron
VP ARRANGEMENTS	Benjamin Nicot TOTAL
SECRETARY/TREASURER	Christa Vindum Vindum Engineering
PAST PRESIDENT	Lesley James Memorial University of Newfoundland
PRESIDENT ELECT	Benjamin Nicot TOTAL
YOUNG PROFESSIONALS DIRECTOR	Alvaro Munoz-Beltran Weatherford Labs
DIRECTOR-AT-LARGE	Douglas Ruth University of Manitoba
INDUSTRIAL RELATIONS LIAISON	Craig Lindsay Core Specialist Services Limited
MARKETING DIRECTOR	Stefano Pruno Weatherford Labs
CHINA DIRECTOR	Hu Dong iRock Technologies
AUSTRALIA/ASIA DIRECTOR	Christoph Arns University of New South Wales
EUROPEAN DIRECTOR	Jules Reed Lloyd's Register
JAPAN DIRECTOR	Satoru Takahashi JOGMEC
MIDDLE EAST DIRECTOR	Moustafa Dernaika Ingrain Rocks
NORTH AMERICAN DIRECTOR	David Potter University of Alberta
SOUTH AFRICA DIRECTOR	Edith Akwaeke Ansett Integrated Services Limited
U.S. DIRECTOR	Eric M. Withjack PTS Laboratories
U.S. DIRECTOR	Will Richardson Chevron

2019 Technical Committee

The SCA Technical Committee is the backbone of the Society of Core Analysts. Its members rank and evaluate the submitted abstracts and the technical quality of the manuscripts. In the review of each individual manuscript and in-depth discussions with the authors, many hours are spent until a manuscript is accepted. In many cases, the interaction between authors and TC members leads to a higher quality of individual manuscripts. As a result, the annual SCA symposia are packed with high-quality presentations. For this, sincere acknowledgement go to this year's Technical Committee members:

Adam Moss
Arjen Cense
Benjamin Nicot
Carl Songergeld
Cyril Caubit
David Potter
Derrick Green
Doug Ruth
Evren Unsal
Eric Withjack
Fabrice Pairoys
Gerald Hamon
Hassan Mahani
Hendrik Roler
Holger Ott
James Howard
John Shafer
John Mills
Jos Maas
Josephina Schembre-McCabe
Kevin Flynn
Marc Fleury
Matthias Appel
Matthias Halisch
Mohammed Ameen
Ole Torsæter
Olga Vizika
Olivier Lopez
Patrick Egermann
Roland Lenormand
Ryan Armstrong
Souhail Yussef
Stefano Pruno
Steffen Berg
Ted Braun
Will Richardson

Technical Achievement Darcy Award Recipient
Ted Braun

Ted Braun obtained a BS degree in Chemical Engineering from Case Western Reserve University in 1976, and immediately began a 34-year career at Exxon Production Research Co., which later became ExxonMobil Upstream Research Co. He spent his first two years researching in-situ combustion as an enhanced oil recovery technique, and the remainder in core analysis. After formally retiring from the company in 2010, he has worked as an independent consultant.

His first assignment was to develop the capability to measure relative permeability by the steady-state method at reservoir temperature and pressure, in order to obtain the most reliable data possible for the Prudhoe Bay field. A series of other research areas followed, each targeting specific data needs of the company:

- Accurately measuring oil saturations and compositions in gas cap cores.
- Defining hysteresis behavior in oil-water relative permeability.
- Restoring the wettability of cores contaminated by oil-based mud.
- Measuring relative permeability for gas condensate reservoirs.
- Measuring shale permeabilities in the nano-Darcy range and their response to mechanical stress.

In addition, he has taught in Exxon in-house training courses and served as a mentor for many of the company's core analysts and reservoir engineers. A frequent topic of advice was to apply the principles of chemical engineering and geochemistry to laboratory core analysis measurements.

Ted has authored or co-authored seven SCA papers and nine SPE papers. He is a lifetime member of the Society of Core Analysts. He has held a series of SCA offices, beginning with the Vice President of Technology for the 2010 Symposium in Halifax, and progressing to president-elect, president, and past-president. Among his colleagues over the years have been four earlier recipients of the SCA's Darcy Award / Technical Achievement Award.

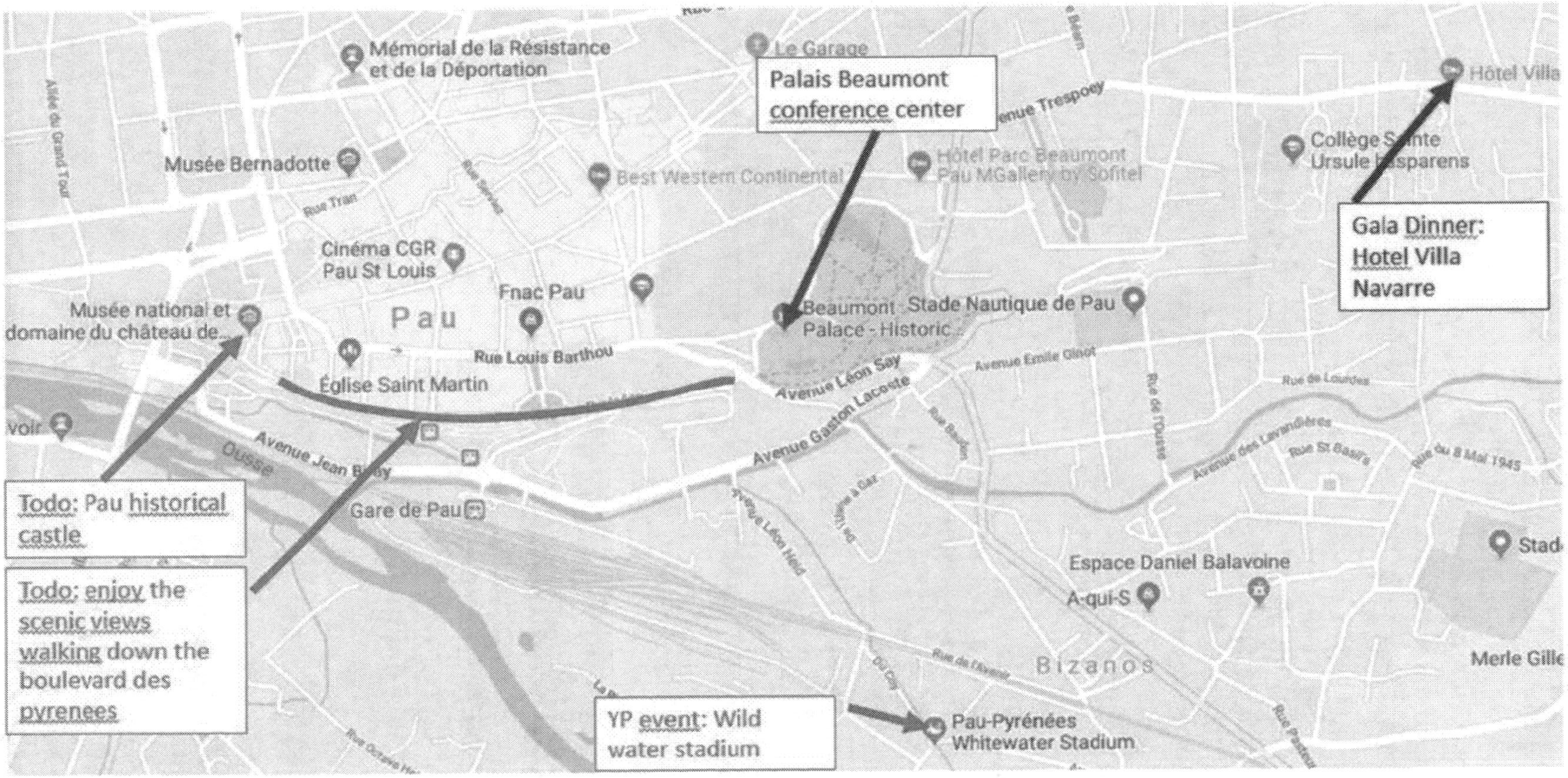
Palais Beaumont conference center
Gala Dinner: Hotel Villa Navarre
Todo: Pau historical castle
Todo: enjoy the scenic views walking down the boulevard des pyrenees
YP event: Wild water stadium
Mémorial de la Résistance et de la Déportation
Le Garage
Hôtel Villa
Musée Bernadotte
Best Western Continental
Hôtel Parc Beaumont Pau MGallery by Sofitel
Collège Sainte Ursule Hasparens
Cinéma CGR Pau St Louis
Fnac Pau
Pau
Musée national et domaine du château de...
Beaumont Palace - Historic...
Stade Nautique de Pau
Rue Louis Barthou
Église Saint Martin
Avenue Léon Say
Avenue Gaston Lacoste
Avenue Emile Gilnot
Rue de Lourdes
Rue de l'Ousse
Avenue des Lavandières
Rue St Basil's
Rue du 8 Mai 1945
Avenue Trespoey
Allée du Grand Tour
Rue Tran
Avenue Jean Biray
Ousse
Gare de Pau
Avenue Léon Heid
Espace Daniel Balavoine
A-qui-S
Bizanos
Merle Gille
Pau-Pyrénées Whitewater Stadium
Rue Pasteur

Golf – Sunday 25th, 11:00 a.m. – 5:00 p.m.

Pau Golf Club is the oldest club in continental Europe. The development of this sport, can be attributed to the founding in 1856 of the Pau GOLF CLUB. During the Napoleonic wars and the Battle of Orthez in February 1814, some regiments of WELLINGTON's army were quartered in Pau.

Among the soldiers, two Scottish officers who were never separated from their golf clubs and took every opportunity to practice their favorite sport. During their travels, they were impressed by the beauty of the countryside around Billère close to the gates of Pau. They decided that the banksz of the mountain stream, with the Pyrenees as a backdrop would make an ideal setting for a golf course

Short Courses – Monday 26th, 9:00 a.m. – 1:30 p.m.

The short courses are on the numerical interpretation of data, and in more detail: on "Numerical History Matching of SCAL Data; Best Practice" and on "Interpretation of Pore Scale Experiments; Alternative Descriptions of Porous Media Flow by Topological Means". A special thanks goes to Jos Maas, Jules Reed, Ryan Armstrong and Steffen Berg for the preparation of the course material and for organizing and presenting the short courses.

Coffee and tea will be available during breaks. Lunch will be provided to attendees.

Short Course is Kindly Sponsored by: Thermo Fisher Scientific

Opening Reception –Monday 26th, 5:30p.m. - 9:00 p.m.

The Opening "Icebreaker" Reception will be held at the restaurant "la belle époque" on the ground floor of the conference center Palais Beaumont. Snacks and drinks will be served to foster a relaxed atmosphere to meet and greet both old and new colleagues.

Dress Code: Business Casual

Technical Sessions – Monday 26th, Tuesday 27th, Wednesday 28th and Thursday 29th, 8 a.m. - 5 p.m.

(With the exception for start at 1:30 p.m. on Monday and end at 4:15 p.m. on Thursday).

Oral presentations: The Symposium will offer 36 oral presentations, 20 minute presentations followed by 5 minutes for discussion.

Poster sessions: The symposium will offer 55 posters, distributed in two poster sessions. Authors of posters will present their work in 2 minutes presentation format on Tuesday (Poster session Even) and Wednesday (Poster Session Odd).

Young Professionals Event – Tuesday 27th, 6:00-11:00 p.m.

Pau-Pyrénées Whitewater Stadium (French: Stade d'eaux-vives Pau-Pyrénées) is the home training facility for the French national canoe slalom team. It was first used to train the French team for the 2008 Summer Olympics in Beijing. It 2009, it was the first of three venues used in the canoe slalom World Cup.[1] It is also a whitewater park for recreational use by the general public.

> A 5,000-m^2 stretch of water,
> A competition course, an arrival pool and an initiation pool

We will be experiencing wild water rafting! Please bring your swimming suit! Wet suits will be provided if necessary. After this high adrenaline experience, we will enjoy a picnic on the riverside with local food and drinks.

As this is a sport event, there will be a 15min walk from the Palais Beaumont.

Visit of the log calibration center – Tuesday 27th, 6:00-9:00 p.m.

Alternative logging is a cost responsible and innovative way to bring logging companies from diverse energetic domains into the oil and gas environment. By boosting the ecosystem of service companies and manufacturers in logging, Total E&P ASD has built a logging calibration facility to help suppliers to boost business opportunities.
Located in Artigueloutan, in the south-west of France, this Calibration Center is open to all oil and gas companies like manufacturers, service companies, IOCs, NOCs, but also to all energetic domains like hydrogeology, geothermic and mining.
The Center provides 29 standards in 18 silos: rocks, fluids, sand pack, aluminum and composite. These standards are used for the characterization of a large number of logging probes and metrological monitoring. The rock standards used in our workshop are carefully prepared following a certified drying and soaking process. Each rock is chosen for its mineralogical, geochemical and petro physical properties, a large range of lithology (11 types) as limestone, dolomite, sandstone…

Buses are arranged, departure from the Palais Beaumont conference center

Awards Gala Dinner – Wednesday 28th, 6:30 - 11:00 pm

The gala dinner will be held in the gardens of the Villa Navarre hotel, located avenue de trespoey, 15 minutes' walk from the Palais Beaumont conference center. There we will enjoy some typical food and drinks from the region.

Dress code: (Smart) Casual dress code.

Gala Dinner is Kindly Sponsored by: iRock Technologies

Accompanying Persons Tours and Activities

Monday, August 26th

5:30 PM - Opening Reception

Tuesday, August 27th

Departing from the Palais Beaumont at 8:00am and Returning at 4:00pm

Morning

A panoramic guided tour of the city by bus followed by a visit to the castle where the "good King Henri" was born. With sections dating back 1000 years and a unique collection of tapestries, the castle is listed as a National Museum.

Lunch "Poule au pot" (chicken stew), the traditional Béarnaise dish. (as an option)

Afternoon

Walking tour of the historic old quarter, around the castle and along the Boulevard des Pyrenees. Continue by bus to the Jurançon vineyards for a tour and some wine-tasting at a renowned winery.

Wednesday, August 28th

6:30 PM – Gala Dinner

Thursday, August 29th

Departing from the Palais Beaumont at 8:00 am and returning at 6:00 pm

Morning

Meeting with your guide. Departure from Pau to Saint-Jean-de-Luz by the A64. Walking tour in Saint Jean de Luz: the traditional charming fishing harbour, its beautiful sandy beach in front of the ocean...

Lunch at the restaurant in a village of the Basque coast.

Afternoon

Panoramic tour of Biarritz.
Continuation to Bayonne, visit of the port city, the cathedral...
Transfer to Pau.

Optional Field Trip – Friday August 30th, 8:15 a.m.- 6:00 p.m.

The Ossau valley is the ideal site for a fieldtrip which deals with the reservoirs and the source rock of the Aquitaine oil and gas fields. The advantage of this option is that the transit time from Pau to the visited sites is short. Moreover, if the weather is good, the scenery is spectacular and allows various comments (Tectonic, glacier geomorphology etc.). However, we have chosen to focus our attention on subject related to cores interpretation showing (among other aspects) remarkable carbonates reservoirs, in two main stops: one in the morning requiring some walking effort, the other in the evening in the village of Arudy.

We propose to visit two sites that do not require special physical conditions. However adapted shoes are recommended, as part of the trip requires a short walk (3-4km) along natural tracks. This field trip should be combined with a light lunch (picnic style) to allow a full day entertainment. Due to the large spectrum of interest it may satisfy a wide range of people. And for those that do not want any more technical knowledge, Pyrenean Nature will be discovered by your visitors, in a relaxed atmosphere.

The field trip will include lunch.

Publication of Proceedings

The proceedings are prepared in USB format and will be given at the Symposium to all registered participants. Additional USBs may be ordered the SCA web site:

www.SCAweb.org

The SCA has decided to no longer carry printed copies of the proceedings.

Exhibition Hours

Sunday: 2:00 p.m. – 7:00 p.m.	*Exhibition Build-Up
Monday: 8:00 a.m. – 5:30 p.m.	*Exhibition Build-up starts at 8am
Tuesday: 8:30 a.m. – 5:00 p.m.	
Wednesday: 9:00 a.m. – 5:00 p.m.	
Thursday: 8:00 a.m. – 4:05 p.m	*Exhibition Break-down can start after the afternoon break and complete by 6pm.

Exhibitors

AMETEK Chandler Engineering – AMETEK Chandler Engineering produces the highest quality instruments and measurement systems for the Oil and Gas Industry. Our portfolio includes Quizix precision pumps used in core flooding (NEW models available!), SCAL, EOR, and other fluid delivery applications, the 6100 Formation Response (formation damage) systems, Core flood/EOR flow systems, custom core flow systems for Steady State/Unsteady State permeability measurements, and the Chandler 3000 Series PVT systems for phase behavior studies.

Bruker microCT - Bruker microCT develops and produces a wide range of high-end microtomography instruments for life science, material research and in-vivo preclinical studies.
The company is market-leader in the field of desktop micro-CT systems, bringing high-resolution 3D images straight to your desk.
In July, Bruker's latest high resolution nanotomography system was launched: the SKYSCAN 2214 features up to 4 X-ray detectors, integrated in a maintenance-free and intuitive to use system.

CYDAREX - Since 2005, CYDAREX provides expertise in the domain of core and cuttings analysis. CYDAREX is a spin-off company of the French Institute of Petroleum (IFPEN), and is now a completely independent company. CYDAREX provides consulting and training services, commercializes laboratory and training equipment, and develops the software CYDAR™.
Teaching: CYDAREX commercializes a series of teaching equipment, designed to demonstrate flooding experiments (permeability measurements, two-phase flow, porous-plate and electrical experiments) in a rapid and safe laboratory setting.

Equipment: DarcyPress™ allows measurements of gas permeability under confining pressure on small rock samples (5 to 10 mm), with permeabilities ranging from 0.1 nanoDarcy to 10 Darcy. DarcyPress™ is ideal for measurements on shales. DARCYLOG™ allows measurements of permeability on drill cuttings, with sample size ranging from 1 to 5 mm, and permeability ranging from 0.05 to 100 mDarcy.
Software: CYDAR™ allows the design and the interpretation of all conventional and special core analysis experiments (MICP, Porous Plate, Rel Perms, centrifugation...). CYDAREX has recently implemented a module for Two-Phase Flow experiments with EOR (Enhanced Oil Recovery) capabilities for polymer flooding, low salinity, and surfactants injection experiments. CYDAREX is introducing an original porous plate device allowing measurements of gas or oil drainage Pc curves in less than one day. The main application is to replace mercury injection experiments for reservoir samples.

DCI Corporation - DCI offers innovative solutions to your core testing requirements. From turnkey systems to system components we can help you with your laboratory needs. DCI designs and manufactures both custom and standard core holders, accumulators, syringe pumps, acoustic separators, core flood systems, electrical resistivity systems, rock mechanics systems and much more. Stop by our booth to see how DCI can help you make better measurements on core properties.

Dassault Systèmes - Dassault Systèmes SIMULIA reveals the world we live in through realistic simulation of product, nature & life. We provide high-value end-to-end industry processes for digital engineering that employ state-of-the-art connected multidisciplinary-multiscale simulation applications. With SIMULIA, customers can reduce testing, increase confidence & quality, and get to market faster using always-available virtual worlds for discovery and testing. As an integral part of the 3DEXPERIENCE platform, SIMULIA applications connect directly to both design data and product requirements allowing simulation to power sustainable innovation at all stages of the product lifecycle. Our global team of simulation experts provides services, support, and training to help customers meet their business goals. www.3ds.com/simulia

FloXlab - Floxlab is an engineering firm specialized in the design and manufacture of an ample range of products encompassing advanced geotechnical testing equipment, high precision syringe pumps and compression frames. Whether a unique lab-scale unit or a complete turnkey system is required, our clients will benefit from our unrivaled depth of technological knowledge and experience as well as unparalleled customer service. With over 90% of our operations lying outside of France, Floxlab systems have become a standard in all major laboratories, with a strong presence in the U.S., Europe, the Middle East, China, and Russia. Our customers would agree that we are surely the world's foremost pump designer and manufacturer. Sectors benefiting from our devices' precision in pressure, volume and flow rate include: Petroleum research, reactant feeding, supercritical fluids, geoscience etc. At Floxlab, our philosophy is continuous improvement through constant market demand monitoring, higher quality material prospection, and meticulous customer feedback collection and analyses.

GeoTek - Geotek specialises in the non-destructive analysis of geological cores. We supply our range of Multi-Sensor Core Logger (MSCL) and X-ray CT systems that use multiple geophysical and geochemical sensors to rapidly and automatically gather measurements on sediment or rock cores. The rugged nature of the equipment makes it suitable for use in either an onshore laboratory/repository environment or onboard survey and drilling vessels. Stop by our booth and discuss our dedicated core plug CT system PlugXcan, and find out about our brand new system BoxScan, specifically designed to analyse your core trays accurately, automatically and efficiently.

Green Imaging Technologies, Inc – Green Imaging Technologies is the industry leader in NMR rock core analysis. Our software is the backbone of the GeoSpec line of NMR Rock Core Analyzers, which are used by all major oil producers, oil service companies and the most active research institutes worldwide. Our customers have access to exclusive, patented measurements including NMR capillary pressure and quantitative saturation profiles. For those that don't have their own laboratory, or service companies who want to add NMR to their product offerings, H2 Laboratories, a subsidiary of Green Imaging, leverages our expertise and patented technologies to offer NMR rock core analysis services both directly to customers, and via industry partners. Our team of experts are completely focused on rock core analysis, and as such have developed relationships with the most respected researchers and experts in the NMR rock core analysis field. At this years' SCA we will be showcasing new applications, including a new method for determining methane isotherms in shales and a technique to reducing sample heating during NMR measurements. Stop by the Green Imaging/Oxford Instruments booth in the exhibit hall to find out what is new in NMR rock core analysis.

HOT Microfluidics GmbH - HOT Microfluidics GmbH (HOT) is the leading provider of turnkey microfluidic solutions for IOR/EOR applications. The pioneering, fast and cost-effective Rock-on-a-Chip approach provides unmatched EOR process visualisation and chemicals screening capabilities and yields an unrivalled understanding of EOR processes, pore-scale oil mobilisation and trapping mechanisms.

Math2Market GmbH - Our vision is to replace laboratory analysis of material properties by material models and software. The mission is to offer a complete solution for Digital Rock Physics (DRP). For that mission, we develop our own software GeoDict and provide it to our clients together with hardware consulting, customized software development and support, and user training. The DRP suite of GeoDict is a purpose-built tool for the specific needs of core analysts. In combination with imaging capabilities, GeoDict enables you to perform the entire DRP workflow in-house, representing an alternative to service providers' solutions. GeoDict comprises a broad portfolio of DRP parameters, provides tools for the automation of workflows, and is offered as a service on request. In combination with its unmatched fast runtimes and low hardware requirements, GeoDict is designed to bridge the gap between "hand-made" and automated generation of DRP data.

Object Research Systems - Founded in 2004 and based in Montreal, Object Research Systems (ORS) Inc., develops advanced 3D visualization solutions used by research centers, engineering groups, and healthcare facilities to process, visualize, and analyze scientific and medical image data.
Our development team operates in compliance with ISO and IEC standards to ensure quality and processes that are reliable and reproducible.
Our partners rely on our products to extract the most value from their imaging system investments, to solve complex industrial and research problems, and to address medical imaging challenges. Our services range from product training to custom development and consulting and our software is deployed by registered users in more than 80 countries.

Oxford Instruments – Oxford Instruments will display the renowned GeoSpec line of NMR rock core analysers, including GeoSpec12 which offers ten times greater sensitivity and 100 times faster measurements on tight rocks and low porosity samples. The GeoSpec range measures standard core parameters such as pore size distributions, BVI, FFI, porosity, and T2 cut-off on a single instrument, and can perform advanced measurements such as capillary pressure and spatially resolved T2 distributions with the exclusive use of Green Imaging Technologies' software. We will also be available to discuss new applications such as gas isotherms, wettability and relative permeability.

PanTerra – PanTerra is an integrated laboratory, geosciences, and engineering consultancy serving the international oil and gas industry for more than 25 years. Our services include conventional and special core analysis, PVT, production chemistry, Enhanced Oil Recovery, subsurface evaluation and modelling, field development, souring studies, engineering and project management services. Capitalising on our in-house expertise, PanTerra also specialises in recruitment and secondment of subsurface professionals and additionally offers a unique blend of E&P learning customised to individual needs. For more information please visit www.panterra.nl or connect via LinkedIn, Twitter and YouTube.

Premier Oilfield Group - WE ARE PREMIER - We believe the data-driven oil and gas industry is here, and that generating and sharing relevant data from rock and fluid samples is the key to more effective and efficient hydrocarbon development.
Premier Oilfield Group integrates reservoir rock and fluid measurements across multiple disciplines to deliver completions and reservoir solutions. Our proven workflows deliver timely, relevant data by doing the right experiments, doing those experiments right, and providing client-ready, fit-for-use results. We generate, aggregate, interpret, and deliver consistent, high-quality data that are of immediate value to oil and gas companies. We've assembled world-class experts, facilities, and platforms to produce that data, make it readily accessible, and help our clients share it effectively. These new ways of doing business set Premier apart from other testing-focused laboratories.

Reactiv'IP - Reactiv'IP offers to public and private research centers, software solutions to automate the analysis of their images. Our solutions allow users to replace a manual inspection or counting operation with a completely automatic operation. Thanks to our algorithms, our solutions can support very large volumes of 2D or 3D imaging data. These accelerations are made possible by the IPSDK library, flagship product of the company. IPSDK is used by a lot of major companies. Now, discover 'IPSDK Explorer' the new IPSDK GUI. This graphical interface will help you to use the IPSDK features effectively. In addition, IPSDK Explorer will automatically generate for you the Python scripts.

Rotunda Scientific Technologies LLC - Rotunda Scientific Technologies LLC provides innovative radiation measurement and protection products to the energy industry. This year we will be demonstrating the JCS GMS312 Rolling Gamma Spectrum Core Analyzer, which is based on the field proven design of the GMS310 Gamma Spectrum Core Logger handheld. Both the GMS310 and GMS312 are designed for core sample analysis (spectral gamma, API and Percent Concentration) in the rugged environment of the exploration site and will be available at our booth for you to evaluate. The latest firmware & software will provide enhancements to your operation. In addition to the GMS310 & GMS312 we offer many other state of the art radiation detection and protection products for use during exploration or nondestructive testing. Several of the radiation detection products are intrinsically safe for potentially explosive atmospheres. You are invited to see these products at our booth #5 during the SCA Annual Meeting. We look forward to meeting you for the first time or seeing you again and catching up.

Spectra-Map - Spectra-Map provides unique mineral analysis services to the upstream O&G sector, using visible and infrared reflectance imaging spectroscopy (IRS). IRS identifies a range of minerals including all clay polytypes, carbonates and sulphates, as well as solid and liquid hydrocarbons. It is well suited to analysis of cores, cuttings and plugs, generating continuous high density mineral data (at mm scale) without sample preparation or damage. Imaging provides the ability to see spatial relationships between different minerals and oils. The company has invested heavily in other spectroscopy techniques in order to expand our range of services, with a view to providing comprehensive non-destructive whole rock mineralogy on sub millimetre to metre scales.

Strata Technology – Strata Technology is one of the industry's leading specialist engineering companies. Strata designs, manufactures, commissions, inspects and maintains bespoke laboratory equipment, skid mounted rigs and pilot plants. Our equipment is used to undertake critical experiments into enhanced oil recovery, drilling operations and other Oil & Gas related research and development activities, where high pressure and high temperature are key requirements. In addition to the bespoke engineering services we offer, Strata's team of highly trained and skilled designers, engineers and technicians have developed a number of standard, CE Marked products including:

- DCP50 Pump – your ideal replacement to the Pharmacia P-500 for core preparation &

flooding applications

- Core Holders – off the shelf or bespoke designed solutions
- Piston Vessels – 0.5 to 4 litres capacity, operating at high pressures and temperatures

We are also the exclusive distributor of Quizix Pumps in the UK and France and the exclusive European distributor of Komax Systems's range of Static In-Line Mixers.

Stratum Reservoir – Stratum Reservoir supports energy resource investment and divestiture strategies worldwide through the analysis and interpretation of rocks and fluids. We specialize in reservoir characterization, laboratory services, and instrumentation designed to deliver scientific insights in upstream, midstream, and mining operations. In addition to software and equipment, we offer expertise in these areas:

Field Services - We accommodate our client's requests during handling and sampling to ensure materials are preserved, shipped, and stored according to specifications.

Laboratory Services - Our laboratories are equipped and expertly staffed to manage and execute your analytical programs.

Interpretive & Advisory Services - We have a wealth of reference data, domain knowledge and extensive experience that guides and informs our interpretation of the various analyses we conduct.

Some of the services we offer include: Routine Core Analysis, Special Core Analysis, Geochemical Analysis & Interpretation, Petrographic Analysis & Interpretation, Express Lab Services, PVT, EOR, and Storage Services

Synergie4 / BRUKER Nano - Synergie4, a French company created in 1994, is specialized in Electron Microscopy and high performance microanalysis systems (EDS, EBSD, μXRF, WDS). We provide installation, training and service on site for all our products since 25 years.
We are the official representative of Bruker Nano and μCT, Hirox, Denton Vacuum and Technoorg Linda:

- MICROANALYSIS (EDS, WDS, μXRF and EBSD) from Bruker Nano
- TABLE TOP ELECTRON MICROSCOPE (SEM) from Hirox
- X-RAY MICRO-CT from Bruker μCT-SkyScan
- COATING SYSTEMS from Denton Vacuum
- ION BEAM POLISHING SYSTEM (SEM-TEM) from Technoorg Linda
- SERVICE and TRAINING

We are continuously looking for the complete satisfaction of our customers by providing the best solutions and the service support.

TESCAN XRE - TESCAN's leading-edge 3D X-ray micro-computed tomography (micro-CT) systems are designed to meet application demands across a variety of fields by facilitating high performance 3D and time-lapse imaging methods. UniTOM, CoreTOM and DynaTOM are demonstrating leadership in dynamic CT, enabling high-speed tomography and materials evolution studies under various in situ sample environments. TESCAN XRE, a subsidiary of TESCAN ORSAY HOLDING (Brno, Czech Republic), was formed via acquisition of XRE NV in 2018 and is based in Ghent, Belgium.

Thermo Fisher Scientific – Thermo Fisher Scientific is the world leader in serving science. Our mission is to enable our customers to make the world healthier, cleaner and safer. Through our Thermo Scientific brand, we help customers accelerate innovation and enhance productivity.

Powerful and built for multi-scale, multi-modal data, providing built-in automated workflows and unique flexibility, Thermo Scientific™ PerGeos Software allows geoscientists and petrophysicists to visualize and analyze 2D/3D digital rock imagery, perform image-based simulations to improve reservoir quality evaluations and receive rapid insight into static and dynamic rock properties that impact production.

Universal Medical Systems - A Merry X- Ray Company– Universal Systems bring over 30 years of industry-specific experience providing computed tomography solutions for special core analysis. Ranging from Micro and Robotic CT for fluid flow, portable compact CT, and rapid multi-slice systems for general characterization. Universal delivers rock-ready, turnkey CT solutions. http://www.universal-systems.com/

Vinci Technologies - Vinci Technologies is a well-established engineering firm specializing in the manufacture of equipment for the Petroleum, mining and geology sectors. We operate in over one hundred countries and our clientele ranges from research universities to the most well-known service companies. The company comprises three business divisions, namely: Exploration & Production, Geophysical Tools and the Pilot Plants. We work in E&P field and the number of products that we offer exceeds 200. These are arranged in five major categories: Conventional Rock Analysis, Advanced Rock Analysis, Fluid Analysis, Geochemistry and Pumps.

Vindum Engineering Inc. - Vindum Engineering Inc. – MANUFACTURER OF PRECISION HIGH PRESSURE PULSE-FREE PUMPS, VALVES AND FLUID FLOW EQUIPMENT

- Vindum VP Series Pumps: High Precision, Continuous Pulse-Free Metering Pumps
  - Latest generation design: lower cost, higher performance
  - Models up to 25,000 psi
  - Ambient Temperature or High Temperature (up to 160C) Options
- CV Valves: Constant-Volume, High-Pressure Valves
  - Air activation
  - High Temperature applications (up to 300C)
  - Gas or Liquids, 2-Way and 3-Way Models
- MV Valves: Modular Manual Valve System
  - Mountable, compact design saves space
  - O-ring seals: replaceable, various seal materials
  - Lower Cost Hastelloy Design
- Acoustic Separators: High Pressure 2-Phase & 3-Phase
  - New signal detection system for increased accuracy
  - Hastelloy or Stainless Steel

Booth Layout

1. Vindum Engineering
2. Vindum Engineering
3. Green Imaging Technologies, Inc.
4. Oxford Instruments
5. Rotunda Scientific Technologies
6. AMETEK Chander Engineering
7. Strata Technology
8. Reactiv'IP
9. DCI Corporation
10. CYDAREX
11. HOT Microfluidics GmbH
12. Dessault Systemes
13. Spectra-Map Ltd
14. Math2Market GmbH
15. Object Research Systems
16. Universal Medical Systems
17. Thermo Fisher Scientific
18. Geotek Limited
19. PanTerra Geoconsultants B.V
20. Bruker micro-CT
21. Vinci Technologies
22. FloXLab
23. Premier Oilfield Group
24. Synergie4 / Bruker
25. TESCAN
26.
27.
28.
29.
30. Stratum Reservoir

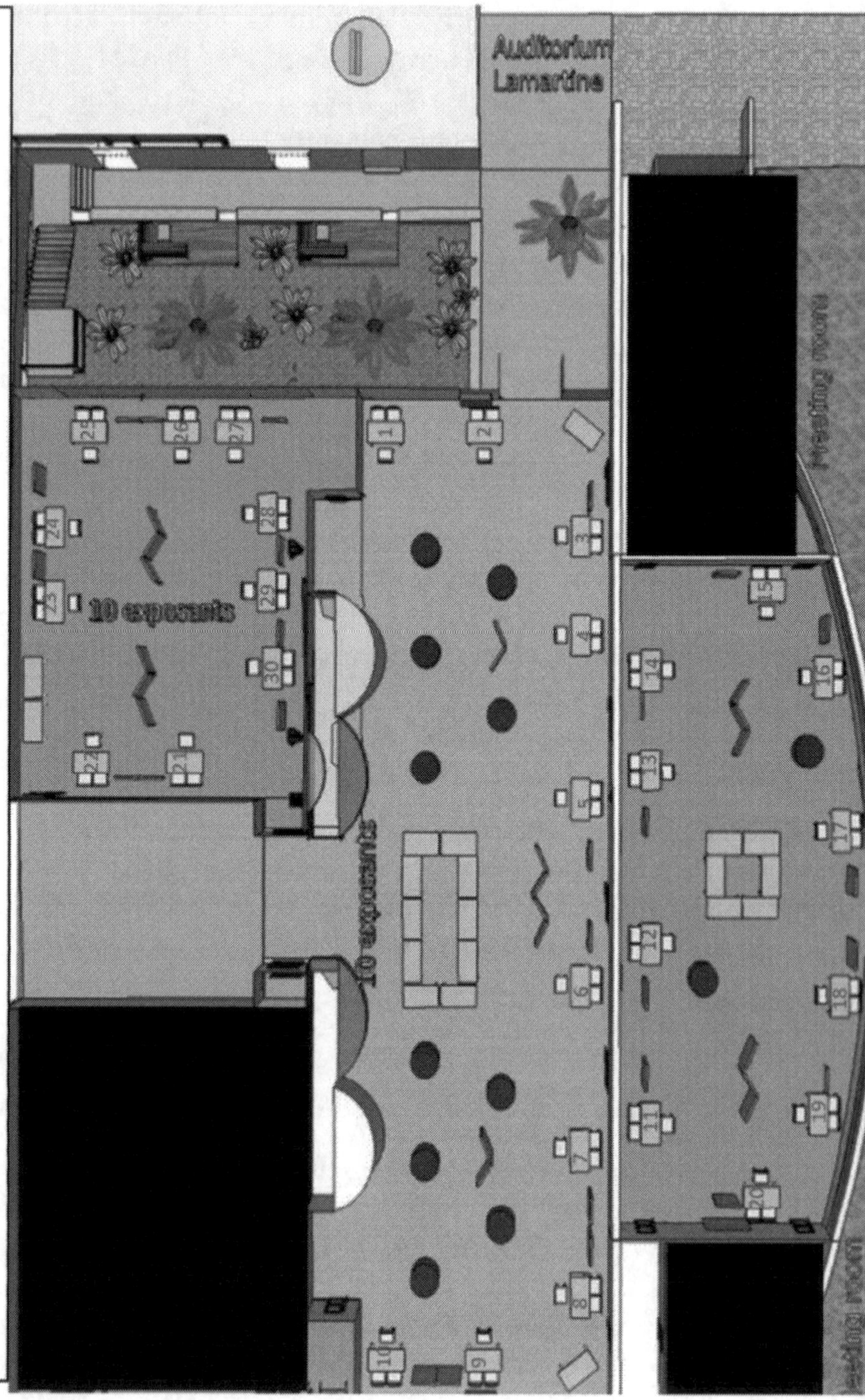

PREFACE

The 2019 International Symposium of the Society of Core Analysts (SCA) was held in Pau, France between August. 26[h] and 30th. The theme for the Symposium in 2018 was *"Core Analysis in a Digital World"*. The symposium was attended by representatives from industry, academia and organizations and vendors. With a large variety of vendors, technical veterans, industry leaders, and rising young professionals, the Annual meeting of the Society of Core Analysts is a great opportunity for professional interaction, to share ideas, innovations, knowledge, best practices, products and services.

The collection of articles published in the SHS Web of conferences consists of 23 articles prepared for oral presentations. Twenty-one (21) proceedings presented in this symposium, published in Petrophysics or other journals, and (47) posters are not included in this collection. Here is the list of articles with full text available in www.scaweb.org and/or other journals.

Papers presented at the 2018 SCA Symposium not submitted / published in other journals

Pore-scale imaging and determination of relative permeability and capillary pressure in a mixed-wet carbonate reservoir rock at subsurface conditions
A. M. Alhammadi, Y. Gao, T. Akai, M. J. Blunt, and B. Bijeljic

Workflow for upscaling wettability from the nano- to core-scales
M. Rücker, W.-B. Bartels, T. Bultreys, M. Boone, K. Singh, G. Garfi, A. Scanziani, C. Spurin, S. Yesufu, S. Krevor, M.. J. Blunt, O. Wilson, H. Mahani, V. Cnudde, P. F. Luckham, A. Georgiadis and S. Berg

The link between microscale contact angle measurements and corescale wettability
C. Sun, J. McClure, M. Shabaninejad, P. Mostaghimi, S. Berg and R. T. Armstrong

Effect of fractures on hot solvent injection in viscous oil: a study using HP-HT micromodel
I. Bondino, G. E. Romero, J.-P. Chaulet, A. Brisset and M. Mujica

Screening of EOR potential on the pore scale by statistical and topological means
H. Ott, A. Kharrat, M. Borji, T. Clemens, and P. Arnold

Estimation of Gas Condensate Relative Permeability using a Lattice Boltzmann Modelling approach
J. Schembre-McCabe, J. Kamath, A. Fager, and B. Crouse

Low permeability measurement on crushed rock: insights
S. Profice and R. Lenormand

Multiphase flow imaging through X-ray microtomography : Reconsideration of capillary end-effects and boundary conditions
F. Nono, P. Moonen, H. Berthet and R. Rivenq

Determination of Critical Gas Saturation by Micro-CT
S. Berg, Y. Gao, A. Georgiadis, N. Brussee, A. Coorn, H. van der Linde, J. Dietderich, F. O. Alpak, D. Eriksen, M. Mooijer-van den Heuvel, J. Southwick, M. Appel and O. B. Wilson

A New Apparatus for Coupled Low-field NMR And Ultrasonic Measurements in Rocks at Reservoir Conditions
P. R. J. Connolly, J. Sarout, J. Dautriat, E. F. May, M. L. Johns

Defining a sample heterogeneity cut-off value to obtain representative Special Core Analysis (SCAL) measurements
J. G. Maas, N. Springer, and A. Hebing

Dielectric Polarisation in Partially Saturated Shales
P. R.J. Connolly, M. Josh, K. O'Neill, E. F. May, M. L. Johns

CT-scan in-situ investigation of waterflood front instabilities during immiscible displacements: effect of viscosity contrast and flow rate
M. Mascle, E. Rosenberg, B. Roboele, E. Kowalewski and S. Youssef

Papers presented in Poster Sessions

Causal Protocols to assess the Viability of Native State or Restored State Preparation
J. Reed, S. Pruno, I. Zubizarreta, R. S. Johansen

Validation of Simulated Relative Permeability Estimates in Very Low Permeability Rocks
N. Drenzek, R. Antler, S. Zhang, J.Howard

Composite Core – An Experimental Approach in Evaluating Ordering Criteria for Individual Cores in Composite
E. Sripal and L. James

Impact of Brine Spontaneous and Forced imbibition on effective permeability in gas shales
D. Dzhafarov and B. Nicot

A new approach to measure the wettability of porous media under different saturation conditions by Temperature Sensitivity of Nuclear Magnetic Resonance Relaxation Time
H. T. Kwak and J. Gao

Wettability - Combine the macroscopic approach to pore-scale analysis
P. Andriamananjaona, P. Moonen, M. Chamerois, and R. Rivenq

Lithologically Controlled Core-Log-Geocell Integration Using Probabilistic Multivariate Clustering Analysis and an Expert System
A. A. Curtis, E. Eslinger, S. Nookala, and F. Boyle

An experimental and digital investigation into the impact of diagenesis above and below the oil-water contact
F. Mujica, L. A. James, C. F. Berg and D. H. C Wilton

Optimizing Magnetic Measurements on Drill Cuttings for Reservoir Characterization
M. Bik and D. K. Potter

Automated Determination of Formation Porosity from Drill Cuttings using Nuclear Magnetic Resonance
J. Billiotte, M. Ammar, K. Bondabou and J. Breviere

Hele-Shaw Setup for Investigation of two-phase flow in fractures
M. Raphaug, P. Bergmo, A. Lavrov, B. Bjørkvik, C. Ringstad

A Review of Effects of Bedding Plane and Anisotropy on Indirect Tensile Strength Test of Rocks
D. Moronkeji, R. Shouse and R. Lew

Pore scale visualization of Improved Sweep Efficiency During Fines Assisted Low-salinity Waterflooding Using Micro-CT Imaging
Y. Wang, M. Yu and F. Hussain

New method for characterising the nano- to macro-scale voidage within black shale and for modelling shale gas recovery efficiency
G. P. Matthews, K. L. Jones and G. M. Laudone

Pore Structure Impact on Polymer Retention in Carbonate Cores
A. M. AlSofi, J. Wang, A. M. Boqmi

Automated Quantitative Micromodel Image Analysis Applied to High- Pressure CO2 Foam Injections
T. L. Føyen, J. Gauteplass, Ø. Eide, M. A. Fernø

Reducing Sample Heating During NMR Measurements
M. J. Dick, D. Veselinovic, and D. Green

Modeling Carbonate Microfractures with the Lattice Boltzmann method
A. Fager, B. Crouse, D. Freed, J. Schembre-McCabe and N. Hurley

Sunday, Aug 25	
11:00 – 5:00	Optional Golf Event – Pau Golf Club

Monday, Aug 26	
8:00 – 5:00	Registration Desk Open

Short Course Program

Numerical History Matching of SCAL Data; Best Practice MONDAY, AUGUST 26, 2019

Numerical interpretation of data	
9:00 – 9:10	Welcome
9:10 – 10:40	Numerical Interpretation of Data...N/A Presenters: Jos Maas & Jules Reed
10:40 – 11:00	Coffee Break Kindly Sponsored by: AMETEK Chandler Engineering
11:00 – 12:30	Interpretation of Pore Scale Experiments; Alternative Descriptions of Porous Media Flow by Topological Means...N/A Presenters: Ryan Armstrong & Steffen Berg
12:30 – 12:35	Closing Remarks
12:35 – 1:30	Lunch Kindly Sponsored by: TOTAL

Short Course Kindly Sponsored by: Thermo Fisher Scientific

Core analysis in a digital world
Technical Program

Monday, August 26, 2018

1:30 - 1:45		Opening Remarks
1:45 – 3:15		Session 1: Core Analysis in a Digital World (1) - Pore Scale Chairs: Oliver Lopez and Josephina Schembre-McCabe
	SCA001	Pore-scale imaging and determination of relative permeability and capillary pressure in a mixed-wet carbonate reservoir rock at subsurface conditions...N/A Amer M. Alhammadi, Ying Gao, Takashi Akai, Martin J. Blunt, and Branko Bijeljic
	SCA002	Density Functional Hydrodynamics in Multiscale Pore Systems: Chemical Potential Drive...1 Oleg Dinariev, Nikolay Evseev, and Denis Klemin
	SCA003	Uncertainty span for relative permeability and capillary pressure by varying wettability and spatiality flow directions utilizing pore scale modelling...11 Thomas Ramstad, Anders Kristoffersen, and Einar Ebeltoft
3:15 – 3:30		Exhibitor Presentations (x5)
3:30 - 3:45		Coffee Break Kindly Sponsored by: AMETEK Chandler Engineering
3:45 – 4:00		Exhibitor Presentations (x5)
4:00- 5:30		Session 2: Improved SCAL Techniques and Interpretation (1) Chairs: Derrick Green and Doug Ruth
	SCA004	Permeability alteration by salt precipitation: numerical and experimental investigation using X-Ray Radiography...23

Olivier Lopez, Souhail Youssef, Audrey Estublier, Jostein Alvestad, Christin Weierholt Strandli

5:30 - 9:00 **Opening Reception "Icebreaker Reception" – Restaurant "La belle Epoque", Palais Beaumont**

Tuesday, August 27, 2018

8:00 – 5:00 **Registration Desk Open**

8:30 - 10:00 **Session 3: Wettability**

Chairs: Steffen Berg and Josephina Schembre-McCabe

SCA007 Workflow for upscaling wettability from the nano- to core-scales...N/A

Maja Rücker, Willem-Bart Bartels, Tom Bultreys, Marijn Boone, Kamaljit Singh, Gaetano Garfi, Alessio Scanziani, Catherine Spurin, Sherifat Yesufu, Samuel. Krevor, Martin. J. Blunt, Ove Wilson, Hassan Mahani, Veerle Cnudde, Paul F. Luckham, Apostolos Georgiadis and Steffen Berg

SCA008 Is contact angle a cause or an effect? – A cautionary tale...58

Douglas Ruth

SCA009 The link between microscale contact angle measurements and corescale wettability...N/A

Chenhao Sun, James McClure, Mehdi, Shabaninejad, Peyman Mostaghimi, Steffen Berg and Ryan T. Armstrong

10:00 – 10:15 Exhibitor Presentations (x5)

10:15 - 10:30 Coffee Break - Kindly Sponsored by: Thermo Fisher Scientific

10:30 – 10:45 Exhibitor Presentations (x5)

10:45 – 12:15 **Session 4: Displacement Mechanisms/EOR/IOR (1)**

Chairs: Jos Maas and Doug Ruth

SCA010 Effect of fractures on hot solvent injection in viscous oil: a study using HP-HT micromodel...N/A

Igor Bondino, Gerardo Emanuel Romero, Jean-Philippe Chaulet, Anne Brisset and Marelys Mujica

SCA011 Screening of EOR potential on the pore scale by statistical and topological means...N/A

Holger Ott, Ahmad Kharrat, Mostafa Borji, Thorsten Clemens, Pit Arnold

	SCA012	Polymer-functionalized Silica Nanoparticles for Improving Water Flood Sweep Efficiency in Berea Sandstones...66 Alberto Bila, Jan Åge Stensen, and Ole Torsæter
12:15 - 1:15		Lunch
1:15 – 2:45		**Poster Session (Even Numbers)** **Chair: Christoph Arns and Ryan Armstrong**
2:45 - 3:45		Coffee and Poster Session Break Coffee Break Kindly Sponsored by: Thermo Fisher Scientific Poster Session Kindly Sponsored by: Green Imaging Technologies, Inc. and Oxford Instruments
3:45 – 4:00		Exhibitor Presentations (x5)
4:00 – 5:30		**Session 5: Pore Scale Imaging and Modelling (1)** **Chairs: Hendrik Roler and Steffen Berg**
	SCA013	Permeability Prediction using multivariant structural regression...78 Matthew Andrew
	SCA014	A fast FFT method for 3D pore-scale rock-typing of heterogeneous rock samples via Minkowski functionals and hydraulic attributes...85 Han Jiang and Christoph H. Arns
	SCA015	Estimation of Gas Condensate Relative Permeability using a Lattice Boltzmann Modelling approach...N/A Josephina Schembre-McCabe, Jairam Kamath, Andrew Fager and Bernd Crouse
6:00 – 11:00 p.m.		Young Professional Event – Pau-Pyrenes, Whitewater Stadium
6:00 – 9:00 p.m.		Visit to the Log Calibration Center

Wednesday, August 28, 2018

8:30 – 5:00	**Registration Desk Open**
8:30 – 10:00	**Session 6: Unconventionals and Shales (1)**
	Chairs: Ryan Armstrong and Jos Maas
SCA016	Low permeability measurement on crushed rock: insights...N/A
	Sandra Profice, and Roland Lenormand
SCA017	Towards Relative Permeability Measurements in Tight Gas Formations...94
	Denis Dzhafarov, and Benjamin Nicot
SCA018	Storing CO2 as solid hydrate in shallow aquifers: Electrical resistivity measurements in hydrate-bearing sandstone...101
	Jarand Gauteplass, Stian Almenningen, and Geir Ersland
10:00 - 10:30	Coffee Break
10:30 – 12:00	**Session 7: Pore Scale Imaging and Modelling (2)**
	Chairs: Oliver Lopez and Souhail Youssef
SCA019	Multiphase flow imaging through X-ray microtomography: Reconsideration of capillary end-effects and boundary conditions...N/A
	Franck Nono, Peter Moonen, Hélène Berthet and Richard Rivenq
SCA020	Local Capillary Pressure Estimation Based on Curvature of the Fluid Interface – Validation with Two-Phase Direct Numerical Simulations...110
	Takashi Akai, Branko Bijeljic, and Martin Blunt
SCA021	Determination of Critical Gas Saturation by Micro-CT...N/A
	Steffen Berg, Ying Gao, Apostolos Georgiadis, Niels Brussee, Ab Coorn, Hilbert van der Linde, Jesse Dietderich, Faruk Omer Alpak, Daniel Eriksen, Miranda Mooijer-van den Heuvel, Jeff Southwick, Matthias Appel, Ove Bjørn Wilson

12:00 - 1:00	Lunch
1:00 – 2:00	**Session 8: Improved SCAL Techniques and Interpretation (2) - NMR** **Chairs: Stefano Pruno and Matthias Appel**
SCA022	Inverted bucket centrifugation with fluorinated oils and its applications to T2 cut-offs...119 Ben Anger, Stefan Hertel, Keith Love, Michael Ehiwario, Matthias Appel
SCA023	A New Apparatus for Coupled Low-field NMR And Ultrasonic Measurements in Rocks at Reservoir Conditions...N/A Paul R. J. Connolly, Joël Sarout, Jérémie Dautriat, Eric F. May, Michael L. Johns
2:00 – 3:30	**Poster Session (Odd Numbers)** **Chair: Lesley James and Jules Reed**
3:30 – 4:30	Coffee and Poster Session Break Coffee Break Kindly Sponsored by: Thermo Fisher Scientific Poster Session Kindly Sponsored by: Green Imaging Technologies, Inc and Oxford Instruments
4:30 - 5:30	**Session 9: Core Analysis in a Digital World (2)** **Chairs: Olga Vizika-Kavvadias and Christopher Arns**
SCA024	Defining a sample heterogeneity cut-off value to obtain representative Special Core Analysis (SCAL) measurements...N/A Jos G. Maas, Niels Springer, and Albert Hebing
SCA025	Digital core repository coupled with machine learning as a tool to classify and assess petrophysical rock properties...125 Vanessa Hébert, Thierry Porcher, Valentin Planes, Marie Léger, Anna Alperovich, Bastian Goldluecke, Olivier Rodriguez and Souhail Youssef
6:30 – 11:00	Gala Dinner – At Villa Navarre Kindly Sponsored by: iRock Technologies

Thursday, August 29, 2018

8:30 – 4:30	Registration Desk Open
8:30 - 10:00	**Session 10: Unconventionals and Shales (2)**
	Chairs: Matthias Appel and Olga Vizika-Kavvadias
SCA026	Using Capillary Condensation and Evaporation Isotherms to Investigate Confined Fluid Phase Behavior in Shales...136
	Elizabeth Barsotti, Evan Lowry, Mohammad Piri, and Jin-Hong Chen
SCA027	Methane Isotherms and Magnetic Resonance Imaging in Shales...142
	M. J. Dick, D. Heagle, D. Veselinovic and D. Green
SCA028	Dielectric Polarisation in Partially Saturated Shales...N/A
	Paul R.J. Connolly, Matthew Josh, Keelan O'Neill, Eric F. May, Michael L. Johns
10:00 - 10:30	Coffee Break
10:30 – 12:00	**Session 11: Displacement Mechanisms/EOR/IOR (2)**
	Chairs: Souhail Youssef and Hendrik Roler
SCA029	CT-scan in-situ investigation of waterflood front instabilities during immiscible displacements: effect of viscosity contrast and flow rate...N/A
	Matthieu Mascle, Elisabeth Rosenberg, Berit Roboele, Espen Kowalewski, and Souhail Youssef
SCA030	Core-scale sensitivity study of CO2 foam injection strategies for mobility control, enhanced oil recovery, and CO2 storage...152
	Zachary Paul Alcorn, Sunniva B. Fredriksen, Mohan Sharma, Tore Føyen, Connie Wergeland, Martin A. Fernø, Arne Graue, and Geir Ersland
SCA031	New laboratory core flooding experimental system for EOR surfactant screening, especially for foam...N/A

Xuesong Li, and Matthias Appel

12:00 – 12:30		**Business Meeting**
12:30 - 1:30		Lunch
1:30 – 3:00		**Session 12: Laboratory Core Analysis**
		Chairs: Stefano Pruno and Matthias Halisch
	SCA032	A new CEC measurement proxy using high-frequency dielectric analysis of crushed rock...N/A
		M. Rebecca Stokes, Z. Elton Yang, Prince Ezebuiro, and Timothy Fischer
	SCA033	Gas Slippage in Partially Saturated Tight Rocks and During Drainage...N/A
		Alexandra Amann-Hildenbrand, Mohammadebrahim Shabani, Thomas Hiller, Norbert Klitzsch, Norbert Schleifer, and Bernhard M. Krooss
	SCA034	Modeling Permeability in Carbonate Rocks...N/A
		Moustafa Dernaika, Shehadeh Masalmeh, Bashar Mansour, Osama Al-Jallad, and Safouh Koronfol
3:00 – 3:30		Coffee Break
3:30 - 4:30		**Session 13: Improved SCAL Techniques and Interpretation (3) - NMR**
		Chairs: Matthias Appel and Derrick Green
	SCA035	Direct Magnetic Resonance Measurement of Average Pore Size..N/A
		Florin Marica, Armin Afrough, Derrick Green, Laura Romero-Zeron, and Bruce Balcom
	SCA036	Two-Phase Fluid Flow Experiments Monitored by NMR...164
		Thomas Hiller, Gabriel Hoder, Alexandra Amann-Hildenbrand, Norbert Klitzsch, and Norbert Schleifer

4:30 **Closing Remarks**

Alternate Orals

SCA037	Transport properties of the Cobourg Limestone: A benchmark investigation...N/A C. A. Davy, Zhazha Hu, A.P.S. Selvadurai, Jop Klaver, M.-C. Willemetz, F Agostini, F. Skoczylas, Jan Dewanckele, Alexandra Amann-Hildenbrand and Roland Lenormand
SCA038	Pore-scale experimental investigation of in-situ wettability and displacement mechanisms governing WAG in oil-wet carbonates...N/A Ziqiang Qin, Maziar Arshadi, and Mohammad Piri
SCA039	High-resolution inline density measurements: insight on multiphase flow and transport phenomena in porous media...173 Jelayne Falat, Adam Fehr, Ali Telmadarreie, and Steven Bryant
SCA040	Pore-scale experimental study of carbonated water injection in an oil-wet carbonate: an improved insight into wettability alteration and displacement mechanisms...N/A Ziqiang Qin, Maziar Arshadi, and Mohammad Piri
SCA041	The digital rock analysis of biogenically induced reservoir heterogeneities in Cretaceous reservoirs of Saudi Arabia...184 Ivan Deshenenkov, and Camilo Polo
SCA042	Novel technique to measure mutual bulk fluid diffusion using NMR 1-D gradient...194 Son Dang, Carl Sondergeld, and Chandra Rai

Thanks to all Symposium Sponsors

iRock Technologies 数岩科技

ThermoFisher SCIENTIFIC

GIT GREEN IMAGING TECHNOLOGIES, INC.

OXFORD INSTRUMENTS *The Business of Science®*

TOTAL

CHANDLER ENGINEERING®

AMETEK® OIL & GAS

VINDUM ENGINEERING, INC.

STRATA TECHNOLOGY LTD

SOCIETY OF CORE ANALYSTS

139 Whittaker Rd • Rusagonis, NB E3B 7X3, Canada

E-Mail: office@scaweb.org • www.SCAWeb.org

edp sciences

Web of Conferences

Statement of Peer review

In submitting conference proceedings to *Web of Conferences*, the editors of the proceedings certify to the Publisher that

1. They adhere to its **Policy on Publishing Integrity** in order to safeguard good scientific practice in publishing.
2. All articles have been subjected to peer review administered by the proceedings editors.
3. Reviews have been conducted by expert referees, who have been requested to provide unbiased and constructive comments aimed, whenever possible, at improving the work.
4. Proceedings editors have taken all reasonable steps to ensure the quality of the materials they publish and their decision to accept or reject a paper for publication has been based only on the merits of the work and the relevance to the journal.

Title, date and place of the conference

The 33rd International Symposium of the Society of Core Analysts
August 26 to August 29 2019
Le Palais Beaumont, Pau, France

Proceedings editor(s):

Holger Ott
Josephina Schembre-McCabe

Date and editor's signature

January 20, 2020

Holger Ott (SCA VP of technology)

Density Functional Hydrodynamics in Multiscale Pore Systems: Chemical Potential Drive

Oleg Dinariev[1], *Nikolay* Evseev[1,*], and *Denis* Klemin[2]

[1] Schlumberger, Schlumberger Moscow Research, 113, Pudovkina Street, Moscow 119284, Russia
[2] Schlumberger, Schlumberger Reservoir Labs, 6350 West Sam Houston Pkwy N, Houston, TX 77041, Houston, USA

Abstract. We use the method of density functional hydrodynamics (DFH) to model compositional multiphase flows in natural cores at the pore-scale. In previous publications the authors demonstrated that DFH covers many diverse pore-scale phenomena, starting from those inherent in RCA and SCAL measurements, and extending to much more complex EOR processes. We perform the pore-scale modelling of multiphase flow scenarios by means of the direct hydrodynamic (DHD) simulator, which is a numerical implementation of the DFH. In the present work, we consider the problem of numerical modelling of fluid transport in pore systems with voids and channels when the range of pore sizes exceed several orders of magnitude. Such situations are well known for carbonate reservoirs, where narrow pore channels of micrometer range can coexist and interconnect with vugs of millimeter or centimeter range. In such multiscale systems one cannot use the standard DFH approach for pore-scale modeling, primarily because the needed increase in scanning resolution that is required to resolve small pores adequately, leads to a field of view reduction that compromises the representation of large pores. In order to address this challenge, we suggest a novel approach, in which transport in small-size pores is described by an upscaled effective model, while the transport in large pores is still described by the DFH. The upscaled effective model is derived from the exact DFH equations using asymptotic expansion in respect to small-size characterization parameter. This effective model retains the properties of DFH like chemical and multiphase transport, thus making it applicable to the same range of phenomena as DFH itself. The model is based on the concept that the transport is driven by gradients of chemical potentials of the components present in the mixture. This is a significant generalization of the Darcy transport model since the proposed new model incorporates diffusion transport in addition to the usual pressure-driven transport. In the present work we provide several multiphase transport numerical examples including: a) upscaling to chemical potential drive (CPD) model, b) combined modeling of large pores by DFH and small pores by CPD.

1 Introduction

In recent years we witnessed an increasing level of capability and acceptance of digital rock as a complementary tool in core analysis. This tendency is following a gradual increase in both resolution of rock imaging techniques and the power of high-performance computing that are the key to unlock the potential in the core analysis by digital rock. As an answer to the growth in the hardware possibilities various modeling methods applicable in digital rock also evolve; see reviews in [1-6]. One of such methods, which the authors are actively developing, is the density functional hydrodynamics (DFH).

DFH is essentially a pore scale multi-phase, multi-compositional approach. As it has been demonstrated in previous publications [7-10], various multiphase compositional problems can be described in the frame of the DFH. The DFH method has found many applications in direct pore-scale modeling of various hydrodynamic and petrophysical phenomena on digital rock models obtained by X-ray micro-CT and SEM [11-21]. The practical usefulness of a standard workflow is very much dependent on the possibility to resolve necessary pore structure adequately, as well as on the rock sample being representative for the considered formation. There are specific ways to meet both of these requirements, but in this work, we discuss the resolution problem only.

In many cases the natural rocks are characterized by a wide range of pore sizes making the choice of X-ray micro-CT resolution very difficult if not altogether impossible. Indeed, if pores of both micrometer and millimeter range coexist within one piece of rock then, at present, it is not always possible to develop 3D model of rock microstructure, which can represent all pores. In situations like this we propose a combined approach, when several 3D digital rock models having different resolutions are integrated into one synthetic model. Combined multiscale approach is also being developed in the frame of pore-network modelling [1], but we follow image-based approach as it is inherent in DFH. High-resolution sub-models with explicitly resolved

* Corresponding author: nevseev@slb.com

small pores are used for direct modeling of transport processes by DFH to obtain effective flow characteristics. Then these results are incorporated into low-resolution model with explicit imaging of large pores while small pores are not necessarily resolved. At this stage the hydrodynamics in large pores is still simulated directly by DFH, while the transport in matrix is simulated using the previously computed effective transport properties. All these operations are done by the DFH pore-scale simulator called DHD (Schlumberger).

In effect, the described procedure represents upscaling from pore-scale DFH equations to macroscopic porous medium equations. As concerned with the single-phase flow, our idea is similar to the one described in [22, 23]. However, in addition to being applicable in general multiphase compositional scenarios, the rigorous upscaling starting with the DFH equations leads to the transport equations, which constitute the significant extension of the Darcy model. The driving force for the fluid flow happens to be gradients of chemical potentials of fluid chemical components that gives the model its name Chemical Potential Drive (CPD). Because change in pressure is directly related to change in chemical potential, this new model is consistent with pressure-drive Darcy approach as will be shown below. On the other hand, the CPD approach covers certain physical phenomena outside the scope of the Darcy model. Here are two examples in relation to the latter assertion. First, there are cases, when definition of pressure in confined fluid is problematic, e.g., when gas mean free path length is comparable to pore size (high Knudsen number), or liquid in pores with significant disjoining pressure; at the same time, the definition of chemical potential is still correct in both said instances. Second, there is an easy to derive exact solution for two-phase water-oil transport velocities in a hydrophilic circular capillary

$$u_w = -(8\eta_{sw})^{-1} R^2 (1-(r/R)^2)^2 G \tag{1}$$

$$u_o = -(8\eta_{so})^{-1} R^2 (2(r/R)^2 (1-(r/R)^2)^2 (\eta_{so}/\eta_{sw}) + (r/R)^4) G \tag{2}$$

where the water and oil velocities u_w, u_o are defined as phase fluxes per capillary cross-section, η_{sw}, η_{so} are the shear viscosities of water and oil, respectively, R is the capillary radius, r is the radius of the central region occupied by oil, and G is the pressure gradient. An important observation following from Eqs. (1), (2) is that the transport of oil is influenced by viscosity of water, because oil, being the central phase, is carried by water. This cross-phase friction is in direct contradiction with the routine concept of phase permeabilities indicating the possible role of cross-terms in phase permeability matrix [24]. At the same time the cross-component influence is inherent in CPD, as it will be demonstrated below. Limitations of the Darcy law are discussed in the literature and deviations from it are observed in both high-accuracy 4D visualization of pore-scale fluid dynamics as well as in numerical simulations of pore-scale flows [25]. One may expect more deviations when dealing with complex fluids.

In Sec. II we give a brief reminder of the DFH equations, as well as a derivation of upscaling from DFH to CPD. Numerical demonstration of combined DFH and CPD models is presented in Sec. III. The overall summary and discussion of results is in Sec. IV.

The summation over repeated indices is implied everywhere. The indices $a, b, c = 1,2,3$ are related to Cartesian coordinates x^a, the indices $i, j, k = 1,...,M$ are related to chemical components in the fluid mixture. We consider isothermal processes, so the temperature is assumed to be fixed, and dependence of certain variables on temperature is omitted. We use short symbols for partial derivatives: $\partial_a = \partial/\partial x^a$ and $\partial_t = \partial/\partial t$, where t is time.

2 Theoretical concepts and equations

2.1 Density functional hydrodynamics

Here we provide only some of the basic definitions necessary to evolve the step from the DFH equations to the CPD ones. A detailed exposition of the DFH can be found in [7-10].

We consider continuum mechanics description of a mixture of M chemical components present inside a spatial region D having volume V_D. The region contains N_{iD} of each type of molecules. To avoid large numbers, the quantities N_{iD} are measured in moles. In case of homogeneous and static mixture we define chemical component molar densities by $n_i = N_{iD}/V_D$. If the mixture is inhomogeneous and possibly evolving, one can define n_i as a dynamic variable at time t and spatial point x^a by establishing a small volume limit, such as $n_i = n_i(t, x^a) = \lim_{V_D \to 0}(N_{iD}/V_D)$. Like elsewhere in continuum mechanics, the small volume limit is understood as the convergent procedure with V_D being small, but still larger than the molecular volume.

By counting the flow rate of molecules through a small area inside the mixture, one can define the component flux $I_{ia} = I_{ia}(t, x^b)$. The component fluxes are used to calculate the mass flux $I_a = m_i I_{ia}$, where m_i is the molar mass of the ith component. By introducing mass density $\rho = m_i n_i$ it is possible to define mass velocity $v_a = \rho^{-1} I_a$. Component flux I_{ia} can be represented as a combination of transport term $n_i v_a$ and diffusion flux Q_{ia}:

$$I_{ia} = n_i v_a + Q_{ia} \tag{3}$$

where by definition diffusion flux does not influence net mass transfer $m_i Q_{ia}=0$. We assume the existence of the Helmholtz energy functional

$$F_D = F_D[n_i] = \int_D \omega dV + \int_{\partial D} f_* dA, \quad (4)$$

$$\omega = f(n_i) + 2^{-1} \nu_{ij}(n_k) \partial_a n_i \partial_a n_j, \quad (5)$$

where ∂D is the boundary surface of the region D (when the region is finite), $f = f(n_i)$ is the bulk Helmholtz energy density of homogeneous mixture, ν_{ij} is the positive-definite symmetric matrix, and $f_* = f_*(n_i)$ is the surface Helmholtz energy density, which is not equal to zero if ∂D is a contact surface with some immobile solid. It is convenient to recollect certain thermodynamic equations involving Helmholtz energy

$$f = \kappa_i n_i - p, \; df = \kappa_i dn_i, \; dp = n_i d\kappa_i, \quad (6)$$

where κ_i is the chemical potential of the ith component, and p is the hydrostatic pressure.

The model in Eqs. (4) and (5) is adequate for description of many important phenomena involving multiphase multicomponent mixtures. Up to now, it was successfully used to simulate multiphase multicomponent phenomena with or without phase transitions, surfactants, and mixtures with solid phases such as gas hydrates or solid particles [9-21, 26].

Now we list the DFH statements necessary to move forward to CPD.

The multiphase compositional transport in case of isothermal flow of fluids with Newtonian rheology is governed by the following equations written in a form of conservation laws for chemical components and momentum

$$\partial_t n_i + \partial_a (n_i v_a + Q_{ia}) = 0, \quad (7)$$

$$\partial_t (\rho v_a) + \partial_b (\rho v_a v_b - p_{ab}) = 0, \quad (8)$$

together with the relations

$$p_{ab} = \sigma_{ab} + \tau_{ab}, \quad (9)$$

$$\sigma_{ab} = (\omega - \Phi_i n_i)\delta_{ab} - \nu_{ij} \partial_a n_i \partial_b n_j, \quad (10)$$

$$\tau_{ab} = \eta_v \delta_{ab} \partial_c v_c + \eta_s (\partial_a v_b + \partial_b v_a - \frac{2}{3} \delta_{ab} \partial_c v_c), \quad (11)$$

$$Q_{ia} = -D_{ij} \partial_a \Phi_j, \quad (12)$$

$$\Phi_i = \kappa_i + 2^{-1} \frac{\partial \nu_{jk}}{\partial n_i} \partial_a n_j \partial_a n_k - \partial_a (\nu_{ij} \partial_a n_j), \quad (13)$$

and subject to the boundary conditions at ∂D, when it is a contact surface with some immobile solid

$$v_a = 0, \quad (14)$$

$$l^a Q_{ia} = 0, \quad (15)$$

$$\frac{\partial f_*}{\partial n_i} - l^a \nu_{ij} \partial_a n_j = 0, \quad (16)$$

where p_{ab} is the stress tensor, σ_{ab} is the static stress tensor, τ_{ab} is the viscous stress tensor, Φ_i is the generalized chemical potential of the ith mixture component, η_v and η_s are nonnegative bulk and shear viscosity coefficients, respectively, D_{ij} is the nonnegative definite symmetric matrix subject to $D_{ij} m_j = 0$, l^a is the internal normal unit vector at the boundary surface ∂D, and δ_{ab} is the Kronecker symbol. We also note that there is the identity $\partial_b \sigma_{ab} = -n_i \partial_a \Phi_i$, which reduces to $\partial_b \sigma_{ab} = 0$ in static equilibrium.

It is possible to derive the differential equation for the total energy (that includes both Helmholtz and kinetic energy) with energy flux J_a and dissipation rate function χ:

$$\partial_t (\omega + 2^{-1} \rho v_a v_a) = -\partial_a J_a + \chi, \quad (17)$$

$$J_a = \frac{1}{2} \rho v_b v_b v_a - \nu_{ij} \partial_t n_i \partial_a n_j - \tau_{ab} v_b + \Phi_i I_{ia}, \quad (18)$$

$$\chi = -\tau_{ab} \partial_a v_b + Q_{ia} \partial_a \Phi \quad (19)$$

In accordance with Eqs. (11), (12), and (22) the inequality $\chi \le 0$ holds. Also, in accordance with Eqs. (14)-(16) the energy flux vanishes at the boundary ∂D: $l^a J_a = 0$. This signifies that in absence of external sources and forces the total energy decreases $\frac{d}{dt} \int_D (\omega + 2^{-1} \rho v_a v_a) dV \le 0$ in consistency with the second law of thermodynamics.

2.2 Upscaling to macroscopic transport in porous medium: chemical potential drive

Now we consider transition from microscopic DFH to macroscopic CPD description of fluid transport in porous medium. We use variables related to microscopic pore-scale description in parallel to similar variables related to macroscopic description. If a is a microscopic variable, then similar macroscopic variable is denoted by symbol $\bar{a}$. The latter variable is usually obtained from the former by using certain averaging procedure. For example, we define component molar density by $\bar{n}_i = \bar{n}_i(t, x^a) = \lim_{V_D \to 0} (N_{iD} / V_D)$, where the pore volume V_D is now sufficiently large to contain representative piece of saturated porous medium. The respective spatial region D encompasses pores belonging to a porous medium region Δ containing both solid skeleton and pores. This porous medium region Δ is supposed to be representative of pore structure in statistical terms but not necessarily identical to the neighbor regions of porous medium. This is different from homogenization approach with periodic geometry of pore system [27]. The volumes of two regions D and Δ are related by $V_D=\phi V_\Delta$, where $\phi=\phi(x^a)$ is porosity. The processes in a porous medium are assumed to be sufficiently slow in respect to the changing macroscopic component molar densities, so that the local distribution of chemical components and phases in pores are close to the equilibrium. Therefore, it is possible to define the macroscopic Helmholtz energy density $\bar{f}=\lim_{V_D\to 0}(F_D/V_D)$ as a function of parameters $\bar{n}_i$ since the equilibrium state of the mixture is determined by the functional (4) together with the components conservation condition $N_{iD}=\int_D n_i dV$. Using $\bar{f}$ it is possible to define macroscopic chemical potentials $\bar{\kappa}_i=\frac{\partial\bar{f}}{\partial\bar{n}_i}$.

Let us study transport through the spatial porous region D. Transport is possible only if there are free subregions of the boundary ∂D , which are not associated with solid surface. We designate $\partial D=\Gamma_f\cup\Gamma_s$, where Γ_f, Γ_s are free and solid boundary, respectively. At the free boundary Γ_f the so-called free boundary condition for velocity is used in the form of $l^b\partial_b v_a=0$ instead of the no-slip condition in Eq. (14) of the DFH. The diffusion transport is generated by gradients of generalized chemical potentials $\partial_a\Phi_i$ (see Eq. (12)). The convective transport is determined by mass velocity, which under current assumptions satisfies the reduced momentum equation

$$\partial_b\tau_{ab}=n_i\partial_a\Phi_i. \tag{20}$$

with mixed no-slip and free boundary conditions. In terms of differential operations Eq. (20) can be represented as follows

$$L_{ab}v_b=-\eta_a=-n_i\partial_a\Phi_i, \tag{21}$$

where the left side is positive elliptic self-adjoint operator in $L_2(D)$ acting on the velocity field. It admits the solution with a Green's function, which can also be represented in operator form

$$v_a=-G^{ab}\eta_b=-G^{ab}(n_i\partial_a\Phi_i). \tag{22}$$

Here the integral operator in G^{ab} is positive definite, and also $G^{ab}=G^{ba}$. Now combining Eqs. (12) and (22) we arrive to the expression for the component flux generated by the gradients $\partial_a\Phi_i$

$$I_{ia}=n_i v_a+Q_{ia}=-(n_i G^{ab}n_j+\delta_{ab}D_{ij})\partial_b\Phi_j. \tag{23}$$

Averaging Eq. (23) over different cross-sections of the region Δ produces the macroscopic flux of components generated by gradients of chemical potentials (Chemical Potential Drive or CPD)

$$\bar{I}_{ia}=\langle I_{ia}\rangle=-K^{ab}_{ij}\partial_b\bar{\kappa}_j, \tag{24}$$

where the transport matrix $K^{ab}_{ij}=K^{ab}_{ij}(x^c)$ is positive definite and satisfies the symmetry conditions $K^{ab}_{ij}=K^{ba}_{ij}=K^{ab}_{ji}$. The component conservation equation together with the CPD transport law in Eq. (24) that constitutes the closed formulation of macroscopic fluid transport model is as follows

$$\partial_t(\phi\bar{n}_i)+\partial_a\bar{I}_{ia}=0. \tag{25}$$

Direct calculations lead to the macroscopic equivalent of Eq. (17)

$$\partial_t(\phi\bar{f})=-\partial_a\bar{J}_a+\bar{\chi} \tag{26}$$

$$\bar{J}_a=\bar{\kappa}_i\bar{I}_{ai}, \tag{27}$$

$$\bar{\chi}=\bar{I}_{ai}\partial_a\bar{\kappa}_i\le 0. \tag{28}$$

2.3 Boundary conditions between DFH and CPD domains

In highly heterogeneous porous material there can be situations when there are adjacent spatial regions where one can use alternatively DFH or CPD. It is necessary to discuss what boundary conditions should be set at the common boundary $\Gamma=\partial D_1\cap\partial D_2$ between the DFH region D_1 and the CPD region D_2. Eq. (16) is still valid at Γ in D_1 , while Eqs. (14) and (15) are not because there can be transport of components through the boundary.

One evident boundary condition follows from conservation of component fluxes

$$l^a I_{ai}=l^a\bar{I}_{ai}. \tag{29}$$

The next condition can be derived from energy conservation at the boundary

$$l^a J_a=l^a\bar{J}_a. \tag{30}$$

Using Eqs. (18), (29), and (30) one can satisfy Eq. (27) by imposing boundary condition for chemical potentials

$$\bar{\kappa}_i=\Phi_i-m_i X, \tag{31}$$

where variable X is determined by the additional boundary equation

$$X\rho l^a v_a+\frac{1}{2}\rho v_b v_b l^a v_a-l^a\tau_{ab}v_b=0. \tag{32}$$

 https://doi.org/10.1051/e3sconf/202014601001

2.4 Examples of CPD models

For the first example we assume that the porous medium is homogeneous and isotropic, and there is single-phase single-component flow. In this case the CPD transport law is equivalent to the Darcy law (see the last equation in (6))

$$\bar{I}_{1a} = -K\partial_a \bar{\kappa}_1 = -K\bar{n}_1^{-1}\partial_a \bar{p}_1 \qquad (33)$$

with CPD transport coefficient K being proportional to the absolute permeability k, $K = k\bar{n}_1^2\eta_s^{-1}$, where coefficient η_s is the shear viscosity.

Second, we consider two-phase two-component mixture modeling water and oil. Here we have the following definitions

$$\bar{n}_1 = \bar{s}_w\bar{n}_w,\ \bar{n}_2 = (1-\bar{s}_w)\bar{n}_o, \qquad (34)$$

$$\bar{f} = \bar{f}(\bar{n}_i) = \bar{s}_w\bar{f}_w(\bar{n}_w) + (1-\bar{s}_w)\bar{f}_o(\bar{n}_o) + \bar{f}_{cap}(\bar{s}_w), \qquad (35)$$

$$\bar{p}_w - \bar{p}_o = \bar{p}_{cap}(\bar{s}_w) = \bar{f}_{cap}'(\bar{s}_w), \qquad (36)$$

where $\bar{s}_w$ is the water saturation, $\bar{f}_w, \bar{p}_w$ are the Helmholtz energy density and pressure of water, $\bar{f}_o, \bar{p}_o$ are the Helmholtz energy density and pressure of oil, and $\bar{f}_{cap}(\bar{s}_w)$, $\bar{p}_{cap}(\bar{s}_w) = \bar{f}_{cap}'(\bar{s}_w)$ are the Helmholtz energy density and pressure for capillary forces. Eqs. (34)-(36) determine variables $\bar{s}_w, \bar{n}_w, \bar{n}_o$ when variables $\bar{n}_1, \bar{n}_2$ are known. The transport model in Eq. (24) is consistent with the generalized Darcy law if there are the following additional conditions:

$$K_{12} = K_{21} = 0,\ K_{11} = k_w\bar{n}_w^2\eta_{sw}^{-1},\ K_{22} = k_o\bar{n}_o^2\eta_{so}^{-1}, \qquad (37)$$

where k_w, k_o are water and oil phase permeabilities. If conditions in Eq. (37) are not satisfied, CPD model can demonstrate some phenomena outside traditional Darcy transport.

3 Numerical examples

To demonstrate application of a workflow based on the joint DFH modeling within resolved porosity and CPD modeling within matrix containing unresolved smaller pores (DFH+CPD workflow) we chose a heterogeneous, but relatively well characterized chalk sample. The sample is grainstone composed of pelloids, skeletal grains and ooids. The grains are cemented by equant calcite spar cement. Intergranular porosity is the dominant pore system, additionally some moldic and micropores are present in partially leached grains. Pore throats are ranging between approximately 1 and 10 microns. Pore throat distribution is skewed to smaller pore throat sizes. Measured gas porosity of the sample is 26.2%, and gas permeability is 19.4 mD.

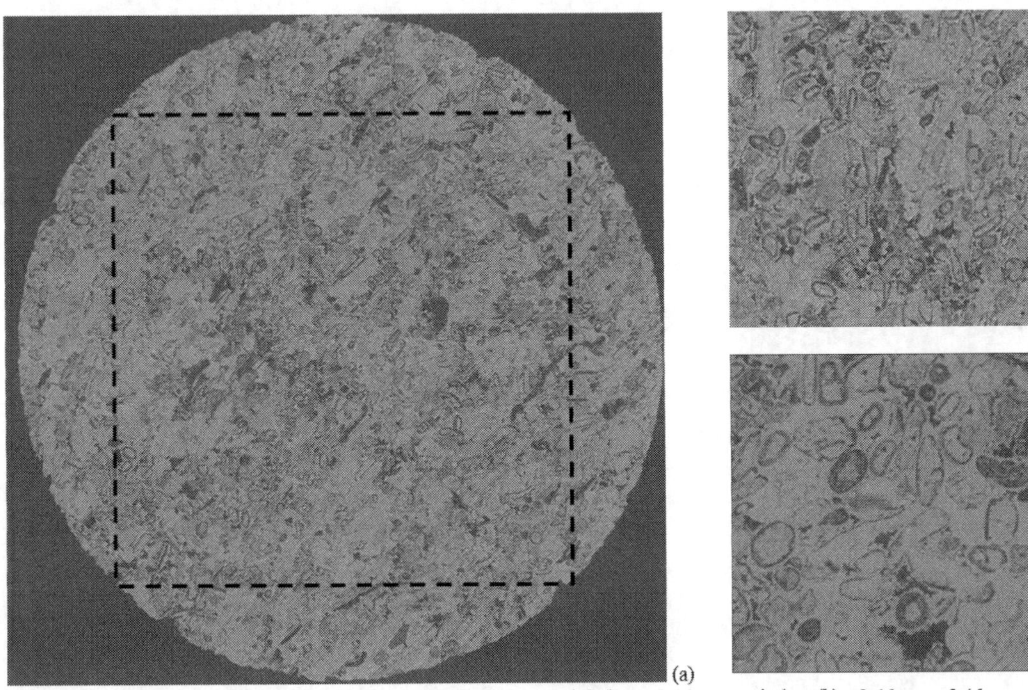
(a) (b) (c)

Fig. 1. Micro-CT grayscale cross-section of the dry chalk (a) 8 mm mini plug at 2.46 um resolution, (b) a 2.46 mm x 2.46 mm fragment of a cross-section at 2.46 um resolution, and (c) a 0.82 mm x 0.82 mm fragment of a cross-section at 0.82 um resolution; the dashed line in (a) shows the position of the maximum size parallelepiped model extracted from the mini plug.

An 8 mm diameter mini plug was scanned with 2.46 um per voxel resolution, and a portion of the mini plug was scanned with higher resolution of 0.82 μm per voxel (Fig. 1).

Two 3D digital rock models (DRM) with these two different resolutions were constructed (Fig. 2). Both models cover the same spatial region with the size 0.82 mm x 0.82 mm x 0.82 mm, but they have different number of cells (voxels). The high-resolution DRM has 1000^3 (1000x1000x1000) cells and the low-resolution DRM has 333^3 (333x333x333) cells so that the cell sizes are 0.82 um and 2.46 um, respectively. In either model some portion of pores fell below resolution. However, it is evident from Fig. 2 that significantly higher proportion of pores remains below resolution in the low-resolution DRM than it does in the high-resolution DRM. The data on the resolved and unresolved porosity is shown in Table 1.

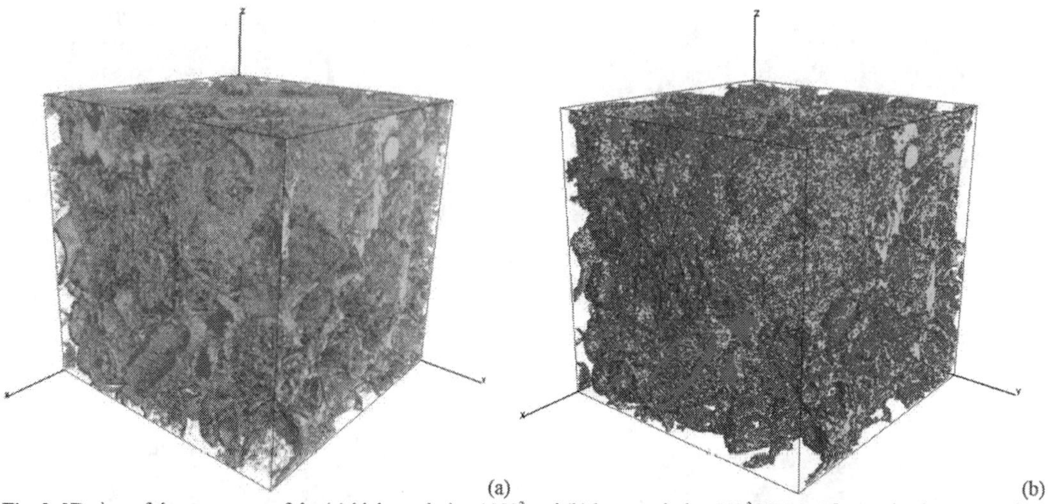

(a) (b)

Fig. 2. 3D view of the pore space of the (a) high-resolution 1000^3 and (b) low-resolution 333^3 DRMs. The resolved pores are shown in green, the unresolved pores (subresolution matrix) are shown in red, and the skeleton is made transparent.

Table 1. The portion of resolved porosity and unresolved porosity (subresolution matrix) within digital rock models of different spatial resolution.

Model	Resolved porosity	Subresolution matrix
1000^3, 0.82 um	0.166	0.073
333^3, 2.46 um	0.091	0.157

In addition to X-ray micro-CT, we have done scanning electron microscopy (SEM) of parts of the sample to reveal the pores below micro-CT resolution down to 50 nm. Using both micro-CT high-resolution (0.82 um) and SEM data we performed digital modeling of MICP experiment and reconstructed pore throat size distribution that is compared to the standard experimental MICP measurements in Fig. 3. The two curves in Fig. 3 match well meaning that most of the controlling pore throats have been adequately resolved on the corresponding scales.

In modeling on the high-resolution DRM, we assumed that the subresolution matrix portion of the model is effectively impermeable for oil at relevant oil-water capillary pressure levels so that only the DFH equations need to be solved within the resolved pore space. On the other hand, in modeling on the low-resolution DRM, we assumed that the subresolution matrix is permeable and therefore the full DFH+CPD simulation needs to be done. To set up a single-phase DFH+CPD simulation it is necessary to provide absolute permeability in the subresolution matrix cells so that the single-phase CPD transport law given in Eq. (33) could be resolved. In order to do this, we first simulated single-phase flow on the high-resolution DRM to obtain velocity distribution (Fig. 4a).

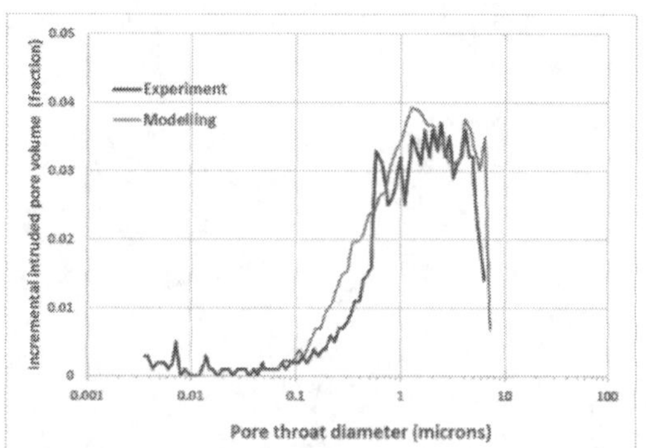

Fig. 3. Reconstructed numerically from the micro-CT and SEM data (green curve) and obtained from the MICP measurements (blue curve) pore throat size distributions.

Then, by interpreting the simulation results in terms of Eq. (33) and linking them with the gray scale distribution from the high-resolution images such as those shown in Fig. 1c, we established a correlation between the gray scale and absolute permeability values. We then applied the same correlation to the low-resolution gray scale

* Corresponding author: nevseev@slb.com

 https://doi.org/10.1051/e3sconf/202014601001

images to inform the low-resolution DRM about the subresolution matrix permeability. The velocity distribution obtained on this 333^3 low-resolution DRM is presented in Fig. 4b. The absolute permeabilities calculated on the high- and low-resolution DRMs are 17.3 mD and 18.2 mD, correspondingly.

The next step was simulating two-phase flow and the steady-state relative permeability experiment. The two immiscible phases were assumed to be water and oil with the properties as follows: $\rho_w = 1000\,\mathrm{kg/m^3}$, $\rho_o = 800\,\mathrm{kg/m^3}$, $\eta_{sw} = 0.001\,\mathrm{Pa{\times}s}$, $\eta_{so} = 0.004\,\mathrm{Pa{\times}s}$, $\gamma_{ow} = 0.03\,\mathrm{N/m}$, where ρ_w, ρ_o are the water and oil mass densities, η_{sw}, ρ_{so} are the water and oil viscosities, and γ_{ow} is the oil-water interfacial tension. Since the sample is carbonate, it was assumed that it has mixed wettability. To model this, we first injected oil into initially water-wet 100% water saturated high-resolution DRM using our standard DFH workflow [11, 14-17, 19-21]. (In all cases here and below injection was arranged in direction from bottom to top as shown in the 3D views.) Then at those places where oil touched the pore walls we switched wettability to moderately oil-wet while leaving the untouched pore walls water-wet. In this way most of the smaller pores as well as the pores isolated by thin pore throats remained water-wet, while the larger pores attained wettability towards oil. The distribution of water and oil at the residual water saturation state is shown in Fig. 5a. Using the obtained mixed-wet wettability distribution we simulated the steady-state relative permeability experiment by injecting oil and water mixture in several steps different in water content in the influx; at the final step, with only water injected, the residual oil saturation was obtained (Fig. 5c).

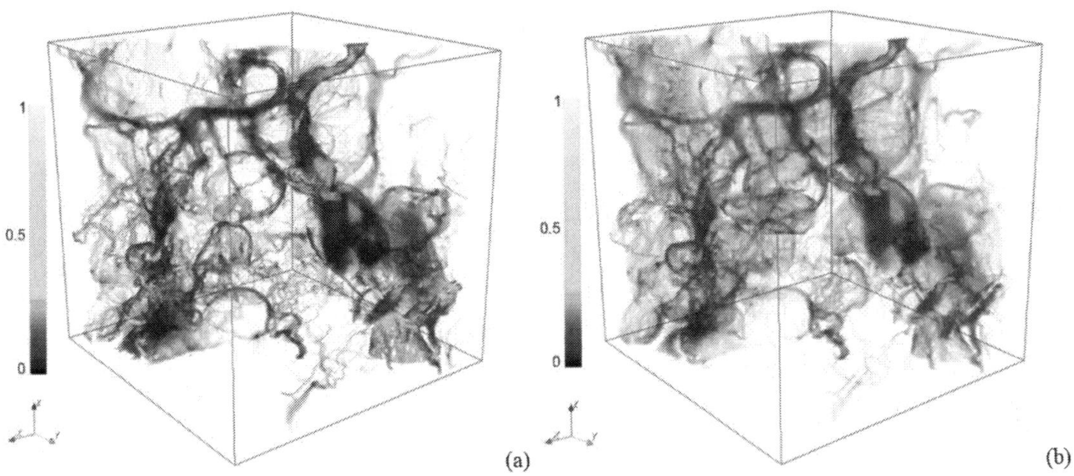

Fig. 4. 3D view of the velocity absolute value distribution in (a) the high-resolution 1000^3 DRM and (b) the low-resolution 333^3 DRM of the same location within the core. The shown distribution was scaled to the [0; 1] range. The velocity corresponds to flow in vertical (z) direction.

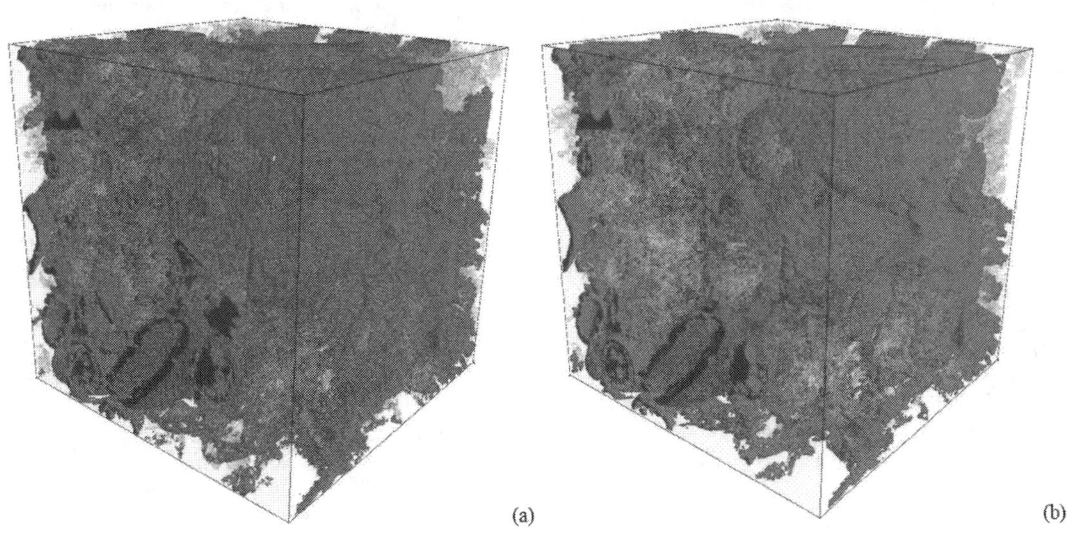

https://doi.org/10.1051/e3sconf/202014601001

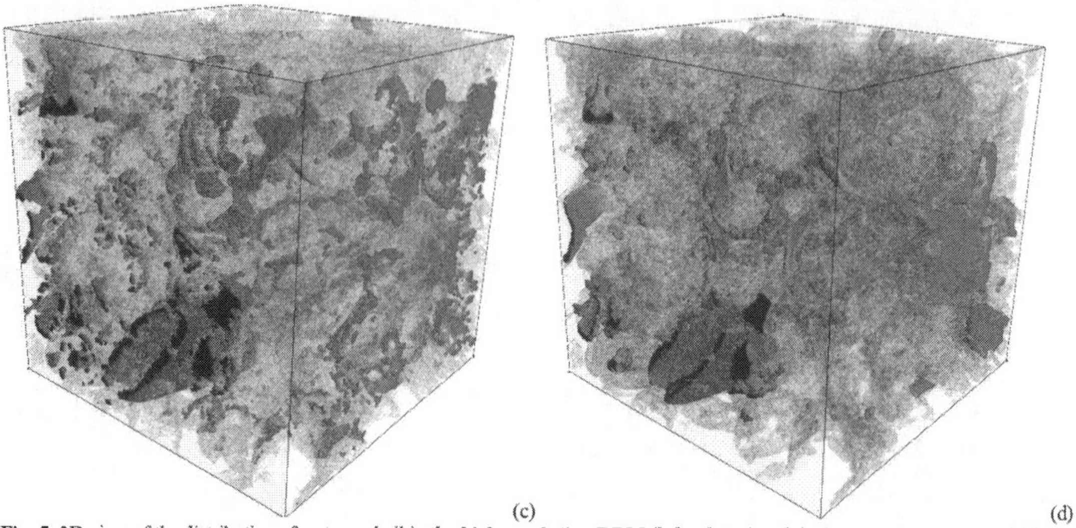

(c) (d)

Fig. 5. 3D view of the distribution of water and oil in the **high-resolution DRM** (left column) and the **low-resolution DRM** (right column) at the initial conditions corresponding to the residual water saturation (a, b), and the final state corresponding to the residual oil saturation (c, d). Water is shown in semitransparent blue, oil is shown in red, and skeleton is completely transparent.

To run DHD+CPD two-phase simulation, it is necessary to provide CPD transport matrix of Eq. (24) in subresolution matrix cells. Alternatively, when dealing with immiscible phases like oil and water that are each described by just one pseudo-component, it is possible to use simplified CPD transport model provided in Eq. (37) and based on the concepts of absolute and relative permeabilities and capillary pressure. We already discussed how absolute permeability was assigned to the subresolution matrix cells of the low-resolution DRM. The relative permeability was assigned in a similar way since all the velocity distribution, pressure, and water saturation for the subresolution matrix cells are known from the simulation on the high-resolution model. The capillary pressure function was constructed using the Young-Laplace equation and then scaled to the pore size distribution correlated by the gray scale values.

The distributions of oil and water obtained in simulation of the two-phase steady-state relative permeability experiment modeled using the DFH+CPD workflow on the low-resolution DRM are presented in Fig.5b,d. The chart in Fig. 6 compares the relative permeability curves obtained on both high- and low-resolution DRMs.

The final example we wish to present here is the simulation of both absolute and steady-state relative permeabilities on another low-resolution DRM with the size of 4.92 mm x 4.92 mm x 2.46 mm approximated with 500 x 500 x 250 cells with the cell size of 9.84 um. This model covers maximum size parallelepiped that can be extracted from the micro-CT scans of the 8 mm mini plug as shown by the dashed line in Fig. 1a. All of the parameters necessary to set up the DFH+CPD model were taken from the previous calculations on the high-resolution DRM as described previously.

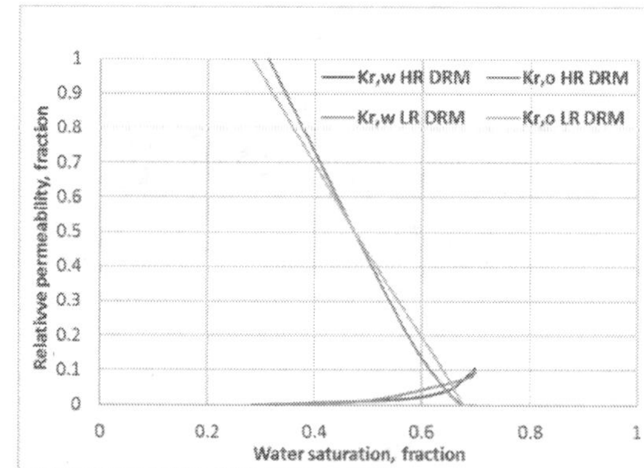

Fig. 6. Comparison of the simulated steady-state relative permeability curves obtained on the high-resolution (orange and blue curves) and low-resolution (yellow and gray curves) DRMs of the same spatial region within the sample.

Simulated absolute permeability of this model is 19.8 mD. The distributions of water and oil corresponding to the residual water saturation and residual oil saturation as obtained during the steady-state relative permeability simulation are shown in Fig. 7. The simulated relative permeability curves are presented in Fig. 8.

In this work, the simulation results were interpreted in consistency with the conditions in Eq. (37) that is with the traditional Darcy transport. However, in complex fluid problems deviations from relative permeability concept are expected, which then must be treated in the frame of the wider CPD model.

 https://doi.org/10.1051/e3sconf/202014601001

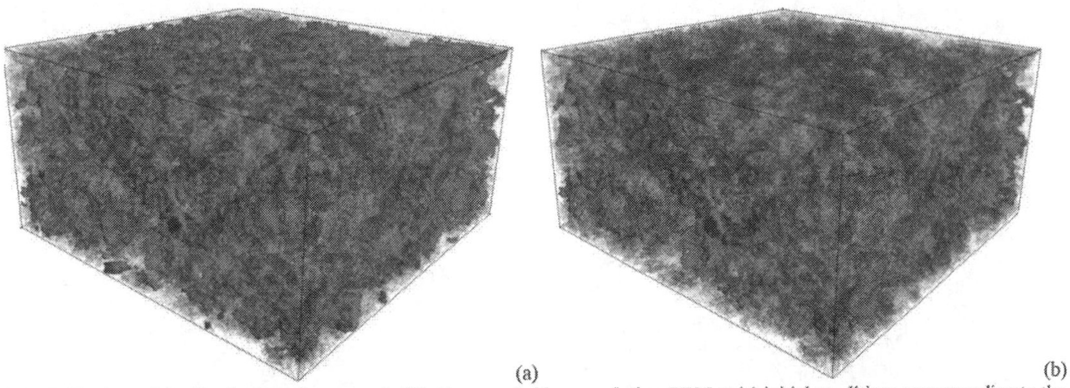

Fig. 7. 3D view of the distribution of water and oil in the upscaled low-resolution DRM at (a) initial conditions corresponding to the residual water saturation, and (b) final state corresponding to the residual oil saturation. Water is shown in semitransparent blue, oil is shown in red, and skeleton is transparent.

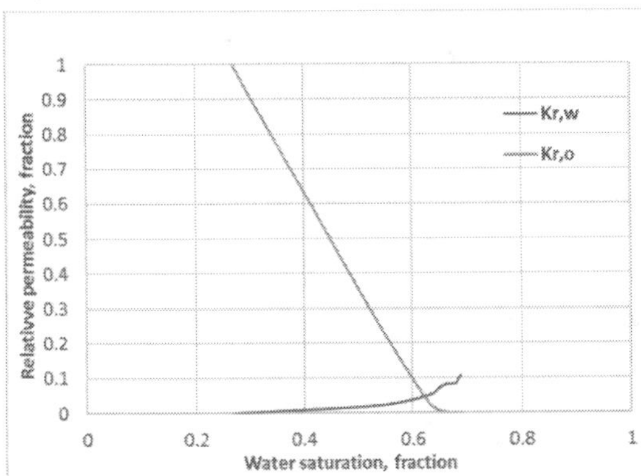

Fig. 8. Simulated steady-state relative permeability curves obtained on the upscaled low-resolution model.

2 Conclusion

We developed a new digital rock workflow applicable for simulation of multiphase flow on multiscale models representing heterogeneous core samples. The workflow called DFH+CPD is based on solving the standard DFH equations within resolved porosity and solving the new CPD equations within unresolved porosity or subresolution matrix. The CPD equations are derived in mathematically rigorous way from the DFH equations averaged over statistically representative ensemble of pores. The new workflow amounts to an effective upscaling procedure allowing multiphase simulation on coarse resolution digital rock models that include information that can be obtained on high-resolution models. The DFH+CPD workflow was implemented within the Schlumberger DHD simulator.

To validate the new workflow, we compared absolute and steady-state relative permeabilities simulation results on the high-resolution model and the low-resolution model of the same region within the heterogeneous carbonate core. Then we demonstrated simulation of the steady-state relative permeabilities on the upscaled digital rock model that covers almost entire 8 mm mini plug.

We thank Schlumberger for permission to publish this work.

References

1. Blunt, M.J., *Multiphase Flow in Permeable Media: A PoreScale Perspective*, (Cambridge University Press, Cambridge, 2017). ISBN 978-1-107-09346-1
2. C.F. Berg, O. Lopez, H. Berland, J. Petrol. Sci. Eng. **157**, 131 (2017)
3. T. Bultreys, W. De Boever, V. Cnudde, Earth Sci. Rev., **155**, 93 (2016)
4. M.J. Blunt, B. Bijeljic, H. Dong, O. Gharbi, S. Iglauer, P. Mostaghimi, A. Paluszny, C. Pentland, Adv. Water Resour. **51**, 197 (2013)
5. V. Joekar-Niasar, M.I.J. van Dijke, S.M. Hassanizadeh, Transp. Porous Med. **94**, 461 (2012)
6. P. Meakin, A.M. Tartakovsky, Rev. Geophys., **47**, RG3002 (2009)
7. O. Dinariev, J. Appl. Math. Mech. **59**, 745 (1995)
8. O. Dinariev, J. Appl. Math. Mech. **62**, 397 (1998)
9. A. Demianov, O. Dinariev, N. Evseev, Can. J. Chem. Eng., **89**, 206 (2011)
10. A.Yu. Demianov, O.Yu. Dinariev, N.V. Evseev, *Introduction to the density functional method in hydrodynamics* (Fizmatlit, Moscow, 2014). ISBN 978-5-9221-1539-1
11. D. Koroteev, O. Dinariev, N. Evseev, D. Klemin, A. Nadeev, S. Safonov, O. Gurpinar, S. Berg, C. van Kruijsdijk, R. Armstrong, M. T. Myers, L. Hathon, H. de Jong, Petrophysics **55**, 294 (2014).
12. R.T. Armstrong, N. Evseev, D. Koroteev, S. Berg, Adv. Water Resour., **77**, 57 (2015)
13. R.T. Armstrong, S. Berg, O. Dinariev, N. Evseev, D. Klemin, D. Koroteev, S. Safonov, Transp. Porous Med., **112**, 577 (2016)
14. N. Evseev, O. Dinariev, M. Hürlimann, S. Safonov, Petrophysics, **56**, 32 (2015)
15. A. Shandrygin, V. Shelepov, R. Ramazanov, N. Andrianov, D. Klemin, A. Nadeev, Yakimchuk, SPE-176629-MS (2015)

16. A. Shandrygin, V. Shelepov, R. Ramazanov, N. Andrianov, D. Klemin, A. Nadeev, S. Safonov, I. Yakimchuk, SPE-182037-MS (2016)
17. O. Dinariev, N. Evseev, Computat. Geosci., **20**, 835 (2016)
18. O. Dinariev, N. Evseev, Phys. Rev., **E95**, 063307 (2017)
19. S.I. Geetan, R.M. MacDonald, D. Klemin, 2461115-MS URTeC Conference Paper (2016)
20. S.I. Geetan, R.M. MacDonald, D. Klemin, 2671283-MS URTeC Conference Paper (2017)
21. R.M. MacDonald, S.I. Geetan, D. Klemin, B. Hill, 2902911-MS URTeC Conference Paper (2018)
22. T.D. Scheibe, W.A. Perkins, M.C. Richmond, M.I. McKinley, P.D.J. Romero-Gomez, M. Oostrom, T.W. Wietsma, J.A. Serkowski, J.M. Zachara, Water Resour. Res., 51, 1023 (2015)
23. C. Soulaine, P.Creux, H.A. Tchelepi, Transp. Porous Med. **127**, 85 (2019)
24. O.Yu. Dinariev, J. Appl. Math. Mech. **69**, 213 (2005).
25. R.T. Armstrong, J.E. McClure, M.A. Berrill, M. Rücker, S. Schlüter, S. Berg, Phys. Rev. **E94**, 043113 (2016)
26. O.Yu. Dinariev, N.V. Evseev, Miner. Eng., **125**, 239 (2018)
27. K.R. Daly, T. Roose. Proc. R. Soc. **A 471**, 20140564 (2015)

Uncertainty span for relative permeability and capillary pressure by varying wettability and spatial flow directions utilizing pore scale modelling

Thomas Ramstad, *Anders* Kristoffersen, and *Einar* Ebeltoft

Equinor ASA, Norway

Abstract. Relative permeability and capillary pressure are key properties within special core analysis and provide crucial information for full field simulation models. These properties are traditionally obtained by multi-phase flow experiments, however pore scale modelling has during the last decade shown to add significant information as well as being less time-consuming to obtain.

Pore scale modelling has been performed by using the lattice-Boltzmann method directly on the digital rock models obtained by high resolution micro-CT images on end-trims available when plugs are prepared for traditional SCAL-experiments. These digital rock models map the pore-structure and are used for direct simulations of two-phase flow to relative permeability curves.

Various types of wettability conditions are introduced by a wettability map that opens for local variations of wettability on the pore space at the pore level. Focus have been to distribute realistic wettabilities representative for the Norwegian Continental Shelf which is experiencing weakly-wetting conditions and no strong preference neither to water nor oil. Spanning a realistic wettability-map and enabling flow in three directions, a large amount of relative permeability curves is obtained. The resulting relative permeabilities hence estimate the uncertainty of the obtained flow properties on a spatial but specific pore structure with varying, but realistic wettabilities.

The obtained relative permeability curves are compared with results obtained by traditional SCAL-analysis on similar core material from the Norwegian Continental Shelf. The results are also compared with the SCAL-model provided for full field simulations for the same field. The results from the pore scale simulations are within the uncertainty span of the SCAL models, mimic the traditional SCAL-experiments and shows that pore scale modelling can provide a time- and cost-effective tool to provide SCAL-models with uncertainties.

1 Introduction

Wettability is a key factor for multiphase flow behaviour in porous media in general, and for reservoir rocks especially. In the origin of the petroleum industry, clastic reservoir rocks where assumed to behave water-wet due to the water-wetting minerals composing the rock and pore surface. Since the 70'ties, several papers [see i.e 1,2] have questioned this assumption by adsorption of oil components on the pore surface due to braking the thin water-film established between the oil-phase and the rock surface [3]. It is now well acknowledged that the wettability of clastic reservoir rocks experiences weakly wetting behaviour and generally no strong preference to neither oil nor water.

Wettability can be determined and studied by different approaches [2-8]. The most common for the petroleum industry have been either measurements of contact angles, θ, and interfacial tension, σ/IFT, [4] or macroscopic experiments and measurements on core samples by the traditionally Amott/Amott-Harvey and USBM methods [9,10,11]. The benefit of the first approach is that the wetting preference is well defined, however the major – and crucial - drawback is that the measurements must be done on a flat surface with a geometry far from the geometry and shape of the pore structure in porous media. Macroscopic measurements are done on core samples exposed to real fluids at reservoir conditions, however the measurements are not regarded as well-suited to define the specific wettability where limited preferences to neither of the phases are experienced – which is the case for most clastic reservoir rocks. Even though the macroscopic approach has its limitation, it is the most used method by the industry due to its pragmatic approach and it provides reliable quality information of the wettability preference.

A huge amount of wettability definitions has been suggested during the years: mixed-wet, intermediate-wet, neutral-wet, weakly water-wet, weakly to strongly water-wet etc. Even though the origin of these definitions was intended to be well-defined, the pragmatic use of the terms is not. When providing SCAL-models for full-field applications, the methods and pragmatic understanding described in ref. [2-8] are generally applied. The core sample is generally cleaned, irreducible water saturation, S_{wir}, is established, the core sample is then saturated with reservoir oil at reservoir conditions – and subsequently aged for two to three weeks at irreducible water saturation. The core is then, by definition,

* Corresponding author: trams@equinor.com

restored to its original wetting conditions. This approach is well established, even though the distribution of the (irreducible) water-film is not known and the local wettability in the pore space, pore throats, body and pore necks are not known.

SCAL data for use in reservoir modelling are commonly obtained through core flooding experiments. Detailed information about wettability is distributed in the core sample, and how that influence the fluid distributions inside the pore space, is limited. Recent advances in X-ray CT technology have made non-destructive imaging of core material readily available in the petroleum industry. High resolution micro-computed tomography (micro-CT) enables both resolving the pore space to a high degree of accuracy and visualize the distribution of fluids at the pore level – before and after a flooding. Such technology provides a valuable tool for interpreting and augmenting SCAL data. Micro-CT images can in addition be segmented into pore scale models that can be used for flow simulations. To optimally utilize the results from this new imaging equipment, a digital workflow that combines experimental imaging with simulations directly on these images will be essential.

In this paper we combine the approach of using the well-defined wettabilities by contact angles and interfacial tension distributed on the porous structure of the rock surface obtained by micro-CT imaging and direct pore scale simulations.

Wettability maps are generated, where wettabilities can be varied for different parts of the pore structure and/or varied from sample to sample, experiencing identical pore structure. This will provide additional and important information how the various definitions of wettabilities influence the dynamic flow behaviour, i.e. relative permeability and capillary pressure.

2 Methods

In this section we present the background and techniques for the different steps in the workflow. Every step provides useful information, but ultimately, we wish to gain direct information about essential SCAL data like relative permeability and capillary pressure for different wettability and orientations.

The total workflow contains the following steps:

- Sample selection and preparation
- High resolution Micro-CT imaging
- Image segmentation
- Direct pore scale flow simulations in the segmented images

2.1 Samples and preparation

End trims from representative core material were selected as starting point for the workflow. The core material has been part of a SCAL program and comes from two different wells in the same formation on the Norwegian Continental Self (NCS).

The end-trims themselves are incomplete and otherwise not suited for flooding experiments. However, they can be imaged in micro-CT and thus provide pore scale images used for further modelling.

Examples of overview scans and sample selection area are shown in Fig. *1*. Coarse micro-CT were first done to evaluate the integrity of the material and find suited parts for sub-sampling. The samples were embedded in epoxy solution before further handling to avoid alterations in the pore space of loosely consolidated grains. After epoxy treatment, micro-plugs with diameter 5 mm were drilled for further imaging. Presently, no confining pressure can be applied which may affect the pore throats and measured permeability.

The plug preparation and scanning were carried out by a FEI HeliScan® micro-CT instrument.

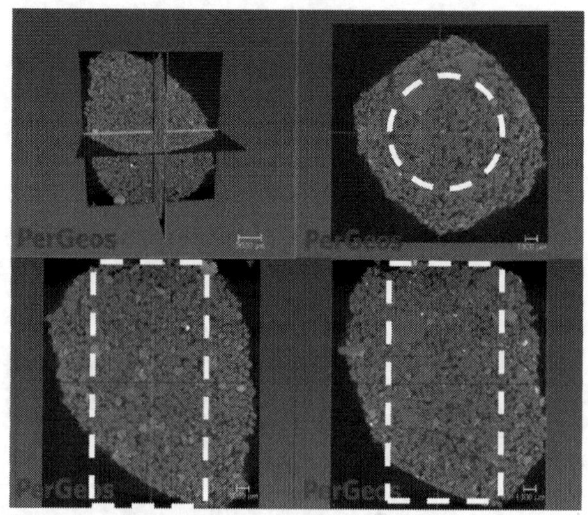

Fig. 1 Overview micro-CT scan of an end-trim sample. The micro-plug selection is marked.

2.2 MCT, segmentation and downsampling

The 5 mm samples were scanned at a photon energy of 100 kV and the resolution was 4.2 µm/pixel.

https://doi.org/10.1051/e3sconf/202014601002

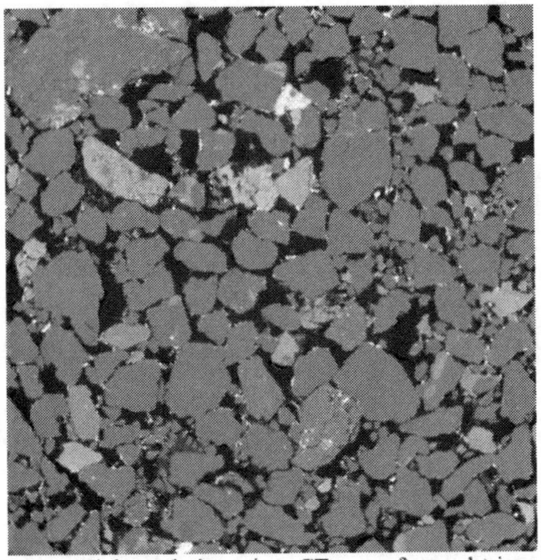

Fig. 2 High resolution micro-CT scan of an end-trim sample.

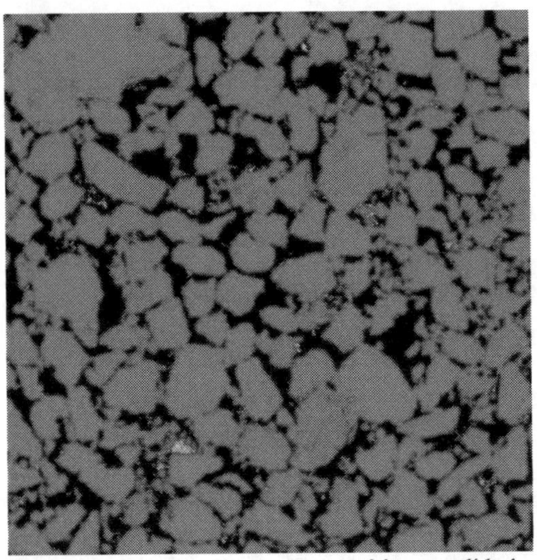

Fig. 3 The image in Fig. *2* segmented into a solid phase (red), pore phase (black) and a sub-resolution porosity phase, which mainly corresponds to clay (green).

The segmentation was performed in the Avizo® software, using a marker-controlled watershed algorithm. An example of a raw image and the corresponding label image is shown in Fig. *2* and Fig. *3*, respectively. The image is segmented into three phases (solid, pore and sub-resolution porosity). The lattice-Boltzmann algorithm inputs a binary map (solid/pore). Therefore, the sub-resolution porosity was assigned to the solid phase. The reason is that the sub-resolution porosity is clay, and we do not expect clay-bound water to contribute to the flow. It should also be noted that the clay content was small in all of the examined samples as can be seen in Fig. *3*.

Lattice Boltzmann modelling is computationally intensive, and it is crucial to limit the size of the grid. Therefore, it was decided to downsample the images prior to the flow simulations. By downsampling to a coarser resolution, it is possible to model a larger physical section of the plug, thus coming closer to a representative elementary volume.

Fig. *4* shows a slice at the original resolution and one that is downsampled. Visually, there seems to be no significant loss of information. To verify this, the relative permeabilities were calculated on the same physical sample at three different resolutions; 7 µm, 10 µm and 15 µm (data not shown). Indeed, the results were equal. It was therefore concluded that the downsampling approach was justified. The length dimensions of the samples are listed in Table 1.

Fig. 4. An image slice at the original resolution of 4.2 µm (top) and the corresponding slice downsampled to a resolution of 15 µm (bottom).

2.3 Direct pore scale simulations

Multiphase flow is simulated directly on the segmented images from the micro-CT scans. We use a multiphase lattice Boltzmann method (LBSIM) for this purpose, and the simulations perform the basis for the SCAL results produced.

The lattice Boltzmann method is a kinetic approach to fluid dynamics and is well documented in the literature [16]. The method is well suited for simulating multiphase flow in complex geometries and fluid interfaces that are the cases for flow in porous media. From such simulations it is possible to obtain capillary pressure and relative permeability that can be compared to different experimental set-ups.

For the multiphase simulations presented in this paper we use the following set-up:

1. Irreducible water saturation is established through primary drainage simulations
2. Wettability in the model is established either as a uniform wettability or through a local wettability distribution on the pore space, i.e. wettability map
3. Water flooding capillary pressure curves are obtained through forced imbibition
4. Relative permeability points are generated from a steady-state set-up. The results are then analysed using the software Sendra® [12] with the end-point obtained from the capillary pressure simulations.

The different steps of the workflow can be repeated with different conditions, and the simulations thus provide a fast tool for sensitivity testing, such as wettability and spatial flow direction.

The forced imbibition capillary pressure simulations are done by injecting water with a fixed pressure drop over the model. When the water saturation has converged, the capillary pressure is recorded, and the applied pressure drop increased. This way we obtain the capillary pressure curve and the end-point residual saturations.

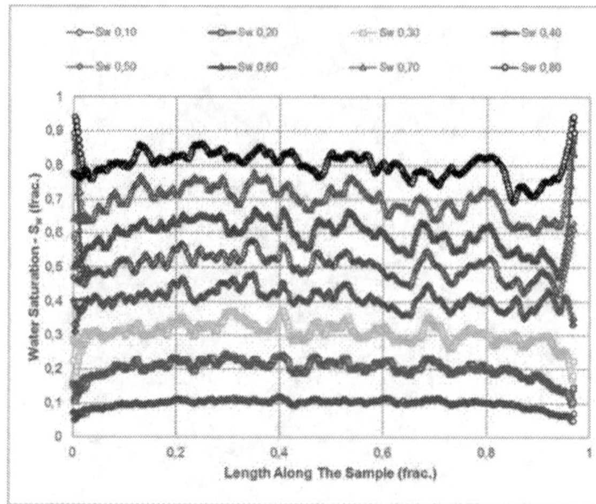

Fig. 5. Water saturation profile at steady state in the simulations.

The steady state relative permeability curves are done with periodic boundary conditions to minimize end-effects. When the effective permeability of each phase has converged, the saturation is changed to preserve the fluid distribution history. This is done to model the proper effect of steady state water flooding with increasing water saturation.

As shown in Fig. 5, the saturation distribution in the sample is close to uniform, and end-effects are relatively small despite the limited physical size of the sample.

Steady state effective permeability values are normalized with the initial oil effective permeability at irreducible water saturations to get relative permeability. The steady state set-up is suited for obtaining the relative permeability in the mid saturation region. However, it suffers from resolution limitations closer to low end-point values. To compensate for this, a curve fit is done combining steady state points with residual points from the capillary pressure simulations.

2.4 Wettability distribution

One of the main aims of this study is to investigate the effects of wettability and to clarify some of the concepts related to wettability distribution. With the use of rock images and modelling, we can set the wettability locally on the grains and investigate the resulting fluid distributions.

In LBSIM the wettability is defined through a scalar affinity value that controls the local wettability. The local affinity values range from 1.0 to -1.0. On a plain surface, this correlation can be translated into a contact angle between the fluid and solid pore walls. However, for complex and rugged grain surface it is difficult to define unique contact angle. Hence, it is common to refer a local wettability of 1.0 to strongly water wet (WW) while a value of -1.0 relates to strongly oil wet (OW). In terms of global wettability for the sample, there are two ways of setting that in the simulations, which we refer to as: Uniform wettability and non-uniform wettability.

For a uniform wettability the local wetting parameter is constant throughout the sample. This is indeed a special case, as there are many local effects that can alter wettability, like e.g. fluid composition and mineralogy. However, a uniform wettability can be justified in cases of a clear wettability trend in the entire sample like strongly WW or strongly OW.

A more sophisticated way of handling the wettability is by distribution of the local wettability parameters in the pore space. Such a control of the local wettability opens for a more in-depth treatment of wettability effects in a setting of non-uniform wettability.

There are many ways to define the wettability in the samples, but in this study, the local wettability parameters are limited to a static binary wettability map as shown in Fig. *6*.

The wettability maps consist of two affinity parameters: OW and WW, and they only apply to the fluid grid voxels that border to solid. The OW and WW are distributed according to the fluid distribution at initial water saturation, S_{wi}. Grid voxels where the oil is in contact with the grain is set to OW while the WW is applied elsewhere.

We chose to set the WW parameter to WW = 1.0 while the OW will vary from 1.0 to -1.0. This to mimic an aging process. Changing wettability though an aging is a slow process that usually takes weeks. Hence, the values of OW and WW are kept static and do not change throughout a simulation.

A resulting fluid distribution is shown in Fig. *7*. We can observe the curvature of droplets and the contact with grains. This is further magnified in Fig. *9* and Fig. *10*. Such fluid distributions can again be compared directly to micro-CT images of flooding experiments (see e.g. ref [19]).

Fig. 6 Binary wettability map where red denotes OW index and blue is WW index. Only voxels that border to solid grains will be affected by the wettability index.

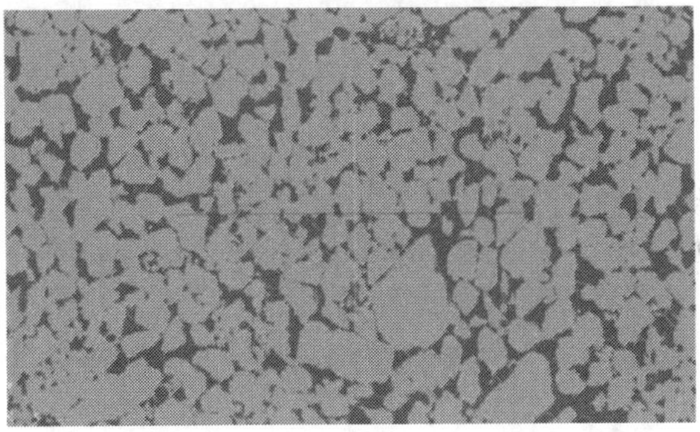

Fig. 7. A resulting fluid distribution from the wettability map in Fig. **6**. The water saturation is $S_w = 0.5$. Water is blue and oil is red.

3 Results

This section provides the results obtained by combining Micro-CT images, spatially pore scale simulations combined with wettability distribution within the porous structure.

3.1 Models and Set-up

We have used the workflow and tools listed in Section 2 to build pore scale models and run two-phase direct flow simulations with LBSIM. The dimensions and properties of the models are listed in Table 1:

Table 1. Basic properties of the digital models

Sample	A	B	C
Size [voxels]	485 x 485 x 1085	485 x 485 x 878	485 x 485 x 878
Resolution [microns]	15	15	15

Dynamic SCAL-experiments utilizing either the steady state- or the unsteady state techniques are generally exposed to capillary end-effects and are not well suited to provide estimates of the residual oil saturation, S_{orw}. A common approach is to combine such flow experiments with comparable centrifuge or porous plate techniques which are well suited to provide estimates of residual saturations [13]. This behaviour is also experienced for steady state pore scale modelling on digital images. Hence, it tends to be unsuited for determination of very low relative permeabilities close to residual saturations, S_{wir} and S_{orw}, both because of low fractional flow and limited resolution. The steady state dynamic simulations are here done at typical capillary numbers of $Ca = 10^{-5}$.

Pore network analogues can reduce some of the problems related to limited resolutions, but these come at the cost of simplifying the pore scale micro structure and reducing dynamic effects for low capillary numbers [18].

Similar approach as for traditional SCAL-experiments have been utilized in the simulations. Capillary pressure with the coherent saturation end-points are determined prior to determination of the relative permeabilities. These latter simulations are performed with steady state conditions and anchored in the saturation end-points, S_{orw}, determined from the capillary pressure simulations.

When the pore scale models are built, it is straight forward to change input parameters related to wettability and sample orientation and re-run simulations from the same starting point, i.e. the configuration after primary drainage. With the data points obtained from capillary pressure and relative permeability simulations, we have used the software Sendra® for further curve fitting.

The simulations have been done on end-trims from core material defining the SCAL-model used for full-field simulations. The obtained curved are then compared to relevant field SCAL model with uncertainty bands [20].

From the samples listed in Table 1 with sensitivity on wettability and orientation for each sample, we obtain a total of 54 relative permeability curves and 9 capillary pressure curves. These are used to in-depth analyse the impact by wettability on SCAL data and the associated uncertainty in interpretation of SCAL data.

3.2 Capillary pressure and end-points

All capillary pressure data points from the simulations are obtained through a forced imbibition set up. The pressure drop over the model is fixed, and when the time dependent saturation profile has converged, both the resulting saturation and pressure are recorded.

Fig. *8* shows the capillary pressure points determined by pore scale modelling and subsequently curve fitted by the LET-correlation for capillary pressure [14]. This provide guidelines for where to anchor residual oil saturation, S_{orw}, for the relative permeabilities shown in the next section. In addition, the capillary pressure curves are also determined.

The forced imbibition simulations were done with uniform wettability distribution in the intermediate wet range. From the curves, there is a relatively long range of water saturations with negative Pc values.

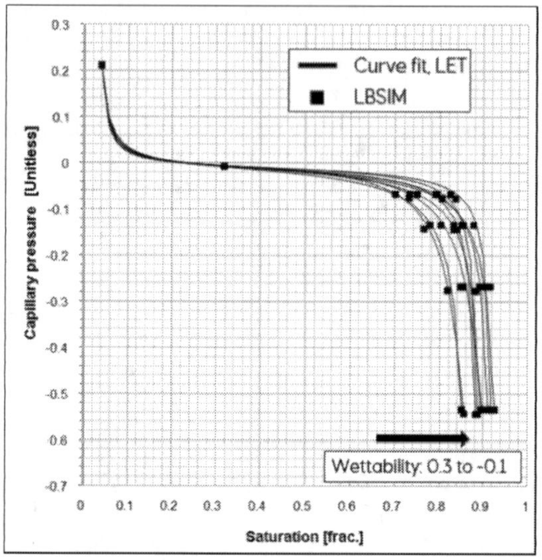

Fig. 8 Capillary pressure curves for all samples with different wettabilities on the intermediate wet range. The end-points are listed in Table 2.

The residual oil saturation, S_{orw}, experiences reliable values for $S_{orw} \in [0.06 - 0.15]$ and are listed in Table 2. These end-points are used in further interpretation of steady state relative permeability curves.

Table 2. End-point saturations from the capillary pressure curves.

Sample	A	B	C
S_{wir}	0.04	0.04	0.04
S_{orw}, OW = -0.5	0,088	0,105	0,149
S_{orw}, OW = 0.3	0,080	0,073	0,101
S_{orw}, OW = 0.5	0,112	0,116	0,144

3.3 Relative permeabilities

Relative permeabilities are obtained by pore scale simulations (LBSIM) with steady state flow conditions.

Each of these samples utilized different wettabilities, and simulations have been done for three spatial directions

A total of six different wettabilities have been used, three cases of uniform wettability and three cases of non-uniform wettability where a binary wettability maps, as described in section 2.4, have been applied. For the uniform wettability cases, the local wettability constant takes the values 0.3, 0.1 and -0.2 giving intermediate global wettabilities. These correspond to the capillary pressure curves shown in Fig. ***8***.

In the cases of the non-uniform wettability the parameter WW (see Section 2.4) has been defined to strictly 1.0 and the OW parameter takes the values 0.5, 0.3 and -0.5. The wettability maps are generated from the S_{wir} distributions, and hence the WW only applies to small pores and corners in pore bodies.

With the OW index taking the value -0.5, the resulting fluid distribution can be observed in detail Fig. ***10***. A wetting configuration that is more towards a global oil wet condition is observed, and water droplets forming on the grain walls are observed.

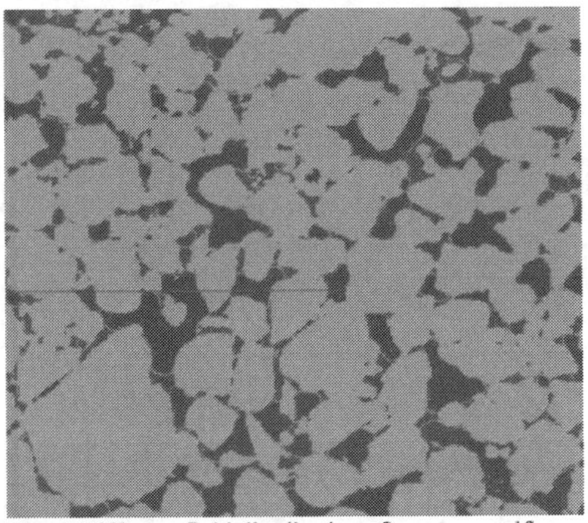

Fig. 9 Different fluid distributions from non-uniform wettability OW = 0.5

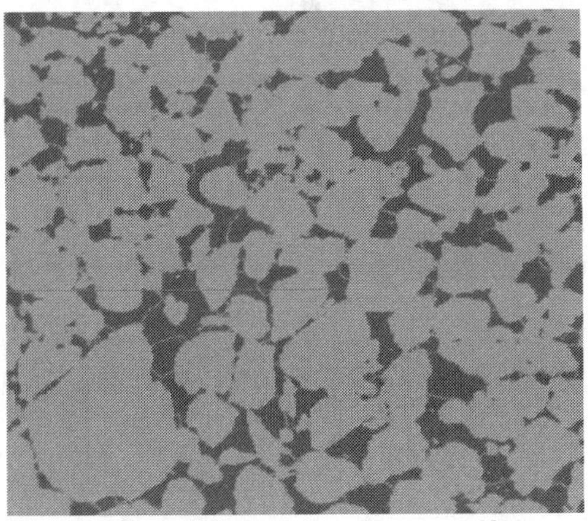

Fig. 10 Different fluid distributions from non-uniform wettability OW = -0.5

With OW = 0.5, the oil resides on the pore bodies as droplets formed by the water wet films and corners, but there is not a large curvature for the interfaces between water and oil in large pores (Fig. ***9***).

These two wettability settings have been compared to the SCAL models in Fig. ***11*** and Fig. ***12***. All relative permeabilities obtained by LBSIM have been curve fitted by the LET-correlation for relative permeability [15]. The observation is that the uniform wettability case produces more symmetric relative permeability curves and the relative permeability of water (K_{rw}) is consistently higher than the SCAL model close to residual oil saturation, see Fig. ***11***. For the non-uniform wet cases the bundle for curves stay within the uncertainty of the SCAL model, $K_{rw}(S_{wir})$ is lower than for the uniform wettability case, albeit on the pessimistic side of the SCAL model, see Fig. ***12***.

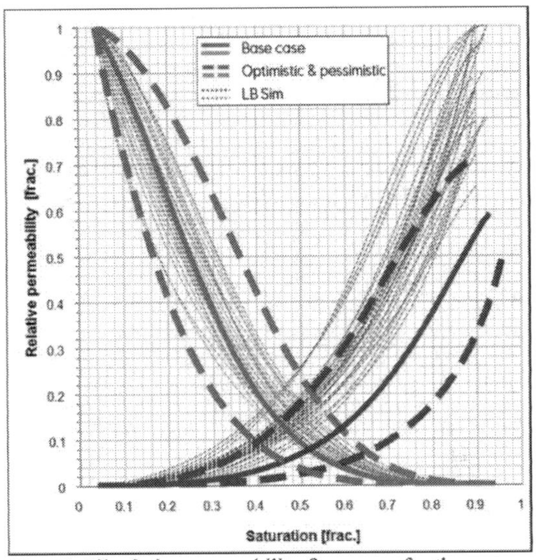

Fig. 11 All relative permeability for curves for the uniform wet cases.

https://doi.org/10.1051/e3sconf/202014601002

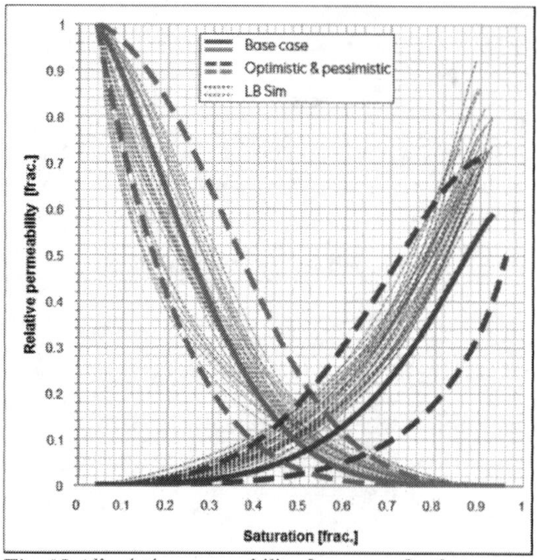

Fig. 12 All relative permeability for curves for the non-uniform wet cases.

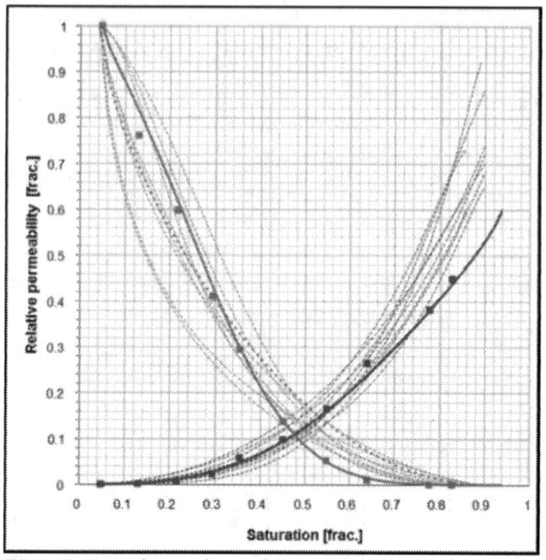

Fig. 13 Experimental steady state relative permeability curves compared to different non-uniform wet conditions and orientations, lin-lin.

In the following we will focus on the sensitivity to wettability using the approach of wettability maps with non-uniform wettability. All results from pore scale modelling non-uniform wet conditions are within the provided uncertainty span for the SCAL-model, even though it can be argued that the water relative permeabilities and oil relative permeabilities are on the pessimistic and optimistic side, respectively. This observation is similar to the results for the comparison to steady state experiments from one individual core - Fig. *13* and Fig. *14* - and is considered for further work when defining the wettabilitiy maps.

As for the individual interpretation the results are within the expected uncertainty for SCAL-experiments and multi-phase flow behaviour in general. The steady state experiments were performed with the following fraction flow of water, f_w = 0.008; 0.06; 0.22; 0.5; 0.78; 0.94; 0.992; 1.

The relative permeablity curves tend to be overestimated for low values - toward the residuals - compared to the experimental results. This is due to limited resolution and finite-size effects in the models and fluid dynamics.

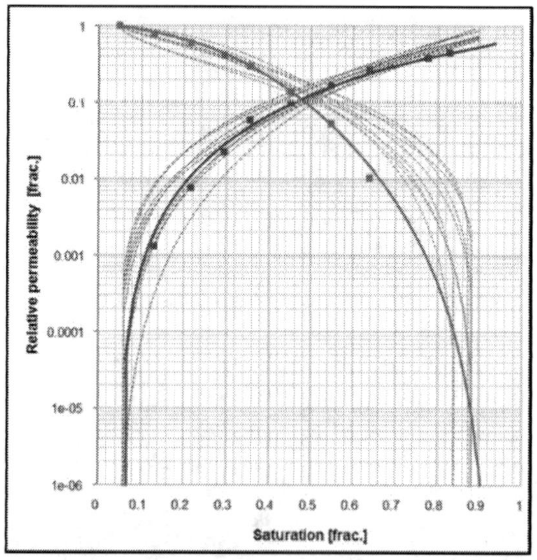

Fig. 14 Experimental steady state relative permeability curves compared to different non-uniform wet conditions and orientations, semilog.

 https://doi.org/10.1051/e3sconf/202014601002

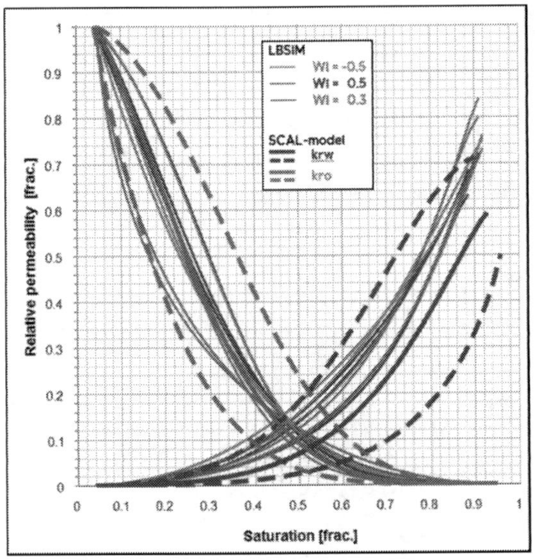

Fig. 15 Wettability sensitivity in relative permeability for Sample A. The data are compared to the SCAL model with uncertainty span (bold lines).

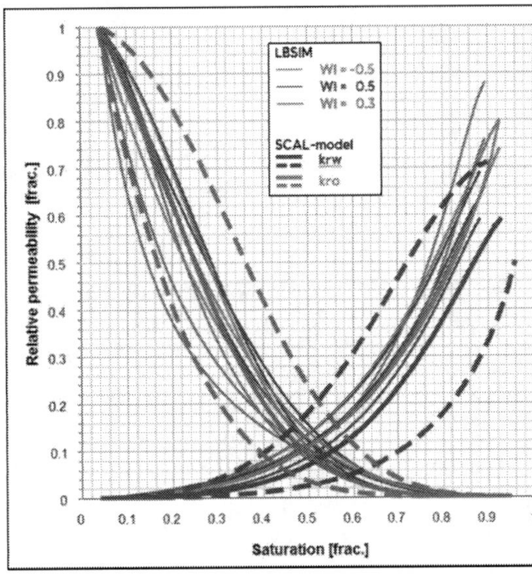

Fig. 17 Wettability sensitivity in relative permeability for Sample B. The data are compared to the SCAL model with uncertainty span (bold lines).

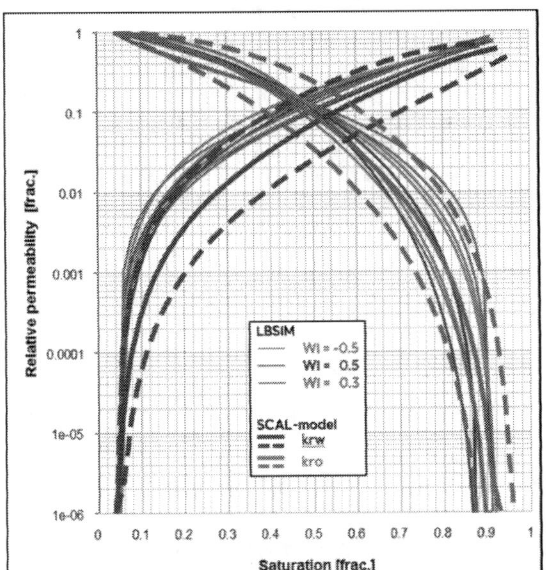

Fig. 16 Wettability sensitivity in relative permeability for Sample A. The data are compared to the SCAL model with uncertainty span (bold lines).

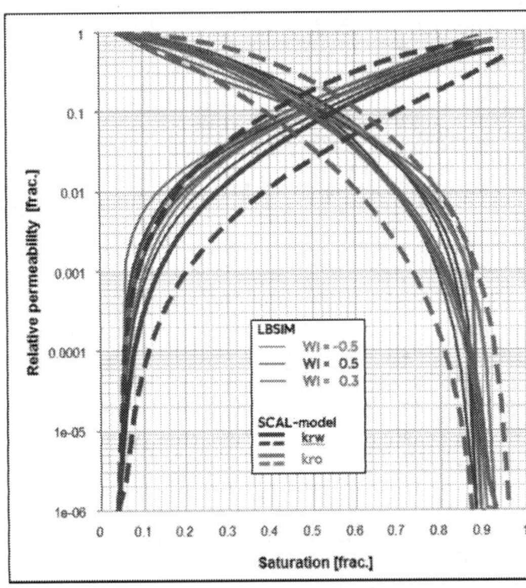

Fig. 18 Wettability sensitivity in relative permeability for Sample B. The data are compared to the SCAL model with uncertainty span (bold lines).

In terms of analysis of uncertainty on relative permeability measurements that owe to wettability, we have plotted the results for each sample in Fig. **16** to Fig. ***20***. The results are differentiated on sample with three different non-uniform wetting conditions. The samples are homogeneous and isotropic. Hence, the

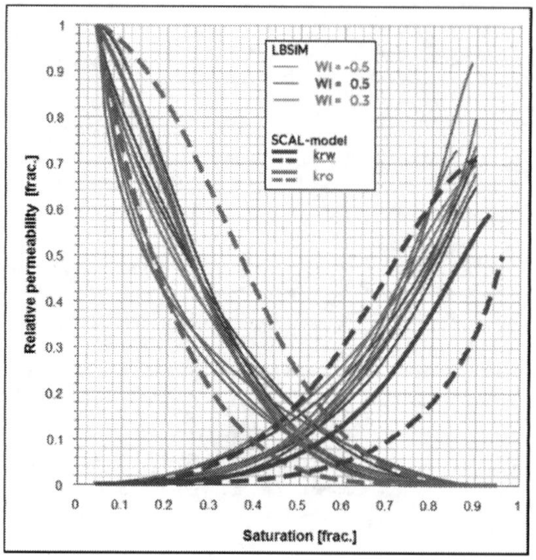

Fig. 19 Wettability sensitivity in relative permeability for Sample C. The data are compared to the SCAL model with uncertainty span (bold lines).

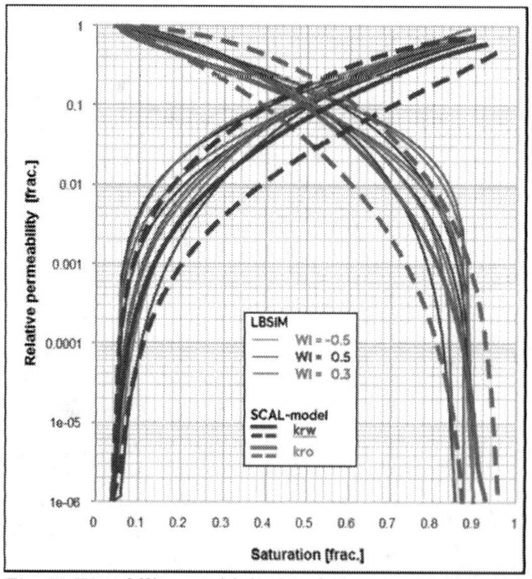

Fig. 20 Wettability sensitivity in relative permeability for Sample C. The data are compared to the SCAL model with uncertainty span (bold lines).

results from orientation should not be systematically different although we expect a certain degree of spreading in the results. Also, the SCAL models with uncertainty span should apply to all samples.

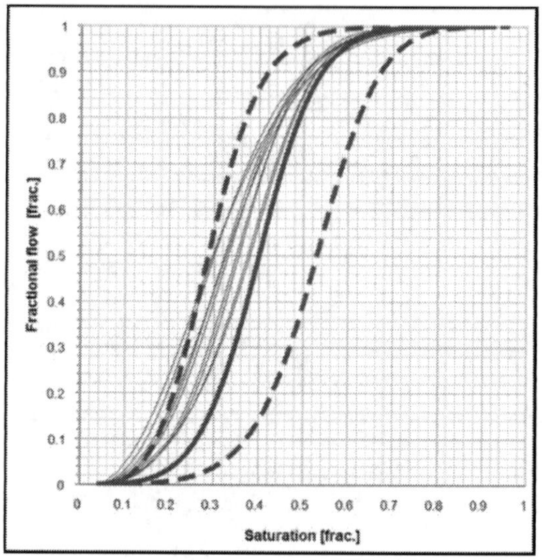

Fig. 21 Fractional flow curves for all non-uniform wettability cases Sample A.

However, we expect to see a certain trend in relative permeability owing to different wettability. With wettability towards more oil wet, the curves tend to be more pessimistic. As the wettability tends more towards the neutral wet regime, the curves are more in agreement with the SCAL model also for the end-points.

Fraction flow curves for all samples are shown in Fig. *21* to Fig. *23*. Most curves fall within the uncertainty span of the SCAL model, but towards more mobile water.

https://doi.org/10.1051/e3sconf/202014601002

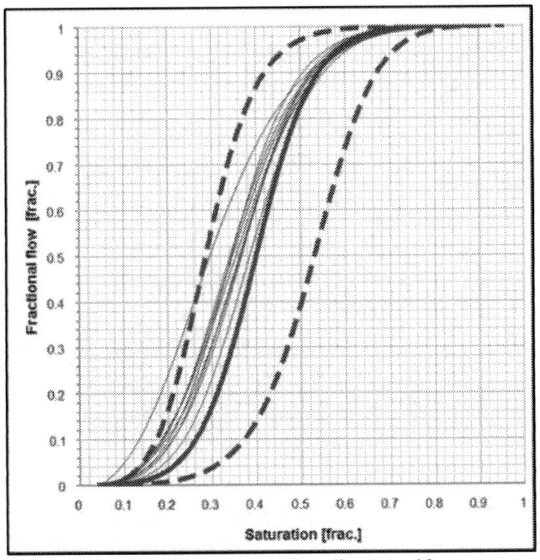

Fig. 22 Fractional flow curves for all non-uniform wettability cases Sample B.

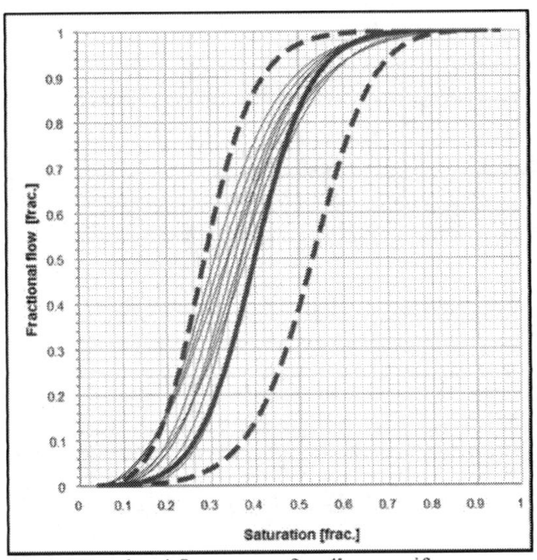

Fig. 23 Fractional flow curves for all non-uniform wettability cases Sample C.

4 Discussion

Pore scale simulations of the individual sample maps the steady state experiment reasonably. It also provides an estimate of the uncertainty span due to spatial- and wettability variations as shown in **Fig.** *11* and Fig. *12*. Even though we have compared with the steady state experiment, it is not - a priori - expected that the results should map each other perfectly. The core samples used for the composite core in the steady state experiments have undergone the common "recommended practice" preparation prior to experimentation – restored state, S_{wir} establishment and aging with live oil. There is no guarantee that this procedure provides the "correct" wettability-state of the core sample prior and during the flow experiment.

The relative permeability to oil and water are slightly over-estimated toward residual oil saturation, S_{orw}, interpreting the water relative permeability, k_{rw}, slightly too pessimistic and the oil relative permeability, k_{ro}, slightly too optimistic compared to the steady state experiment. However, this is within the expected uncertainty for SCAL-experiments and multi-phase flow behaviour in general.

It needs to be emphasized, that the digital models are restrained by the resolution of the micro-CT images. Low relative permeability values close to the saturation end-points and residuals tend to be over-estimated. It is a general short-coming of direct simulation techniques to handle very low fractional flows. There will be a certain amount of finite size effects linked to fluid-fluid interfaces and fluid-solid interactions. In general, these will lead to noise and small-scale currents that contribute relatively stronger to small fractional flow rates. Nevertheless, steady state set-ups are in general not the best candidate for end-point estimations and a combination with additional centrifuge experiments/set-up is advised.

With a detailed mapping of local wettability, we can reproduce expected trends, but also categorize properties on the pore space that are considered "Black Box" in ordinary core flooding experiments. More sophisticated wettability maps should preferably be considered for further work. A wettability map that distribute various in-situ contact angels throughout a mixed-wet pore structure with local varieties depending upon pore-shapes [19]. Further the traditional classifications of wettability - intermediate-wet, neutral-wet, weakly water-wet, weakly to strongly water-wet etc. - can be distributed by defining in-situ contact angels and interfacial tensions to further study variations of realistic wettabilities where none of the fluids have strong wetting preferences.

5 Conclusion

In this paper the following conclusions are drawn:

- LBSIM combined with wettability mapping and spatial directions provides reliable uncertainty span for the relative permeability curves.
- LBSIM and wettability mapping has the potential to provide basic understanding of mechanisms that influence the flow behaviour.

- Non-uniform wettability defined from irreducible oil distribution leaves corners and small pores water wet. Such wettability distribution lowered the water relative permeability curves compared to a uniform wettability distribution. A non-uniform wettability distribution is considered a more realistic scheme.

- Uniform wettability with a weak overall wetting preference produces almost symmetric oil and water relative permeability curves and thus a more pessimistic case compared to SCAL models.

- For a fruitful and possible concluding approach, we need to properly define contact angle and interfacial tension for wettabilities.

References

1. Morrow, N.R. (editor): "Interfacial phenomena in petroleum recovery", 1991, Marcel Dekker Inc., New York.
2. Radke, C.J., Kovscek, A.R. and Wong, H.: "A Pore-Level Scenario for the Development of Mixed Wettability on Oil Reservoirs", SPE 24880, presented at the 67th Annual Technical Conference and Exhibition of the Society of Petroleum Engineers, Washington. DC, October 4-7, 1992.
3. Anderson, W.G.: "Wettability Literature Survey – Part 1; Rock/Oil/Brine Interactions, and the Effects of Core Handling on Wettability" JPT (Oct. 1986), 1125-44.
4. Anderson, W.G.: "Wettability Literature Survey – Part 2; Wettability Measurements" JPT (Nov. 1986), 1246-62.
5. Anderson, W.G.: "Wettability Literature Survey – Part 3; The Effects of Wettability on the Electric Properties in Porous Media" JPT (Dec. 1986), 1371-78.
6. Anderson, W.G.: "Wettability Literature Survey – Part 4; The Effects of Wettability on Capillary Pressure" JPT (Oct. 1987), 1283-1300.
7. Anderson, W.G.: "Wettability Literature Survey – Part 5; The Effects of Wettability on Relative permeability" JPT (Nov. 1987), 1453-68.
8. Anderson, W.G.: "Wettability Literature Survey – Part 6; The Effects of Wettability on Waterflooding" JPT (Dec. 1987), 1605-22.
9. Amott, E.: "Observations Relating to the Wettability of Porous Rock", Trans., AIME (1959) **216**, 156-62.
10. Boneau DF and Clampitt RL: "A Surfactant System for the Oil-Wet Sandstone of the North Burbank Unit" Journal of Petroleum Technology 29, no. 5 (May 1977): 501–506.
11. Donaldson EC, Thomas RD and Lorenz PB: "Wettability Determination and Its Effect on Recovery Efficiency" SPE Journal 9 (March 1969): 13–20.
12. Sendra User guide 2018.5, ProRes AS, 2019.
13. Sylte, A., Ebeltoft, E. and Petersen, E.B.: "Simultaneous Determination of Relative Permeability and Capillary Pressure Using Data from Several Experiments". Paper SCA 2004-08 presented at the International Symposium of the Society of Core Analysts, Abu Dhabi, UAE, 5-9 October 2004.
14. Lomeland, F., and Ebeltoft, E.: "A New Versatile Capillary Pressure Correlation". Paper SCA 2008-08 presented at the International Symposium of the Society of Core Analysts, Abu Dhabi, UAE, 29 October – 2 November 2008.
15. Lomeland, F., Ebeltoft, E., and Thomas, W.H.: "A New Versatile Relative Permeability Correlation". Paper SCA 2005-32 presented at the International Symposium of the Society of Core Analysts, Toronto, Canada, 21-25 August 2005.
16. T. Ramstad, I. Nasiru, C. Nardi, P. E. Øren, "Relative permeability calculations from two-phase flow simulations directly on digital images of porous rocks", Transport in Porous Media **94**, 487 - 504 (2012)
17. Lopez, O., Mock, A., Skretting, J., Pedersen EB., Øren, P., Rustad, AB., "Investigation into the Reliability of Predictive Pore-Scale Modeling for Siliciclastic Reservoir Rocks", Proceedings of the 2010 International Symposium of Core Analysts, SCA2010-23 (2010)
18. Ruspini, LC., Farokhpoor, R., Øren, P., "Pore-scale modeling of capillary trapping in water-wet porous media: A new cooperative pore-body filling model". Adv. Water Resources, (2017), **108**, 1-14.
19. Lin, Q, Bijeljic, B., Krevor, S.C., Blunt, M. J., Berg, S., Coorn, A., van der Linde, H., Georgiadis, A., Wilson, O.B.: "A New Waterflood Initialization Protocol for Pore-Scale Multiphase Flow Experiments". Paper SCA 2018-32 presented at the International Symposium of the Society of Core Analysts, Trondheim, Norway, 27-30 August 2018.
20. Ebeltoft, E., Lomeland, F., Brautast, A., Haugen, Å.: "Parameter Based SCAL - Analysing Relative Permeability for Full Field Application", reviewed Proceedings of the 2014 International Symposium of the SCA, Avignon, France, September 8-11, 2014.

Permeability alteration by salt precipitation: numerical and experimental investigation using X-Ray Radiography.

Olivier Lopez[2], *Souhail Youssef*[1], *Audrey Estublier*[1], *Jostein Alvestad*[2], *Christin Weierholt Strandli*[2]

[1]IFP Energies Nouvelles, Rueil-Malmaison France
[2]Equinor, Trondheim, Norway

Abstract. The injection of a gas phase through a water saturated porous medium can reduce the water saturation not only by displacement mechanisms but also by evaporation mechanisms. In the presence of brine, this process can induce salt crystallization and precipitation within the porous medium with a risk of permeability alteration. In the field of gas production and storage, the occurrence of such a phenomenon can have detrimental consequence on the well productivity or injectivity. In this work, we investigated experimentally and numerically the effect of dry gas injection on salt precipitation and permeability impairment. State of the art equipment designed for high throughput coreflood experimentation was used to capture the dynamic of salt migration using X-Ray radiography. A set of experiments have been conducted on a sample of Bentheimer sandstone (10mm in diameter and 20 mm in length) as well as a two layers composite sample with a significant permeability contrast. Experiments were conducted using Nitrogen and KBr brine with different boundary conditions (i.e. with and without capillary contact). Results showed that salt precipitation results from the interplay of different parameters, namely pressure gradient, brine salinity, capillary forces and vapor partial pressure. Experimental observations indicate that in the case of dry gas injection, salt systematically precipitates but permeability alteration is observed only if a capillary contact is maintained with the brine. We built a 2D flow model integrating two-phase Darcy flow, capillary forces, salt effect on vapor partial pressure, dissolved salt transport, as well as the different PVT equilibria needed to describe properly the systems. Once calibrated, the model showed good predictability of lab scale experiment and thus can be used for parametrical study and upscaled to the well bore scale.

1. Introduction

In the field of natural gas production as well as gas storage in aquifer, salt deposition is considered as a potential risk for gas wells. In Northern Germany, field evidences of mainly halite precipitation have been reported during gas production [1]. A synthetic review of gas well productivity impairment by salt plugging can be found in [2]. Injection of a gas phase through a water saturated porous medium can reduce the water saturation not only by displacement mechanisms but also by evaporation mechanisms. To understand the phenomena of drying induced by gas production, Mahadevan *et al.* [3] did an extensive experimental and theoretical study to show that injection of vapor saturated gas (i.e., humid gas) in a rock sample can induce drying [3-4]. The origin of the drying is that the water mole fraction in the gas phase at equilibrium depends on pressure. In the context of gas storage in aquifer, experiments have been conducted by [5-6] where dry gas was injected in brine saturated sandstone samples with different salt concentrations in the brine. They showed that the drying rate is proportional to the gas velocity and that any dissolved salt in the water can be transported by capillary flows induced by drying and can even accumulate near the injection surface. The permeability can then be reduced by pore clogging. They conclude that all these mechanisms must be included in models to calculate the near wellbore permeability and porosity changes with time, to correctly determine well injectivity.

A first attempt to model the drying process in a porous media at the laboratory scale was proposed by Mahadevan *et al.* [4]. The authors considered two different models, one to model the first stage of the immiscible displacement, and another during the evaporation regime. The first part is described by a standard two-phase Darcy model. The second accounts for compressibility-driven drying and capillarity effects. The two models were not coupled. The evaporation model was then extended in [7] to consider the impact of capillary-driven liquid film flow on the evaporation rates as well as the rate of transport of salt through those films. TOUGH2 family of reservoir simulators was also used by [8-9] to simulate two-phase flow of a gas mixture in equilibrium with a sodium chloride brine, including the modeling of halite precipitation/dissolution. The simulations were conducted for near well bore model in the context of CO_2 injection at laboratory scale.

In this study, we present a comprehensive experimental investigation of the effect of gas flow in a sandstone sample initially saturated with water or brine. A set of experiments have been conducted on a Bentheimer sandstone sample (10mm in diameter and 20 mm in length) as well as on a two layers composite sample of different permeability (high and low) with the aim of investigating the effect of salt precipitation on permeability impairment at different boundary conditions. Experiments were then simulated using an in-house advanced compositional model (CooresFlow).

2. Materials and methods

2.1. Experimental set-up

The experimental set-up used in this study is shown in **Fig. 1**. Nitrogen was injected at a constant pressure using a downstream pressure regulator. Its flow rate was monitored by a Coriolis mass flow meter Brine solution was injected using an Isco Pump. The different lines were connected to an X-ray transparent hassler type cell equipped from each side with three connections (inlet, outlet and static line for the pressure transducer). A data acquisition system was used to monitor the pressure, mass flow and temperature. Finally, the set-up was installed in dedicated radiography facility equipment for in-situ saturation monitoring.

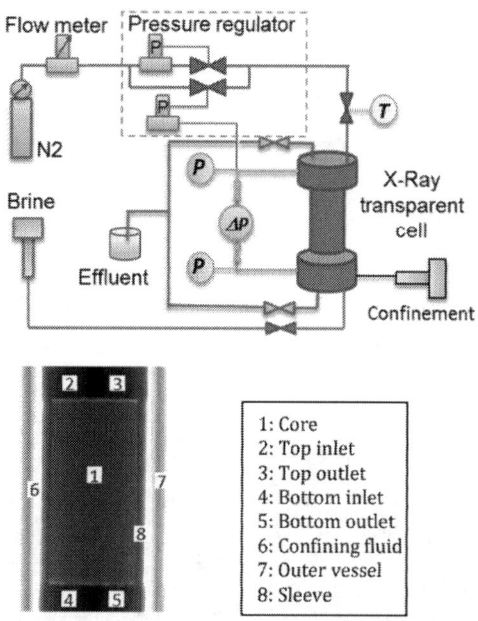

Fig. 1 Experimental set-up and radiograph showing the sample shape and the basic components of the Hassler injection cell.

During fluid injection, the sample was imaged using an X-ray radiograph equipment. The system consists of an X-ray source, a flat panel and a data processing unit for computation, visualization and data analysis. The radiographs were captured with a resolution a 20 µm at a rate of one frame per second. Radiographs were then cropped to show only the sample (see zone 1 in Figure 1). The stack of images was processed to monitor water saturation and salt concentration evolution during gas injection [10]. Indeed, image gray levels are proportional to the transmitted X-ray intensity which is defined by the Beer Lambert's law in the case of a multicomponent material as:

$$I_t = I_o\, e^{-(\mu_{bt} X_b + \mu_g X_g) + \alpha}$$

where I_o and I_t are respectively the incident and the transmitted X-ray intensity, μ_{bt} and μ_g are respectively the brine and the gas attenuation coefficients, X_w and X_g are respectively the brine and the gas portions in the linear path of the X-ray and α is the attenuation of the structure. Considering two references states, 100% brine saturated and dry images the above equation can be written as:

$$I_b = I_o\, e^{-\mu_{bo} X + \alpha} \text{ and } I_g = I_o\, e^{-\mu_g X + \alpha}$$

where X is the linear path of the X-ray that corresponds to the total porosity and μ_{b0} is the initial brine attenuation coefficient.

By construction we obtain:

$$X_b + X_g = X$$

Afterward, normalizing the X-ray intensity I_t by both the 100% brine and the dry case, and considering $\mu_b \gg \mu_g$ we obtain:

$$\frac{\ln\left(\frac{I_b}{I_0}\right) - \ln\left(\frac{I_t}{I_0}\right)}{\ln\left(\frac{I_b}{I_0}\right) - \ln\left(\frac{I_g}{I_0}\right)} = \frac{-\mu_{b0} X + \mu_{bt} X_b}{-\mu_{b0}\, X}$$

When salt concentration is constant we have:

$$\mu_{bt} = \mu_{b0} \text{ and } \frac{X - X_b}{X} = S_g$$

where S_g is the gas saturation. If the salt concentration changes, μ_{bt} will vary with an upper limit equal to the KBr attenuation coefficients. If we normalize the X-ray transmitted intensities (i.e. relative X-ray absorption) by the mean intensity of a dry sample with precipitated salt, we can assess qualitatively the variation of salt quantity compared to the initial distribution.

2.2. Experimental procedure

A set of experiments has been conducted using the set-up described above. The experiments have been conducted on the Bentheimer sandstone sample and the Bentheimer/Molière composite core. Four steps have been applied to investigate the effect of salt on the drying dynamics and the effect of capillary contact on salt accumulation and permeability alteration. the sequence of injection consisted in:

1-Saturating the sample with the wetting phase by injecting the water (or brine) from the bottom inlet and opening the top outlet (the bottom outlet and top inlet are closed). To ensure a full saturation, we have imposed a back pressure of 4 bar and monitored the saturation by X-ray.

2-Lowering the pore pressure to 1 bar, opening the bottom outlet and top inlet and closing the top outlet. The water injection is maintained allowing a sweeping of the bottom face of the sample.

3-The gas is then injected at a constant pressure. The variation of the saturation and the local concentration of salt is monitored by the X-ray radiographs and the gas flow rate is monitored by the Coriolis flow meter. The

https://doi.org/10.1051/e3sconf/202014603001

injection is stopped once no more variations are observed on the radiograph and the flow rate.

4-The sample is finally cleaned by injecting deionized water until the reference X-ray absorption is reached.

2.3. Sample properties and fluids

The experiments were conducted using two consolidated water-wet sandstones: Bentheimer, and Molière. Bentheimer sandstone is a high porosity (23%) permeable (1350 mD) rock composed mainly of quartz; while Molière sandstone is a medium porosity (14%) with very low permeability (0.05 mD) composed of quartz and pore filled clay.

In the three first experiments a Bentheimer core of 1 cm in diameter and 2 cm in length was used. In the last experiment and to reproduce a lateral capillary contact parallel to the flow direction, a composite sample was shaped using Bentheimer and Molière sample. The plugs were first half cut in the longitudinal direction using a diamond saw. The two half pieces were then gathered by using a nonwoven fabric reinforced by a wet ceramic powder (see **Fig. 2** a) to improve capillary contact. The Bentheimer half core is made shorter than the Molière one to avoid outlet capillary contact with the distributor (see **Fig. 2**.b).

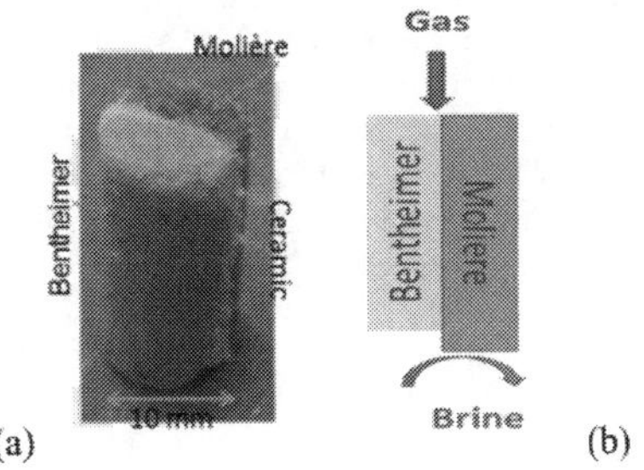

Fig. 2 (a) Photographs of the composite core composed of Bentheimer and Molière half cores contacted with nonwoven fabric reinforced by a wet ceramic powder. (b) Sketch of the composite sample during fluids injection.

To perform the different experiments, we have used dry nitrogen and 100 g/l KBr brine. KBr salt was chosen as it has a high electronic density allowing a good X-ray contrast with water and gas. The solubility limit of KBr in water at 20°C is 560 g/l.

3. Experimental observations

As pointed out in the introduction, drying and salt precipitation result from the interplay of different parameters. To separate the effect of these parameters, we have conducted a non-exhaustive number of experiments that focus on the effect of salt on permeably alteration with three different boundary conditions: no capillary contact at the outlet face, capillary contact at the outlet by brine seeping and lateral capillary contact using a composite sample as described in the previous section. In the following, we discuss the impact of these parameters.

3.1. Effect of salt

To study the effect of salt on permeability alteration, we have conducted two experiments where we have first injected dry gas in a Bentheimer sample initially saturated with deionized water (Exp 1). The purpose of this first experiment is to observe the drying dynamics without the interference of the salt. The inlet relative gas pressure was set to 100 mbar. The pressure drop was chosen to be above the entry pressure of 70 mbar and corresponds to a pressure gradient of 5 bar/m which is representative of near well bore pressure gradient. **Fig. 3** shows the evolution of the mean gas saturation within the sample as well as the mass flow rate of the gas for Exp 1.

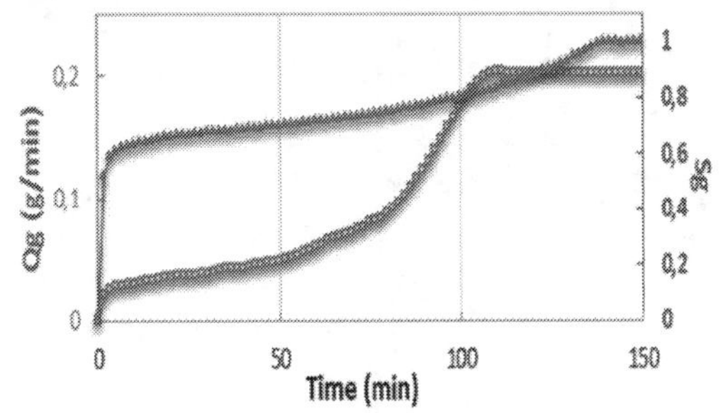

Fig. 3 Exp 1 Gas saturation (red curve) and mass flow rate (blue curve) evolution as a function of time for gas injection in sample initially saturated with deionized water.

Four regimes can be distinguished:

- In the first few minutes (between 0 and 10 min) the gas drains the sample until it reaches the residual water saturation that was established to 40%. **Fig. 4** shows the local saturation of the sample after 10 min. The longitudinal profile as well as the 2D saturation map show a typical drainage profile with a quite constant saturation on the first 2/3 of the sample (around 30% of water) followed by an increasing water saturation toward the outlet due to capillary end effects.
- The next regime (between 10 and 70 min) exhibits a smooth increase in the gas saturation that reaches 74% as well as in the flow rate that establishes at 35% of the final rate (cf. **Fig. 3**). The profile at 70 min displays a constant shift of the initial profile. This observation can be explained by an apparent homogeneous drying all along the sample which means that the equilibrium between the gas and the water is not instantaneous. Or by an equilibrium reached between the water evaporation and capillary forces that tends to redistribute liquid from regions of high-water saturation to region of low-water saturation.
- Following this second regime that ends at 70 min and at water mean saturation of 26% we observe a noticeable increase in the gas flow rate until 110 min as well as in the gas saturation until 140 min. This change in the drying rate corresponds to the progression of a discontinuity and a sharp front on the saturation profile and 2D map. This front propagates through the sample

toward the outlet allowing a subsequent increase in the flow rate until it reaches a plateau.

- The final stage corresponds to the dry state (above 140 min). It is to be noticed that the flow rate reaches its limit before the gas saturation.

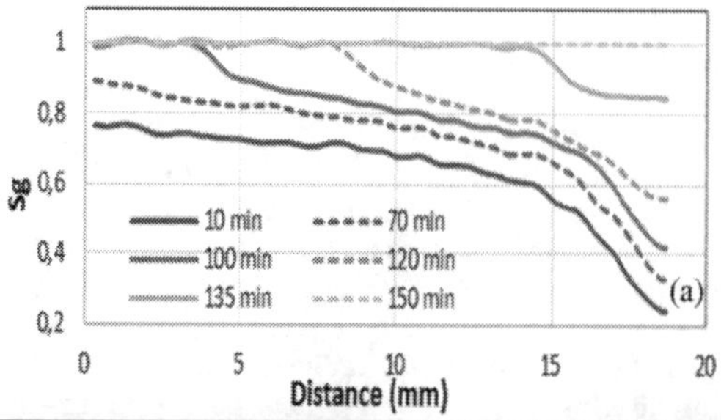

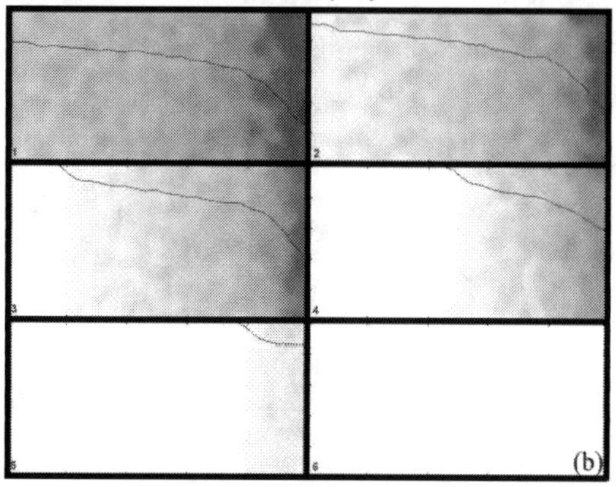

Fig. 4 Exp 1 evolution of gas saturation (a) 1D profile et (b) 2D map as a function of time (gas is injected from left) in sample initially saturated with deionized water.

This first experiment shows that injecting a dry gas in a porous media containing only deionized water allows to evaporate the total content of water without altering the permeability of the sample. The observation of the saturation profiles shows that the drying effect takes place first at the inlet of the sample and then propagates throw a sharp front downstream. The two first regimes involved in this particular case are most likely an instantaneous phase equilibrium between gas and water coupled with a capillary equilibrium generated by the gas flow and the water saturation.

The drying scenario would probably be first the gas vaporizes water at the entrance of the sample creating a capillary disequilibrium and decreasing the water saturation. This results in water redistribution by counter-current imbibition to compensate for the water evaporation at the inlet. Once the mean water saturation reaches a minimum (26%), the capillary disequilibrium cannot be compensated anymore. The evaporation front appears and propagates in the sample until the total drying of the sample.

In a second experiment (Exp 2) dry gas was injected in the same sample initially saturated with KBr brine at a relative gas pressure of 150 mbar. **Fig. 5** shows the evolution of the gas flow rate as a function of time as well as the relative X-ray absorption that was normalized by the final stage of Exp 3. The flow rate exhibits the same behavior as for the first experiment (i.e. a first increase corresponding to the drainage phase, a smooth increase slope followed by a more pronounced rise till flow rate reaches a constant plateau corresponding to the dry state). The relative X-ray absorption that corresponds in this particulate case to the mean gas saturation shows an asymptotic behavior with a smooth increase of the gas saturation. The final flow rate did not show any evidence of permeability alteration due to salt precipitation.

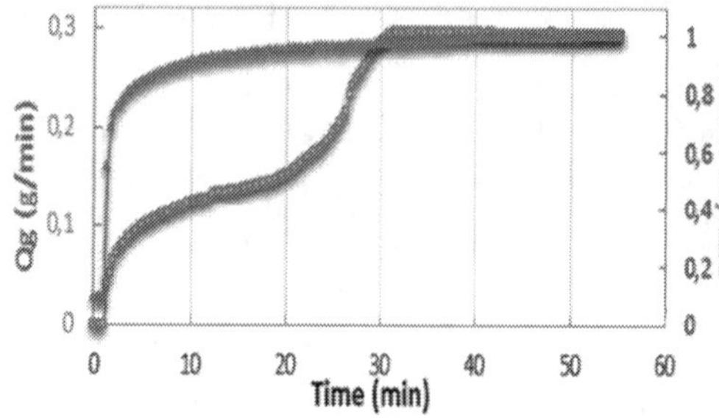

Fig. 5 Exp 2 X-ray relative absorption (red curve) and gas mass flow rate (blue curve) evolution as a function of time for gas injection in sample initially saturated with 100 g/l KBr brine.

The observation of the local profiles shows a more complex behavior (see **Fig. 6**). As the relative X-ray absorption was normalized by the gray level of the final sate, the profile can be qualitatively interpreted as the relative variation in salt localization (i.e. when the profile goes over or under one, it means that the quantity of salt decreases or increases respectively). Thus, we can observe that the salt concentration slightly decreases in the first half of the sample while it considerably increases at a specific location situated 5 mm from the outlet, corresponding to the location of the capillary end effects.

This behavior can be explained by the competition between drying and capillary rise. Actually, as drying decreases the saturation from the inlet toward the outlet as shown by case 1, capillary forces tend to equilibrate the saturation by displacing brine toward the inlet. As a consequence, the salt concentration locally increases and once it reaches the solubility limit, it precipitates.

The measured gas flow rate shows that, even if salt has precipitated, it did not affect the permeability. This could be explained by the low quantity of salt initially present in the sample after the drainage. An estimation of the initial volume of dissolved salt after the drainage gives a value 0.003 cc which corresponds to 0.01PV (PV is the pore volume). This reduction of the pore volume does not seem sufficient to reduce the absolute permeability. To confirm this observation, a 3D images of the sample was taken at initial dry state and after salt precipitation. **Fig. 7** shows a slice picked at the level of maximum salt precipitation. We can distinguish almost no difference between the two images apart few bright points corresponding to solid salt in some locations.

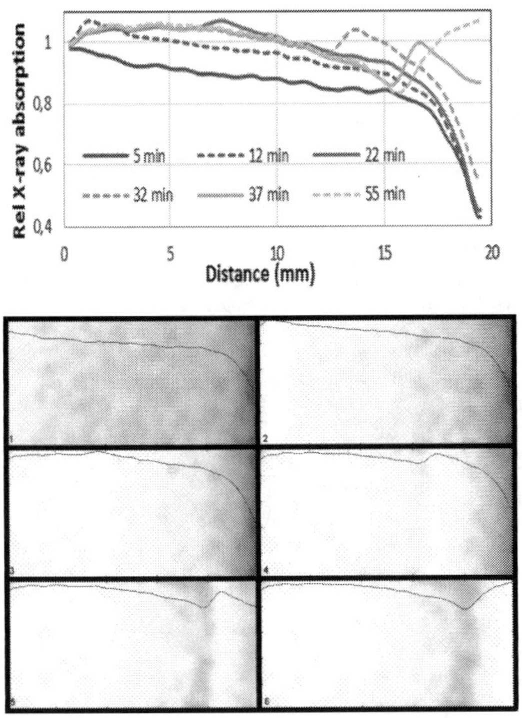

Fig. 6 Exp 2 evolution of X-ray relative absorption (a) 1D profile and (b) 2D map as a function of time (gas is injected from left) in sample initially saturated with 100 g/l KBr brine.

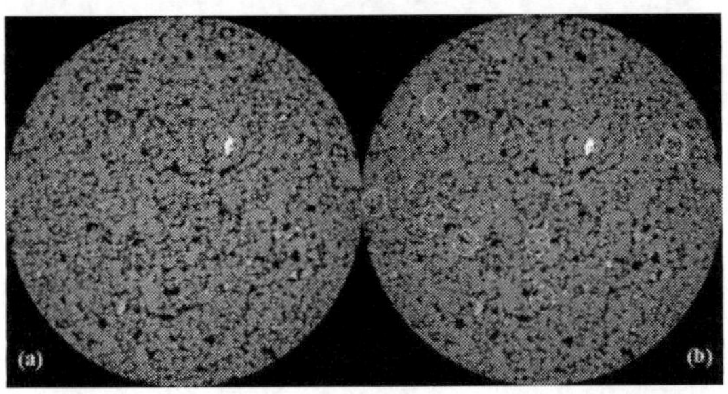

Fig. 7 2D slices picked at the maximum salt precipitation location (4 mm from the outlet) (a) at the initial dry state and (b) at the final state of the experiment corresponding to a dry state with precipitate salt. Some locations where salt has precipitated are indicated by circles.

3.2. Outlet capillary contact effect

To investigate the capillary effect, dry gas was injected in a sample initially saturated with a 100g/l KBr brine while contact with brine was maintained by sweeping the bottom face of the sample at a relative gas pressure of 150 mbar. **Fig.8** shows the evolution of the gas flow rate and the relative X-ray absorption as a function of time. In this case, we observe a completely different behavior than in the previous cases. After an initial increase due to drainage, the gas flow rate experiences successive decline and rise cycles with increasing amplitude before it reaches zero after six hours of injection. The relative X-ray absorption shows a constant decrease which corresponds to an increase of the sample density. This observation suggests an accumulation of additional salt within the sample that is supplied by the brine in contact with the bottom face of the sample.

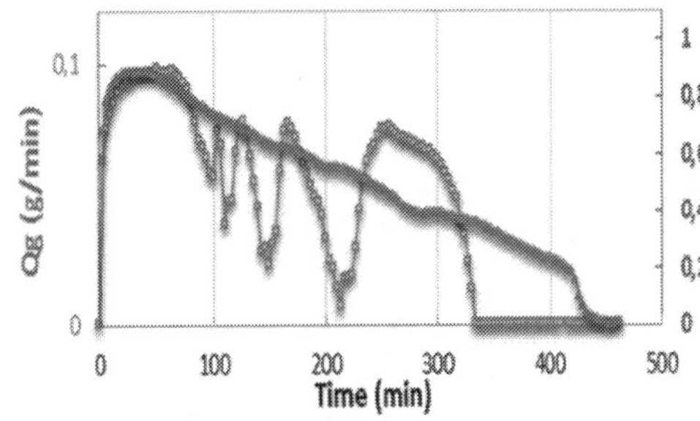

Fig. 8 Exp 3 X-ray relative absorption (red curve) and gas mass flow rate (blue curve) evolution as a function of time for gas injection in sample initially saturated with 100 g/l KBr brine and with outlet capillary contact.

The profiles illustrated in **Fig. 9** give more insight into the mechanisms that take place during this experiment. We observe first a local decrease in the relative X-ray attenuation at 8 mm from the inlet at 63 min that suggests an increase in salt concentration. The advent of this salt strip is correlated with a decrease in the gas flow rate and corresponds most probably to salt precipitation that alters permeability to gas. We observe then an alternation of dilutions and precipitation corresponding respectively to the increase and the decline of the gas flow rate with an intensification of the relative X-ray attenuation along the salt strip. Finally, we observe a displacement of the salt strip during the successive cycles of dilution precipitation toward the inlet of the sample. At the end of the experiment the salt strip is located 5 mm far from the inlet.

The observed phenomena can be explained by the competition between drying and capillary rise. Indeed, the gas flow dries out the sample, which is manifested by a decrease in the water saturation. However, as we have an active capillary contact, capillary suction compensates for the decrease of water saturation to maintain a capillary equilibrium, which in turn leads to an increase of local salt concentration. At one point the salt concentration is sufficiently high to precipitate salt and altering the gas permeability. As the gas flow rate declines, the capillary rise becomes predominant, allowing the dilution of the precipitated salt and a regeneration of the gas permeability. The alternation of this instable equilibrium leads to an increase of the precipitated salt after each stage till a point of no return is reached where the volume of precipitated salt is sufficiently high to stop the gas flow and the capillary suction.

At the end of the experiment a 3D image of the sample was taken to evidence salt precipitation. **Fig. 10** shows a slice picked up at the maximum salt precipitation location. The accumulation of solid salt is clearly observable all over the porous space of the sample (white part). The solid salt plugs the pores and is responsible of the permeability alteration.

 https://doi.org/10.1051/e3sconf/202014603001

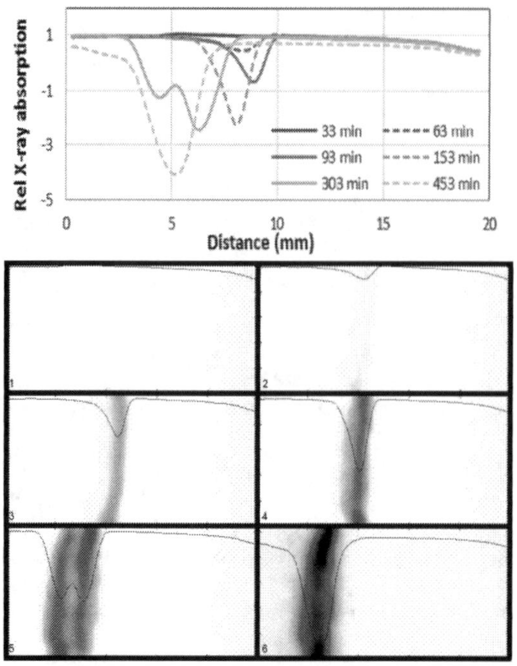

Fig. 9 Exp 3 X-ray relative absorption (a) 1D profile and (b) 2D maps (gas is injected from left) as a function of time for dry gas injection in sample initially saturated with 100 g/l KBr brine and with outlet capillary contact.

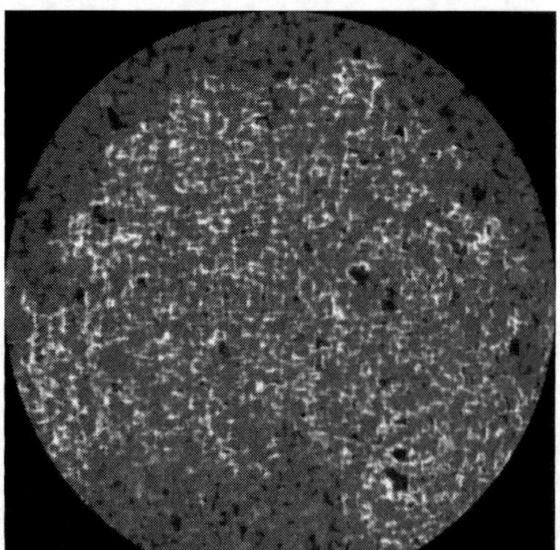

Fig. 10 2D slice picked at the maximum salt precipitation location (5 mm from the inlet) at the final state of the experiment. Light gray level corresponds to solid salt that has precipitate due to gas drying out.

3.3. Lateral capillary contact effect

In this last experiment (Exp 4), we have injected dry nitrogen in the composite sample initially saturated with 100 g/l KBr brine. The inlet gas pressure was set to 450 mbar (the outlet was opened to atmosphere). **Fig. 11** shows the evolution of the gas flow rate as a function of time as well as the relative X-ray absorption that was normalized by the finale stage within the Bentheimer half core. We can observe that when gas is injected, the relative X-ray absorption and the gas flowrate rise rapidly to a maximum that correspond probably to the residual water saturation state. Following this first stage, we note a monotonous decrease in the relative X-ray adsorption which means salt accumulation within the sample. As consequence, the gas flow rate decreases which corresponds to a progressive permeability alteration.

Fig. 12 shows the 1D and 2D locale relative X-ray absorption. The Bentheimer half core is positioned on the upper part of the images. The first image at 26 min shows that when gas is injected, a preferential path is created throw the Bentheimer sample that goes to Swi whereas the Molière half core remains fully brine saturated. The second image shows the advent of salt precipitation that is located almost at the middle of the sample and near the Bentheimer and sleeve boundary. However, the shape of the precipitate salt zone is completely different from the one observed in the previous experiment. Indeed, we do not observe a strip perpendicular to the gas flow direction but rather an arrow like spot. The next images show that salt precipitates and accumulates in parallel to the flow direction. As the Bentheimer half plug is only in contact with the Moliere one, this observation confirms that the lateral capillary contact is active and can provide enough salt to plug completely the permeable zone.

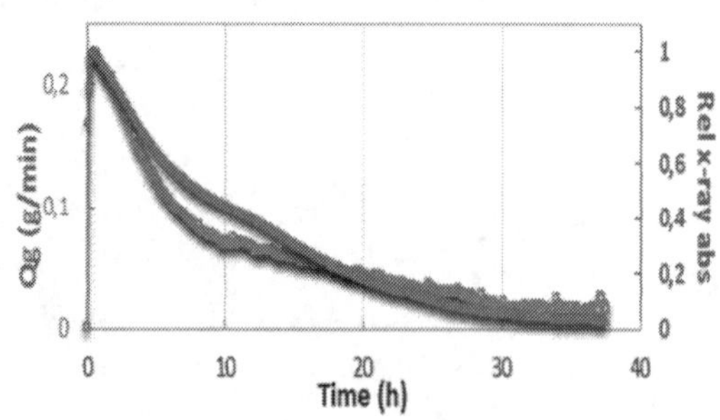

Fig. 11 Exp 4 X-ray relative absorption (red curve) and gas mass flow rate (blue curve) evolution as a function of time for gas injection in sample initially saturated with 100 g/l KBr brine and with outlet capillary contact.

To better observe the shape of the salt precipitate zone, a 3D image was taken at the end of the experiment. Fig. 13 shows 2D sagittal and axial images of the sample. We can clearly see the solid salt barrier that cause permeability alteration. It is interesting to note that this barrier lies in diagonal to the Bentheimer sample and extend on the two third of the length. This shape reveals the complexes interplay of the different mechanisms involved in salt precipitation.

In this case, the drying front is opposed not only to parallel capillary forces coming from the capillary end effect in the Bentheimer half core (as described in Exp 2) but also to perpendicular capillary forces resulting from the contact with the Moliere half core. This configuration can then explain the shape of the precipitated salt that results from the competition between the two perpendicular drying fronts.

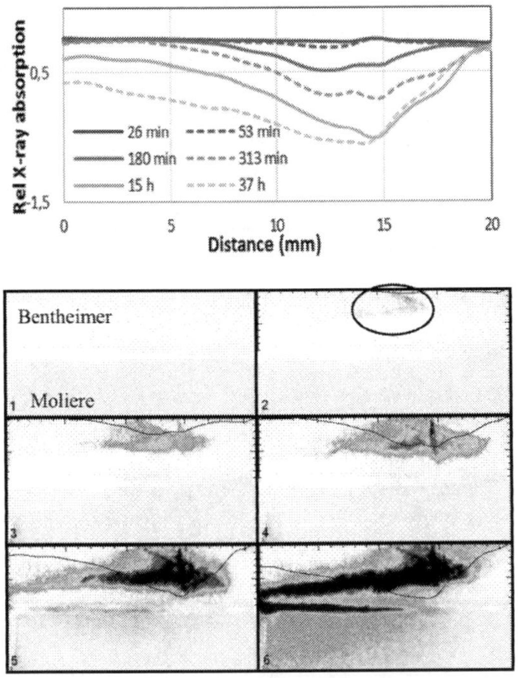

Fig. 12 Exp 4 X-ray relative absorption (a) 1D profile and (b) 2D maps (gas is injected from left) as a function of time for dry gas injection in sample initially saturated with 100 g/l KBr brine and with lateral capillary contact.

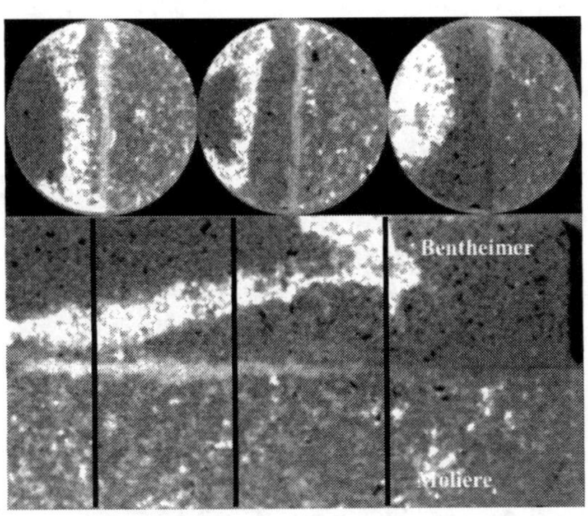

Fig. 13 2D sagittal slice (bottom) and 2D axial slices picked at three different salt precipitation location at the final state of the experiment. Bright zone corresponds to solid salt that has precipitated due to gas drying out.

4. Numerical interpretation

In this study we used CooresFlow an in-house simulator designed for reactive flows in porous media. It was developed to model multiphase transport processes coupled with fluid-rock interactions, for various applications including CO_2 underground storage as well as enhanced recovery of hydrocarbon reservoirs.

4.1. General characteristics

CooresFlow is a THC (thermal hydro-chemistry) simulator that consider the following phenomena:

- multiphase flow in porous media, with viscous and capillary forces,
- transport of chemical components by advection, diffusion and dispersion,
- transfers within the fluids or on the surface of the rock governed by local equilibrium,
- chemical equilibrium or kinetic reactions to describe geochemical exchanges,
- dynamic modification of the porosity and permeability of the porous medium over time.

In the case of gas production or storage, it allows to consider:

- the phenomena of mass transfer between liquid water and water vapor (drying),
- the transport and precipitation of dissolved species (salts),
- the coupling between porosity and permeability changes.

The mathematical formulation used in the simulator is a generalization of the compositional formulation based on material and energy balance equations and chemical equilibrium equation (for more see [11])

4.2. Physical and numerical model

In this study we consider a multiphase compositional system composed of four phases: two fluids (water and gas) and two solids (rock and salt). As the system is compositional, each phase can contain one or several species. In the studied case, the species are:

- H2Ow and KBrw (the dissolved salt) in water,
- N2g (nitrogen) and H2O vapor (H2Og) in gas,
- Precipitated salt (KBrs) in salt.
- The Rock phase (Quartz) considered inert in this study (no exchange with other phases)

The rock sample is modelled by a vertical parallelepiped of 100 cells in Z direction. Its sizes are 0.86 cm in X and Y directions and 1.98 cm vertically. Initially, the sample is saturated with either deionized water or a 100 g/l KBr brine. The temperature is constant within the sample and does not evolves with time. The pressure is also initially constant and equal to the atmospheric pressure. The boundary conditions were defined in accordance to the experimental design.

The thermodynamic model computes the density and viscosity of the fluids. The brine density evolves with the salt concentration (KBr molar fraction) using a linear relation based on experimental data from the Detherm 2015 database, for a temperature of 25°C. In the same way, the brine viscosity evolves with the salt concentration by interpolating linearly experimental data of the Detherm 2015 database for a temperature of 25°C. The gas properties are assumed constant since the pressure is closed to 1 bar and the humidity rates are low (<2%). We consider a gas density of 1.2 kg/m3 and viscosity of 1.758 10^{-5} Pa.s.

In this work, two phase transfers are modelled: Water evaporation and condensation, and salt dissolution and precipitation. The salt dissolution/precipitation is considered instantaneous. It is modelled by an equilibrium constant with a linear dependency on temperature. The equation is based on the experimental data of the Detherm 2015. The water evaporation/condensation is also considered instantaneous. This hypothesis is confirmed by a Peclet number higher than 1 that means the water advection dominates the water diffusion in gas. The model is defined by a table of equilibrium constants with a dependency to temperature, pressure and salt concentration. This table is based on experimental data from the Carnot and Detherm 2015 databases.

4.3. Flow through drying modelling

To generate the petrophysical model, we have first considered a homogeneous sample with a constant porosity of 23% and permeability to water of 1450 mD. Relative permeabilities and capillary pressure were then estimated by history matching the mean saturation profiles and gas flow rate end point from Exp1. The resulting relative permeability to water and to gas as well as water/gas capillary pressure are illustrated in **Fig. 14**. It is to be noted that the resulting absolute permeability to gas was estimated to 2450 mD which is much higher than the water permeability. This can be explained by the inertial effect . This effect was not considered in this study. Irreducible water saturation was estimated to 12%. In the domain of 0 to 12% of water saturation, the relative permeability to gas is found to be constant while the capillary pressure increases slightly.

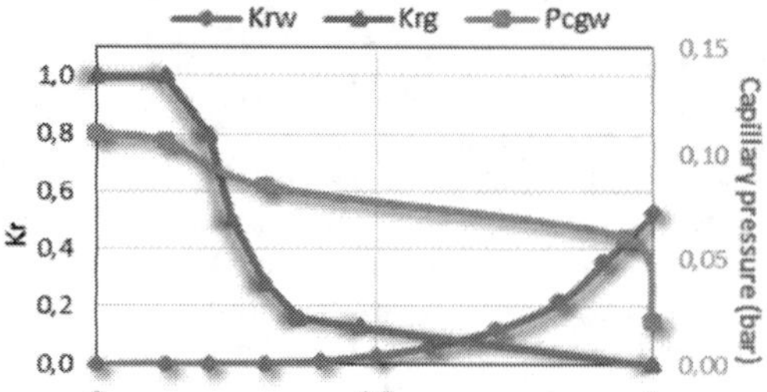

Fig. 14 Water (Krw), gas (Krg) relative permeability and Gas-water capillary pressure (Pcgw) versus water saturation

By comparing the simulation results with the experimental data in the case of Exp 1 (see **Fig. 15**), we observe that the simulated system exhibits the same behavior than the experimental case. Indeed, we observe the same behavior of the gas rate with an initial important increase corresponding to the drainage phase, then a smoother increase followed by a greater increase during the drying phase within the sample until the gas rate reaches a plateau

The model succeeds to fit properly the mean saturation and gas flow rate end point as well as the major part of saturation profiles. However, differences are observed on the gas flow rate dynamic. This discrepancy can be explained by the uncertainties on capillary effects and saturation at the outlet boundary that induce a difference on the gas saturation profiles near the outlet. Nevertheless, the reactive transport model succeeds to simulate the same physical phenomena such as the drying effect, the capillary counter-current, the disappearance of a phase with similar characteristic times.

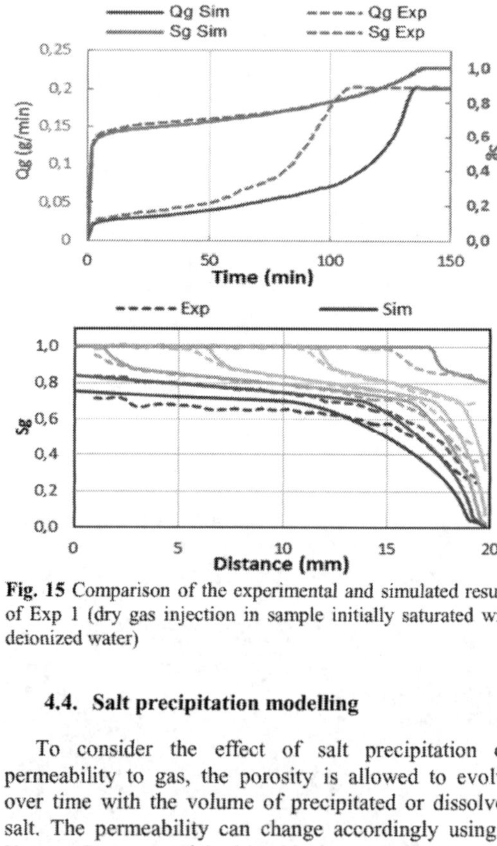

Fig. 15 Comparison of the experimental and simulated results of Exp 1 (dry gas injection in sample initially saturated with deionized water)

4.4. Salt precipitation modelling

To consider the effect of salt precipitation on permeability to gas, the porosity is allowed to evolve over time with the volume of precipitated or dissolved salt. The permeability can change accordingly using a Kozeny Carman model with critical porosity:

$$k = k_0 \left(\frac{\emptyset - \emptyset_c}{\emptyset_0 - \emptyset_c}\right)^a \left(\frac{1 - \emptyset_0 + \emptyset_c}{1 - \emptyset + \emptyset_c}\right)^b$$

Where k and $\emptyset$ are respectively dynamic permeability and porosity, k_0 and $\emptyset_0$ are respectively initial permeability and porosity and $\emptyset_c$ is a critical porosity beyond which permeability is null. Exponents a and b as well as $\emptyset_c$ were history matched using data from Exp 3.

https://doi.org/10.1051/e3sconf/202014603001

Best match was found with 0.08 for $\emptyset_c$, 2.5 for a and 1.5 for b. Experiment 2 was simulated using the same model parameters defined in experiment 1 apart from the initial saturation where the salt concentration was set to 100 g/l and the relative gas pressure at the inlet of 150 mbar.

Fig. 16 shows the evolution of the simulated (continuous line) gas flow rate, mean gas saturation within the sample and the inverse of the normalized KBr quantities. Experimental data (dashed line) were added for comparison. Simulated gas flow rate shows the same behaviour as for previous case with also a slower dynamic than the experimental data due probably to the same reason mentioned previously. However, comparison of simulated gas saturation and relative X-Ray attenuation is not straightforward as X-Ray attenuation is sensitive not only to water saturation but also to salt quantities. Still the two quantities decrease monotonically which agrees with the increase of the X-Ray attenuation. We also observe that the salt quantity reaches a plateau after about 20 min indicating that no more salt is evacuated from the sample which mean that no more water convection is active. This time corresponds also to a change in the flow rate slop indicating a change in the water transport mechanism which becomes dominated by the evaporation and drying mechanisms.

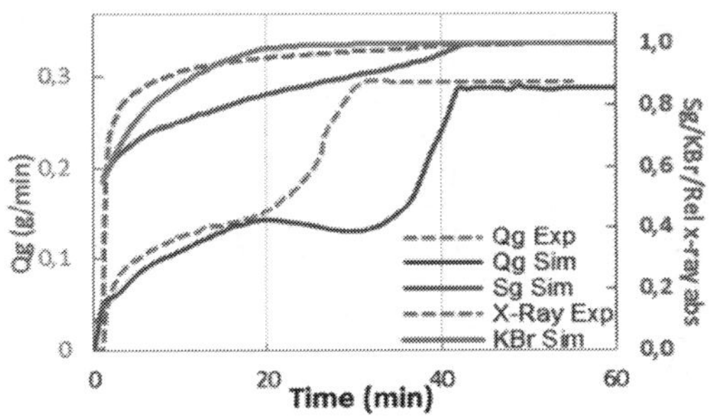

Fig. 16 Comparison of the experimental and simulated results of Exp 2 (dry gas injection in sample initially saturated with 100 g/l brine)

Simulation results show also a similar KBr salt profile evolution with the experimental data as well as a salt accumulation near the outlet of the sample. **Fig. 17** shows four different profiles taken at four different time and corresponding to (a) experimental X-Ray attenuation normalized by the final mean X-Ray attenuation, (b) the KBr quantity along the sample normalized by the final one, (c) dissolved KBr concentration and (d) gas saturation along the sample. For each date, we observe on the KBr quantity a front similar to the experimental data that separates the precipitated-salt zone from the dissolved-salt zone. This front has a characteristic shape that is induced by the combination of an increasing water saturation (cf. **Fig. 17**.d) with a decreasing KBr concentration (cf. **Fig. 17**.c) that reaches a maximum at the interface between the dry zone and the wet zone. The front moves towards the outlet at the same velocity as the evaporation front. It is to be noted that simulation shows no evidence of overconcentrated zone (i.e. zone where concentration is greater than solubility limit). At 55 min, the sample is completely dried and all the KBr salt is precipitated. Within the sample, the KBr amount corresponds to the amount that was included in the residual water saturation (after 20 min). But near the outlet boundary, we observe an accumulation of salt. It results from the initial capillary end effect and very close to the outlet, from the competition between the drying effect and the capillary re-imbibition.

This salt precipitation occurs when the solubility threshold is reached. It reduces the porous volume in proportion with the solid KBr density. The final average porosity was estimated at 0.228 with a minimum of 0.225 which is very closed to the initial 0.23 value. The initial salt amount of the sample is not sufficient to drastically reduce the porosity. This porosity decrease induces a small permeability decrease that has a little effect on the gas rate. The average permeability to gas was 2000 mD compared 2450 mD at the initial state.

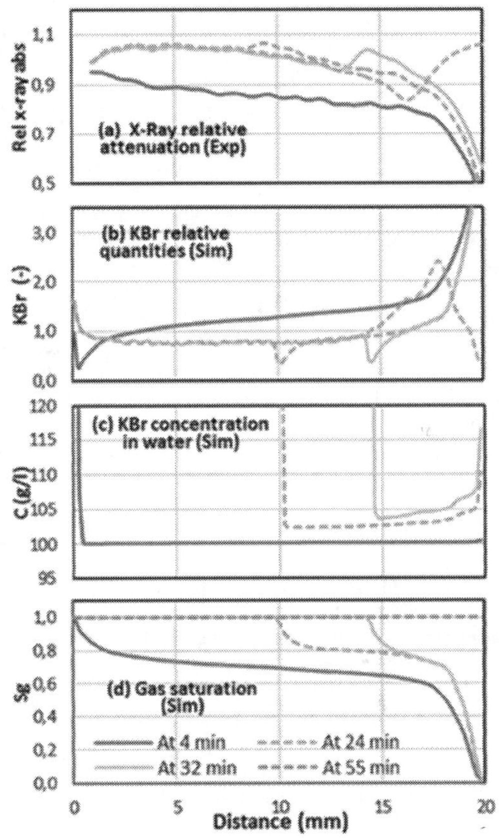

Fig. 17 Comparison of (a) experimental X-Ray attenuation profiles normalized by the final mean X-Ray attenuation and simulated profiles of (b) the KBr quantity along the sample normalized by the finale one, (c) the dissolved KBr concentration and (d) the gas saturation along the sample at four different date for Exp 2.

4.5. Capillary contact modelling

To simulate capillary contact effect, the numerical model was modified by adding an infinite water reservoir in contact with the outlet boundary to reproduce Exp 3 boundary condition. Water was set free to flow through the outlet in the two directions. In the two cases dry gas is injected in a sample initially saturated with a 100 g/l KBr brine.

Fig. 18 (a) shows that the simulated gas flow rate for the third experiment (Exp 3) was comparable to the experimental one. Indeed, we observe four successive cycles of gas flow rate increase and decrease corresponding drainage/dissolution (increasing flow rate) and drying/precipitation phase (decreasing flow rate). The gas rate decreases for the last time at 327 min and reaches a value close to zero.

To get more insight in the cycling behavior, we have reported in **Fig. 18** (b) the simulated mean gas saturation within the sample, the precipitated solid KBr quantities normalized by the maximum KBr quantity and the KBr concentration in water normalized by the initial concentration. Experimental data corresponding to the X-Ray normalized attenuation was added for comparison. The KBr curve shows clearly that the quantity of precipitate salt is increasing over time during each cycle with alternation of stagnation intervals that correspond to gas rate local minima. During those stagnation periods, we observe also a decrease in gas saturation as well as a slight increase in the brine concentration that is probably due to solid salt dissolution. However, as we do not observe any significant decrease in the solid KBr quantity curve, this means that during the stagnation period, an equilibrium is reached between dissolution and precipitation reaction. Salt is dissolved at the contact of brine and precipitate at the drying front.

The simulated KBr normalized quantity profiles at four different date are illustrated in **Fig. 19**. They show where salt forms a deposit. The four times correspond to the four gas local minimum rates. At these times, the salt amount is mainly located in the first 7 mm of the sample with a tendency to move towards the inlet over time. It is characterized by one or several peaks with an increasing base. At these zones, the porosity and the permeability are altered and reach a minimum value of respectively 9% and 1-10 mD. All these figures clearly show a correlation between the increase of the salt amount, the permeability alteration and the decrease of the gas rate.

The following process can explain the different observations made above. As in the case 2, the dry gas injection locally induces a decrease of water saturation due to water evaporation and a capillary re-imbibition from the high-water saturation zone to the low-water saturation zone to compensate the water loss. While the porous medium dries, the brine salt concentration increases until it reaches the solubility threshold. Then, it precipitates with an impact on the porosity and the permeability. The gas front on the saturation profile continues to propagate towards the outlet, leaving behind it, a dried porous medium containing a small amount of precipitated salt. At a specific location defined by the competition between the drying effect and the capillary re-imbibition, a salt accumulation starts, fed by the capillary contact, and alters the petrophysical properties. In that case, the gas rate strongly decreases until the capillary re-imbibition becomes preeminent. Then, the brine supplied by the capillary counter-current partially dissolves the precipitated salt. Indeed, the salt dissolution quickly saturates the brine that cannot dissolve an additional amount of salt. Consequently, the porosity and the permeability are also partially restored, and the gas rate increases again, until a maximum reduction. This is the end of the first cycle. Then, the process starts again with altered petrophysical conditions that impact the competition between the drying effect and the capillary re-imbibition. It explains the displacement of the salt accumulation observed in **Fig. 19**. The cycling process continues and stops when the pressure forces dominate the capillary forces and block the capillary flow.

In conclusion, the simulations results show a similar gas rate evolution with the experimental data as well as a salt accumulation that behaves in the same way. The reactive transport model succeeds to simulate the drying effect and the salt precipitation associated to a petrophysical properties change. The dynamic permeability model used in this study reveals adapted to model the main effects of salt precipitation.

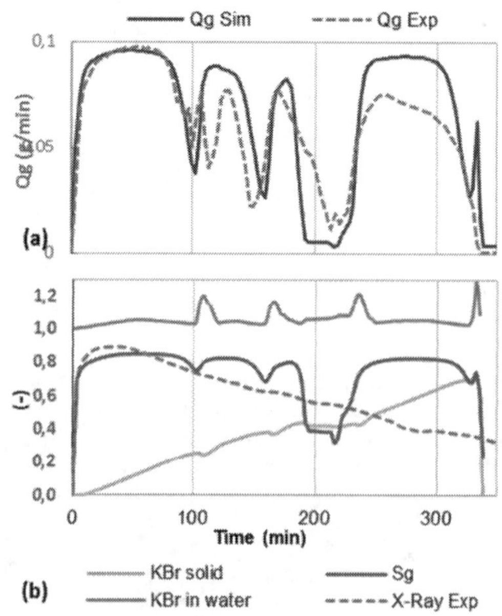

Fig. 18 Comparison of (a) experimental and simulated gas flow rate, (b) experimental X-Ray attenuation normalized by the final mean X-Ray attenuation and simulated evolution of (b) solid (precipitated) KBr quantity normalized by the finale total KBr quantity, the dissolved KBr concentration and the mean gas saturation Exp 3.

 https://doi.org/10.1051/e3sconf/202014603001

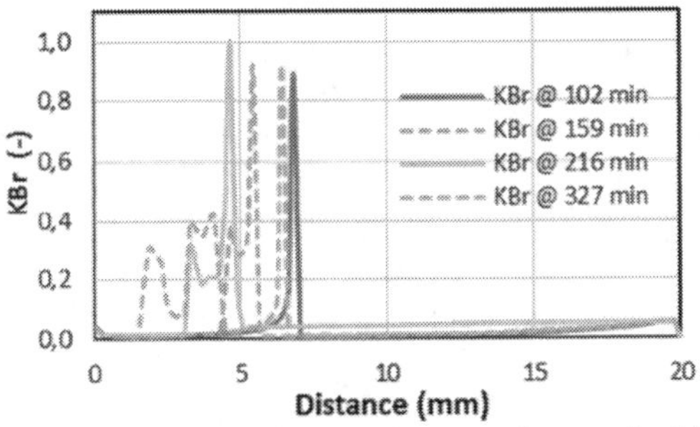

Fig. 19 KBr quantity profile along the sample normalized by the maximum quantity at four different date for Exp 3.

In the case of Experiment 4 a two-layers mode parallel to the pressure gradient, (see **Fig. 2** (b)) was built with high permeability contrast. In this case outlet capillary contact was only maintained for the low permeability layer. **Fig. 20** shows the evolution of the simulated gas flow rate for Exp 4. We observe the same behavior the experimental case (see **Fig. 11**). The curve strongly increases due to the drainage phase then quickly decreases probably due to an alteration of the permeability. However, and as for experiment 1 and 2, the simulated flow rate exhibits a faster dynamic that can also be explained in this case by the uncertainty on the capillary contact description.

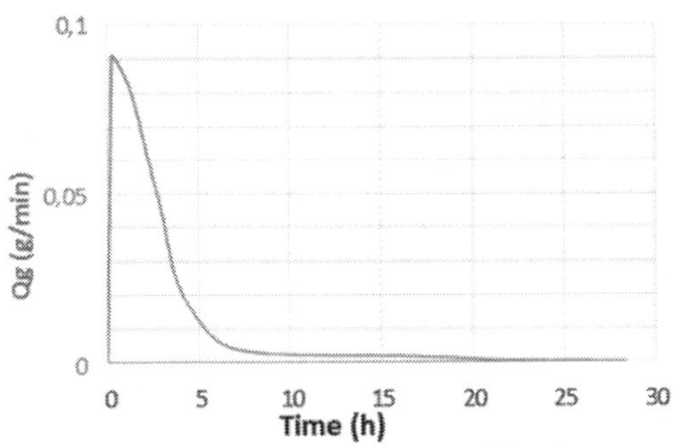

Fig. 20 Simulated gas flow rate in the case of Exp 4.

Within the property maps shown in **Fig. 21**, we observe the result of the longitudinal capillary contact effect. The third experiment has shown that the capillary contact at the bottom of the sample induces a salt deposit that is perpendicular to the fluid flow. In the case of Exp 4, the result is different. The re-distribution of brine from the high-water saturation zone to the low-water saturation zone mainly occurs near the contact surface between the two porous media, that induces a 2D profile of salt deposit that spread over a diagonal in the Bentheimer half core. As in the previous case, the brine feeding induces a salt accumulation that strongly alters the porosity (decrease from 23% to 8.06%) and the permeability (decrease from 1350 mD to 1.2 mD). The sample is plugged in 7 hours.

This simulation case shows that the plugging of a permeable rock can be caused by the combination of the drying and a brine supply coming from a low-permeability rock that contains enough volume of mobile brine.

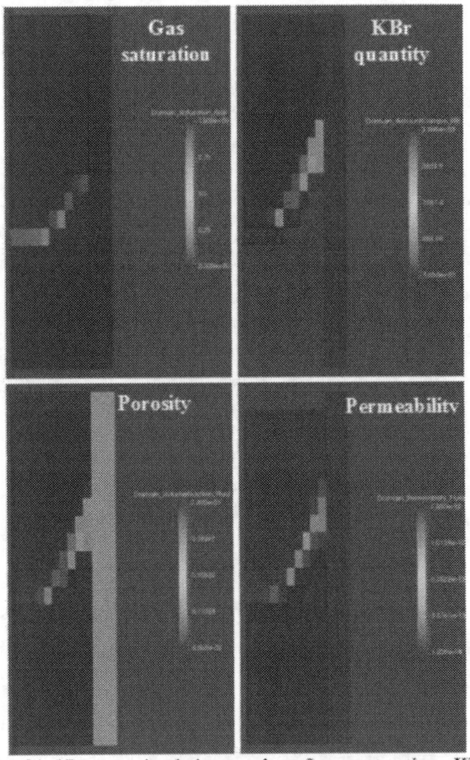

Fig. 21 2D map simulation results of gas saturation, KBr quantity, porosity and permeability within the sample at 25 hours for Exp 4. The low permeable zone representing the Moliere is on the left zone of the maps.

5. CONCLUSIONS

In this study, we have shown the importance and interplay of different parameters that govern brine drying and salt precipitation in porous media subject to dry gas injection. When dry gas is injected in a sample saturated with brine, the water is dried out near the inlet first and then a drying front propagate toward the outlet. As a consequence, salt precipitates in the porous media and can lead to pore plugging. The pore plugging will either completely block the pore space or heavily reduce the rock permeability depending on the boundary conditions.

Indeed, the experimental study showed that dry gas injection combined with a capillary water contact can trigger salt precipitation and possibly block the pore space. These experiments establish the basis for simulation models that can be used to integrate the different physical phenomena that take place during water displacement, evaporation and possibly precipitation.

Results from the experimental study have been modeled using an in-house simulator to interpret the drying and salt precipitation mechanisms. The governing equations consider various coupled mechanisms like flow-through drying, capillary forces, salt displacement, salt precipitation and porosity/permeability alteration. This reactive transport model succeeded in simulating the main effects observed in experimental cases:

- drying of the sample by dry gas injection, and salt precipitation
- salt accumulation supplied by capillary re-imbibition until plugging of the sample,
- successive alteration and improvement cycles of the petrophysical properties due to precipitation and dissolution of salt.

Future work includes:

- improving the simulation approach by integrating
  - modeling of hysteresis in the drainage/imbibition curves of relative permeability and capillary pressure,
  - modeling of the capillary pressure evolution with porosity,
  - modeling of the thermal transfer due to evaporation,
- investigating the effect of other parameters like pressure gradient, salt concentration and/or gas relative humidity to assess the induced risk of salt precipitation on permeability alteration,
- adaptation and application of the model to near well and reservoir scale problems. In particular, radial flow conditions imply rapidly changing pressure gradients close to wellbore, contrary to the uniform pressure gradient characterizing linear flow in a core.

6. References

1. Kleinitz W., Dietzsch G., Köhler M. (2003) Halite scale formation in gas-producing wells; Trans. IChemE, 81(A), pp352-358.
2. Aquilina, P. (2012) Impairment of Gas Well Productivity by Salt Plugging: A Review of Mechanisms, Modeling, Monitoring Methods, and Remediation Techniques, paper SPE 158480 prepared for presentation at the SPE Annual Technical Conference and Exhibition, San Antonio, Tx, 8-10 October.
3. Mahadevan J., Sharma M.M., Yortsos Y.C. (2006) Flow-Through Drying of Porous Media; AiChE Journal ; 52(7); pp2367-2380
4. Mahadevan J., Sharma M.M., Yortsos Y.C. (2007) Water removal from porous media by gas injection: experiments and simulation; Transp Porous Media; 66; pp287-309
5. Peysson Y., Bazin B., Magnier C. Kohler E., Youssef S. (2011) Permeability alteration due to salt precipitation driven by drying in the context of CO2 injection, Energy Proceedia (4) p4387-43941
6. Peysson Y., André L, Azaroual M. (2013) Well injectivity during CO2 storage operations in deep saline aquifers 1: Experimental investigation of drying effects, salt precipitation and capillary forces, Int. Journal of Greenhouse Gas Control, 22, pp 291-300
7. Le D., Hoang H., Mahadevan J. (2009) Impact of capillary-driven liquid films on salt crystallization, Transp Porous Med; 80; pp229-252
8. Giorgis T., Carpita M., Battistelli A. (2007), 2D modeling of salt precipitation during the injection of dry CO2 in a depleted gas reservoir, Energy Conversion and Management, 48, pp1816-1826
9. André L., Peysson Y., Azaroual M., (2014) Well injectivity during CO2 storage operations in deep saline aquifers – Part 2: Numerical simulations of drying, salt deposit mechanisms and role of capillary forces, Int. Journal of Greenhouse Gas Control, 22, pp301-312
10. Youssef S., Mascle M., Peysson Y., Vizika O. (2017) CAL-X: An X-Ray Radiography tool for high throughput coreflood experimentation. Applications in the EOR context. To be presented at Int. Sym. of the Society of Core Analyst Vienna, Austria
11. Gassara O., Estublier A., Garcia B., Noirez S., Cerepi A., Loisy C., Le Roux O., Petit A., Rossi L., Kennedy S., Brichart T., Chiquet P., Luu Van Lang L., André Duboin F., Gance J., Texier B., Lavielle B., Thomas B., (to be submitted), The Aquifer-CO2 Leak project: numerical modeling for the design of a CO2 injection experiment in the saturated zone of the Saint-Emilion (France) site.

Steady-State Two-Phase Flow in Porous Media: Laboratory Validation of Flow-Dependent Relative Permeability Scaling

Marios S. Valavanides[1,*], *Matthieu* Mascle[2], *Souhail* Youssef[2] and *Olga* Vizika[2]

[1]Civil Engineering Department, University of West Attica, 12243 Attica, Greece
[2]IFP Energies Nouvelles, Rueil-Malmaison, France

Abstract. The phenomenology of steady-state two-phase flow in porous media is recorded in SCAL relative permeability diagrams. Conventionally, relative permeabilities are considered to be functions of saturation. Yet, this has been put into challenge by theoretical, numerical and laboratory studies that have revealed a significant dependency on the flow rates. These studies suggest that relative permeability models should include the functional dependence on flow intensities. Just recently a general form of dependence has been inferred, based on extensive simulations with the DeProF model for steady-state two-phase flows in pore networks. The simulations revealed a systematic dependence of the relative permeabilities on the local flow rate intensities that can be described analytically by a universal scaling functional form of the actual independent variables of the process, namely, the capillary number, *Ca*, and the flow rate ratio, *r*. In this work, we present the preliminary results of a systematic laboratory study using a high throughput core-flood experimentation setup, whereby SCAL measurements have been taken on a sandstone core across different flow conditions -spanning 6 orders of magnitude on *Ca* and *r*. The scope is to provide a preliminary proof-of-concept, to assess the applicability of the model and validate its specificity. The proposed scaling opens new possibilities in improving SCAL protocols and other important applications, e.g. field scale simulators.

1. Introduction

The conventional use of saturation as the independent variable in two-phase flow in *porous media* (PM) is based on the oversimplifying assumption that disconnected fluidic elements of the *non-wetting phase* (NWP) (ganglia and droplets) do not move with the average flow but remain stranded in the pore medium matrix. This situation arises when flow conditions of 'relatively small values' of the capillary number are maintained. In those cases the disconnected NWP fluidic elements block part of the available flow cross-section by a fraction analogous to the average saturation and effect a relative reduction of the permeability of both the NWP and the *wetting phase* (WP). Nevertheless, there is ample experimental evidence that disconnected flow is a substantial and sometimes prevailing flow pattern [1-8].

Treating relative permeabilities as functions of the saturation is inefficient in providing a correct and specific-enough description of the process across the domain of all possible flow conditions. The issue is extensively discussed in [9]. A particular value in saturation does not necessarily imply that a unique disconnected structure (arrangement) of the NWP will settle-in. Disconnected structures of the NWP can be coarsely described by a spectrum of population density distributions, extending from distributions of 'many-and-small' fluidic elements (droplets and small ganglia) to 'fewer-and-larger' fluidic elements (small and large ganglia). For any one of those cases, the corresponding superficial velocity of the disconnected NWP would not necessarily attain the same value. The latter would be the result of the 'negotiation' between the two factors inhibiting the transport of each phase, i.e. viscosity and capillarity, over the mass and momentum balances. Moreover, this 'negotiation' takes place within the 'regulatory framework' conformed by two critical 'stakeholders' or 'regulators', namely the particular structure of the PM and the externally imposed, flow conditions.

The term 'structure of the PM' comprises some appropriate, modeling ensemble of geometric /topologic characteristics of the pore network (dimensionality, genus, tortuosity, pore geometry etc.) as well as a set of physicochemical characteristics of the N/W/PM system, pertaining to- or associated with- *wettability* or to the favoring of interstitial, *emulsion* type flows. Wettability may be accounted by the values of the *static* contact angles of the N/W (NWP/WP) menisci against the pore walls. In that context, wettability conforms to the combination of the NWP, WP and PM surface characteristics and tripartite physicochemical affinities, as well as to *hysteresis*, expressed by the difference between the *dynamic receding* and *advancing* contact angles, pertaining to the motion of N/W menisci during drainage or imbibition of the WP within pores [10, 14].

*Corresponding author: marval@uniwa.gr

Flow conditions have an adverse effect on the momentum balance and, in particular, on the relevant magnitude of the Stokes flow resistances (due to bulk viscosity of the NWP and WP) and to Young-Laplace resistances (due to the contact forces between the N/W menisci and the pore walls). The former depend on local (micro /pore scale) velocity gradients (rates of deformations), in contrast to the latter that are relatively independent of rates of displacement of the menisci. From an energy point of view, the power dissipation due to bulk viscosity scales with the square of the local velocity gradients, whereas the capillarity-induced dissipation scales with the rates of displacement of the menisci. At relatively low values of the flow intensity (superficial velocity) viscosity effects are insignificant to capillarity effects; as flow intensity is progressively increased, viscosity takes over from capillarity and the flow progressively mutates from capillarity-dominated to transient capillary-viscosity- to viscosity-dominated characteristics [9].

As a consequence, for any two different cluster configurations (any two different population density distributions) of disconnected phase, the effective permeabilities of the WP and the NWP would differ and, therefore, the corresponding values of the relative permeability to the NWP would be different. Yet, those different cluster configurations (population distributions) and different values of the relative permeability could -in general- correspond to the same saturation value. In such situations, a universal, saturation-dependent description of the flow would be weak.

In addition, during core analysis, saturation is measured indirectly and cannot be externally imposed directly; it is only through control of a combination of pressure difference and/or flow rate of the NWP or WP that the system will attain an average saturation. If one wants to consistently and systematically describe the process in the entire flow regime (extending across extreme values in the capillary number and the viscosity ratio), one has to consider those variables that describe the externally imposed conditions and contain macroscopic *kinematic* information, e.g. the superficial velocity of each phase or, equivalently, the capillary number and the flow rate ratio. Moreover, saturation cannot adequately (or uniquely) describe the flow conditions. This is because saturation alone brings no definite input to the momentum balance, therefore it is questionable if it can provide any information on the *kinetics* of the macroscopic flow.

Observations of single phase flows within pore networks confirm that the macroscopic pressure gradient scales linearly with the superficial fluid velocity (Darcy's law). This seems to be a quite trustworthy modeling consideration in the case of two-phase flow as well, but only when very high superficial velocities are considered and capillary forces are negligible. However, at moderate/low velocities, when capillary forces are comparable to viscous forces, the macroscopic pressure gradient does not scale linearly with the flow rate. Experimental studies on steady-state two-phase flows in glass beads [3, 5, 6], in glass bead columns [11], as well as in sand-pack columns [12], revealed that the non-linear relation between pressure gradient and flow rate can be described by generic power laws with different exponent values. The discrepancy between the values of the scaling exponents is attributed to differences associated with dimensionality of the pertinent variables, measurements pertaining to different flow conditions, dimensionality of the NWP/WP/PM system etc. Therefore, it is worth the effort to examine if these 'different' observations can be integrated in a universal power law relating appropriate, dimensionless variables of the process.

To this end, a first attempt in deriving a universal scaling functional form describing the flow dependency of relative permeability was recently proposed [13]. It is based on the results stemming from systematic, extensive simulations spanning 5 orders of magnitude in both the capillary number and flow rate ratio, implementing the DeProF model algorithm, build around a hybrid, true-to-mechanism, stochastic scale-up model for steady-state two-phase flows in pore networks [14].

The objective of the present work is three-fold:

(a) to examine the applicability of the proposed scaling on the basis of using typical R/SCAL measurements;

(b) to verify the existence, uniqueness and form of the locus of critical flow conditions, an inherent characteristic of such processes in terms of energy efficiency (see next section and [9]);

(c) to examine the potential of the proposed scaling and the underlying theoretical framework, in revealing the flow structure and/or network structure characteristics of any set of R/SCAL measurements (R/SCAL forensics).

To do so we have used a set of SCAL measurements on a typical sandstone core. Data were collected over a grid of specially selected flow conditions, spanning across up to 5 orders of magnitude on *Ca* and up to 10 orders on *r*. The laboratory study was performed independently, i.e. prior to the publication of the recently proposed flow-dependent relative permeability scaling; in that sense it can be characterized as a "blind study". With respect to (b), the existence of critical flow conditions (CFCs) has been already verified in an extensive review of published R/SCAL relative permeability diagrams [15]; nevertheless, verification of the universal form of the locus of CFCs proposed in [9] was lacking a laboratory verification using the same core and across a broad domain of flow conditions (*Ca* and *r*).

The paper deploys in 3 sections. We start by providing the basic modeling assumptions for concurrent, immiscible, steady-state two-phase flows in porous media and present the basic structure of the proposed, flow-dependent relative permeability scaling. We continue with describing the SCAL experiments we have tapped-on (methods and materials). Then, we present the results of the laboratory examination and associated raw data calculations. Finally, we discuss "points of agreement/disagreement" between scaling predictions and actual measurements and we draw conclusions on the applicability of the proposed scaling or its potential exploitation in SCAL technology.

2. Basic Theoretical Background

Consider the simultaneous, one-dimensional concurrent flow of a non-wetting phase (NWP, index n) and a wetting phase (WP, index w) across a porous medium control surface, $\tilde{A}$, at flow rates equal to $\tilde{q}_n$ and to $\tilde{q}_w$ respectively. (A tilde ~ is used to indicate a dimensional variable, no tilde indicates a dimensionless variable.) The phenomenological fractional flow Darcy relations that describe the steady-state, fully developed process, whereby the pressure gradient, $(\Delta\tilde{p}/\Delta\tilde{z})$, is common in both fluids ([9], Appendix I), are given by eqn (1)

$$\tilde{U}_i = \frac{\tilde{q}_i}{\tilde{A}} = \frac{\tilde{k}}{\tilde{\mu}_i} k_{ri} \left(\frac{\Delta\tilde{p}}{\Delta\tilde{z}}\right) \quad , \quad i = n, w \tag{1}$$

In one-dimensional flow under steady-state conditions, especially when the flow is fully developed, the common pressure gradient in both phases can be expressed in reduced form as

$$x = \frac{\Delta\tilde{p}/\Delta\tilde{z}}{(\Delta\tilde{p}/\Delta\tilde{z})^{1\Phi}} = \frac{\Delta\tilde{p}/\Delta\tilde{z}}{\tilde{\gamma}_{nw} Ca/k} = \frac{\Delta\tilde{p}}{\Delta\tilde{z}} \frac{\tilde{k}}{\tilde{\mu}_w \tilde{U}_w} = \frac{1}{k_{rw}} \tag{2}$$

where

$(\Delta\tilde{p}/\Delta\tilde{z})^{1\Phi}$ is the pressure gradient corresponding to an equivalent saturated one-phase flow (1Φ) of the WP at the same superficial velocity ($\tilde{U}_w$).

Ca is the capillary number, conventionally defined as

$$Ca = \tilde{\mu}_w \tilde{U}_w / \tilde{\gamma}_{nw} \tag{3}$$

$\tilde{\mu}_w$ is the viscosity of the WP and $\tilde{\gamma}_{nw}$ is the NWP/WP interfacial tension.

In eqn (2) the actual pressure gradient is divided by the pressure gradient of an equivalent one-phase flow of water at superficial velocity equal to $\tilde{U}_w$ (the 2nd component of the product). Note that, by definition, it is straightforward to verify that the reduced pressure gradient is essentially the inverse of the relative permeability of the WP.

The set of superficial velocities in eqn (1) may be appropriately reduced and replaced by a set of dimensionless variables, namely, the capillary number, *Ca*, as defined previously in eqn (3), and the N/W flow rate ratio,

$$r = \tilde{q}_n/\tilde{q}_w = \tilde{U}_n/\tilde{U}_w \tag{4}$$

At steady-state conditions in one-dimensional flow, the flow rate ratio, r, is equal to the mobility ratio, r_λ. The equivalence, expressed in the first part of eqn (5), is a conjecture resulting from direct flow analysis (see Appendix I in [9]).

$$r = \frac{U_n}{U_w} \equiv \frac{1}{\kappa}\frac{k_{rn}}{k_{rw}} = r_\lambda \quad \Leftrightarrow \quad k_{rn} = \kappa r k_{rw} \tag{5}$$

where

$$\kappa = \tilde{\mu}_n/\tilde{\mu}_w \tag{6}$$

is the NWP/WP bulk viscosity ratio.

Eqn (5) is very useful, as we may easily recover the r_i value pertaining to any pair i of measured $\{k_{rn}, k_{rw}\}_i$ values from a conventional (saturation dependent) relative permeability diagram. We may also recover a relative permeability by knowing the values of the other relative permeability and the flow rate ratio. Therefore, switching to the approach of flow-dependency may be readily applicable as all conventional relative permeability diagrams can still be used.

The conjecture that both fluids share a common pressure gradient, eqn (1), is also based on experimental evidence. In particular, eqn (5), i.e. the direct result of assuming a common pressure gradient, is verified by experiment. When measured values of flow rate ratio are plotted against calculated values of mobility ratio, they align on a straight line. Ample experimental evidence is provided in a recent work whereby ~180 relative permeability diagrams pertaining to steady-state conditions for a variety of N/W/PM systems have been reviewed [15]. The conjecture proves to be valid also in the present work, see the diagrams in **Fig. 7**. Because of the common pressure gradient [eqn (1)] and the equivalence between flow rate ratio and mobility ratio [eqn (5)], relative permeability curves intersect at a fixed value of the flow rate ratio, the so-called cross-over flow rate ratio value, r_x. The latter is reciprocal to the viscosity ratio, $r_x = (1/\kappa)$, as provided by eqn (5). This inherent characteristic of steady-state flows is universally observed in all relative permeability vs flow rate ratio diagrams ([9], [15] and **Fig. 7**).

In the line of the present work we will also analyze the results from an energy efficiency perspective. In this context, we will use the reduced (normalized) measure of the energy efficiency of the process, considered as the flow rate of the NWP per unit of total hydraulic power spent, or equivalently, provided externally to the N/W/PM system (say by the "pumps"), to maintain two-phase flow (an essentially dissipative process) at any set of externally imposed flow conditions, *Ca* and *r*. The associated energy utilization factor, or energy efficiency index, f_{EU}, can be readily calculated in terms of macroscopic measurements [9], as

$$f_{EU} = \frac{k_{rn}}{\kappa(r+1)} = \frac{r k_{rw}}{r+1} = \frac{r}{x(r+1)} \tag{7}$$

For every fixed value of the capillary number, *Ca*, there exists a single value of the flow rate ratio, r^*, for which the energy efficiency index, f_{EU^*}, attains a maximum value [9, 15]. Moreover, for every N/W/PM system, a unique locus of energy efficiency maxima is formed, $r^*(Ca)$. Flow conditions matching the $r^*(Ca)$ locus are called *critical flow conditions* (*CFC*s). Critical flow conditions can be measured and identified by R/SCAL. The efficiency index provides a strong flow analysis and characterization tool [9].

Closing this passage, we need also to address the issue of selecting and using the flow rate ratio, r, instead of the fractional flow of the WP, f_w, as one of the two independent variables of the process – the other being the capillary number, *Ca*. In the core analysis community the applicable standard is to refer to one of the fractional flow of the WP, f_w. Switching between r and f_w (or f_n) is readily provided through the

 https://doi.org/10.1051/e3sconf/202014603002

transformation $f_w = r/(1+r)$. The use of the flow rate ratio, r, instead of the fractional flow, f_w, as independent variable, has the advantage of a more convenient description of the sought physical process, especially (or at least) around the critical flow conditions, whereby the identification of the critical flow rate ratio, $\log r^*(Ca)$, becomes more evident and obvious. The implications in using f_w, or f_n, instead of r are extensively discussed in [15]. As a general observation all energy efficiency diagrams $(\{r, f_{EU}\}_i)$ show a more uniformly smooth trend irrespective of the particular values of the system parameters and flow conditions examined in that review.

2.1 Prediction of Steady-State Relative Permeabilities

The mechanistic model *DeProF* for immiscible, steady-state two-phase flow in pore networks may be used to predict the reduced macroscopic pressure gradient, x, given the flow conditions and system properties. The model is based on the concept of decomposition in prototype flows, hence the acronym *DeProF*. It takes into account the pore-scale mechanisms and the sources of non-linearity caused by the motion of interfaces, as well as other complex, network-wide cooperative effects, to estimate the conductivity of each class of pore unit cells in a statistical sense. It implements effective medium theory with appropriate expressions for pore-to-macro scale consistency for NWP and WP mass transport, to derive an implicit algebraic relation invoking the macroscopic pressure gradient, the capillary number, Ca, the flow rate ratio, r, the viscosity ratio, κ, the dynamic advancing and receding contact angles, (θ_A, θ_R) (wettability), and a set of parameters that describe the geometrical, topological and structural characteristics of the pore network, $\mathbf{x}_{pm}$ [14].

Using the *DeProF* model, one can obtain the solution to the problem of steady-state two-phase flow in porous media in the form of the following transfer function,

$$x = x(Ca, r; \kappa, \theta_A, \theta_R, \boldsymbol{x}_{pm}) \tag{8}$$

Implementation of the DeProF model is possible when analytical expressions of the pore network geometry are plausible and general expressions for the cell conductivities can be calculated for all possible flow configurations. In general, this is not the case for real porous media, but only for a certain class of model pore networks with analytically tractable pore geometry. Yet, one can benefit by scanning entire domains of flow conditions at relatively short times and therefore get a systematic and consistent overview of the behavior of the flow across orders of magnitude on flow conditions. In that context, DeProF can be used as a virtual,large core SCAL simulator.

In a recent work, extensive simulations implementing the *DeProF* algorithm have been carried out to derive maps that describe the dependence of the flow structure on the independent flow variables, namely the capillary number, Ca, and the flow rate ratio, r, or, equivalently, to the reduced, superficial velocities of NWP and WP [14]. The simulations span 5 orders of magnitude in Ca $(-9 \leq \log Ca \leq -4)$ and r $(-2 \leq \log r \leq 2)$ over sufficiently fine steps. Fluid systems with various viscosity ratios (9 in total), have been examined.

Indicative DeProF model predictions of the reduced pressure gradient, x, are presented in **Fig. 1** for a typical NWP/WP/PM system with viscosity ratio $\kappa = 1.5$. The diagrams furnish the projections of the $x_{ij}(Ca_i, r_j)$ predicted values on many constant-logCa_i planes in (a) and constant-logr_j planes in (b). In that context, markers are connected into iso-Ca and iso-r groups.

Referring to **Fig. 1**(a), at the high-end of the log$r$ domain, all curves pertaining to constant-Ca values tend to bundle and align asymptotically to the straight (dashed) line. The corresponding functional form, eqn (9), states that at sufficiently large Ca values the reduced pressure gradient becomes a linear function of the flow rate ratio, with the linearity constant equal to the viscosity ratio, $\kappa = 1.5$.

$$\text{As } r \to \infty, \quad \log x = \log\kappa + \log r \quad \Leftrightarrow \quad x = \kappa r \tag{9}$$

(a)
(b)

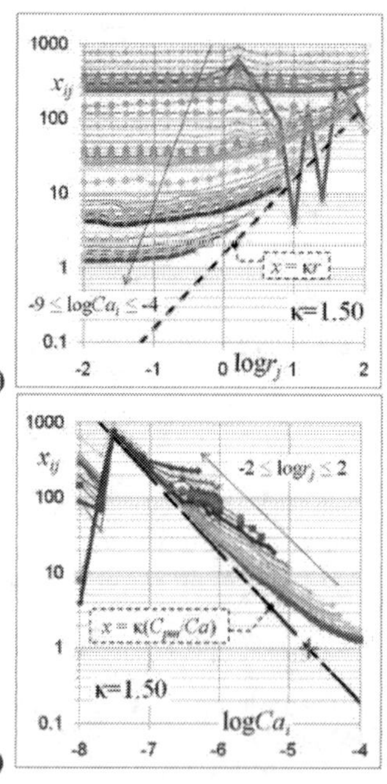

Fig. 1 Reduced pressure gradient values, x, for different values, **(a)** of the capillary number, Ca, and **(b)** of the flow rate ratio, r. Both diagrams pertain to a typical value of the viscosity ratio, $\kappa = 1.5$ function

We may observe a similar trend in **Fig. 1**(b). At the low-end of the Ca domain, all iso-r curves tend to bundle and align asymptotically to a straight line, also illustrated with a dashed line. Now the functional form is given by the expression, up to a constant value, C_{pm}, associated to the N/W/PM system [13].

$$\log Ca \ll 0, \ \log x = \log\kappa - \log C_{pm} - \log Ca \Leftrightarrow x = \kappa/(C_{pm} Ca) \tag{10}$$

2.2 Universal Functional Form of Relative Permeabilities

Based on these simulations it was possible to derive a universal function that can describe the dependence of the DeProF model predicted values $x_{ij}(Ca_i, r_j)$ of the reduced pressure gradient for different values of the capillary number, *Ca*, and the flow rate ratio, *r*, by the universal scaling form [13],

$$x(\log Ca, \log r) = A(\log Ca) + \kappa r \quad (11)$$

where the functional $A(\log Ca)$ may be determined by any fitting procedure. The set of (A_i, Ca_i) values is plotted on the diagram of **Fig. 2**. The cross-markers are lined-up against two straight line asymptotes; one with negative inclination and the other coinciding with the horizontal axis; these meet at an oblate angle.

We may fit the set of ($\log Ca_i$,$\log A_i$) values by a function of the form,

$$\log Ca = \log Ca_0 + C_0/(\log A)^2 - C_3 \log A \quad (12)$$

where $-1/C_3$ is the gradient of the inclined asymptote, and $\log Ca_0$ is the abscissa of the intersection of the two asymptotes and coefficient C_0 is a measure of the overall distance (approach) to the two asymptotes (the higher this value, the greater the distance of the apex of the swarm of markers to the asymptotes' intersection).

The coefficients values pertaining to the particular NWP/WP/PM model system (examined in the DeProF simulations) are: $C_3 = 1.1474$, $\log Ca_0 = -4.3$, and $C_0 = 0.003$. The thick red line is the curve $\log A(\log Ca)$ that has been fitted to the raw data $\log A_i$ using eqn (12). Note, eqn (12) is an implicit function of $\log A$ in terms of $\log Ca$. In particular, it is a cubic equation of $\log A$, which can be solved analytically to provide an explicit analytical expression $\log A(\log Ca/Ca_0\,; C_0, C_3)$.

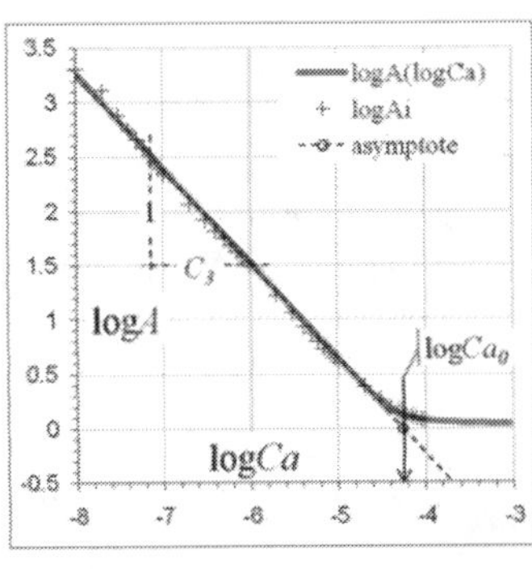

Fig. 2 Kernel function values *A*, in terms of *Ca*. Cross-markers indicate values ($\log A_i$, $\log Ca_i$) of the kernel function used in eqn (11), calculated to fit the DeProF predicted data, **Fig. 1**(b). These have been fitted by the thick red line, using eqn (12). The black dashed line is an asymptote with gradient ($-1/C_3$).

Using the analytical expressions from eqs (11) and (12) with the aforementioned coefficient values, we may plot the reduced pressure gradient in terms of *Ca* and *r*, $x(\mathrm{Ca}, r)$. Comparing the diagrams of **Fig. 2** to those of in **Fig. 1**, we observe that the DeProF predicted values $x_{ij}(Ca_i, r_j)$ are recovered with great specificity, and that the trend at extreme flow conditions, maintains its physically-true characteristics (NWP flow is decoupled to the WP flow). Similar plots for other systems can be found in [13].

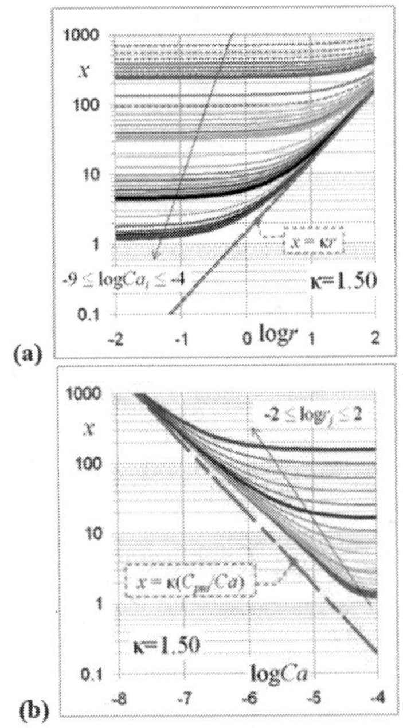

Fig. 3 Plots of the reduced pressure gradient values, *x*, in terms of the flow conditions, *Ca* and *r*, based on eqs (11) and (12); **(a)** iso-*Ca* curves; **(b)** iso-*r* curves.

We also need to stress here that the particular form/shape of the kernel function $\log A(\log Ca)$ produced in Fig. 2, as well as the form of eqn (12), come as a result of fitting the $\log A_i$ values produced by the DeProF model simulations of a NWP/WP/PM system of intermediate wettability. Yet, the system we examined in the laboratory is water wet. To this end, it needs to be investigated if we may implement the same type of functional form for the kernel function [as eqn (12)] or we will have to derive a different one.

3. Laboratory Materials and Methods

A set of SCAL measurements was ran on a typical sandstone core, over a grid of specially selected flow conditions (88 in total), spanning across ~6 orders of magnitude on *Ca* and ~6 orders on *r*.

3.1 Rock samples and fluids

Sample used in this work is water-wet outcrop sandstone from a Clashach quarry, with porosity of 13% and mean permeability of 672 mD. Sample was cored with a diamond core drill with 1 cm internal diameter and cut to a length of 2 cm. After being cored, the sample was dried in an oven for at least 48 hours at a temperature of 60 °C.

Two-phase flow measurements were conducted using synthetic dodecane oil and brine at 5 g/l NaCl (Sodium Chloride) for the low to intermediate capillary number. Viscosity at 20°C were respectively for oil and brine 1.5 cp and 1.07 cp. Interfacial tension was measured at $\tilde{\gamma}_{nw} = 50\text{mN/m}$. For intermediate to high capillary number, i.e. for 1 g/lt of SDBS (sodium dodecyl-benzene-sulfonate) was added to the brine to lower the interfacial tension to 1% of the original, i.e. to $\tilde{\gamma}_{nw} = 0.5\text{mN/m}$. According to Graue *et al* [16], Clashach sandstone exhibits an Amott index of 0.45 which is typical of strongly water wet rock. During our experiments we have checked that the sample do not spontaneously imbibe oil.

3.2 Experimental set-up and procedure

Two phase flow measurements were conducted using the CAL-X set-up that has been recently designed for high throughput core-flood experimentation [17]. It is composed of an X-ray radiography facility, a fully instrumented multi-fluid injection platform and a dedicated X-ray transparent beryllium core holder. The local and averaged saturation S_w is derived from the radiographs, captured every 10 seconds, using Beer-Lambert law for multi-material.

Steady State Method (SSM) was used to derive relative permeability. The SSM method involves the injection of both fluids at the inlet of the plug while measuring the differential pressure and the saturation within the core. Eqn (1) can be solved analytically if capillary pressure gradient is constant and end effects are neglected. This condition can be assessed when fluids are injected at high flow rate. Yet, these assumptions as well as laminar flow (i.e Reynolds number less than 10) have to be confirmed. Laminar condition is verified since a linear relation between flow rate and pressure drop is respected in mono-phasic injection. To validate the constant capillary pressure gradient and negligible capillary end effect, we verify that the saturation at steady state condition is uniform along the sample, see example on **Fig. 4**.

The core has been systematically flooded, by successive cycles of drainage and imbibition at practically constant values of the nominal capillary number, N_{Ca}, defined over total flow and average viscosity as

$$N_{Ca} = (\tilde{U}_n + \tilde{U}_w)(\tilde{\mu}_n + \tilde{\mu}_w)/(2\tilde{\gamma}_{nw}) \tag{13}$$

The diagram in **Fig. 5** illustrates a map of the broad range of scanned flow conditions – spanning across 5 orders of magnitude over Ca and 10 orders of magnitude over r. Markers connected with lines group successive cycles of drainage/imbibition, at nominal capillary number values, N_{Ca}, as indicated in the legend.

Note, Ca is defined by eqn (3) whereas N_{Ca} is defined by eqn (13). Therefore,

$$N_{Ca}/Ca = (1 + r)(1 + \kappa)/2 \tag{14}$$

In total, 14 cycles of drainage and imbibitions have been run and 88 measurements have been taken. The source data measurements are presented in Table 1.

Cycle	Type	$\langle N_{Ca} \rangle$ [×10-6]	k [md]	WP	NWP	IFT [mN/m]	q_n [cc/min]	q_w [cc/min]	S_w [-]	Δp [mbar]
01	1st Dr	0.54	620	5 g/l NaCl	Dodecane	50	0.00001	0.097	0.98	11
							0.002	0.088	0.67	64
							0.048	0.044	0.55	98
							0.105	0.00001	0.28	80
02	1st I	0.54					0.0934	0.00001	0.28	80
							0.073	0.021	0.4	124
							0.048	0.045	0.49	153
							0.004	0.091	0.51	199
							0.00001	0.097	0.52	192
03	2nd Dr	0.56					0.00001	0.097	0.555	191
							0.028	0.091	0.545	189
							0.046	0.046	0.492	164
							0.072	0.021	0.452	130
							0.092	0.004	0.389	97
							0.096	0.00001	0.27	82
04	1st Im	5.55	823				1	0.00001	0.19	174
							0.929	0.044	0.345	407
							0.736	0.246	0.42	587
							0.488	0.491	0.46	670
							0.239	0.736	0.48	722
							0.04	0.933	0.51	682
							0.00001	0.974	0.53	653
05	2nd Dr	5.58					0.00001	0.974	0.54	650
							0.043	0.933	0.52	695
							0.246	0.736	0.5	713
							0.501	0.49	0.473	712
							0.746	0.245	0.441	629
							0.946	0.046	0.384	400
							0.99	0.00001	0.23	180
06	1st Dr	5.60	521				0.00001	0.998	0.984	97
							0.046	0.932	0.75	403
							0.245	0.735	0.65	561
							0.502	0.491	0.574	630
							0.748	0.237	0.51	612
							0.944	0.046	0.415	435
							0.996	0.00001	0.225	243
07	1st Im	5.57					0.996	0.00001	0.225	243
							0.942	0.045	0.39	430
							0.743	0.244	0.464	726
							0.496	0.49	0.505	871
							0.232	0.733	0.525	876
							0.041	0.93	0.55	828
							0.00001	0.982	0.561	760
08	2nd Dr /	5.56					0.00001	0.982	0.572	760
							0.041	0.932	0.561	811
							0.236	0.734	0.533	945
							0.486	0.49	0.504	890
							0.742	0.243	0.46	817
							0.947	0.043	0.387	500
							0.994	0.00001	0.19	245

Table 1 The 88 source data measurements taken for each of the 14 cycles of drainage and imbibitions type injections of the core floods. The panel on the left is for high-IFT set-up, the panel on the right for low-IFT set-up.

Cycle	Type	$\langle N_{Ca} \rangle$ [×10-6]	k [md]	WP	NWP	IFT [mN/m]	q_n [cc/min]	q_w [cc/min]	S_w [-]	Δp [mbar]
09	1st Dr	55.17	693	5g/l NaCl + 1 g/l SDBS	Dodecane	0.5	0.00001	0.98	0.98	86
							0.0044	0.093	0.58	58
							0.048	0.047	0.385	183
							0.099	0.00001	0.14	40
10	1st Im	55.35					0.099	0.00001	0.14	40
							0.094	0.004	0.15	576
							0.074	0.022	0.211	390
							0.049	0.048	0.311	180
							0.005	0.093	0.413	56
							0.00001	0.098	0.431	28
11	2nd Dr /	56.02					0.00001	0.098	0.43	28
							0.005	0.093	0.421	32
							0.026	0.073	0.41	51
							0.05	0.047	0.35	105
							0.074	0.022	0.25	385
							0.096	0.004	0.172	746
							0.104	0.00001	0.165	82
12	1st Dr	558.12	823				0.00001	0.988	0.974	86
							0.041	0.937	0.9	108
							0.236	0.74	0.729	230
							0.493	0.494	0.47	506
							0.74	0.247	0.234	1170
							0.94	0.048	0.143	1449
							0.99	0.00001	0.122	250
13	1st Im	558.11					0.98	0.00001	0.122	255
							0.937	0.0439	0.125	1572
							0.739	0.239	0.191	1340
							0.494	0.493	0.41	520
							0.247	0.744	0.624	270
							0.048	0.935	0.738	134
							0.00001	0.994	0.783	105
14	2nd Dr	558.69					0.00001	0.995	0.787	105
							0.049	0.935	0.746	133
							0.236	0.739	0.639	267
							0.493	0.494	0.437	530
							0.74	0.247	0.2	1350
							0.935	0.048	0.111	1599
							0.99	0.00001	0.11	280

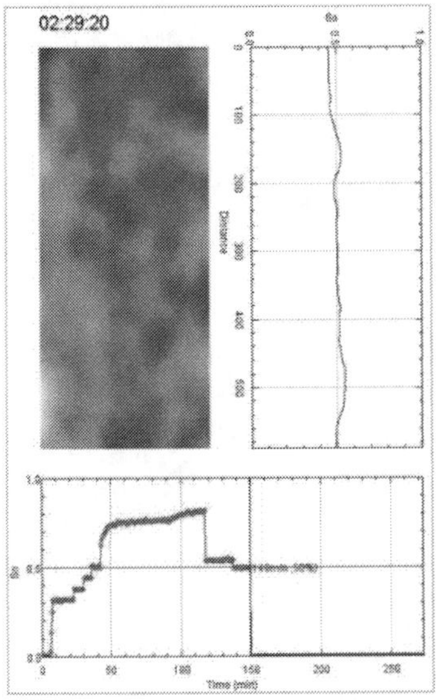

Fig. 4 Example of real-time saturation monitoring: 2D saturation map (upper left) and corresponding saturation profile (upper right); evolution of mean saturation as a function of time (low central)

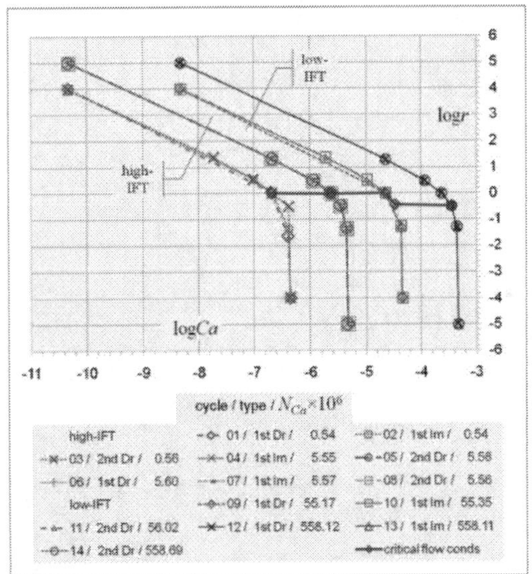

Fig. 5 Mapping of the 88 imposed flow conditions scanned during the 14 core-floods. Markers connected with lines are grouped into successive cycles of drainage/imbibitions, at constant values of the nominal capillary number, N_{Ca}, as indicated in the legend.

4. Results and discussion

Pairs of NWP and WP relative permeability values $\{k_{rn}, k_{rw}\}_i$, were computed using eqn (1) for each measurement i (i = 1, 88). These were then drawn into standard relative permeability diagrams in terms of the WP saturation, S_{wi}, for each one of the 14 different core flood cycles (or iso-N_{Ca} groups). These diagrams are not presented here because of space limitation. For the record, they have been merged into a single diagram, see **Fig. 6**(a). No evidence of grouped correlations between k_{rn}, k_{rw} and S_w is observed (as expected). Yet, the picture becomes more clear when values of relative permeabilities, $\{k_{rn}, k_{rw}\}_i$, are plotted against corresponding values of the flow rate ratio, r_i. The diagram in **Fig. 6** (b) is the result of merging the corresponding source diagrams in **Fig. 7**.

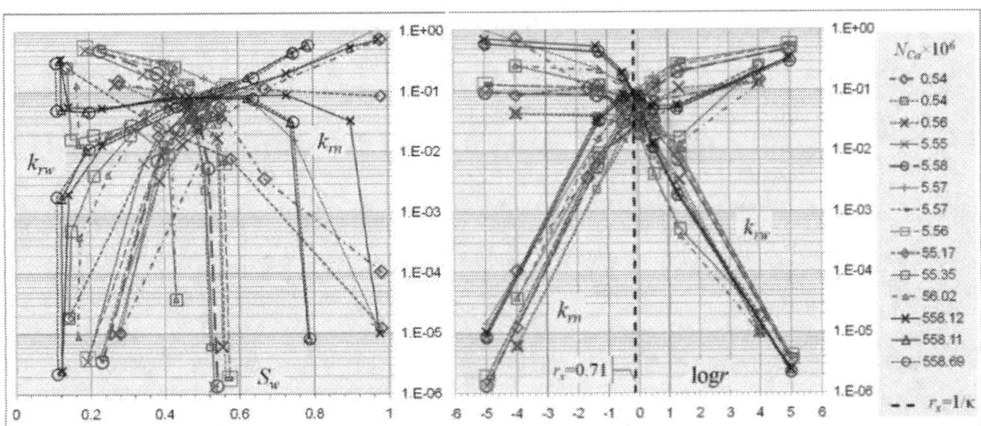

Fig. 6 Core-flood measured relative permeabilities, k_{rn}, k_{rw}, against corresponding mean values of WP saturation, S_w. **(a)** and flow rate ratio values, log$r$ **(b)**. In both diagrams, a total of 88 markers connected with lines are grouped into 14 successive cycles of drainage /imbibition, at constant values of the nominal capillary number, N_{Ca}, as indicated in the legend.

https://doi.org/10.1051/e3sconf/202014603002

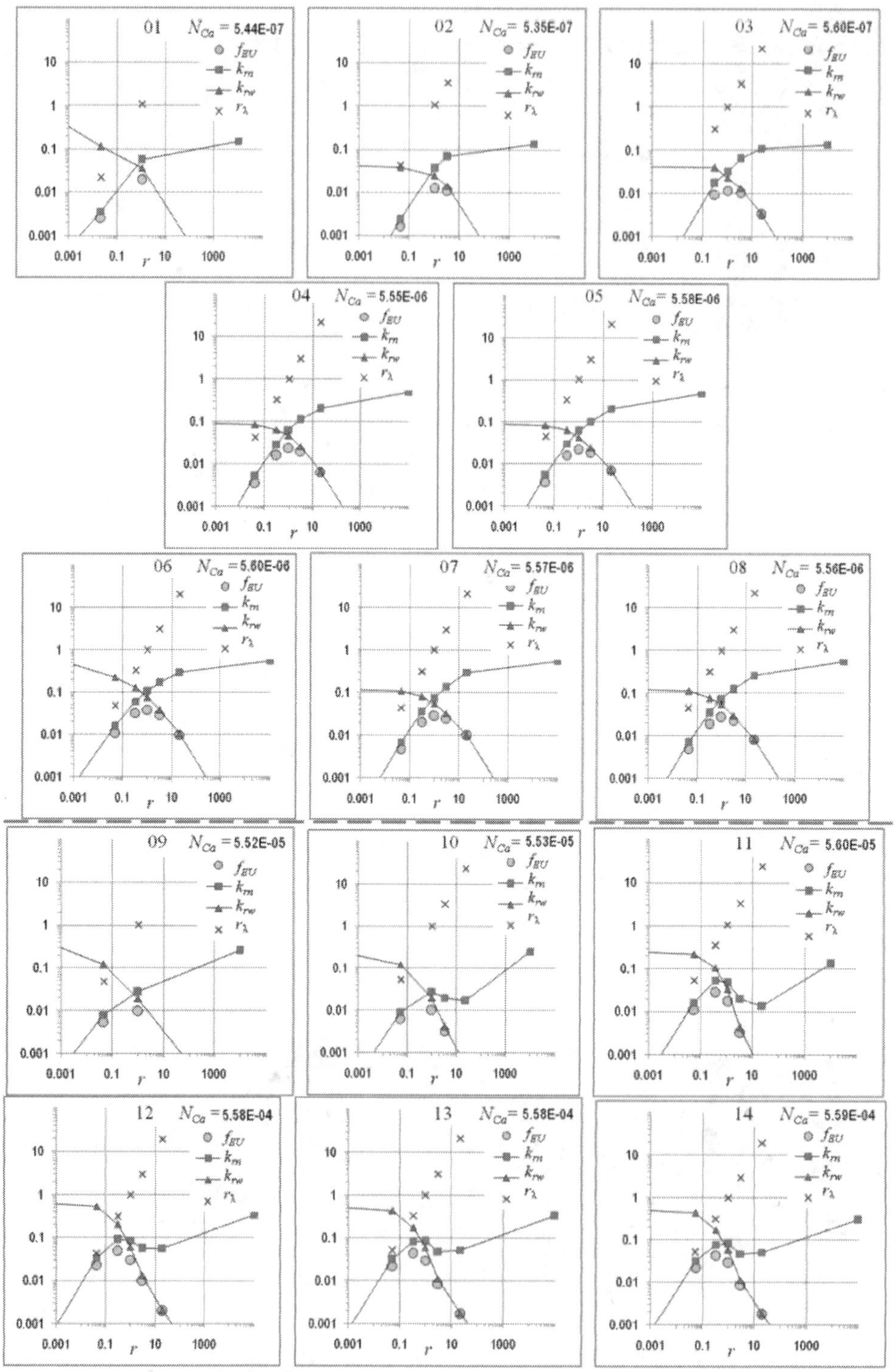

Fig. 7 Measured values of relative permeabilities, k_{rw} (▲) and k_{rn} (■), mobility ratio, r_λ (×) and energy efficiency, f_{EU} (●) against flow rate ratio values, r, for the 88 different steady-state flow conditions imposed during the 14 core floods (see **Fig. 5**). Diagrams 01-08 (above the dashed line) pertain to high IFT (50 mN/m) core-floods whereas 09-14 to low-IFT (0.5 mN/m) floods.

 https://doi.org/10.1051/e3sconf/202014603002

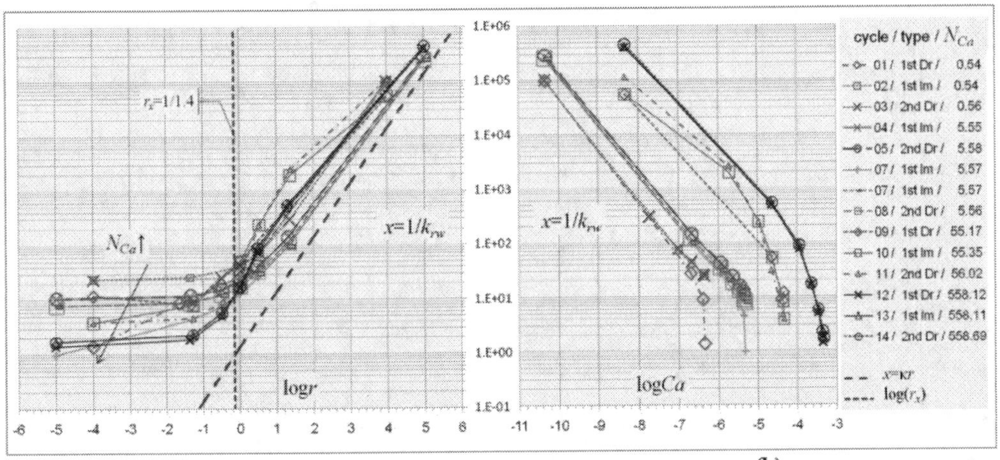

(a) **(b)**

Fig. 8 Reduced pressure gradient values, x, measured during the core-floods. Data markers are organized in iso-N_{Ca} groups. **(a)** Dashed inclined line indicates the asymptotic line for flow decoupling provided by eqn (9). **(b)** Data group markers are clustered into two formations, each pertaining to a value of the interfacial tension (IFT = 50 mN/m (left cluster) and 0.5 mN/m (right cluster).

4. Results and discussion

Pairs of NWP and WP relative permeability values $\{k_{rn}, k_{rw}\}_i$, were computed using eqn (1) for each measurement i ($i = 1, 88$). These were then drawn into standard relative permeability diagrams in terms of the WP saturation, S_{wi}, for each one of the 14 different core flood cycles (or iso-N_{Ca} groups). These diagrams are not presented here because of space limitation. For the record, they have been merged into a single diagram, see **Fig. 6**(a). No evidence of grouped correlations between k_{rn}, k_{rw} and S_w is observed (as expected).

Yet, the picture becomes more clear when values of relative permeabilities, $\{k_{rn}, k_{rw}\}_i$, are plotted against corresponding values of the flow rate ratio, r_i. The diagram in **Fig. 6** (b) is the result of merging the corresponding source diagrams in **Fig.** 7.

Note, in the latter diagrams (**Fig. 7**), measured values of the actual flow rate ratio, r [eqn (4)], and the calculated mobility ratio, r_λ [eqn (5)], are aligned along a straight line with gradient 1. Therefore, the conjecture of common pressure gradient in both fluids, as well as the transformation expressed in eqn (5), because of the underlying equivalence between r and r_λ, are also verified. Note, in **Fig.** 6(b) and **Fig.** 7, all pairs of NWP and WP relative permeability curves intersect at the, so-called, cross-over flow rate ratio value, $r_x = (1/\kappa)$, eqn (5). That value is indicated by the black short-dash line at $\log r_x = -\log(\kappa)$ in **Fig. 6** (b).

Reduced pressure gradient values, x, calculated as the reciprocal to the WP relative permeability, k_{rw}, [eqn (2)], are presented in **Fig. 8**. The diagram in **Fig. 8** (a) presents data values in terms of log$r$, classified into 14 iso-N_{Ca} groups. The general trend is similar to that provided by the proposed functional form, eqs (11) and (12), **Fig. 1** and **Fig. 3**. Note, with increasing r, the asymptotic trend aligns to the dashed inclined line to a virtually decoupled flow state, as provided by eqn (9).

In **Fig. 8** (b) reduced pressure gradient values, x, in terms of logCa are also presented in iso-N_{Ca} groups. The groups are clearly clustered into two formations (or "islands"). Each pertains to core floods performed with the same fluids but under different interfacial tensions (Section 3.1). The cluster on the left pertains to an interfacial tension (IFT) value $\tilde{\gamma}_{nw} = 50\text{mN/m}$, while the cluster on the right (higher N_{Ca}) to an IFT reduced by 1%, $\tilde{\gamma}_{nw} = 0.5\text{mN/m}$. On the high-$Ca$ region the trend is different when compared to that observed in the diagram of **Fig. 1**(b) [also **Fig. 3** (b)]. Specifically, the mild and progressive drop of the x values, as flow conditions shift to high-Ca regimes, contradicts the smooth but very steep (abrupt) drop of the pressure gradient measured in the core-floods (**Fig. 8**). Moreover, we may distinguish two similar behaviors. In the high-IFT set-up (left cluster, groups 01- 08) the drop in x observed at increasing values of Ca is steeper when compared to the low-IFT set-up. In that same set-up (groups 09-11 and 12-14) the drop in x is progressive, yet its onset is practically independent of the N_{Ca}. We will comment on this later on.

Using eqn (11), for every pair $\{x, r\}_i$, $i = 1,88$ of the measured values, we may calculate a corresponding A_i value for each flow condition imposed during the core-floods. The A_i values are plotted in terms of logCa_i values on the diagram in **Fig. 9**. Comparing with **Fig. 2**, we observe the same differences in the trend (drop) of the A_i values at high-Ca regimes. The trend in **Fig. 9** is similar to that observed in **Fig. 8**(b). Note, also, data markers of the high-IFT set-up are aligned within a narrow band in the low-Ca regime, in contrast to the high-IFT set-up where markers are aligned to a wider band. The gradients of these bands in the low-Ca regime, could be considered as a "system identity" (compare with **Fig. 2**).

The overall trend indicates a drop of the effect of capillary forces at high-Ca regimes. The cause may probably be associated to interstitial, emulsion type flow, setting-in at high-Ca flow conditions.

Fitting the A_i data using the functional form expressed by eqn (12) does not succeed in capturing the trend at the high-*Ca* regime. An expression describing more consistently the trend at the high-*Ca* regime is

$$\log A = A_0\left(1-\frac{|\log(Ca/Ca_0)|}{|\log(Ca_m/Ca_0)|}\right)^{\frac{1}{n}} - C_g\log\frac{Ca}{Ca_0} \qquad (15)$$

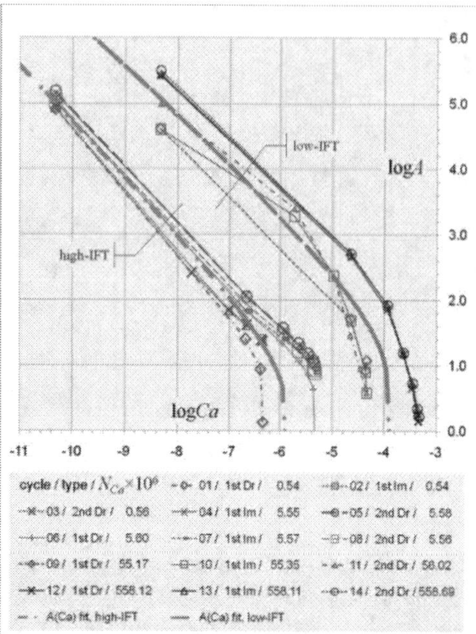

Fig. 9 The log*A* values in terms of log*Ca*, calculated from the core-flood data using eqn (11). The continuous orange dash-dotted line fits high-IFT data, the blue dashed line fits low-IFT data, both using eqn (15).

The orange dash-dotted and the blue dashed lines fit the high-IFT and low-IFT data (**Fig. 9**). Coefficient values of the two fitting curves are presented in **Table 2**.

Table 2. Values of the log*A* fitting expressions coefficients

IFT	A_0	$\log Ca_m$	$\log Ca_0$	C_g	n
High	1.7	-13	-5.9	0.78	5
Low	2.4	-11	-3.9	0.65	5

We may now compare the dissimilar trends at high-*Ca* regimes between the DeProF model predictions for a system of intermediate wettability, eqn (12), and the water wet system examined in the core-floods, eqn (15). To reveal any latent systematic trend and eventually get a better image of the interstitial flow structure, we have further investigated energy efficiency aspects of the core-floods within the particular N/W/PM system.

The diagram in **Fig. 10**(a) provides the map of the *critical flow conditions* (CFCs) identified from the core floods. The diagram is actually a blow-up of the central part of the diagram in **Fig. 5** -focusing on the layout of the CFCs. The line connecting the CFCs attains a clearly formed S-shape, consistent to the $r^*(Ca)$ shape of the CFC locus on the universal energy efficiency map illustrated in [9] (cf. reference Figure 6 and 7). The dashed red horizontal line indicates the *nominal* value of the critical flow rate ratio for fully viscous flow conditions (as $Ca \to \infty$). Provided the viscosity ratio value of the examined N/W system [9], this CFC *nominal* value is estimated to be $r^*_\infty = 1/\sqrt{\kappa} = 0.845$ ($\log r^*_\infty = -0.073$). The *actual value* attained within the lower part of the S-shape form, indicates an asymptotic trend ($Ca \to \infty$) towards $r^*_\infty \cong 0.32$. ($\log r^*_\infty = -0.496$).

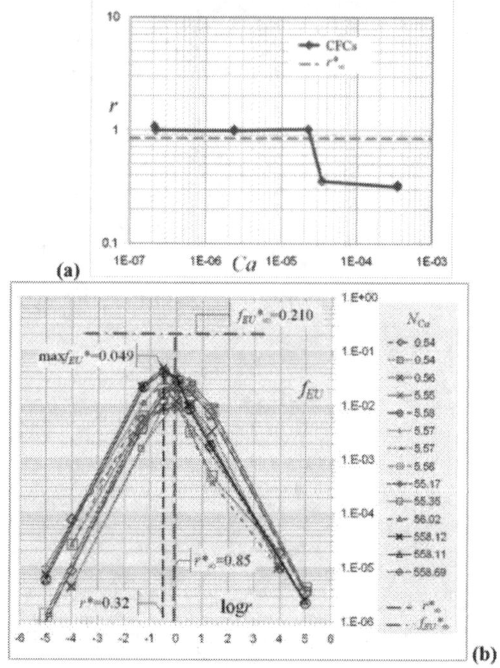

Fig. 10 Energy efficiency characteristics of the examined core-floods. **(a)** Detected critical flow conditions for various values of the capillary number, $r^*(N_{Cai})$, attaining an S-shape form. The diagram is a blow-up of the diagram in **Fig. 5**. **(b)** Energy efficiency values attained in terms of the flow rate ratio, *r*, grouped in iso-N_{Ca} core floods.

We may further examine the energy efficiency characteristics of the core floods, presented in the diagram of **Fig. 10**(b). The dash-dotted red horizontal line indicates the *nominal* value of the maximum attainable energy efficiency for the particular N/W system, the "ceiling of efficiency", toping at $f^*_{EU\infty} = 1/(1+\sqrt{\kappa})^2 = 0.2096$ as $Ca \to \infty$ (cf. eqn (30) in [9]), expected to be met at $r^*_\infty = 1/\sqrt{\kappa} = 0.845$ ($\log r^*_\infty = -0.073$). Yet, the energy efficiency of the examined core floods reached an *actual* maximum value, $\max f^*_{EU} = -0.049$, at $r \cong 0.32$ ($\log r = -0.496$). It is obvious that the *actual* values of energy efficiency attained at large *Ca*, are rather low when compared to the nominal, maximum values pertaining to the particular N/W system.

Overall, the observed differences are associated to high-*Ca* flow conditions. This strange behavior may be correlated to two factors favoring the setting of emulsion type flow: wettability (the examined system is 100%

water wet) and emulsification, affecting the structure of the interstitial flow(s). The aforementioned nominal values ($r^*_\infty, f^*_{EU\infty}$) are estimated considering the ideal case of minimization of capillarity effects at very large *Ca* values (see eqn (30) and Appendix II in [9]). In that modeling approach, capillarity effects are mainly attributed to the net Young-Laplace resistances at the contact lines, when menisci migrate downstream, by taking into account the hysteresis between the dynamic advancing and receding contact angles of menisci. If the flow intensities (superficial velocities) are large, power dissipation due to viscosity resistances prevails over power dissipation due to capillarity hysteresis –as the former depends quadratically on flow intensity, whereas the latter linearly. In that situation the flow is decoupled of capillarity effects and a nominal value of the flow rate ratio maximizing the energy efficiency can be estimated analytically (assuming a rather oversimplifying modeling hypothesis of negligible menisci resistances, [9]), as $r^*_\infty = 1/\sqrt{\kappa}$.

Nevertheless, in reality, e.g. during core floods in R/SCAL measurements, the NWP may disconnect in tiny droplets. (This is also taken into account in the DeProF model simulations albeit the system is of intermediate wettability.) The onset of emulsification is not only due to the reduction of the interfacial tension between the NWP and the WP, but also due to the pore scale flow dynamics and the pore network characteristics [18]. In general, there are two classes of emulsion flows in porous media depending on the size of the drops compared to the capillary throat diameter. For emulsions with smaller drops, the extra-pressure drop does not vary with capillary number and it is a function of the viscosity ratio, dispersed phase concentration and drop size distribution. For emulsions with larger drops, these may partially block the capillaries, leading to higher extra pressure differences at low capillary numbers [1, 19]. Likewise, part of the NWP droplet population may pass through larger throats without contacting the pore walls. In that case a significant fraction of the NWP flows free of contact-line resistances. The associated energy cost is reduced, allowing for a shifting of the critical flow rate ratio to lower values in order to reduce also the power dissipation due to the bulk viscosity of the NWP [9].

Moreover, in water-wet systems, the hysteresis effects are pronounced due to the large differences occurring between receding and advancing dynamic contact angles that, in turn, burden the flow with substantial capillarity-induced resistances at the low-*Ca* regimes. Nevertheless, there is a cut-off limit in pore scale velocities above which lubricating-films are formed and capillarity induced resistances may drop significantly. This is expressed as the steep drop in reduced pressure gradient values pertaining to the examined core-floods systems.

5. Conclusions

We have performed a preliminary assessment/validation of the specificity and applicability of a new flow-dependent relative permeability scaling law that takes explicitly into account capillarity effects. Flow dependency is associated to the flow intensities of the NWP and WP or, equivalently to the capillary number of the WP, *Ca*, and the NWP/WP flow rate ratio, *r*. The proposed scaling is build around a kernel function, log*A*(log*Ca*), that provides a "transitioning bridge" between the decoupled character of the flow at the two extreme flow regimes (low-*Ca*/low-*r* to high-*Ca*/high-*r* flows). The validation was based on a set of 14 core-floods of a typical water-wet sandstone and for a variety of flow conditions spanning across 5 orders of magnitude on *Ca* and 10 orders of magnitude on *r*.

The results suggest that it is possible to derive and implement universal, true-to-mechanism functions of flow-dependent relative permeabilities. The critical step here is to guess or data-fit an appropriate form of the kernel function, log*A*(log*Ca*). The latter is associated to the complex character of the NWP/WP/PM system rheology manifesting itself across the entire *Ca* domain.

The specificity /applicability of the proposed scaling is substantial and deserves further investigation, mainly towards understanding the universal structure of the kernel function log*A*(log*Ca*). Because of the structure of the scaling, eqn (11), the mode of decaying of log*A* at high-*Ca* regimes, whether expressed by eqn (12) or eqn (15), is equally applicable and depends on the physicochemical properties of the N/W/PM system, e.g. wettability. The critical property of the kernel function should be a monotonous decay at the high-*Ca* regime.

Implementing the proposed flow-dependency analysis it was possible to get a consistent and well structured view of the flow in the examined core-floods. The calculated values of reduced quantities (reduced pressure gradient, kernel function) show a *consistent dependency on flow conditions* and can be used to reveal the underlying flow mechanisms. In particular, considering the physicochemical properties of the two fluids, the results suggest that emulsion type flow settles-in above a threshold value of the capillary number. This threshold value depends on the flow conditions and the interfacial tension between the two fluids.

Reviewing the present work in terms of reaching its objectives, we may conclude:

(a) Implementation of the proposed scaling is possible using typical R/SCAL measurements; moreover it was possible to rely on superficial measurements only, i.e. flow rates of the two phases, pressure drop across the core length, IFT, absolute permeability.

(b) the existence, uniqueness and form of the locus of critical flow conditions, $r^*(Ca)$, was confirmed. The S-shape type, confined between two asymptotic lines, r^*_0 and r^*_∞ as proposed in [9] is also confirmed. Differences between nominal (expected) and actual (measured) values, of r^*_∞ and $f^*_{EU\infty}$, are justified on the basis of the induced emulsification of the flow occurring at very high *Ca* values. In order to reveal the S-shaping in greater detail, in follow-up studies the flow conditions should be finely scanned /tuned in proximity to the transition zone of the CFC locus.

(c) By implementing the proposed scaling and considering the underlying mechanisms of the interstitial flows, it was possible to explain the flow behavior as described by the macroscopic measurements. That was possible by examining the structure of the WP relative permeability, k_{rw}, and of the kernel function A, without any need to examine the flow at the pore scale using sophisticated equipment. Therefore, the proposed scaling could potentially serve as an additional tool for quality assessment of measurements (R/SCAL forensics).

Based on the results of this preliminary assessment, it would make sense to consider designing and deploying a systematic laboratory to investigate and reveal the effects of wettability, pore network structure, core size, etc. on the form of $A(\log Ca)$. The scope would be to improve R/SCAL protocols, and develop more robust and accurate macroscopic scale models.

The authors thank the reviewers for their constructive comments. MSV acknowledges partial support by the Research Council of Norway through its Centres of Excellence funding scheme, project number 262644.

References

1. C.T. Tsakiroglou, D.G. Avraam, A.C. Payatakes, "Transient and steady-state relative permeabilities from two-phase flow experiments in planar pore networks", *Adv. Wat. Res.*, vol. 30, pp. 1981–1992, 2007, doi:10.1016/j.advwatres.2007.04.002
2. V.R. Guillen *et al.*, "Capillary-driven mobility control in macro emulsion flow in porous media", *Int. J. Multiphase Flow*, vol. 43, pp. 62-65, 2012, doi:10.1016/j.ijmultiphaseflow.2012.03.001
3. K.T. Tallakstad *et al.*, "Steady-State Two-Phase Flow in Porous Media: Statistics and Transport Properties", *Phys. Rev. Let.*, vol. 102, 074502, pp. 1-4, 2009, doi:10.1103/PhysRevLett.102.074502
4. A. Georgiadis *et al.*, "Pore-scale micro-computed-tomography imaging: Nonwetting-phase cluster-size distribution during drainage and imbibition", *Phys. Rev. E*, vol. 88, 033002, pp. 1-9, 2013, doi:10.1103/PhysRevE.88.033 002
5. O. Aursjo *et al.*, "Film flow dominated simultaneous flow of two viscous incompressible fluids through a porous medium", *Frontiers in Physics*, vol. 2, no. 63, pp. 1-9, 2014, doi:10.3389/fphy.2014.00063
6. S.S. Datta, T.S. Ramakrishnan, D.A. Weitz, "Mobilization of a trapped non-wetting fluid from a three-dimensional porous medium", *Phys. Fluids*, vol. 26, 2014, 022002, 2014, doi:10.1063/1.4866641
7. R. Oughanem *et al.*, "A Multi-Scale Investigation of Pore Structure Impact on the Mobilization of Trapped Oil by Surfactant Injection" *Transp. Porous Media* vol. 109, pp. 673-692, 2015, doi:10.1007/s11242-015-0542-5
8. R.T. Armstrong, J.E. McClure, M.A. Berrill, M. Rücker, S. Schlüter, S. Berg, "Beyond Darcy's law: The role of phase topology and ganglion dynamics for two-fluid flow", *Phys. Rev. E*, vol. **94**, 043113, 2016, doi: 10.1103/PhysRevE.94.043113
9. M.S. Valavanides, "Review of Steady-State Two-Phase Flow in Porous Media: Independent Variables, Universal Energy Efficiency Map, Critical Flow Conditions, Effective Characterization of Flow and Pore Network", *Transp. in Porous Media*, vol. 123, no. 1, pp. 42-99, 2018, doi:10.1007/s11242-018-1026-1
10. R.E. Hinkley, M.M. Dias, A.C. Payatakes, "On the motion of oil ganglia in porous media", *PhysicoChemical Hydrodynamics*, vol. **8**, no.2, pp. 185-211, 1987
11. S. Sinha *et al*, "Effective Rheology of Two-Phase Flow in Three-Dimensional Porous Media: Experiment and Simulation", *Transp. in Porous Media*, vol. 119, no. 1, pp. 77-94, 2017, doi:10.1007/s11242-017-0874-4
12. C.D. Tsakiroglou *et al.*, "Steady-state two-phase relative permeability functions of porous media: A revisit", *Int. J. Multiphase. Flow*, vol. 73, pp. 34–42, 2015, doi:10.1016/j.ijmultiphaseflow.2015.03.001
13. M.S. Valavanides, "Universal, Flow Dependnet Relative Permeability Scaling for Steady-State Two-Phase Flows in Porous Media", SCA2018-066, in *Int. Symp. of the Society of Core Analysts*, Trondheim, Norway, Aug. 27-30, 2018, pp.1-9
14. M.S. Valavanides, "Oil Fragmentation, Interfacial Surface Transport and Flow Structure Maps for Two-Phase Flow in Model Pore Networks. Predictions Based on Extensive, DeProF Model Simulations", *Oil Gas Sci. Tech.*, vol. 73, no.6, pp. 1-36, 2018, doi:10.2516/ogst/2017033
15. M.S. Valavanides, E. Totaj, M. Tsokopoulos, "Energy efficiency characteristics in steady-state relative permeability diagrams of two-phase flow in porous media", *J. Pet. Sci. Eng.*, vol. 147, pp. 181-201, 2016, doi:10.1016/j.petrol.2016.04.039
16. A. Graue *et al*, "Imaging Fluid Saturation Development in Long-Core Flood Displacements", SPE 17438-PA, in *Proc. SPE 58th California Regional Meeting*, Long Beach, California, 1988, pp. 363-372, doi:org/10.2118/17438-PA
17. S. Youssef, M. Mascle, O. Vizika., "High Throughput Coreflood Experimentation as a Tool for EOR Project Design", SPE-190166-MS, in *SPE Improved Oil Recovery Conf.*, Tulsa, Oklahoma, USA, 14-18 April, 2018
18. E. Unsal, M. Broens, R.T. Armstrong, "Pore Scale Dynamics of Microemulsion Formation", *Langmuir*, vol. 32, pp. 7096-7108, 2016, doi: 10.1021/acs.langmuir.6b00821
19. S. Cobos, M.S. Carvalho, V. Alvarado, "Flow of oil–water emulsions through a constricted capillary", *Int. J. Multiphase Flow*, vol. 35, pp. 507-515, 2009, doi: 10.1016/j.ijmultiphaseflow.2009.02.018

https://doi.org/10.1051/e3sconf/202014603003

Improved method for complete gas-brine imbibition relative permeability curves

M. Ben Clennell[1,*], *Cameron* White[1], *Ausama* Giwelli[1], *Matt* Myers[1], *Lionel* Esteban[1], *Michael* Cullingford[2], *William* Richardson[3], *Gavin* Ward[2], *Matt* Waugh[2], *Scott* Cole[2], *Ashley* Hunt[2] and *Peter* Bright[2§]

[1]CSIRO Energy, Australian Resources Research Centre, Kensington, Western Australia 6076
[2]Chevron Australia, QV1, 250 St George's Terrace Perth, Western Australia, 6000.
[3]Chevron Energy Technology Centre, 3901 Briarpark Drive, Houston, Texas 77042 USA

Abstract. Standard test methods for measuring imbibition gas-brine relative permeability on reservoir core samples often lead to non-uniform brine saturation. During co-current flow, the brine tends to bank up at the sample inlet and redistributes slowly, even with fractional flow of gas to brine of 400:1 or more. The first reliable Rel Perm point is often only attained after a brine saturation of around S_w=40% is achieved, leaving a data gap between S_{wirr} and this point. The consequent poor definition of the shape of the Rel Perm function can lead to uncertainty in the performance of gas reservoirs undergoing depletion drive with an encroaching aquifer or subjected to a water flood. We have developed new procedures to pre-condition brine saturation outside of the test rig and progress it in small increments to fill in the data gap at low S_w, before continuing with a co-current flood to the gas permeability end-point. The method was applied to series of sandstone samples from gas reservoirs from the NW Shelf of Australia, and a Berea standard. We found that the complete imbibition relative permeability curve is typically 'S' shaped or has a rolling over, convex-up shape that is markedly different from the concave-up, Corey Rel Perm curve usually fitted to SCAL test data. This finding may have an economic upside if the reservoir produces gas at a high rate for longer than was originally predicted based on the old Rel Perm curves.

1 Introduction

When using standard, best-practice SCAL testing methods for measurement of imbibition gas-brine relative permeability on reservoir core samples, e.g., as recommended by McPhee et al. [1], it is very difficult to establish a uniformly progressing brine saturation. This difficulty in establishing stable co-current flow conditions leads to poor definition of the shape of the Rel Perm function at lower brine saturations when gas has the highest mobility.

Typically, during co-current flow with forced imbibition of brine along with gas, the saturation within the sample does not advance uniformly along the sample length but will bank up at the inlet end of the sample (Figure 1). The brine largely builds up by spontaneous imbibition from the restricted supply available and owing to low k_{rw} it takes hours or days for brine to redistribute along the sample length, even with high fractional flow of gas to brine of 400:1 or more. Moreover, the first reliable co-current gas-brine relative permeability point (stable differential pressure and uniform saturation) is often only attained after a brine saturation of S_w=40% or more is achieved, leaving a significant data gap between S_{wirr} and this point (Figure 2).

The data in Figure 2 can be interpolated with a concave-up Corey function curve [2], but this may be incorrect. The lack of valid data can lead to questionable inputs to (and outputs of) reservoir simulation models and flow on to uncertainty in the performance of gas reservoirs undergoing depletion drive with an encroach-

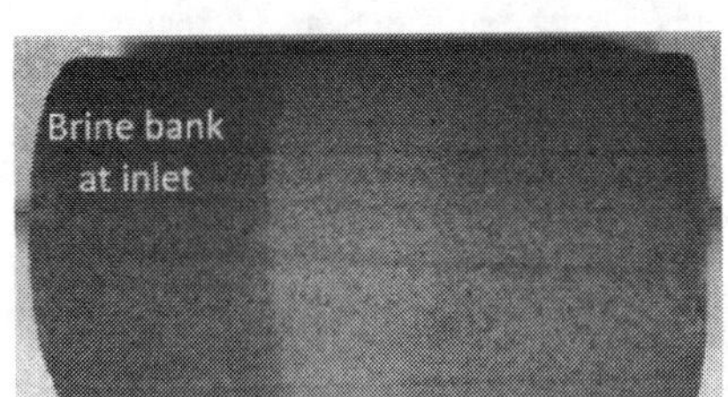

Fig. 1. Berea sandstone sample taken from test rig at the early stage of an imbibition Rel Perm measurement using standard, co-current SCAL methods. A large brine bank has formed at the inlet that blocks the permeability to gas.

* Corresponding author: Ben.Clennell@csiro.au §Present Address: BHP Petroleum, 125 St George's Terrace, Perth W.A. 6000.

https://doi.org/10.1051/e3sconf/202014603003

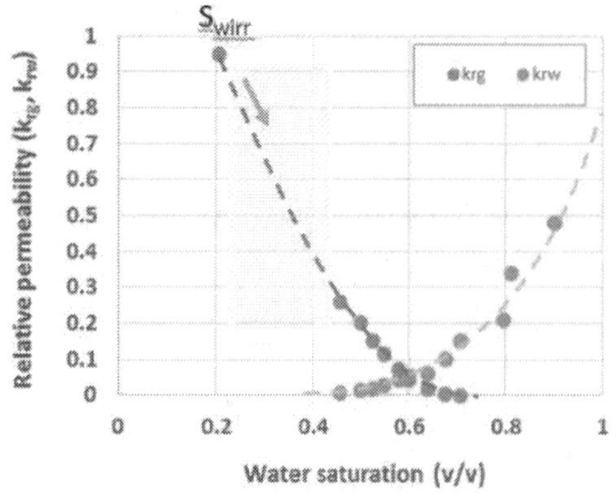

Fig. 2. Example of a good quality, but incomplete imbibition relative permeability curve. There is a large data gap (yellow) in the gas Rel Perm data between the gas-only flow point at S_{wirr} and the first co-current flow point at S_w=0.43. Data from [3].

-ing aquifer or subjected to a water flood.

The general problem of incomplete imbibition Rel Perm curves available for conventional gas field development planning has been pointed out previously at this forum [2] but it has not so far been resolved in the form of a more practical measurement method that can obtain accurate and complete Rel Perm curves at low to intermediate S_w.

In 2016 CSIRO was engaged by industry partners to investigate improvements to Special Core Analysis (SCAL) test methods for the purpose of reducing uncertainties in the development of major gas fields in the Northwest Shelf of Australia. During the ensuing research project, CSIRO worked with the operating company and in-house domain experts to devise and test new laboratory methods. We were able to collect a series of much more complete imbibition gas-brine relative permeability curves on several sandstone samples of moderate to excellent quality. The methods were applied to five NWS gas reservoir sandstones and also to a Berea sandstone, a well characterized "standard" rock, which produced similar results.

2 Materials and Methods

Experiments were conducted on reservoir sandstone samples ranging in dry gas permeability from 10 mD to around 670 mD, and with porosity ranging from 17% to 23% (see Figure A1). The reservoir sandstones are all quartz arenites (Q > 78%) with 6-10% feldspar and 3-4% kaolinite as the only clay mineral. The Berea sandstone outcrop sample used was quartz rich (Q >90%) and well sorted with dry gas permeability of 665 mD and porosity of 21.8%. Core plugs of around 50-60 mm length and 38 mm in diameter were cleaned prior to measurement, and then saturated with synthetic formation brine.

The samples were drained to irreducible brine saturation using a petrophysical centrifuge up to either 100 psi or 250 psi air-brine capillary pressure. The imbibition Rel Perm experiments were commenced from close to the value of irreducible brine saturation S_{wirr}, which in the case of the Berea sample and the most permeable reservoir sample, was around 2%, for the other reservoir samples S_{wirr} ranged from 10 to 15%.

Samples were loaded into the CSIRO x-ray CT monitored core flooding rig (Figure 3). The system uses accurate differential (Keller, Switzerland) and absolute (Quartzdyne, USA) pressure transducers capable of resolving a very small pressure drop across the sample, and gas (Teledyne ISCO, USA) and brine (Quizix, Chandler USA) pumps with high accuracy and a wide range of allowable flow rates (0.001 to 120 cc/min). Flow into and out of the sample is by means of two porous end stones that are removable for cleaning and for controlling their initial brine content. The pressure drop for flow calculations is across the sample only, with the end stones bypassed by thin peek tubes and attached to the differential pressure transducer (Figure 4).

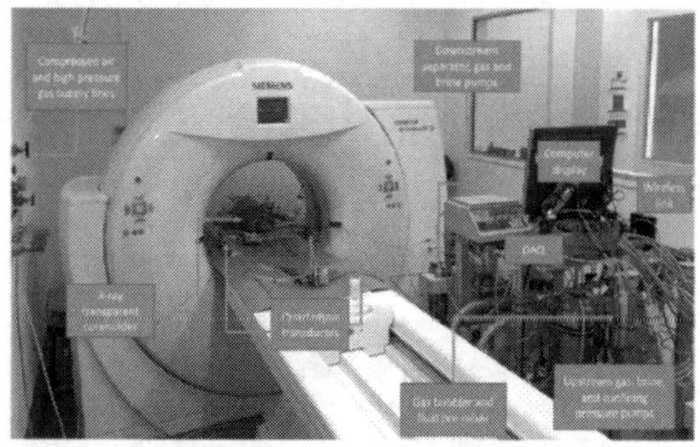

Fig. 3. CSIRO x-ray CT monitored core flooding rig.

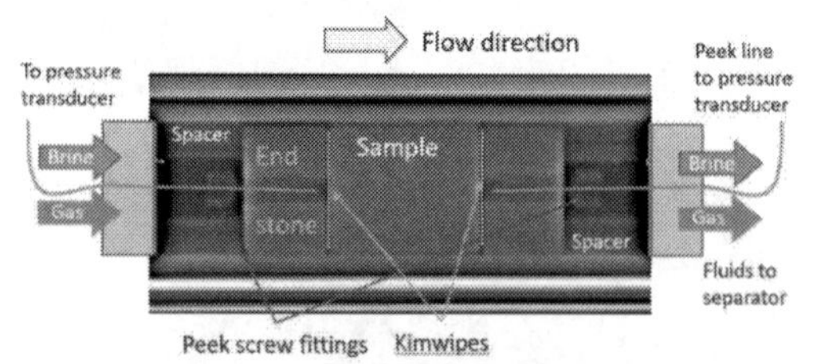

Fig. 4. Configuration of the core holder with removeable end stones made of Bentheimer sandstones that served as flow distributors for gas and brine. Pressure drop is measured only across the sample itself via a thin PEEK tube. Single ply tissue was used for capillary continuity at the ends of the sample.

In all the tests reported here a confining pressure of 2500 psi and a pore pressure of 500 psi were used. The gas used was nitrogen rather than methane for safety reasons. A sodium iodide-based brine was used to enhance x-ray contrast, though the overall composition was tested to be compatible with any clays in the rocks. The saturation computations from x-ray CT numbers used calibrations from dry and brine-saturation samples loaded to the same net overburden pressure of 2000 psi used in the test. Moreover, we also used intermediate points of known saturation before applying back pressure, to better constrain the calibration. The saturation from XCT is determined to within 3%, with errors from inexact repositioning of the sample, and drift in beam energy.

When using the standard technique of high gas fraction/ low brine fraction co-current flow to introduce the water in sample S22B, we encountered the same problem that has been reported previously. The brine entering the rock accumulated at the inlet and did not distribute uniformly until S_w of over 50% had been achieved. By this stage, the gas was nearly immobile. The relative permeability is poorly defined (Figures 5 & 6).

In these figures the valid Rel Perm data points are shown in solid colour. The data point at S_w=49% was collected when pressure was stable and the flow rates into and out of the sample were equal. However, the x-ray images (Figures A7-A9) show that the saturation profile was not even at that time. A brine bank formed at the inlet had not yet dissipated (Figure A7 image 2). The high brine saturation blocked gas flow through the sample, so that its relative permeability was only 2.2 mD. This banked point is labelled with a black border. During continued co-current flow at 400:1 gas to brine ratio the bank spread out through the sample (in fact there was locally, some secondary drainage of excess brine) and a valid point was eventually achieved at S_w=53%, with k_{rg} = 7.6 mD.

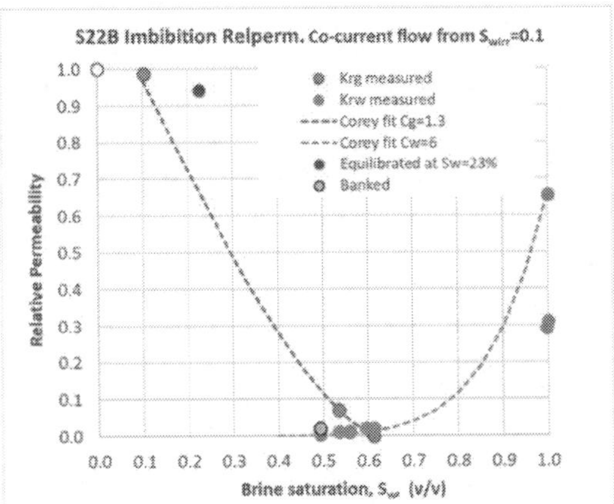

Fig. 5. Relative permeability curve for reservoir sandstone of good quality, (dry gas permeability 107 mD), collected using standard co-current test method. A Corey function with small exponent C_g= 1.3 can be used to fit the valid data points but there is a large gap between k_{rg} at S_{wirr}, around 103 mD, and the next point where the brine was seen on the x-ray CT to be evenly distributed along the length of the sample.

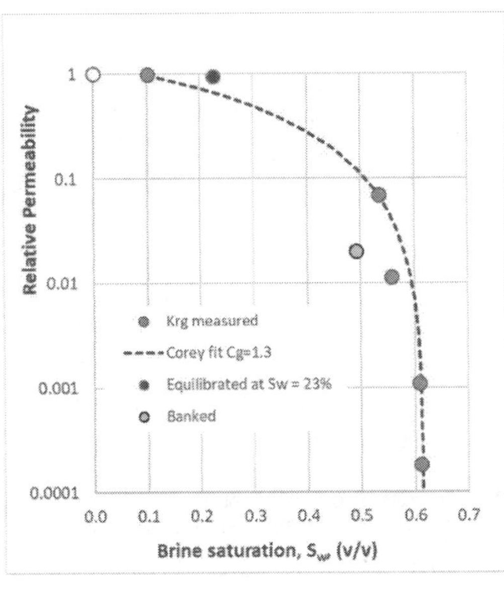

Fig. 6. Gas-only Rel Perm data in Figure 5, plotted in an expanded logarithmic scale. Points at S_w=0.23 (imbibed) and 0.49 (banked) are discussed in the text.

To overcome the problems described above, we developed simple procedures to create a more uniform brine saturation in the low S_w region, but otherwise retain the desirable features of steady-state Rel Perm procedures established as best practice and outlined, e.g. in McPhee et al. [1]. The new step is to control the early imbibition process in small increments of 5-10% S_w outside of the test cell. We achieved this by blotting a known amount of brine slowly into the rock plug from all sides with lint-free laboratory tissue (green Kimwipe™), wrapping the sample in the tissue followed by a layer of polyethylene cling film and allowing it to imbibe and redistribute evenly. This took several hours for permeable samples) while several days was found necessary for the least permeable samples. Several samples were treated in parallel so that they could be loaded sequentially into the test rig and then each advanced further. This means that test rig throughput remains high.

The use of removeable end stones made from highly uniform Bentheimer sandstone meant that the brine saturation in the sample could be pre-equilibrated with the flow boundary materials, also outside of the rig, for each particular sample. We used hand-tightened 'G' clamps to press the end stones against the test samples, while imbibition of the inlet-sample-outlet composite took place. We used mouldable wax film (Parafilm™) wrapped around the polyethyene cling film to keep the blotting tissue firmly in place around the samples and stones, and the whole assembly was stored horizontally in the refrigerator.

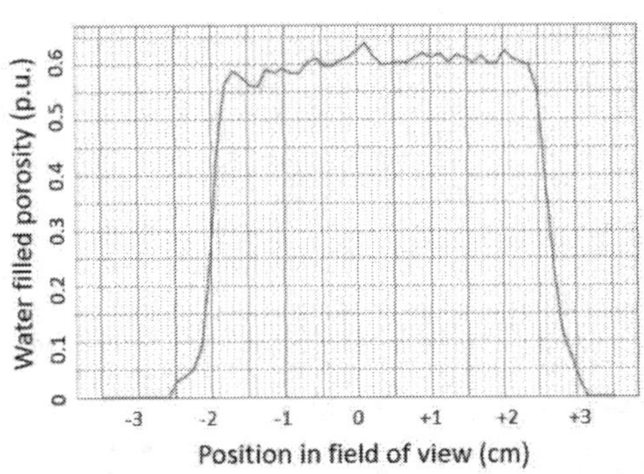

Fig. 7. Example of NMR saturation profile through core plug of high permeability sandstone after blotting with tissue and equilibration. Tissue was removed prior to NMR scan acquisition. Good uniformity in water filled porosity along the core length shows that an even saturation of brine through the sample has been achieved.

Uniform brine saturation throughout the rock was verified using NMR T_2 spectroscopy and profiling along the axis of the cling film-wrapped sample (Figure 7).

The equilibrated samples were loaded into the x-ray imaging flow rig and confining pressure and back pressure were applied slowly. Figure A2 shows an example of a high permeability sandstone together with its end stones loaded into in the test rig.

With the blotting procedure, a series of data points was collected to fill in the data points from the starting point, at or close to S_{wirr}, up to intermediate brine saturations.

At low S_w, before brine became substantially mobile, we injected gas-only, at four or more rates, to define accurate gas relative permeability values that we also found to be highly repeatable. That is, we avoided co-current flow to avoid banking and we used humidified gas to prevent drying of the samples. Using multiple rates, the linearity of the flow (obeying Darcy's Law) could be confirmed, and any offsets in the transducer values (instrument drift and capillary effects not indicative of the flow-induced pressure drop; e.g. see [5], [6]) could be tracked. Generally, we found that capillary effects close to the inlet or outlet upset the stability of the pressure records to such an extent that the data was unusable (Figure A3). Therefore, we developed a procedure to bleed out any brine droplets that were trapped in the inlet of the PEEK tubes and were able to collect very stable pressure readings resolving pressure drops smaller than 0.02 psi. Figure A4 shows an example of a good quality record in a high permeability sample where very small differential pressure signals can be resolved.

Once saturation had advanced sufficiently for brine to be detectably mobile, we progressed into co-current flows with the fractions of gas and brine adjusted to maintain near stable, but slowly advancing S_w. We note that commonly, onset of brine mobility coincided with non-linearity of the pressure response to increased flow rate: in these cases, we used the linear portion of the curve at lower gas flow rates to compute the relative permeability.

The use of paired syringe pumps at inlet and outlet allowed close control of both inflow and outflow brine rates, such that true steady state conditions were verified. We define a point as "valid" when flow rates in and out of the sample are equal, the differential pressure is steady, and the saturation is uniform to within five percent. Points that are invalid or suspect are shaded grey on the plots.

Ideally, the brine saturation should be completely uniform and, in many cases, S_w did vary by less than 2% along the sample length. We found that in the highly permeable sandstones, the brine was easily moved within the sample during loading and pressurization in the rig. As S_w was increased beyond about 50%, brine became more mobile, and there was a tendency to see some end effects so that brine either accumulated at, or retracted from, the sample boundaries: if this led to a saturation range greater than 5% within the sample we consider it invalid. Note that "invalid" points happen to lie on the curves eventually fitted through the valid data points.

The x-ray images of the experiments taken in the medical CT system included the end stones and this gave us advanced warning of potential saturation changes that could be balanced with a higher or lower brine flow rate. In this way, the progression of saturation was made as smooth as possible and the imbibition was controlled to be monotonic (meaning without any regressions in water saturation owing to too-rapid drainage of brine out of the sample). Even so, the experiments are quite challenging to perform as uneven brine saturation can still occur due to brine accumulating at the inlet (banking, as described above, owing to spontaneous imbibition from the upstream stone) or the outlet end of the sample (likely due to poor capillary continuity between the sample and the downstream stone).

3 Results

We present representative results from our study for three samples representing low, intermediate and good quality gas reservoir types.

In Figures 5 and 6 we illustrate the effect of resetting the saturation of Sample S22B from S_{wirr} and advancing

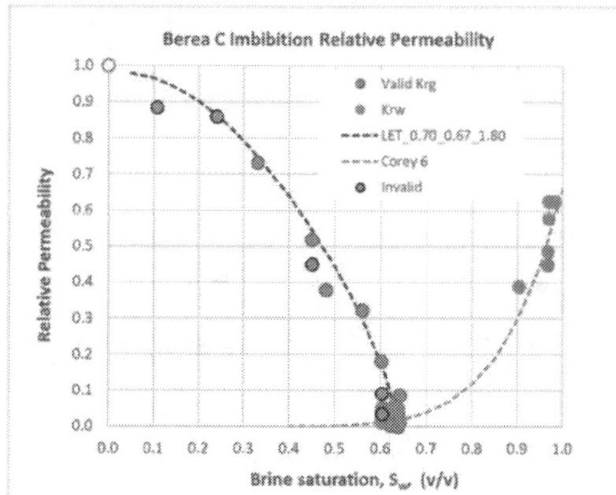

(a)

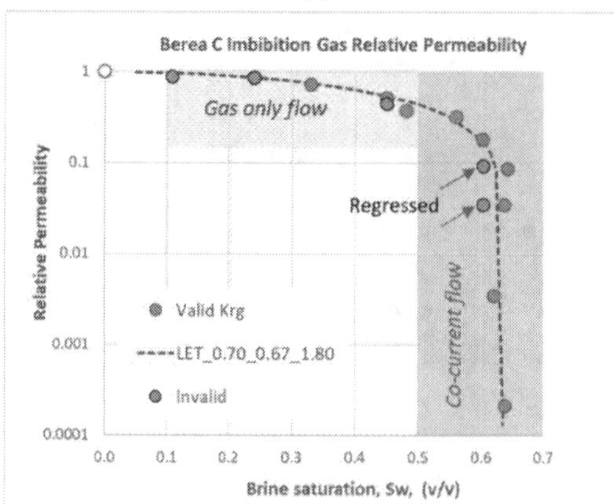

(b)

Fig. 8. Gas relative permeability (orange dots are valid points) and brine relative permeability (blue dots) for Berea C sample. The red dashed line is the LET function fit to gas permeability. Brine data in (a) fit a Corey function. In the expanded, logarithmic-scale plot (b), the location of two stable and equilibrated points that lie below the main Rel Perm curve are labelled. They were acquired after a period of regression in the brine saturation during co-current flow.

S_w to ~23 % using the blotting method to collect just one additional equilibrated imbibition Rel Perm point within the gap region. This point is shown in darker solid colour, and it lies well above the Corey curve fitted through the other points acquired earlier with the standard methodology. The Corey fit was allowable when there was a large data gap, but the new point makes this type of curve impossible. By collecting more points with sequential blotting, we can trace a more complete curve towards S_w values of around 50% brine when the co-current data are again reliable.

Figure 8 shows the full results for Berea sandstone, which has a dry gas permeability of around 665mD. The evolution of brine distribution for this sample is shown in Figure A5. This sample was challenging owing to high water mobility, which was near the limit of causing gravitational segregation during horizontal core floods. Two points at around 11 and 24% brine saturation showed more brine in the middle part of the core than at the ends, and therefore are coloured grey in Figure 8.

These two values were collected immediately after loading the samples into the rig, from an original state that did have uniform brine distribution, in equilibrium with the end stones. We presume that application of the gas pressure from both ends of the sample caused some retraction of the brine towards the middle of the core plug.

Figure 9 shows the gas relative permeability curve for the poorest quality reservoir sample, S26B (porosity 17%, dry gas permeability of 9.4 mD). The first point was collected at S_w=25%, following centrifugation to 100 psi. The general shape of the Rel Perm curve is similar to the Berea and S22B but with slower fall-off and a gas end point at higher water saturation (S_w=73%, S_{gr}=27%).

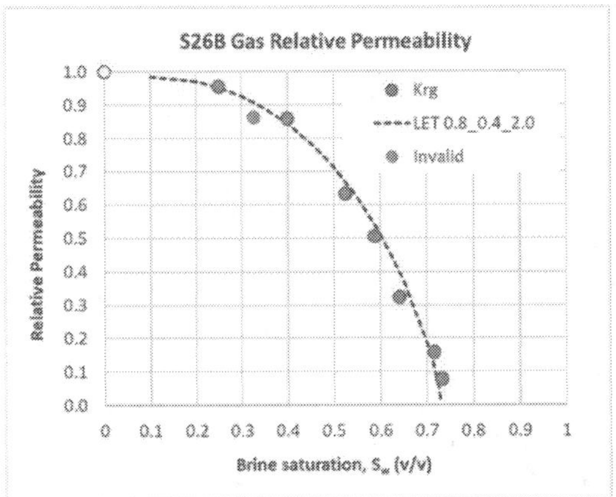

Fig. 9. Gas relative permeability (orange dots are valid points) for sample S26B. The red dashed line is the LET function fit.

For five reservoir sandstones and for the Berea "standard" rock we found that the shape of the more complete gas relative permeability curve, now extending through to low S_w values, is convex upwards and does not show the concave-up or "Corey" type profile that is conventionally expected. The gas Rel Perm function may have an 'S' shaped form or simply roll over to a sharp drop-off without tapering at high S_w. The experimental gas Rel Perm curves can be fitted with the LET function [7], while the water Rel Perm curves still show a classic Corey function shape, as expected from established theory for strongly water wet conditions. The properties and LET fitting values for three representative samples are summarised in Table 1.

We note that during the co-current test periods, we saw several of the phenomena mentioned by Cense et al. [2] for gas-brine imbibition experiments, such as the development of a gradient in brine saturation along the sample near the permeability end point owing to greater compression of trapped gas bubbles at the inlet end than at the outlet end, and also a small amount of "carry-over" of gas in solution or as disconnected bubbles. We therefore used recommended procedures not allowing the differential pressure to exceed 50 psi (10% of the back pressure applied). Usually ΔP was less than 5 psi.

When regressions in saturation occurred, the gas relative permeability for the subsequent point lay below the curve for progressive imbibition, suggesting that additional blocking of gas pathways by displaced brine was occurring. The "invalid" grey points labelled with arrows between S_w=0.4 and S_w=0.64 in Figure 8b were acquired after such episodes of regression in saturation during the co-current flow period in the Berea C sample.

4 Discussion and Implications

Our findings question the adequacy of current "best practice" relative permeability testing methods and reporting in SCAL programs for gas reservoirs. While every Rel Perm point reported may be valid, the limited data available are often fitted through the "gap" at intermediate S_w values using a Corey-type curve that may not be applicable. (Figures A7 to A9 show how this gap came about in sample S22B using co-current flow at 1000:1 gas to brine ratio). Therefore, completing the curve in this region can reduce the uncertainty in what the permeability trajectory during brine imbibition will be.

The results that we obtained for a range of sandstones from good and poor-quality reservoirs suggest that the shape of gas-brine imbibition relative permeability curves should not be assumed to have a concave-up shape, at least for water wet systems such as those investigated in this research project. Considering that in a water wet rock, the brine has a strong affinity for surfaces, it is not surprising to us that the brine does not start to have a marked effect on the gas permeability until it has built up to a level where it blocks the gas pathways in the centres of pores and throat regions. Once the pathways start to block up however, the permeability to gas drops more precipitously in a process akin to the removal of conducting bonds in a percolating network.

Sigmoidal-shaped [imbibition] relative permeability curves for oil-brine and gas-brine system have been reported in the literature, e.g. [8-12], hence the derivation of fitting functions such as LET [7] to describe them. The shape is likely to relate to the pore structure and the "roll-over" shape that has been termed "reverse hysteresis" is sometimes cited as being typical of a sand-pack rather than a heavily consolidated rock [13]. In simple network simulations of two-phase displacements, the shape is typical of low aspect ratio pore structures, and well-connected topology [14,15]. We note that the porosity of the samples that we tested is from ~17%-22%, and microscopic examination shows moderate consolidation and quartz cementation, but in general that sandstones all have an "open" and well-connected type of pore structure.

The Berea sandstone sample that we tested here, is homogeneous in the along-core direction, and looks uniform on visual inspection but has discernible lamination in the x-ray images (Figure A6), whose contrast increases at intermediate brine saturations when the finer layers imbibe more brine relative to the coarse layers. Thus, for Berea C, there are likely to be some relatively unrestricted channels for gas flow at intermediate brine saturation, that remain free flowing until those laminae with large pore apertures also become invaded by brine. This would act to maintain high gas of Rel Perm to a larger value of S_w and generate a "roll over" curve shape thereafter. However, we should point out that all the other samples were chosen for their homogeneity and the overall shape of the permeability decline in laminated Berea C is rather similar to that measured for the highly uniform sample S26B.

Other factors such as interfacial tension, wettability and flow rates affecting the viscous force balance are likely to be as important as the pore structure in controlling the Rel Perm curve shapes. Kjosavic et al. [16] suggest that strong water wetting can produce a sigmoidal shape of Rel Perm curves and leads to an increase in hysteresis between drainage and imbibition processes. A decrease in the exponent describing the non-wetting phase tortuosity (which depends on pore structure, IFT and wettability) is shown by Gerhard and Keuper [13] to generate a convex-up shape for the first imbibition curve and for imbibition scanning curves: physically this implies that the gas flow pathways are relatively simple and dominant with the water phase forced to the margins of the pore space. More sophisticated measures of non-wetting phase continuity and connectedness now being made possible with x-ray microCT image analysis at pore scale may help to put these measures on a more quantitative and predictive footing [17]. Indeed, digital rock simulations with realistic physics can already predict a range of Rel Perm curve shapes, and they can produce output continuously across the full range of saturations [18,19].

Our results apply to the water wet conditions tested in the gas-brine system having high interfacial tension (around 70 mNm^{-1}) in quartzo-feldspathic sandstones with a small proportion of clay, and moderate to high permeability. The prevailing conditions of our tests are representative of the interior of the reservoir (far from a well) and involve small capillary numbers (ratio of viscous (drag) forces to capillary (surface) forces of around 10^{-6} or 10^{-7}. At higher velocity, and capillary number $>10^{-5}$, the viscous forces and momentum transfer will change the shape of the gas Rel Perm curve, which is likely to straighten as the gas can more effectively push brine out of the way. Therefore, prediction of gas flow dynamics close to the wellbore would require a different style of lab test and would most likely lead to differently shaped Rel Perm curves than the examples reported here.

Considering the true sigmoidal or roll-over profile of the gas Rel Perm curves at low capillary number, one can see that the gas flow rates predicted for the earlier stages of brine invasion into the reservoir (away from the near wellbore region) as it is depleted or swept, will be higher than would be predicted by defaulting to the concave-up "Corey" shape. (Figure 5 shows an example a good Corey curve fit to limited data). The Corey and LET type curves only merge towards the tail end of gas mobility, when the relative permeability is less than 10 percent of the starting values at S_{wirr}. Therefore, one would predict that higher gas flow rates are possible in the reservoir during the early stages of water ingress, before a steeper drop-off in k_{rg}. The consequences for reservoir planning are likely to be more important for a water drive situation and be less important for reservoirs only undergoing depletion.

This new understanding of the true gas-brine relative permeability curve shapes may influence the development concept for gas reservoirs, the placement of wells and even the ultimate economics of a project. At the least, having complete and reliable relative permeability curves as inputs will improve the accuracy of reservoir simulation models and increase the level of confidence in their predictions of reservoir performance over time.

Acknowledgements

Chevron Australia and joint venture partners Exxon Mobil Australia and Shell Australia are thanked for funding this research. Dr Ali Saeedi from Curtin University provided important guidance in setting up this project, and advice as to experimental methods. Dr Lincoln Paterson and Dr Tara La Force from CSIRO are thanked for providing internal peer reviews. P. Egermann and K. Flynn are thanked for thoughtful reviews that improved the quality of the paper. This research utilized the x-ray CT and Advanced Core flooding facilities of the National Geosequestration Laboratory (ngl.org.au).

References

1. C. McPhee, J. Reed and I. Zubizarreta, *Core Analysis: A Best Practice Guide*. Elsevier (2015).
2. A. Cense, J. Reed and P. Egermann. "SCAL for gas reservoirs: a contribution for better experiments". SCA 2016-023 (2016).
3. D. Maloney, K. Doggett and R. Gilliland. "New procedure and apparatus for combined reservoir condition gas-brine relative permeability, capillary pressure and electrical property measurements". SCA-9427 (1994).
4. A. T. Corey. "The interrelation between gas and oil relative permeabilities". Producers Monthly, **19**, 38-41, (1954).
5. R. Pini, and S. Benson. "Simultaneous measurement of capillary pressure and relative permeability drainage curves". Water Resources Research, **49**, 3516-3530, (2013).

6. R. Gupta, and D. Maloney. "Intercept Method—A novel technique to correct steady-state relative permeability data for capillary end effects". SPE171797-PA (2014)
7. F. Lomeland, E. Ebeltoft, and W.H. Thomas, "A new versatile relative permeability correlation". SCA 2005-32. (2005).
8. J. Osoba, J. Richardson, J. Kerver, J. Hafford, and P. Blair. "Laboratory Measurements of Relative Permeability". Petroleum Transactions AIME, 192, 47–56 (1951).
9. J. Batycky and F. McCaffery. "Low interfacial tension displacement studies". Petroleum Society of CIM, Calgary, Canada, Paper 78.29.26. 1978.
10. Raimondi, P. "Distribution of the oil phase obtained upon imbibition of water". *Society of Petroleum Engineers Journal*, 49–55 (1964)
11. J. Colonna, F. Brissaud, and J.L. Millet. "Evolution of capillarity and relative permeability hysteresis". Society of Petroleum Engineers Journal, **12**, 28–38 (1972).
12. R. Fulcher, E. Turkay and C. Stahl. "Effect of capillary number and its constituents on two-phase relative permeability curves". SPE 21170-PA (1985)
13. J. Gerhard and B. Keuper. "Relative permeability characteristics necessary for simulating DNAPL infiltration, redistribution and immobilization in saturated porous media". Water Resources Research **39**, W1213 (2003)
14. G. Jerauld, and S. Salter, S. J. "The effect of pore-structure on hysteresis in relative permeability and capillary pressure". Transport in Porous Media, **5**, 103–151 (1990).
15. P. Valvatne and M. Blunt. "Predictive pore-scale modelling of two-phase flow in mixed-wet media". Water Resources Research **40**, W07406, (2004).
16. A. Kjosavik, J. Ringen, and S. Skjaeveland. "Relative Permeability Correlation for Mixed-Wet Reservoirs". SPE Journal, **7**, 49–58 (2007).
17. L. Andersson, S. Schlueter, and D. Wildenshild. "Defining a novel pore-body to pore-throat 'Morphological Aspect Ratio' that scales with residual non-wetting phase capillary trapping in porous media". Advances in Water Resources, **122**, 251–26. (2018)
18. Schembre-McCabe, R. Salazar-Tio, and J. Kamath, 2012. "Two examples of adding value through Digital Rock Technology". SCA 2012-081 (2012)
19. R. Farokhpoor, E. Westphal, N. Idowu, P. Oren and B. Fletcher. "Gas-water steady-state relative permeability determination with two approaches; experimental and digital rock analysis, strengths and weaknesses". SCA2016-12. (2016)

Table 1. Properties of the three representative sandstones, and their LET function parameters for the gas Rel Perm [7]

	Porosity p.u.	Dry Gas Perm. mD	L,E,T for k_{rg}
S22B	20.2	107	2.2, 0.09, 1.5
S26B	17.1	9.4	0.8 ,0.4 ,2.0
Berea C	21.8	665	0.7, 0.67, 1.8

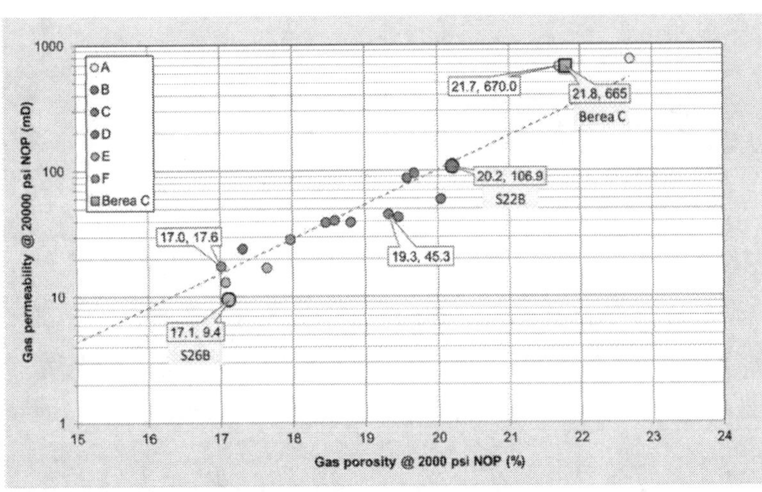

Fig. A1. Dry nitrogen porosity and permeability of the five tested samples from NWS gas field, measured at 2000 psi net overburden pressure. The set of samples represents the range of routine core analysis values of medium to good reservoir quality. The Berea C sandstone has a porosity and permeability very similar to the highest permeability sample among those tested from the reservoir set. The three samples for which relative permeability results are presented in this paper are highlighted in yellow.

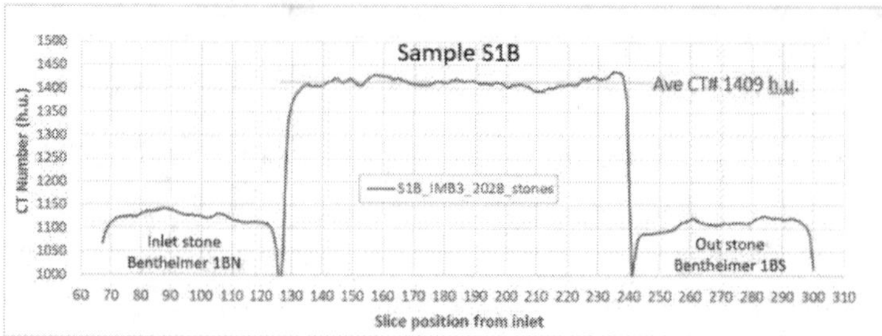

Fig. A2. Profile of average X-ray CT numbers through the assembly of inlet stone-sample-outlet stone mounted in the test rig, prior to commencing measurements. The sample has a uniform brine saturation of 30% and is in capillary equilibrium with the stones. The dip in CT number at the boundary marks the position of the single ply Kimwipe tissue used to maintain good continuity of the brine phase between the end faces of rocks.

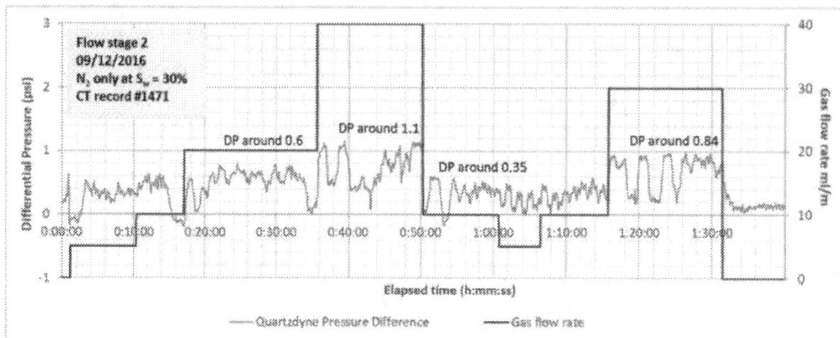

Fig. A3. Example of a completely unstable pressure record during multiple rate gas flow experiments in sample S22B. This problem was rectified by bleeding off any brine droplets that had entered the PEEK tubes connected to the flow transducers.

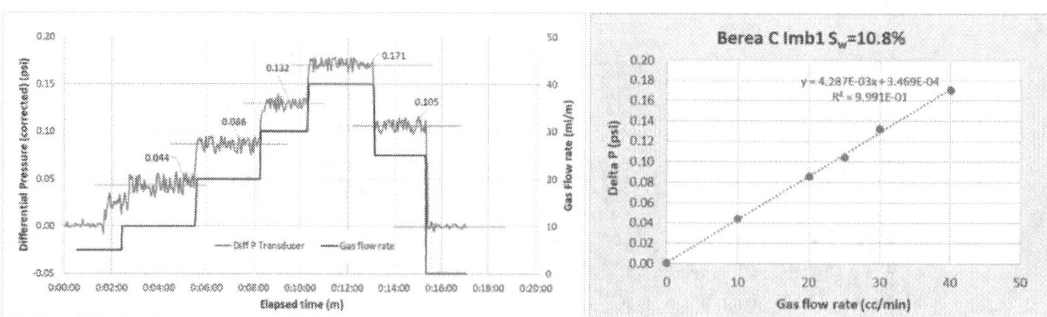

Fig. A4. Example of stable flow-pressure record in sample Berea C, at multiple gas flow rates and a plot showing linearity of the Darcy flow regime. The static periods of no flow show the very high resolution and low noise level of the differential transducer. The fluctuations around the mean pressure values during gas flow are real "flutter" in the multiphase flow process.

 https://doi.org/10.1051/e3sconf/202014603003

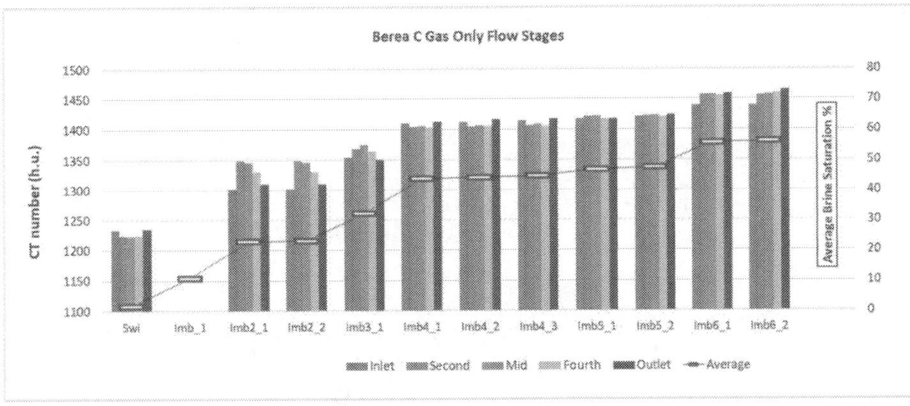

Fig. A5. Saturation distribution through the Berea C sample during tests with gas-only flow. The average CT values for 5 regions running from inlet through to outlet are shown here for simplification, although the data were collected with a slice thickness of around 0.5mm, as seen in the following figure. For Imb 1, x-ray CT data was not collected, but the brine distribution was measured before loading into the rig, when it was uniform, and then after testing, blow-down and extraction, when the brine was found to have retracted towards the mid-point of the plug. For the two CT scans in Imb2, brine again was initially uniform as determined by NMR profiling, prior to loading into the rig, but retracted away from the core ends. This is seen very slightly in the third imbibition stage, and thereafter the brine distributions were highly uniform apart from small end effects, as seen in the following figure.

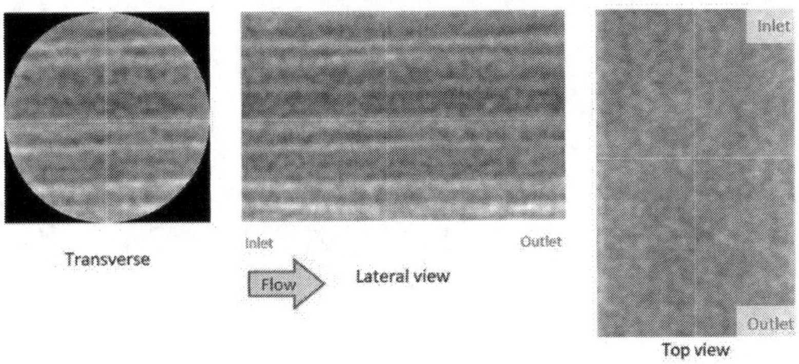

(a)

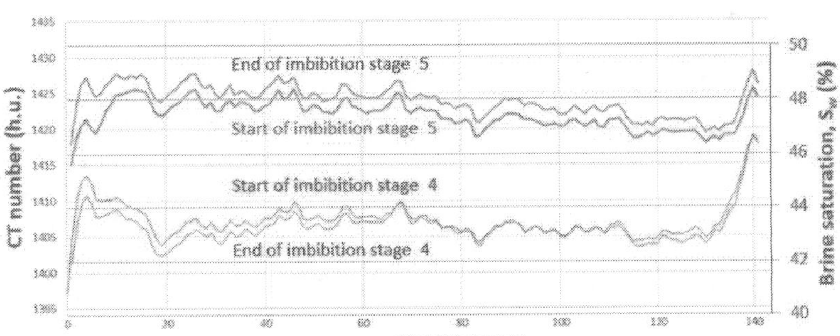

(b)

Fig. A6. (a) X-ray CT images of Berea C Sample during imbibition stage 4, at around 43% average brine saturation. The distribution of brine appears to be very uniform. The horizontal laminations are well defined, and the contrast between them is highest at intermediate saturations owing to the finer laminae having a higher brine content than the coarser-grained laminae. (b) Detail of x-ray profiles showing average CT number variations from inlet to outlet in imbibition stage 4, and the subsequent stage after a further increment of blotting. Small capillary end effects are evident that do not seem to have affected flow or pressure data. While there was essentially no change in brine saturation during stage 4, there was a small net transfer of brine into the sample from the end stones during stage 5, even during gas-only flow, but this amounted to an overall increase in saturation of less than 1 percent.

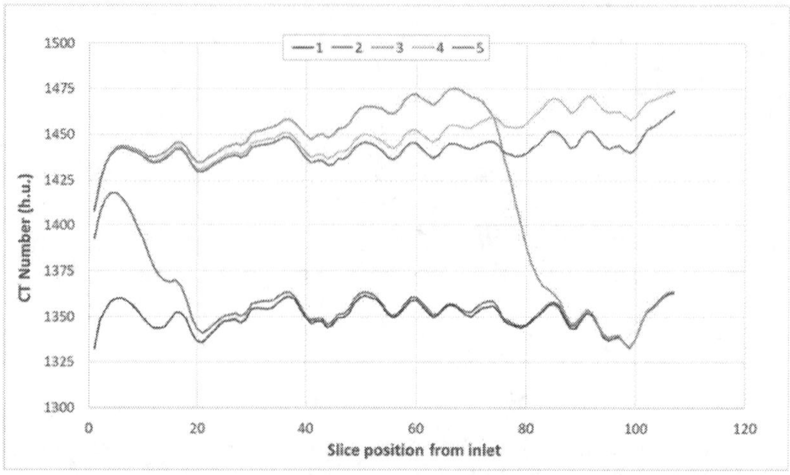

Figure A7. CT value evolution during brine imbibition stages from S_w=22% in sample S22B, during co-current flow at 1000:1. The initial uniform distribution of brine in Image 1 becomes banked at the inlet (2), and a large front progresses two thirds of the way into the sample (3) before reaching the outlet and dissipating (4) and becoming nearly uniform again by Image 5. This figure illustrates the development of the significant gap in relative permeability data for situations with an even brine saturation when using co-current flow, even at high gas rate (60 ml/min) and very low brine fraction.

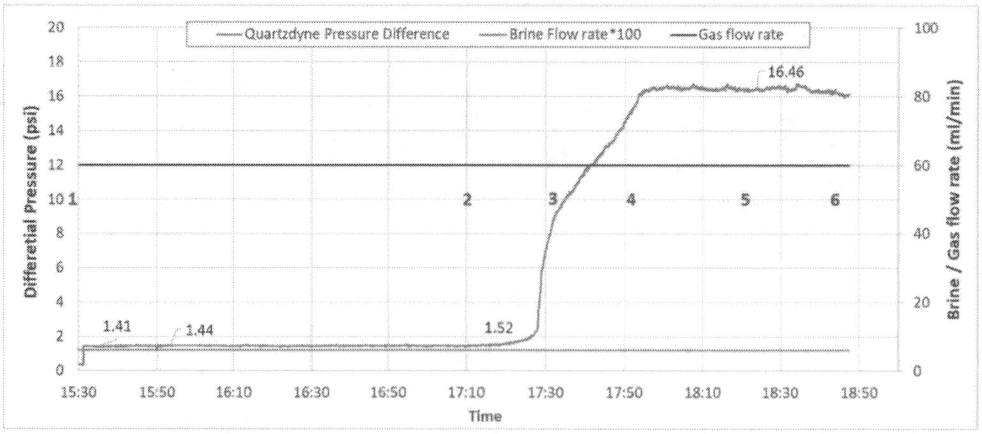

Figure A8. Pressure record for initial gas flow in sample S22B commencing from S_w=22%, followed by high gas fraction co-current flow (1000:1 ratio) that matches to the saturation history shown in the previous figure. There are no good Rel Perm data points between an initial gas permeability of 103 mD, and the drop right down 2.2 mD after the brine spreads out at an average brine saturation of 49% (Image 5 in the previous figure, banked point in Figures 5 and 6).

https://doi.org/10.1051/e3sconf/202014603003

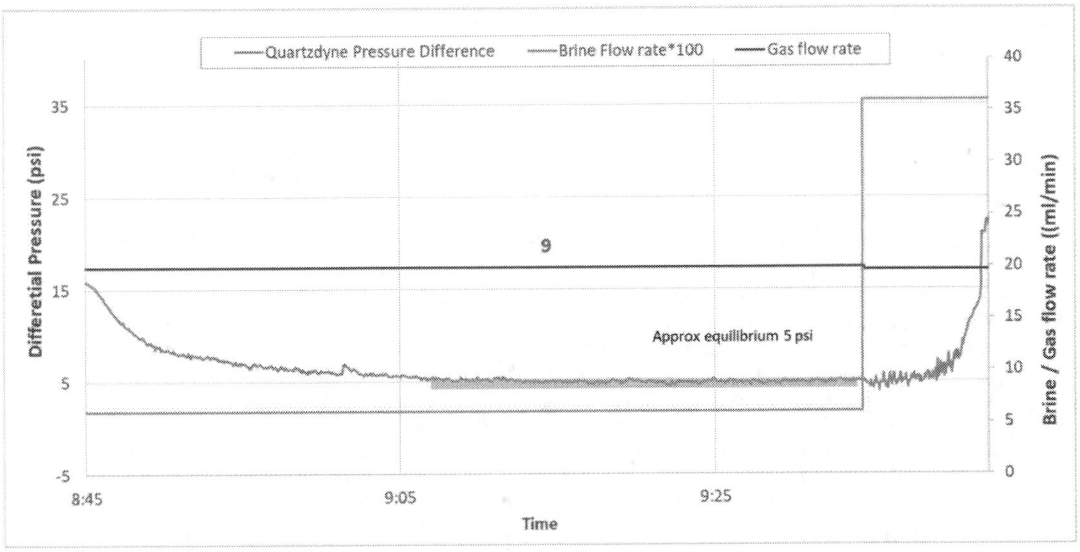

Figure A9 After the sample S22B was left overnight, co-current flow was recommenced at a ratio of 333:1. The permeability to gas recovered to 7.6 mD and flow was very stable. Brine saturation was uniform and had progressed slightly to 53% (the next valid Rel Perm point in Figures 5 and 6). A relatively small redistribution of the brine in the sample significantly affected its gas permeability, as can be seen from the drop off in differential pressure from the peak of 16.5 psi down to 5 psi during this flow stage.

Is contact angle a cause or an effect? – A cautionary tale

Douglas Ruth

University of Manitoba, Winnipeg, Manitoba, Canada

Abstract. The most influential parameter on the behavior of two-component flow in porous media is "wettability". When wettability is being characterized, the most frequently used parameter is the "contact angle". When a fluid-drop is placed on a solid surface, in the presence of a second, surrounding fluid, the fluid-fluid surface contacts the solid-surface at an angle that is typically measured through the fluid-drop. If this angle is less than 90°, the fluid in the drop is said to "wet" the surface. If this angle is greater than 90°, the surrounding fluid is said to "wet" the surface. This definition is universally accepted and appears to be scientifically justifiable, at least for a static situation where the solid surface is horizontal. Recently, this concept has been extended to characterize wettability in non-static situations using high-resolution, two-dimensional digital images of multi-component systems. Using simple thought experiments and published experimental results, many of them decades old, it will be demonstrated that contact angles are not primary parameters – their values depend on many other parameters. Using these arguments, it will be demonstrated that contact angles are not the cause of wettability behavior but the effect of wettability behavior and other parameters. The result of this is that the contact angle cannot be used as a primary indicator of wettability except in very restricted situations. Furthermore, it will be demonstrated that even for the simple case of a capillary interface in a vertical tube, attempting to use simply a two-dimensional image to determine the contact angle can result in a wide range of measured values. This observation is consistent with some published experimental results. It follows that contact angles measured in two-dimensions cannot be trusted to provide accurate values and these values should not be used to characterize the wettability of the system.

1 Introduction

This paper will discuss various considerations of the interface between a gas, a liquid, and a solid surface. The fundamental principle of a force balance between the gas and the liquid, if they are separated by a curved surface, is that the difference in pressure across that surface, ΔP, is given by

$$\Delta P = \sigma_{lg}\left(\frac{1}{R_1}+\frac{1}{R_2}\right) \qquad (1)$$

Here σ_{lg} is the interfacial tension between the gas and the liquid, and R_1 and R_2 are the two radii of curvature of the surface. If a drop of fluid is placed on a flat, solid surface, a balance of forces requires that (Young's equation)

$$\sigma_{lg}\cos\theta = \sigma_{sg} - \sigma_{sl} \qquad (2)$$

Here σ_{sg} is the interfacial tension between the solid and the gas, σ_{sl} is the interfacial tension between the solid and the liquid, and θ is the angle that the liquid-gas surface makes with the solid surface, termed the "contact angle". If the solid is a circular capillary tube, then the pressure difference across the surface is termed "capillary pressure", P_c, and simple rules of geometry lead to the expression

$$P_c = \frac{4\,\sigma_{lg}\cos\theta}{\delta} \qquad (3)$$

Here δ is the diameter of the tube. These three equations are all that is required to demonstrate the characteristic behaviors of contact angles.

Generally, Equation 2 is used to determine the wettability of a three component (liquid-gas-solid) system. Specifically, if θ when measured through the liquid is found to be much less than 90°, the system is said to be strongly "liquid wet"; a typical liquid wet system is water/ air/ glass. If θ is near 90° the system is said to be "neutrally wet". If θ is much greater than 90°, the system is said to be "gas wet"; a typical gas wet system is mercury/ air/ glass. For very simple cases, such as a drop of water on a horizontal solid surface, this definition of wettability is universally accepted. Unfortunately, this definition has become dogmatic and it has been extended to mean that the contact angle directly defines the wettability in all situations. In particular, with the advent of high resolution imaging methods, apparent contact angles can be measured *in-situ* in both static and dynamic situations. In the present paper, published results and simple thought experiments are used to demonstrate that contact angles are not always characteristic of wettability. Furthermore, it is demonstrated that direct measurement of *in situ* contact angles is difficult without sophisticated interpretation methods.

This paper is not a condemnation of the ability to determine contact angles on the microscopic scale – it is a cautionary statement that such determinations must be done with great care. However, this paper does demonstrate that interpreting wettability from direct contact angle measurements is not always a meaningful activity.

2 The Contact Angle and Wettability

It is a tacit assumption that if a contact angle can be measured, then the wettability of a system can be determined. However, there are many cases where this assumption is obviously wrong. One of these cases is demonstrated in the classic monograph by Craig [1]. His Fig. 2.3 clearly demonstrates that even in the case of a static contact angle measurement, it can take many hundreds of hours to obtain an equilibrium contact angle. The illustration is for an oil/ water/ solid surface and it could be argued that such systems are susceptible to complex chemical factors. However, even this simple case provides a cautionary note. Modelling of displacement processes generally utilize capillary pressure data that is time-independent. But if the contact angle is time dependent, then it follows from Equation 3 that the capillary pressure must be time dependent. Therefore, all displacement processes are potentially time dependent, an issue rarely considered when analyzing such processes.

A second case of interest is that of a drop on a surface, when the surface is gradually tilted from the horizontal. This case has been studied since at least the late 1930s [2]. A more recent paper by Krasovitski and Marmur [3] provides a succinct description of this case. What they found is that the drop distorts but does not move until the plate is tilted to a critical angle, the "slip angle". Assuming, a gas/ water/ solid system, when the drop distorts, the contact angle at the up-tilt end of the drop, θ_u, decreases and the contact angle at the down-tilt end of the sample, θ_d, increases. Using the conventional interpretation, this would mean that simply tilting the surface causes the up-tilt end of the system to become more water-wet while the down-tilt end of the system moves toward neutral-wettability

It is found both experimentally and theoretically that the slip angle, α, is related to the two contact angles by the relation

$$\sin\alpha = C\,\sigma_{lg}(\cos\theta_u - \cos\theta_d) \qquad (4)$$

Here C is a constant that depends on the densities of the fluids, gravitational acceleration, and drop shape. Krasovitski and Marmur also document that once the drop begins to move, the values of θ_u and θ_d do not remain constant. An argument can be made that these changes are due to the fact that the up-tilt contact line is receding over a surface which was in contact with water, while the down-tilt contact line is advancing over a surface that was in contact with gas, and that this difference accounts for the difference between "advancing contact angles" and "receding contact angles". However, at tilt angles less than the slip angle, this argument does not apply because the drop has not yet moved.

Next, consider a water/ air/ solid system in a vertical capillary tube with one end immersed in a tray of liquid and the other open to air. The well-known behavior is that water will rise in the tube until a force balance obtains between the hydrostatic and capillary forces. The pressure in the water at the water-surface in the tray must be equal to the pressure in the air at this point. Therefore, this force balance requires that

$$\frac{4\,\sigma_{lg}\cos\theta}{\delta} = (\rho_l - \rho_g)\,g\,h \qquad (5)$$

Here $\rho_l - \rho_g$ is the difference between the densities of water and air, g is the acceleration of gravity, and h is the height of the liquid/ air interface in the capillary tube relative to the height of the liquid/ gas interface in the tray.

Now consider what would happen if the tube is gently removed from the tray and a second water/ air surface is created at the bottom of the tube. The pressure below the bottom water/ air surface differs from the pressure above the top water/ air surface by $\rho_g\,g\,h$. Therefore, the hydrostatic pressure of the water in the tube just makes up for the capillary pressure at the upper surface and the pressure difference at the bottom water/ air surface must remain at zero. But this implies that the bottom surface is flat and that θ is 90°. Using our conventional definition of wettability means that the top surface is strongly water-wet and the bottom surface is neutrally-wet.

Continuing this line of reasoning, various amounts of water can be forced from the tube to obtain force balances that would require

$$\frac{4\,\sigma_{lg}}{\delta}(\cos\theta_{top} - \cos\theta_{bottom}) = (\rho_l - \rho_g)\,g\,h \qquad (5)$$

Therefore, by implication, the contact angle at the bottom water/air surface would have values given by the equation

$$\theta_{bottom} = \arccos\left(\cos\theta_{top} - \frac{\delta\,(\rho_l - \rho_g)\,g\,h}{4\,\sigma_{lg}}\right) \qquad (6)$$

This equation implies that wettability at the bottom of the tube varies continuously with height of water.

The above description is an approximation to what would actually happen experimentally. The bottom of the tube includes a flat surface which complicates the situation. Water can spread across this surface resulting is a convex-up surface that increases the pressure at the bottom surface, hence the pressure at the top surface. However, the conclusion remains the same – the contact angle at the top surface cannot be the same as the contact angle at the bottom surface or the tube will drain and experimentally it is found that the tube does not drain.

A variation on the above reasoning can be made for a horizontal tube that is spun in a centrifuge. The equation describing the capillary pressure in this system is

$$P_{co} - P_{ci} = (\rho_g - \rho_l)\frac{\omega^2}{2}(r_o{}^2 - r_i{}^2) \qquad (7)$$

Here r is the radius of rotation, the subscript i denotes the variables at the fluid interface closest to the center of rotation, o denotes the variables at the interface farthest from the center of rotation, and ω is the angular velocity. At zero angular velocity, the pressures in the water at the two ends of the tube are the same. As the angular velocity is increased, the equilibrium condition requires that the capillary pressures at the two ends of the tube must differ. Using Equation 3, this difference can be expressed in terms of contact angles:

$$\frac{4\,\sigma_{lg}}{\delta}(\cos\theta_o - \cos\theta_i) = \left(\rho_g - \rho_l\right)\frac{\omega^2}{2}({r_o}^2 - {r_i}^2) \tag{8}$$

We can conclude from this that the contact angle for the interface farthest from the center of rotation varies with angular velocity; hence, following conventional reasoning, the wettability at this point, is a function of angular velocity which is not true.

There are also many reports in the literature of how the contact angle at the interface of two fluids in a tube differ with the velocity of the interface. It is well known that the problem of a moving contact angle cannot be solved analytically because a singularity arises in the equations at the contact point of the two fluids and the solid surface. This singularity is generally treated by assuming that there is "slip" at the contact line. Based on this model, the classic paper by Cox [4] showed that the apparent contact angle was related to the static contact angle as a function of the capillary number. This behavior was verified experimentally in such works as Fermigier and Jenffer [5]. It is found that, above a capillary number of approximately 10^{-3}, the apparent contact angle can vary significantly, going from static values of 30° to values approaching 180°. In terms of wettability, these contact angles imply a variation from strongly water wet to strongly oil wet. Therefore, for dynamic displacements in capillary tubes, the contact angle cannot be used to characterize the wettability. If this is true for flow in capillary tubes, it must also be true for flow in porous media.

3 In-Situ Measurements of Contact Angles

There has been a recent interest in measuring contact angles in-situ by means of micro-imaging techniques (*e.g.* Andrew *et al* [5], Held *et al* [6]). However, similar to the measurement of porosity by means of two-dimensional images, in-situ measurement of contact angles are problematic. This will be illustrated by using the case of an interface in a vertical capillary tube.

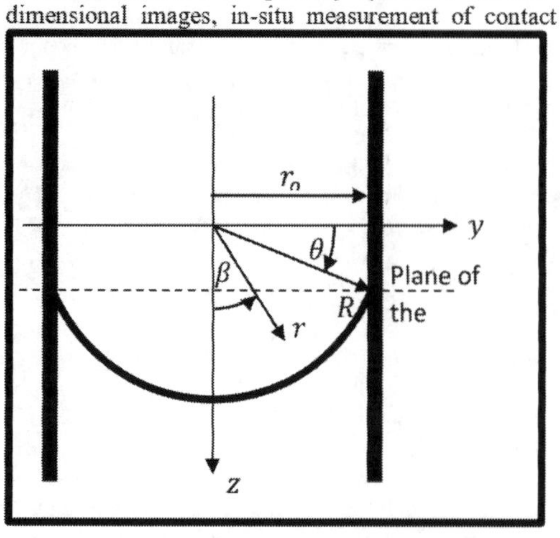

Fig. 1a Nomenclature for an interface in a vertical tube (elevation view).

Fig. 1 shows the nomenclature for an interface in a vertical tube. The interface is assumed to have a spherical shape with a radius of curvature, R. Two coordinate systems will be used: a Cartesian system with coordinates x, y, z and a spherical coordinate system with coordinates r, α, β. The origin of both coordinate systems is the center of curvature of the surface. Here, z is assumed positive downward.

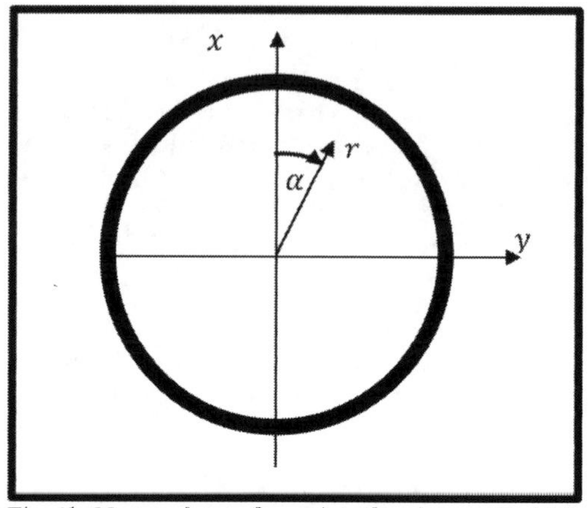

Fig. 1b Nomenclature for an interface in a vertical tube (plan view).

The equations of the surface in the Cartesian coordinate system in terms of the spherical system are

$$x_s = R\sin\beta\cos\alpha \tag{9a}$$

$$y_s = R\sin\beta\sin\alpha \tag{9b}$$

$$z_s = R\cos\beta \tag{9c}$$

The equations for the wall in Cartesian coordinates in terms of the spherical system are

$$x_w = R\sin\theta\cos\alpha \tag{10a}$$

$$y_w = R\sin\theta\sin\alpha \tag{10b}$$

$$z_w = z \tag{10c}$$

It is next assumed that the interface is cut by a plane, as shown in Fig. 2, such that the cross-section of the interface can be examined in a two-dimensions plane, $u - w$. (This plane is meant to represent the observation plane in a two-dimensional imaging experiment.) The equations for this plane in terms of the Cartesian coordinate system are

$$x_p = u \tag{11a}$$

$$y_p = y_e - w\cos\gamma \tag{11b}$$

$$z_p = z_e + w\sin\gamma \tag{11c}$$

The equations for the surface in the two-dimensional plane in terms of the spherical system can be found by combining Equations 9 and 11. The result is

$$u = R \sin\beta \cos\alpha \quad (12a)$$

$$w = \frac{R\cos\beta - z_e}{\sin\gamma} \quad (12b)$$

$$w = \frac{y_e - R\sin\beta\sin\alpha}{\cos\gamma} \quad (12c)$$

The two equations for w require that

$$(\cos\beta - z'_e)\cos\gamma = (y'_e - \sin\beta\sin\alpha)\sin\gamma \quad (13)$$

where y'_e is y_e/R and z'_e is z_e/R.

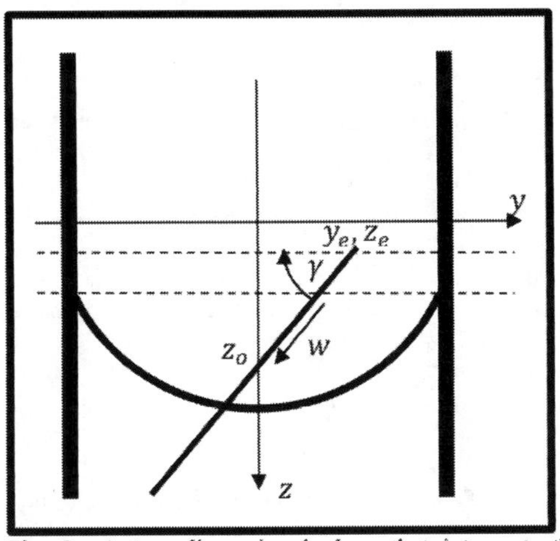

Fig. 2 A two dimensional plane that intersects the surface and the wall.

The two-dimensional surface can be characterized by the point where the plane crosses the z- axis, z_o, and the angle the plane makes with the horizontal, γ. This will allow the calculation of the coordinates where the plane crosses the surface in the cross-section shown in Fig. 2. Provided that the plane intersects the plane of contact, the minimum β-value, denoted by β_1, will always be $\pi/2 - \theta$ (the point where the surface contacts the wall). For this condition,

$$w = \frac{R\sin\theta - z_e}{\sin\gamma} = \frac{y_e - R\cos\theta}{\cos\gamma} \qquad \tan\gamma > \frac{z_o}{r_o} \quad (14)$$

If the intersecting plane does not intersect the plane of contact, then the β-value can be found using Equation 13 and setting the α-values to $\pi/2$; this β_1-value is given by

$$\sin\beta_1 = (z'_e\cos\gamma + y'_e\sin\gamma)\sin\gamma - \cos\gamma\sqrt{\left(1 - (z'_e\cos\gamma + y'_e\sin\gamma)^2\right)} \quad (15)$$

The second point where the plane intersects the surface, β_2, is also given by Equation 13. Again setting the α-values to $\pi/2$, this time the solution is

$$\sin\beta_2 = (z'_e\cos\gamma + y'_e\sin\gamma)\sin\gamma + \cos\gamma\sqrt{\left(1 - (z'_e\cos\gamma + y'_e\sin\gamma)^2\right)} \quad (16)$$

Furthermore, Equation 13 can be used to generate α-values that correspond to β-values

$$\alpha = \sin^{-1}\left(\frac{y'_e\sin\gamma + z'_e\cos\gamma - \cos\beta\cos\gamma}{\sin\beta\sin\gamma}\right) \quad (17)$$

The shape of the surface in the intersecting plane can therefore be calculated for various β-value between the minimum and the maximum.

For the wall, the coordinates in the intersecting plane become

$$u = R\cos\theta\cos\alpha \quad (18a)$$

$$w = \frac{y_e - R\cos\theta\sin\alpha}{\cos\gamma} \quad (18b)$$

The shape of the wall in the intersecting plane can therefore be calculated for any α–value. For the wall, α–values range from 0 to 2π.

A special case is when the intersecting plane is vertical. Then the equation for the plane is

$$x = R\cos\beta\cos\alpha \quad (19a)$$

$$y = y_o \quad (19b)$$

$$z = z_p \quad (19c)$$

the intersection of the surface in the plane is

$$R\cos\beta\sin\alpha = y_o \quad (20a)$$

$$y_p = R\sin\beta \quad (20b)$$

the maximum β-value is

$$\beta = \cos^{-1}\left(\frac{y_o}{R}\right) \quad (21)$$

and the α–values are given by

$$\alpha = \sin^{-1}\left(\frac{y_o}{R\cos\beta}\right) \quad (22)$$

The apparent contact angle can be found from

$$\left(\frac{dw}{du}\right)_{surface} - \left(\frac{dw}{du}\right)_{wall} = \tan\theta_a \quad (23)$$

evaluated at the wall. This is a complex derivative best done numerically.

4 Typical Interface Shapes

Fig. 3 through 9 show typical interface shapes for planes that intersect the interface at various positions. These figures were calculated assuming that $\theta = 30°$. In each figure, the left panel shows a schematic of the intersection and the right panel shows what would be observed in the intersecting plane, that is, the plane of observation in an imaging experiment.

Fig. 3 shows the case where the plane intersects the interface horizontally. The conclusion that could be drawn from the image is that gas is flowing as an encapsulated phase and that the water is perfectly wetting with an apparent contact angle of $\theta_a = 0$. In this case, the intersection of the plane with the wall is circular.

Fig. 4 shows the case where the plane intersects the interface before it intersects the wall. The conclusion that could be drawn from the image is that gas is flowing as an encapsulated phase, that the water is perfectly wetting with an apparent contact angle of $\theta_a = 0$, and perhaps water is trapped in a pore space located at the bottom of the image. In this case, the intersections of the plane with the wall is elliptical and the plane with the interface is circular.

Fig. 5 shows the case where the plane intersects the surface at the point where the interface contacts the wall. In this case, $\gamma \cong 20°$ and the apparent contact angle that would be measured in the image is $\theta_a = 0$. In this case, the intersections of the plane with the wall is elliptical and the plane with the interface is circular.

Fig. 6 shows the case where the plane intersects the surface at the point such that $= 45°$; the apparent contact angle that would be measured in the image is $\theta_a = 24.5°$. In this case, the intersection of the plane with the wall is elliptical.

Fig. 7 shows the limiting case where the plane is vertical,$\gamma = 90°$; the apparent contact angle that would be measured is the true contact angle $\theta_a = 30°$. In this case, the intersection of the plane with the wall is elliptical; however, the ellipse is infinitely extended such that the wall appears as two vertical lines.

Fig. 8 shows the case where the plane is vertical, $\gamma = 90°$, but is offset from the z-axis half the distance to the wall. In this case the apparent contact angle that would be measured is $\theta_a = 33.7°$. The wall still appears as two vertical lines; however, the spacing between the lines is less than for the previous case. This could be interpreted to mean that the image is for a capillary tube with a smaller diameter than in the previous case.

Fig. 9 shows the case where the plane is vertical, $\gamma = 90°$, but is offset from the z-axis by 95% of the distance to the wall. In this case the apparent contact angle that would be measured is $\theta_a = 61.6°$. The wall still appears as two vertical lines; however, the spacing between the lines is much less than for the previous case. This could be interpreted to mean that the image is for a capillary tube with a much smaller diameter than in the previous case.

The implications of the last two cases is obvious. Simply examining the images to obtain apparent contact angles, it could be concluded that the contact angles for smaller sized capillary tubes increase and approach $\theta_a = 90°$ in the limit. This would suggest that smaller capillary tubes are neutrally wet while larger tubes are water wet. This conclusion is, of course, specious.

In summary, using intersecting planes for various positions, it is possible to measure apparent contact angles anywhere in the range $0° \leq \theta_a \leq 90°$. This conclusion is completely supported by the work of Andrew *et al* [6]. They used what they called a "resampling plane" that could be reoriented relative to the grain surface. They found that the measured angle could vary from approximately 10° to 110° (their Fig. 10) by repositioning the resampling plane.

The above analysis considers the very simple example of a straight, circular capillary tube. Real porous media have flow passages that diverge, converge, twist and turn. An analysis of such passages is beyond the scope of the present paper. However, it is obvious that the measurement of true contact angles in such passages would be much more complex than that for the simple model considered here.

5 A Way Forward

It should be no surprise that the contact angle must be an effect, not a cause. Examining Young's equation, there are four parameters $\sigma_{lg}, \theta, \sigma_{sg}$, and σ_{sl}. Basic principles of functions requires that only three of these variables can be independent. The interfacial tensions are assumed to be properties of the substances, related to how various species of molecules interact across a surface. They can therefore be assumed to be the independent variables. This implies that the contact angle is the dependent variable and

$$\theta = f\left(\sigma_{lg}, \sigma_{sg}, \sigma_{sl}\right) \quad (24)$$

Furthermore, Young's equation depends on a static force balance and other force terms must affect that balance if they are present. It is unreasonable to expect the values of the interfacial tensions to change with the additional forces; hence, the contact angle must change. In fact, many theoretical studies utilize a modified Young's equation that includes body forces and fluid stresses.

I emphasize that this paper is not a condemnation of the ability to determine contact angles on the microscopic scale – it is a cautionary statement that such determinations must be done with great care. The work of Andrew *et al* [6] shows how the measured contact angle varies depending on how the plane of investigation is oriented. Their work could be taken one-step further. Using multiple images, a three-dimensional interface model could be reconstructed and this model could then be interpreted to determine a unique contact angle that characterizes the system, assuming that such an angle exists. Based on observations made in the Andrew *et al* work, such a process would require much more detailed images than they acquired. They found that there was insufficient accuracy in the digital data to determine contact angles computationally and they resorted to a manual process. This would be impractical if reconstructions of interfacial surfaces are performed, particularly in a dynamic experiment. It would appear that one more

https://doi.org/10.1051/e3sconf/202014603004

level of sophistication in image analysis must be achieved before contact angles can be systematically interpreted from micro-imaging experiments. There is evidence that this type of work is being pursued (AlRatrout *et al* [8]).

6 Conclusions

The present work supports the following conclusions:

1. Contact angles are dependent variables, functions of fluid/ solid properties (interfacial tensions), displacement velocities, history (time, hysteresis, and externally applied body forces and viscous stresses) – they are effects not causes.
2. Wettability can be defined dependably by the contact angle only for the static case with no externally imposed forces. Contact angles measured in dynamic and complex systems may not be good indictors of wettability.
3. It is not possible to measure accurately a contact angle in a simple two-dimensional image. Such measurements would require detailed reconstruction and interpretation of the three-dimensional interface.

7 References

1. F.F.Craig, jr., *The Reservoir Engineering Aspects of Waterflooding*, (SPE Monograph Series, **3**, 1971).
2. G.Macdougall, and C.Ockrent, "Surface energy relations in liquid/solid systems. I. The adhesion of liquids to solids and a new method of determining the surface tension of liquids", PRSA **180**, 981, 151-173, (1942).
3. B.Krasovitski and A.Marmur,"Drops down the hill: Theoretical study of limiting contact angles and the hysteresis range on a tilted plate", Langmuir, **21**, 3881-3885, (2005).
4. R.G.Cox, "The dynamics of spreading of liquids on a solid surface. Part 1. Viscous flow", JFM, **168**. 169-194, (1986).
5. M.Fermigier and P.Jenffer, "An experimental investigation of the dynamic contact angle in liquid-liquid systems", JCIS, **146**, 1, 226-241, (1991).
6. M.Andrew, B.Bijeljic, and M.J.Blunt, "Pore-scale contact angle measurements at reservoir conditions using X-ray microscopy", AWR, **68**, June, 24-31, (2014).
7. R.Held, N.Schleifer, L.Genolet, and A.Fogden, "Added insight from image-based wettability characterization", E3S Web Conference, **89**, (International Symposium of the Society of Core Analysts, 2018).
8. A.AlRatrout, A.Q.Raeini, B.Bijeljic, and M.Blunt, "Automatic measurement of contact angle in pore-scale images", AWR, **109**, 158-269, (2017).

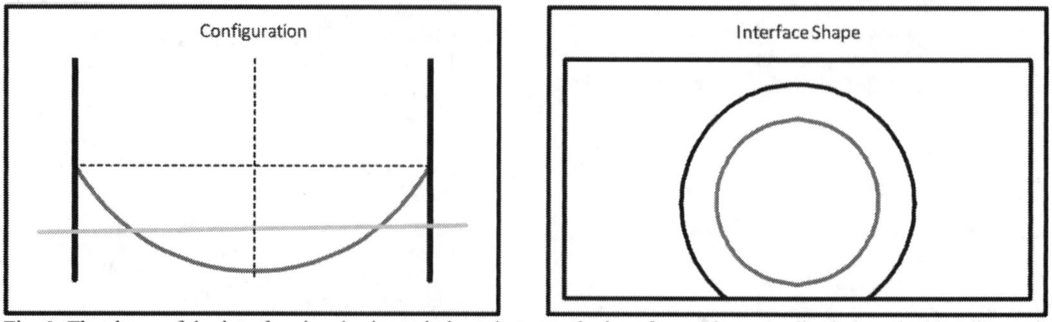

Fig. 3 The shape of the interface in a horizontal plane that cuts the interface. The water lies below the blue line and the air above.

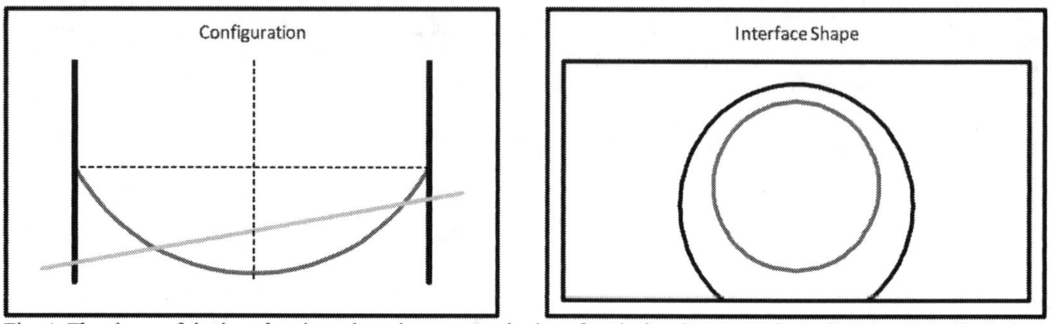

Fig. 4 The shape of the interface in a plane that crosses the interface before it crosses the wall. The water lies below the blue line and the air above.

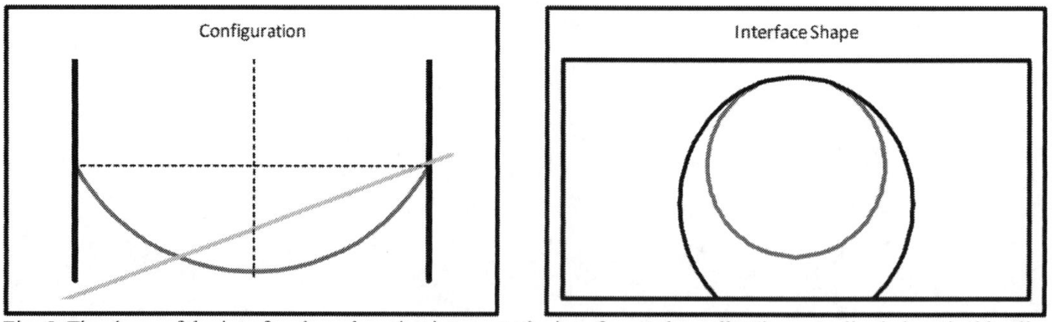

Fig. 5 The shape of the interface in a plane that intersects the interface at the wall. The water lies below the blue line and the air above.

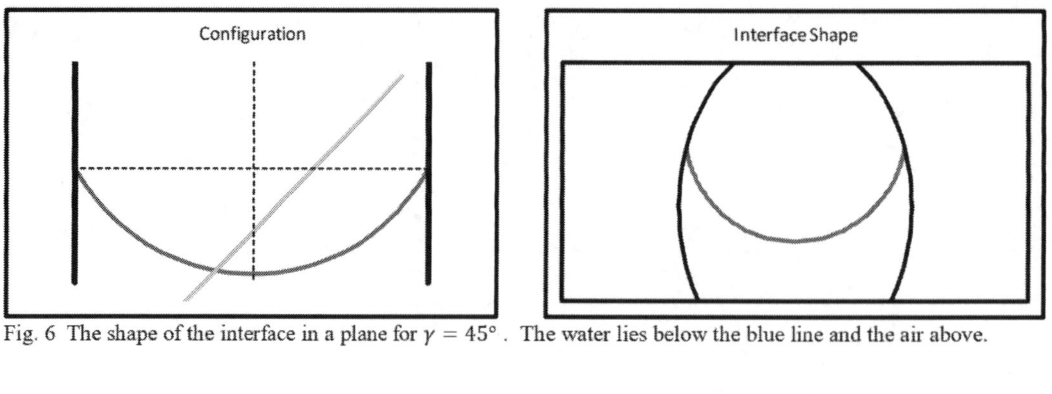

Fig. 6 The shape of the interface in a plane for $\gamma = 45°$. The water lies below the blue line and the air above.

https://doi.org/10.1051/e3sconf/202014603004

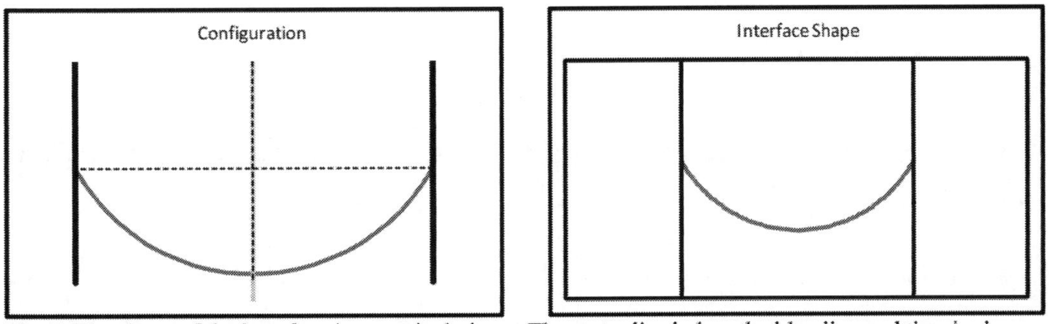

Fig. 7 The shape of the interface in a vertical plane. The water lies below the blue line and the air above.

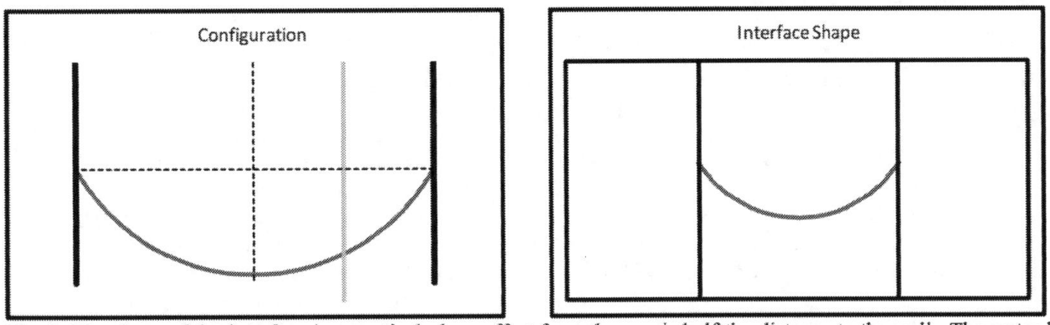

Fig. 8 The shape of the interface in a vertical plane offset from the z-axis half the distance to the wall. The water lies below the blue line and the air above.

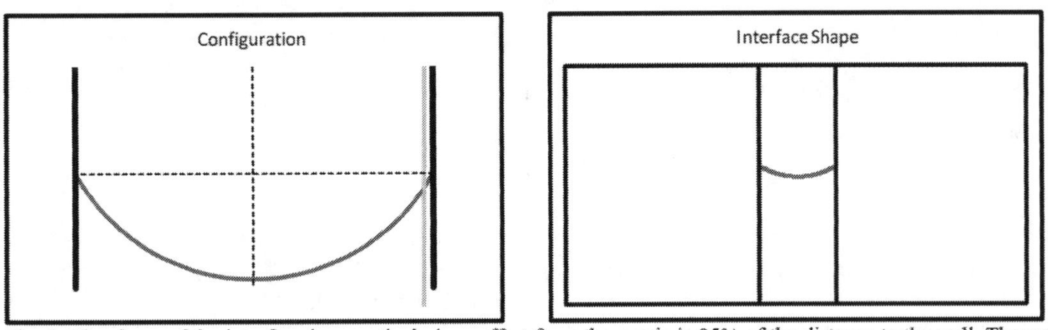

Fig. 9 The shape of the interface in a vertical plane offset from the z-axis is 95% of the distance to the wall. The water lies below the blue line and the air above.

 https://doi.org/10.1051/e3sconf/202014602001

Polymer-functionalized silica nanoparticles for improving water flood sweep efficiency in Berea sandstones

Alberto Bila[1,*], *Jan* Åge Stensen[1,2], and *Ole* Torsæter[1]

[1]PoreLab Research Center, Department of Geoscience and Petroleum-Norwegian University of Science and Technology (NTNU), S. P. Andersens veg 15a, 7031, Norway

[2]Sintef Industry, S. P. Andersens veg 15a, 7031, Norway

Abstract. Extraction of oil trapped after primary and secondary oil production stages still poses many challenges in the oil industry. Therefore, innovative enhanced oil recovery (EOR) technologies are required to run the production more economically. Recent advances suggest renewed application of surface-functionalized nanoparticles (NPs) for oil recovery due to improved stability and solubility, stabilization of emulsions, and low retention on porous media. The improved surface properties make the NPs more appropriate to improve microscopic sweep efficiency of water flood compared to bare nanoparticles, especially in challenging reservoirs. However, the EOR mechanisms of NPs are not well understood. This work evaluates the effect of four types of polymer-functionalized silica NPs as additives to the injection water for EOR. The NPs were examined as tertiary recovery agents in water-wet Berea sandstone rocks at 60 °C. The NPs were diluted to 0.1 wt. % in seawater before injection. Crude oil was obtained from North Sea field. The transport of NPs though porous media, as well as nanoparticles interactions with the rock system, were investigated to reveal possible EOR mechanisms. The experimental results showed that functionalized-silica NPs can effectively increase oil recovery in water-flooded reservoirs. The incremental oil recovery was up to 14% of original oil in place (OOIP). Displacement studies suggested that oil recovery was affected by both interfacial tension reduction and wettability modification, however, the microscopic flow diversion due to pore plugging (log-jamming) and the formation of nanoparticle-stabilized emulsions were likely the relevant explanations for the mobilization of residual oil.

1. Introduction

Oil production rates from existing fields are declining, and the occurrence of new discoveries has been scarce [1, 2]. Therefore, improving the recovery rates from mature oil fields is a priority for oil companies. The oil recovery factor from oil fields is typically around 20-40% of OOIP [3], the remaining oil is in some cases amenable for EOR. After conventional secondary water flooding, residual oil dwells in the reservoir pores as capillary-trapped and/or by-passed oil [2, 4]. Mobilization of the capillary-trapped oil requires increasing the viscous forces or decreasing the capillary forces [4, 5]. Applying sufficient pressure gradients between wells when injecting viscous fluid to increase viscous forces is the major challenge [2], as it can fracture the reservoir; hence, decreasing the capillary forces becomes an assertive option. This can be achieved through the reduction of interfacial tension (IFT) between the reservoir fluids. The by-passed oil or the oil located in unswept areas of the reservoir can be mobilized by pore plugging and flow diversion of injectants [4-7].

To this end, research suggests that properly designed nanoparticles (NPs) can change the physical or chemical behavior of an oil reservoir, thereby improving microscopic sweep efficiency of water flood. This is because NPs have small diameter size (1-100 nm) and large surface area-to-volume ratio. These properties enhance the mobility and surface activity of NPs, particularly at high temperature, making them suitable to modify fluid-rock properties and increase oil recovery [8, 9]. Furthermore, NPs can easily travel through and reach untouched zones of a reservoir with no severe impact to the rock permeability [1, 7, 9].

Thus far, silica NPs are the most researched nano-materials because of their natural occurrence in sandstone formations and low cost [1, 10]. Additionally, silica NPs can be synthesized and surface functionalized to meet particular reservoir conditions [8, 9]. Promising oil recovery results have been reported in the literature due to silica nanoparticles injection [10-17] under a variety of experimental conditions. That is, the use of silica NPs of variable size, composition, surface functionalities, etc.,

* Corresponding author: alberto.bila@uemep.uem.mz

suspended in aqueous solutions with diverse ionic strength [10-16] or in non-aqueous solutions [15]. The concentration of the NPs in the base fluid is also varied including injection flowrate, temperature, evaluation criteria of oil recovery, etc. This, in turn, makes it difficult to adequately grasp the oil recovery mechanisms of silica nanoparticles. Aside from i) IFT reduction and wettability alteration, which are the two most suggested mechanisms [11, 15, 16, 18], NPs improve oil recovery by ii) structural disjoining pressure [26], iii) formation of emulsions [19-21], and iv) log-jamming effect [5-7, 13, 22-24]. In some circumstances, more than one mechanism is associated with oil recovery.

The shortcoming of using typical NPs is associated to their aggregation and agglomeration, especially at harsh reservoir conditions. Therefore, recent advances suggest renewed application of surface-functionalized NPs for EOR to overcome some of the problems encountered with bare nanoparticles. Silica NPs, which have polymers covalently attached to its surface forms a new class of nanomaterials that may be superior to bare NPs for EOR due to improved properties such stability, stabilization of emulsions, low retention on porous media, etc. [25, 26]. These are known as polymer-functionalized/coated NPs. Few studies on their application for EOR have shown to be suitable for reservoir conditions [27]. Ponnapati, et al. [28] conducted injection experiments with polymer-grafted silica NPs in Berea sandstone. The NPs could mobilize residual oil and yield 7.9% of OOIP. The authors speculated that NPs clog the pores and increase local pore pressure to remove oil trapped in larger pores. Behzadi and Mohammadi [29] reported that polymer-coated silica NPs can modulate oil and water IFT and change the wettability of the oil-wet glass micromodel to more water-wet, resulting in higher EOR than unmodified silica nanoparticles. Experiments conducted by Choi, et al. [30] found that, when injecting polymer-coated silica NPs into water-wet sandstones, 74.1% oil was recovered, which was comparable to water flood (68.9%) and 72.7% of OOIP from unmodified NPs. The authors argued that modified silica NPs increased oil recovery by lowering the injection pressure relative to unmodified ones, which was associated to the ordering of NPs in the wedge film between oil and rock surface, rendering more hydrophilic surface. Recently, Bila, et al. [17] evaluated various polymer-coated silica NPs for EOR application in neutral-wet cores. The authors reported an incremental oil recovery up to 5.2% of OOIP after water flood. Like Choi, et al. [30], they reported pressure decrease with NPs injection. The oil recovery was attributed to the interfacial tension reduction and wettability alteration to more water-wet condition.

Based on the above studies, polymer-functionalized NPs can offer a pathway for EOR, however, more studies are needed to improve understanding of the underlying EOR mechanisms. This work evaluated the efficacy of polymer-functionalized silica NPs for EOR in water-wet reservoirs and identified the main mechanisms. Flood tests were conducted at high temperature with nanofluids prepared at 0.1 wt. % synthetic seawater.

2. Experimental materials

2.1 Silica NPs and synthetic seawater

Four types of silica NPs were used. The particles were spherical and hydrophilic with surface functionalized with polymer molecules. The polymer coating aimed at providing a protective layer to the NPs and to prevent the NPs from attracting each other in solution (i.e. steric repulsive forces). These NPs are referred to as polymer-functionalized/coated silica nanoparticles hereafter. The main component of the NPs was silicon dioxide, other components such as aluminum oxide (Al_2O_3) and mixed oxides (MOX) were present. The NPs are special research and development (R&D) products from Evonik Industries and were supplied to us as AERODISP®, which is AEROSIL® particles in liquid solution. The properties of the NPs suspended in distilled water, as received, are given in **Table 1**.

Table 1. Properties of silica NPs dispersed in distilled water.

NF	Basis	wt. %	Size (nm)
02-3	SiO_2 (sol gel cationic)	38.6	107
02-4	SiO_2 (sol gel anionic)	26.0	32
02-6	SiO_2/Al_2O_3/MOX	21.6	218
02-8	SiO_2/Al_2O_3/MOX	25.6	145

The concentrated solutions of the NPs were diluted to 0.1 wt. % in synthetic North seawater (SSW), here referred to as nanofluid (NF). The number is used to identify the nanofluid type. The composition of prepared SSW by wt. % was NaCl (74.4), KCl (1.85), $NaHCO_3$ (0.57), Na_2SO_4 (10.62), $CaCl_2{\cdot}6H_2O$ (4.24), $MgCl_2{\cdot}6H_2O$ (8.25) and $Sr_2{\cdot}6H_2O$ (0.07). The total dissolved salts were ≈38,318 ppm. The density, viscosity and pH of SSW was 1.008 g/cm^3, 0.53 cP and 7.97, respectively. The nanofluid density was 1.007-1.009 g/cm^3 and the viscosity ranged from 0.51-0.67 cP. All measurements were performed at 60 °C using Anton Paar Density meter and Anton Paar Rheometer, respectively.

2.2 Oleic phase

A North Sea reservoir crude oil was used in this work. It was a light oil (30 °API, 6 cP at 60 °C) with 71.57 wt. % saturates, 20.81 wt. % aromatics, 7.44 wt. % resins and 0.18 wt. % of asphaltenes. The crude oil was filtered twice through a 5-µm Millipore filter to remove any suspended particles that can block the lines and the pores spaces and or change the oil composition. Normal decane with density of 0.73 g/cm^3 and viscosity of 0.92 cP at 20 °C was used for wettability experiments.

2.3 Core characterization and preparation

Ten core plugs initially at water-wet conditions were used in this work. The cores were drilled from the same block of Berea sandstone; their mineral composition was measured with X-ray diffraction and were composed of 93.7 vol % quartz, 5 vol % of Microcline (Alkali feldspar) and 1.3 vol % Diopside. The core plugs were prepared to have similar dimensions of 3.8 cm diameter and length of 10 cm. They were cleaned with methanol through Soxhlet extractor and then dried at 60 °C for 2-3 days. Afterwards, gas porosity and permeability were measured on dried core plugs. The next step was to evacuate the core plugs for 2 hours and saturate them with SSW at 100 mbar vacuum pressure for 2-3 hours. The saturated cores were left soaked in the same SSW for at least 10 days for ionic equilibration with the rock constituents. The weight of the wet and dried core plugs was used to calculate the porosity and pore volume (PV), and these data are given **Table 2**.

Table 2. Properties of Berea sandstone core plugs.

Core	Porosity (%)	K_{abs}[1] (mD)	PV (ml)
L1	18.7	404	21.4
L2	17.4	537	21.8
L3	15.9	460	17.6
L4	16.2	411	17.9
L5	17.1	367	18.9
L6	16.8	333	18.5
L7	18.4	384	20.3
L8	18.3	184	20.3

[1]Klinkenberg corrected permeability.

2.4 Core flooding equipment

Fig. 1 presents a schematic of core flooding rig with its main components labelled. It utilizes an injection pump, three cylinders containing oil, synthetic sea water and nanofluid each. All cylinders were assembled vertically inside a temperature-controlled oven.

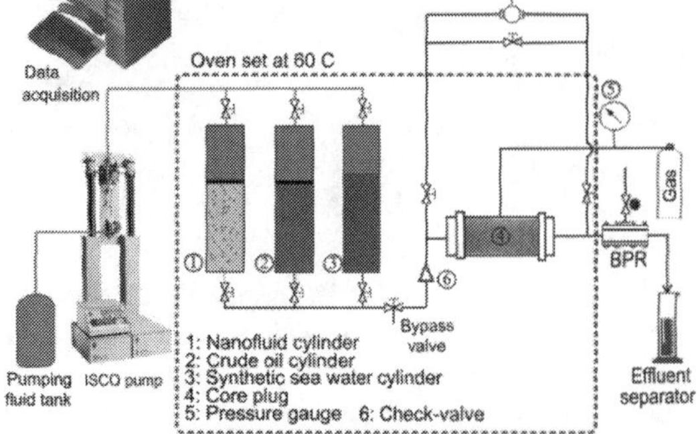

Fig. 1. Schematic of core flooding apparatus.

The core was loaded in the core-holder and oriented horizontally at a confining pressure held within 18-22 bar. A check valve and a backpressure regulator (BPR) were used to prevent back-flow of produced fluids and maintain the pore pressure constant during the core flooding experiments. The BPR wrapped with a heating plate was set to 5-bar pressure. The flooding experiments were conducted at 60 °C with the core oriented horizontally.

3. Experimental methodology

3.1 Core flooding

Eight flooding tests were conducted with nanofluids as tertiary EOR-agents (parallel tests). First, the 100% SSW saturated core plug was injected with fresh SSW for 1-2 PVs to ensure a complete removal of gas bubbles. Second, the drainage process was conducted by successively increasing the crude oil injection flowrate (0.5, 1.5 to 3 ml/min) for 15 PVs. This procedure ensured that there was no SSW production and the irreducible water saturation (S_{wir}) was achieved. The direction of the injection was reversed after half of the total PVs to even the distribution of the fluids in the core. This procedure also established the OOIP reported in **Table** 3. It is worth note that the drainage step was conducted at ambient conditions.

Before the experiments, the flooding system was heated while injecting crude oil at low rate of 0.02 ml/min until the temperature stabilized at 60 °C. Then, water (SSW) flood followed at constant flowrate of 0.2 ml/min until there was no oil production for 1-2 PVs. Thereafter, the flowrate was increased ten-fold (bump rate) for ≈1PV to overcome the capillary end-effects. In the following step, the injection was continued with nanofluid at 0.2 ml/min until there was no more oil production for 2-4 PVs. Then, the flowrate was bumped for ≈1PV. During the flood experiments, produced oil was collected every ¼ PV and corrected for the flooding system dead volume. When the production was occurring at low pace, a camera with automated capturing was used to record the production in a graded line over time, while the total volume was being collected in a large effluent separator. The recorded pictures were then analyzed to measure the amount of oil produced. This oil was compared to the total volume of oil produced in the large effluent separator. The differential pressure (dP) and oil recovery were recorded versus PVs injected, and the residual oil saturation (S_{or}) was calculated for each flooding stage.

3.2 Interfacial tension measurement

The interfacial tension (IFT) between crude oil and SSW or nanofluids was determined with pendant drop and spinning drop techniques at 60 °C. For the pendant drop method, Kruss drop shape analyzer 100 assembled with a J-shape syringe-needle of 1.0047 mm of inner diameter was used to dose the oil drops. With the oil drop hanging from the needle in the bulk phase, the measurements were taken every 20 seconds until static IFT value was reached.

The IFT was calculate using Young-Laplace model. For the spinning drop method, a SVT20N (Data Physics) spinning video tensiometer was used.

3.3 Amott-wettability test

Wettability of the cores was evaluated before and after nanofluid flooding at ambient conditions using Amott-test. Each core plug was set at S_{wir} before placing it in the Amott cell filled with SSW. The SSW was allowed to imbibe spontaneously into the core displacing oil over time. The oil production was recorded stepwise until the equilibrium was reached and the total amount denoted as V_{o1}. The remaining mobile oil (V_{o2}) was forcibly displaced by core flooding method, i.e. by injecting SSW at high flowrates (1-3 ml/min). Then, the core was removed from the core-holder and dipped in the Amott cell filled with oil to assess whether the oil could spontaneously displace water phase. At equilibrium, the amount of SSW produced was recorded as V_{w1}. The remaining mobile SSW in the core (V_{w2}) was displaced by injecting oil at high flowrates. The oleic phase was *n*-decane.

The wettability index (WI) was is the difference between the displacement-by-water ratio (water index, $I_w = V_{o1}/(V_{o1}+V_{o2})$) and displacement-by-oil ratio (oil index, $I_o = V_{w1}/(V_{w1}+V_{w2})$). A Categorization of wettability based on WIs is given by ref. [31].

4. Experimental results

4.1 Oil recovery

Eight core flood tests were conducted at 60 °C. Two core plugs were used for each nanofluid type to reproduce the results and reduce experimental errors. To overcome the capillary end-effects and ensure that any additional oil recovery was a result of nanofluid effect, the flow rate was increased from 0.2 to 2 ml/min at the end of the floods (procedure described in section **3.1**). The oil recoveries achieved by water flood varied from 47.2 to 56.1% of OOIP. The incremental oil recovery due to nanofluid flooding ranged from 7.0 to 14% of OOIP. The production of first oil due to nanofluid injection was occurred from 1.5 to 4 PVs. This was found to depend on NP type, and it was also affected by core properties. Significant oil production was observed after injection of several PVs of nanofluid. An example of oil recovery profile (water and nanofluid flood) at low- and high flow-rates is given in **Fig. 2** as function of PVs (for NF02-4 and NF02-6). Crude oil was produced as an oil-in-water emulsion, especially when the flowrate was increased, during the flood of nanofluids NF02-3 and NF02-4. These samples (NF02-3 and NF02-4) could increase oil recovery without having significant effect on the injection pressure relative to the base case, i.e. the differential pressure remained at the level of water flood (see **Fig. 7a**) throughout the duration of the process.

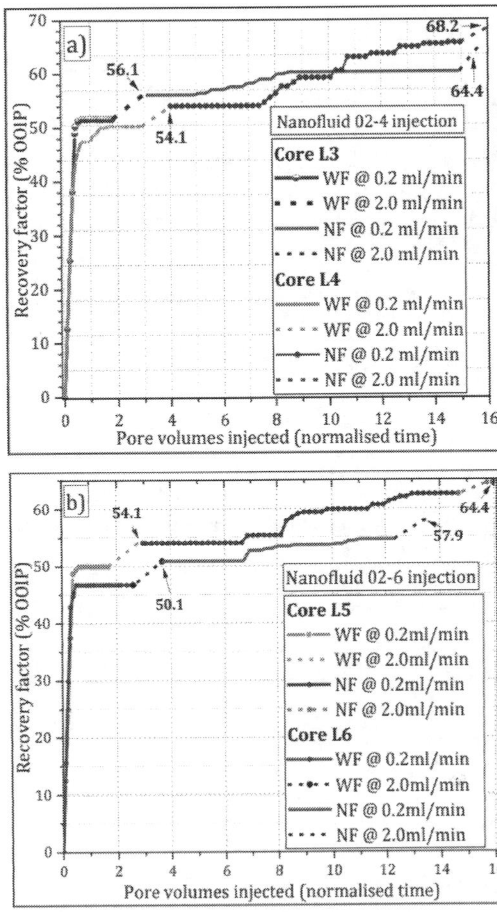

Fig. 2. Oil recovery factors (RFs) versus PVs injected: a) NF02-4 and, b) NF02-6 injection. In both cases, the first oil production was observed after 3 PVs of nanofluids injection.

In contrast, injection of NF02-6 and NF02-8 significantly increased water flood pressure (see **Fig. 7b**); no visible signs of formation of emulsions were observed, even when the flowrate was increased. At the end of the flooding, a filtered NP "cake" was observed at the core inlet for both NF02-6 and NF02-8, showing physical filtration of large NPs and plugging of the pores during the floods. **Table 3** summarizes the main results used to evaluate the nanofluid samples.

Table 3. Oil recovery factors (expressed as % of OOIP) achieved at each injection rate. The RF_1 and RF_2 represent the oil recovery factors at low flowrate (0.2 ml/min) and high flowrate (2 ml/min), respectively.

NF	Core	S_{wir} (%)	OOIP (ml)	Water flooding			S_{or1} (%)	Nanofluid flooding			S_{or2} (%)	E_D (%)	Total RF
				RF_1	RF_2	RF_t		RF_1	RF_2	RF			
02-3	L1	24.7	16.1	44.7	2.5	47.2	39.8	8.6	1.4	9.9	33.6	15.6	57.1
	L2	32.2	14.8	43.9	4.1	48.0	35.3	7.0	4.1	11.1	27.7	22.0	59.1
02-4	L3	10.5	15.7	51.7	4.4	56.1	39.3	15.3	3.0	8.3	31.9	19.0	64.4
	L4	10.6	16.0	50.3	3.8	54.1	41.1	11.6	2.5	14.1	28.5	31.0	68.2
02-6	L5	15.5	16.0	50.0	4.1	54.1	38.8	8.4	1.9	10.3	30.1	22.4	64.4
	L6	16.7	15.4	46.8	4.2	51.0	44.3	4.7	2.3	7.0	35.1	21.0	58.0
02-8	L7	18.8	16.5	49.4	4.2	53.6	37.7	8.2	0.9	9.1	30.0	20.4	62.7
	L8	18.7	16.5	52.4	2.7	55.2	36.5	7.3	3.0	10.3	28.1	23.0	65.5

The oil displacement efficiency was evaluated by equation:

$$E_D = \left[1 - \left(\frac{S_{or2}}{S_{or1}}\right)\right] \times 100\% \quad (1)$$

Here, S_{or_1} and S_{or_2} represent residual oil saturation at the end of water- and nanofluid-flooding (including low and high flowrate injection), respectively.

4.2 Interfacial tension

The interfacial tension (IFT) between crude oil and NF02-3 and NF02-4 was measured with both pendant drop and spinning drop methods. Nanofluids NF02-6 and NF02-8 were very sensitive at high temperature; the solutions immediately precipitated at 60 °C and became opaque, thus making it impossible to use the pendant drop method for IFT analysis. The spinning drop was used instead. **Fig. 3** shows a crude oil drop hanging from a needle within solution of sample NF02-6. Increasing temperature from at 22 °C (**Fig.** 3a) to 60 °C (**Fig.** 3b) caused the NPs to self-assemble at the oil and water (o/w) interface. The formed layers of NPs at the interafce eventually settled to the bottom of the cuvette due to gravity force.

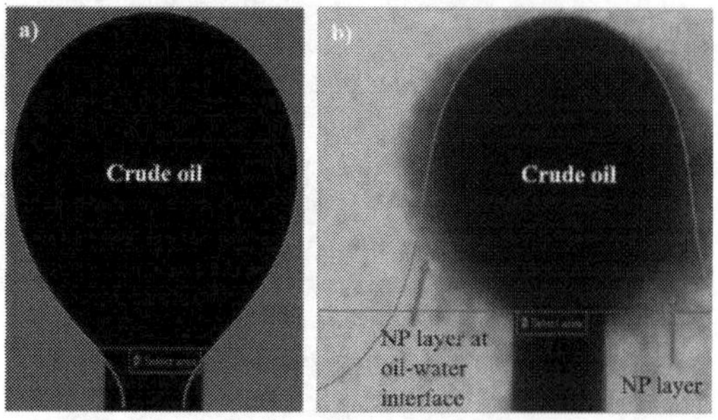

Fig. 3. Crude oil drop suspended from a needle in NF02-6: a) at 22 °C, and b) at 60 °C, NPs self-assembled at the w/o interface.

This phenomenon was viewed thanks to high-resolution pendant drop camera. The measured values of IFT are presented in **Table 4**.

Table 4. Interfacial tension values measured at 60 °C.

Fluid	Interfacial tension (mN/m)	
	Pendant drop	Spinning drop
SSW	11.0	10.0
NF02-3	3.5	3.1
NF02-4	3.3	2.9
NF02-6	-	6.8
NF02-8	-	4.7

The spinning drop was also used to confirm the IFT values obtained from pendant drop method for samples NF02-3 and NF02-4. **Fig. 4** shows a crude oil droplet dipped in the capillary tube and spun at 7,000 rpm and at 60 °C. Due to the high rotational speed, there were no visible aggregates of NPs and the IFT could be measured for all samples.

Fig. 4. Shape of the crude oil drop inside the capillary tube filled with NF02-6 and spun at 7,000 rpm for IFT analysis using STV20 spinning drop video tensiometer.

4.3 Core wettability

Initial wettability of the cores was evaluated using two reference cores. The average wettability index was, WI = 0.86, indicating strongly water-wet condition. The cores used afterwards were considered water-wet. This assumption was made because all core plugs were drilled

* Corresponding author: alberto.bila@uemep.uem.mz

from the same block and followed the same preparation procedures. After nanofluid flooding was completed, each core (at S_{or}) was flooded with *n*-decane at high flowrates (1-3 ml/min) until residual water saturation was achieved. Afterwards, the core was dipped into the Amott cell following the procedure described in section **3.3**. **Fig. 5** shows oil recovery from SSW spontaneous imbibition (SI) during 30-days test. As shown in **Fig. 5**, the rate of water imbibition decreased slightly in NP treated cores compared with that in original ones (not injected with NPs). Initially, oil was produced from all faces of the core plugs. After 24 hours, the production declined, and it was

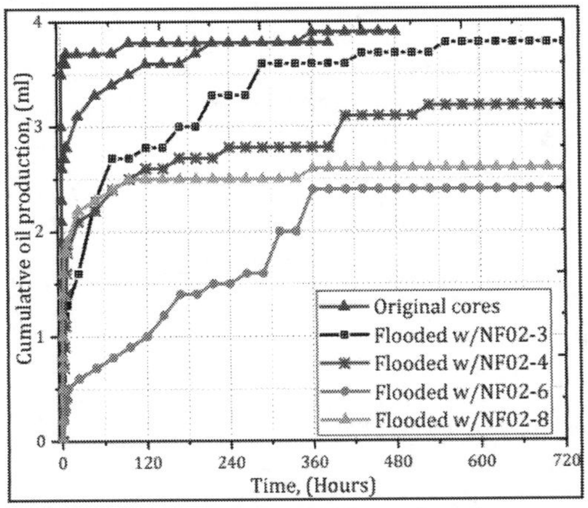

Fig. 5. Water spontaneous imbibition performed on cores flooded with nanofluids (at ambient conditions).

occurring from the top. The produced oil remained on the top surface of the cores and detached from it after several days or by gently shaking the Amott cells. These events are depicted by sudden increase in oil recovery at some points in **Fig. 5**. At equilibrium, the measured water indexes (I_w) varied from 0.5 to 0.89. In the second cycle of spontaneous imbibition, oil did not imbibe into the core or displace water, implying that Amott oil index is zero, I_o= 0. Therefore, the Amott-test was not completed, and Amott water indexes were used to assess the effect of NPs on the rock surface.

4.4 Permeability and porosity

After nanofluid flooding, the cores were cleaned with toluene and methanol for several days with Soxhlet extraction and dried at 60 °C for 2-3 days. Then, the gas permeability and porosity were measured. The results are presented in **Table 5**. The objective was to determine and quantify the changes of initial values of permeability and porosity as result of NPs adsorption or retention during nanofluid flood. In **Table 5** negative values indicate permeability or porosity impairment.

5. Discussion of the results

5.1 Nanofluid stability analysis

The stability of nanofluids at 0.1 wt. % was analyzed via: i) particle size distribution (PSD), ii) zeta potential, and iii) sedimentation rates of nanoparticles.

Malvern Zetasizer instrument was used to obtain PSD and zeta potential of NP solutions. After setting the temperature at 60 °C, the average particle diameter size for NF02-3 was 94 nm and 32 nm for NF02-4. These values are within the range claimed by manufacturer (i.e. 107 nm for NF02-3 and 32 nm for NF02-4), which suggest that NPs were stable in seawater solution, at least for the measured period. The measurements of the particle size for samples NF02-6 and NF02-8 resulted in high values of polydispersity index, suggesting growth of particle size. It was therefore not suitable to characterize the particle size with dynamic light scattering technique. Measurements of zeta potential of NPs gave unreliable values, for unknown reasons.

An analysis of sedimentation rate of NPs was conducted to verify particle size measurements. To this end, nanofluids were placed in the oven at 60 °C and monitored daily. It was observed that samples NF02-6 and NF02-8 were rapidly rendered unstable at 60 °C, which is consistent with observation made while measuring the IFT. Large NPs structures were formed and deposited to the bottom of the ampules within one-hour storage time. Stability of nanofluids "NF02-3" and "NF02-4" was achieved for up to four days. After this storage-time, they also started to precipitate and gradually settling out of solution, forming large aggregates. **Fig. 6** shows NPs in the aggregated form at the bottom of the ampules (white). Interesting that some polymers were seen flocculating in NF02-4 solution. The precipitation of polymers is thought to be due to the divalent cations in aqueous solution (e.g. Mg^{2+}, Ca^{2+}) and high temperature [26, 32]. Consequently, the desired repulsion forces between the NPs in the solution vanished.

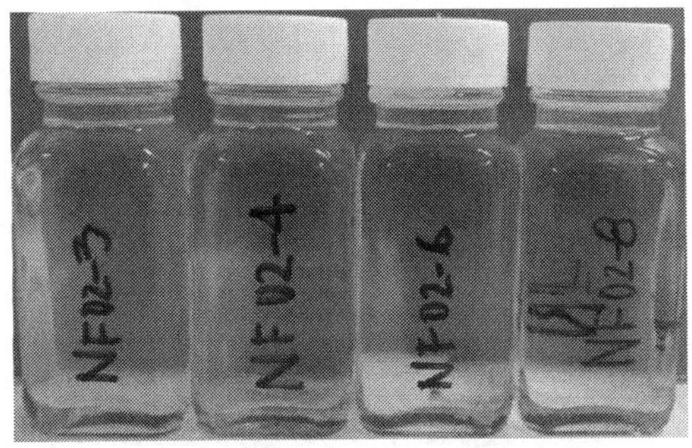

Fig. 6 Visual stability analysis of nanofluids. The white color at the bottom of the ampule shows NPs in the aggregated form.

Table 5. Permeability and porosity measured before and after nanofluid flooding on core plugs.

Core	Injected nanofluid	Permeability (mD)			Porosity[2] (%)		
		Before	After	% difference	Before	After	% difference
L1	02-3	404	350	-13	19	17	-10
L2		537	452	-16	20	16	-19
L3	02-4	460	446	-3	17	15	-11
L4		411	370	-10	18	16	-15
L5	02-6	367	232	-37	18	13	-25
L6		331	246	-26	18	14	-25
L7	02-8	384	292	-24	18	15	-18
L8		265	193	-20	20	17	-16

[2]Measured with Helium porosimeter

Based on sedimentation test, it was concluded that NF02-3 and NF02-4 were stable for approximately 4 days whereas NF02-6 and NF02-8 became unstable at 60 °C. The short-term stability of nanofluids would reduce the transport of NPs through porous media and their surface functionalities, thus lowering the economic feasibility at field scale. This draws the conclusion that the surface modification needs to be improved to achieve a long-term stability over a long injection (expected at field-scales). The improvement should not only focus on the NPs stability, but also on the surface chemistry and other possible interactions that may occur between NPs and an oilfield system for better oil recovery.

5.2 Evaluation of oil recovery

Evaluation of tertiary oil recovery not only requires the reproducibility of experimental results, but also the proper determination of the waterflood S_{or} to ensure that any additional oil is solely due to the EOR fluid. To achieve this, twin core plugs were used for each nanofluid type; a bump rate was applied, for 1 PV, after waterflood no longer produced oil for 2 PVs during low rate.

The results showed that nanofluids at a concentration of 0.1 wt. % can mobilize residual oil, but it was delayed compared to breakthrough of nanofluids. The arrival of first oil at the core outlet varied among the samples. It occurred within 1.5 to 4 PVs. The production due to NF02-6 (**Fig. 2b**) and NF02-8 was occurred at late injection times compared to other two samples NF02-3 and NF02-4 (**Fig. 2a**). Most of oil production occurred in the form of small droplets that merged with the oil in the separator. In **Fig. 2**, the points at which oil recovery increases over time shows the ability of NPs to progressively mobilize residual oil. Large oil production occurred at the expense of large PVs of injected nanofluid, and the late occurrence is partly attributed to differences between the viscosities of nanofluids and crude oil. Additionally, the production was affected by water flood stage and variations in core properties. Therefore, the nanofluids to mobilize the trapped oil, an extra energy was needed. This energy was achieved by physicochemical interactions between NPs and rock system, i.e. deformation and breakage of oil drops and/or NPs blocking enough pores to build up enough pressure within the pores to detach and mobilize the oil in adjacent pores.

Table 3 reports the overall oil recoveries achieved at the end of water- and nanofluid-flooding. The oil recovery achieved by water flood varied from 47.2% to 56.1% of OOIP after the flowrate was bumped. When the nanofluid system was injected at low rate, the incremental oil recovery varied from 4.7% to 11.6% of OOIP. Increasing the flowrate tenfold, for ≈1 PV, the ultimate recoveries reached 7.0% to 14.1% of OOIP. The oil recovery during the bump rate is shown in **Fig. 2** by the "dotted" lines. This was because it was difficult to record the amount of oil produced per step during bump rate, as it was a quick and short process. Additionally, oil was produced as small droplets and could easily be measured at the end of injection. Note that the "dotted" lines in **Fig. 2** do not indicate a linear production of oil. The oil production on a bump flood was unexpected as the residual oil is trapped in the pores where high capillary forces prevail. This oil would resist flow even if it is flooded at higher capillary numbers than those present at trapping [33]. In this work, the produced oil indicated that capillary stability was perhaps not reached during low rate and/or the oil trapped in the thin gap between the end core plugs was produced at high flowrate. Future work is recommended to water flood enough to attain the equilibrium before applying EOR fluid injection and mitigate capillary end-effects.

The twin cores produced comparable results; the variation was below 5% of OOIP. Upscaling the present results for field scale injection is still a challenge. Therefore, one cannot directly compare core scale results to large fields, it may appear economically unfeasible because of the cost of NPs. Research suggests that to better simulate the injections at field-scale, large PVs of nanofluids must be injected at core scale [34]. The results presented in **Table 1** and **Table 3** show that the highest incremental oil recovery was achieved with the smallest NPs size. The increase in oil recovery with decreasing NP size is consistent with the IFT reduction, see **Table 4**; small size NPs were efficient in lowering o/w interfacial

* Corresponding author: alberto.bila@uemep.uem.mz

tension, resulting in higher oil recovery than the large sized NPs. These results are in agreement with previous findings reported by others [12, 19], but are also in contrast to the positive correlation observed between the NP size and oil recovery reported by Aurand, et al. [22].

The NPs at 0.1 wt.% had no plausible effect on the viscosity of the SSW, thus unable to improve the mobility ratio. The maximum differential pressure was achieved from NPs of large size. This supports the notion that increasing NP size decreases the permeability [12], and reduces the surface functionalities of the NPs at the oil-water-rock interfaces [35, 36], resulting poor oil recovery.

In summary, the present work proved that polymer-functionalized silica NPs can EOR efficiently in water-wet waterflooded reservoirs.

5.3 The EOR mechanisms of nanoparticles

This work reports promising results on the application of polymer-functionalized silica NPs for EOR. It was noted that various recovery mechanisms can be operating for oil recovery, depending on NP type. This implies that more studies are needed to quantify the contribution of each mechanism involved. The following sections discuss and suggest the underlying oil recovery mechanisms of silica NPs based on our experimental results.

5.3.1 Interfacial tension and wettability alteration

Microscopic sweep efficiency can be improved by decreasing capillary forces through lowering oil and water IFT. When injecting NPs, the major expectation is that the differential pressure across the core exceed the capillary forces trapping the oil. This can be achieved if the NPs adsorb to the o/w interface and decrease IFT. Capillary number, N_c, describes the balance between the oil mobilizing forces (viscous forces) and the oil trapping forces, and it provides a mean to evaluate how effective the NPs are for EOR.

$$N_c = \frac{\mu v}{\gamma} \quad (2)$$

Here, μ and v are the viscosity and velocity of the displacing fluid, respectively, and γ is the interfacial tension between the displacing and displaced phases.
The N_c for water flooding calculated in this work is of order of 10^{-6}. It is generally recognized that the critical capillary number for the onset of mobilization of residual oil in water-wet reservoirs is of orders of $\approx 10^{-5}$ [7, 37] above which, a complete mobilization of residual may take place. This condition can be met if any additive to the injection fluid, such as NPs, can decrease the IFT between oil and water down to 10^{-3} mN/m [38].

In this work, it was observed that silica NPs reduced the o/w interfacial tension. In **Table 4**, one can see that the reduction is modest, this is the IFT values are not low enough to produce enough N_c for significant mobilization the residual oil. Similar conclusion was reached by other researchers [22, 39]. Furthermore, the results indicate poor surface activity performance of NPs at the o/w interface. The reasons for that might be i) the high temperature that increased the degree of collision between the NPs in solution and the aggregation tendency at the o/w interface (see **Fig. 3**); ii) the accumulation (or coating) of polymer molecules on the particle surface could have prevented or reduced the amount of NPs that can move to the oil-water interface [40, 41], thus diminishing particle surface activity. This can occur even under favorable adsorption energy [42]; and iii) the surface modification was lost due to the precipitation of polymers in seawater due to multivalent ions. Furthermore, the aluminum present in composition of the particles (NF02-6 and NF02-8) would form sulphates in solution and prompting the formation of precipitates.

The oil recovery due to NPs relies on both fluid-fluid and fluid-rock interactions. These interactions can modify the fluid distribution and flow-ability of the fluids within the reservoir. To investigate fluid-rock interactions induced by NPs, we explored the rate of water imbibition and the Amott indexes. From **Fig. 5**, one can see that the rate of water imbibition on NPs treated cores was as high as in original cores (not injected with NPs) from the beginning of the tests, which resulted in significant amount of oil recovery. The oil production was observed from all core faces showing a contribution of both gravity and capillary forces during wettability alteration [43]. Furthermore, in **Fig. 5**, we see a sudden increase in oil recovery over time from all cores indicating a progressive alteration of rock wettability due to NP exposure on the rock surface. The NPs could destabilize oil films on the surface, rendering the pore spaces more water-wet. At later stage, the oil was produced from top of the cores due to gravity forces. The measured Amott water indexes ranged from 0.5 to 0.86. There is some variation in the rate of water imbibition and Amott indexes in treated cores compared with original ones and it can be attributed to the retained NPs on the pores and of differences in core properties.

The overall results suggest that wettability was affected by nanofluids to water-wet condition. These results were expected because of hydrophilic nature silica NPs and exposed to a substrate of like surface charge (Berea sandstone). During nanofluid flooding, NPs are forcibly confined to the pre-existing wedge formed between the three-phase water-oil-rock contact line, where they arranged themselves in a well-ordered layer and create an additional disjoining pressure [44], and developing new hydrophilic surface on the pores. The results are in agreement with previous studies [13, 45] that studied the wettability alteration using hydrophilic silica nanoparticles. Owing to small effects of NPs on IFT and wettability alteration, the polymer-functionalized silica NPs studied in this work are not expected to increase oil

https://doi.org/10.1051/e3sconf/202014602001

recovery by solely reducing the IFT and/or by affecting the wettability. However, the irreversible adsorption of NPs to the o/w interface still plays a predominant role for stabilization of emulsions [20, 25, 46]. The emulsion formation is discussed in the Section **5.3.2**.

5.3.2 Transport of NPs through porous media

Characterization of transport in the reservoir rocks has two components: the retention and the mobility of the nanoparticle dispersion. The former quantifies the fraction of injected NPs that survive and reach the target zone. The latter defines the operating conditions (e.g. injection pressure and/or flowrate) to bring the injected NPs through the desired pathway and time to the target location [47]. If the differential pressure (dP) across the core is decreasing throughout EOR flooding, it may indicate emulsification of oil [19] otherwise, it indicates microscopic flow diversion due to pore plugging [4, 6]. This information can lead to an understanding of the EOR mechanisms of NPs flooding.

During the flood of nanofluids NF02-3 and NF02-4, it was observed that dP increased slightly for ≈1 PV, then it decreased and levelled to the reference waterflood value (**Fig.** 7a for NF02-4). The slow rise of dP is thought to be caused by physical confinement of the particles at the core entrance. Then, the pressure levelled relative to the base case as the injection advanced, indicating the ease with which the NPs were propagating through the pore spaces without clogging or causing significant damage to the rock permeability. The reduction of core absolute permeability varied from 3% to 16%, which confirmed low particle retention on the pore walls. The low injection pressure is likely a manifestation of the role the polymer coating on the NP surface played to improve particle mobility on the pores. Accordingly, this allowed the NPs to stick more to the oil/water interface. From this, we could infer the generation of in-situ emulsions with nanofluid flooding also prompted by flow dynamics and high temperature. This was confirmed by increasing the flowrate and oil-in-water emulsions could be visible in the effluent separator. Increasing the flowrate provided an extra energy required to break up oil phase, thus allowing more NPs to adsorb at the oil-water interface [21]. The generation of emulsions concurs with the reduction of IFT, especially for NF02-3 and NF02-4 (with the lowest IFT reduction among the studied samples). The formed emulsion droplets merged upwards with the oil in the effluent separator within one hour. Earlier studies that investigated the EOR effect of polymer-coated silica NPs, such as those used in this work, have also highlighted the formation of nanoparticles-stabilized emulsions as a mean by which the NPs increased oil recovery increase [13, 20, 25]. It is most likely that samples NF02-3 and NF02-4 improved oil recovery by increasing oil mobility due to generation of stabilized oil-in-water emulsions.

The contrasting results were observed during the injection of samples NF02-6 and NF02-8, little to no emulsions were observed, even at high flowrate. The injection of these samples was accompanied with continuous increased pressure (see **Fig. 7b**). After 5 PVs of nanofluid injection, dP showed high degree of oscillation until the tests were stopped. At the end of the floods, a filtered "cake" of NPs was formed at the core inlet (see the inset in **Fig. 7b**), indicating the occurrence of physical filtration and blockage of the pores.

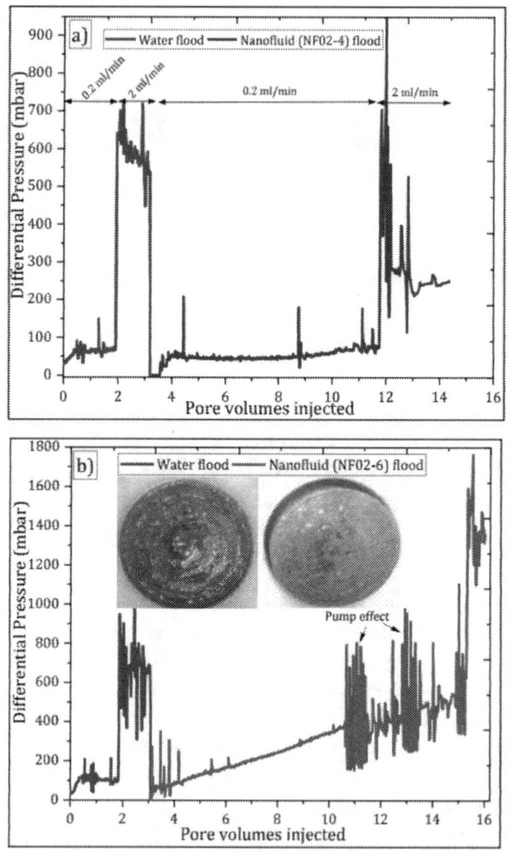

Fig. 7. Differential pressure profiles as a function of PVs of injected fluids: a) NF02-4 injection and b) The inset is the NPs "cake" formed at the core inlet during the injection of NF02-6.

The aggregation at core entrance was expected because were unstable and creating favorable conditions for blockage of small pores from start of the injection. This is likely the primary reason for pressure increase. Consequently, the permeability reduction was in the range of 20% to 37% and was higher than that on cores injected with samples NF02-3 and NF02-4. The pressure pattern also indicated that pore plugging was also happening inside the core until an extreme pore blockage was reached. This likely lead to the diversion of the injection water and mobilization of bypassed oil in the adjacent pores. **Fig. 7b** also shows periods of an abrupt variation of the pressure. These events supported the notion that NPs were further aggregating within the core; as a result, the pressure was redistributed within the pores to assist the injected water to reach the by-passed oil in the adjacent pores and push forward to the core outlet. The pressure was reduced by displacement of the oil and detachment of weakly adsorbed NPs on the pores. This led to the restoration of the normal flow of water. The events of sudden pressure drop in **Fig. 7b** reflect such migration behavior of NPs. The blockage continued in the available pore channels. Due to the high pressure generated with advancing of the injection and subsequent confinement of NPs in the pores, the oil was detached from the surface and produced before the pores were relocked. However, the pressure increase was not always accompanied by oil production.

The primary recovery mechanisms for NF02-6 and NF02-8 is possibly through microscopic flow diversion also referred to as log-jamming effect [6, 7, 22]. According to Spildo, et al. [4] and Skauge, et al. [6] this phenomenon can be explained by mass difference between the NPs and water. The NPs in aqueous solution carry water molecules, which decreases their flow ability, hence causing the particles to accumulate at the pore entrance. This concurs with our results because the NPs were aggregating in SSW solution before injection.

Due to the synergistic effect of IFT reduction and wetting effect, the high dP is believed to have contributed in the mobilization of both by-passed and capillary-trapped oil. Quantifying the contribution of pore blocking and fluid flow diversion mechanisms in the mobilization of oil based on differential pressure is complex because pressure fluctuation.

5.4 Effect of NPs on permeability and porosity

To evaluate the influence of NPs on rock properties, permeability and porosity were evaluated and compared with initial values. The results are presented in **Table 5**. The permeability and porosity decreased relative to initial values. The cores injected with nanofluids containing small-sized NPs diameter (NF02-3 and NF02-4) resulted in the lowest permeability and porosity reduction in the range of 3% to 16% and from 10% to 19%, respectively. For these samples the particle retention may be due to chemical adsorption of NPs owing to their small size [11] and composition. The slightly high values of porosity reduction suggested that adsorption and blockage due to NPs were relevant in the large pores. As expected, large size particles gave the largest permeability and porosity impairments. The permeability reduction was from 20% to 37% and from 16% to 25% for the porosity. Because the NPs aggregated in SSW, significant permeability and porosity reductions were expected. It is likely that the measurements were affected by core preparation steps for wettability evaluation post nanofluid flood.

6. Conclusions

In this work, a series of flooding experiments were conducted to investigate to unleash the EOR potential of four different types of polymer-functionalized silica NPs on water-wet Berea sandstone cores at 60 °C. The underlying EOR mechanisms of the NPs were also investigated. The following conclusions were obtained:

- The core flooding studies demonstrated that silica nanoparticles coated with polymer chains can effectively mobilize residual oil and increase oil recovery. The incremental oil recovery ranged from 7 to 14% of OOIP;
- The IFT between water and oil was decreased with modified silica NPs, but the reduction is not in orders of magnitude to contribute significantly for the remobilization of residual oil;
- The nanoparticles caused variation of the Berea sandstone surface structure, thus affecting the wettability towards water-wet condition;
- Aside from IFT reduction and wettability alteration, nanofluids NF02-3 and NF02-4 increased the mobility of trapped oil by creating oil droplets emulsions, while pore blocking, and microscopic flow diversion was relevant mechanism for samples NF02-6 and NF02-8.

Further work includes improving evaluation procedure of oil recovery as well as recovery mechanisms of the nanoparticles in neutral-wet reservoirs.

Acknowledgements: The authors gratefully acknowledge Evonik Industries for providing the nanoparticles used in this work and the Research Council of Norway for their financial support through PoreLab, Centre of Excellence, project number 262644.

References

[1] X. Sun, Y. Zhang, G. Chen, and Z. Gai, "Application of Nanoparticles in Enhanced Oil Recovery: A Critical Review of Recent Progress," *Energies,* vol. 10(3), 2017.

[2] A. Muggeridge *et al.*, "Recovery rates, enhanced oil recovery and technological limits," *Philosophical transactions. Series A, Mathematical, physical, and engineering sciences,* vol. 372(2006), 2014, pp. 20120320-20120320.

[3] I. Sandrea and R. Sandrea, "Recovery factors leave vast target for EOR technologies," *Oil & Gas Journal,* vol. 105(41), 2007, pp. 44-47.

[4] K. Spildo, A. Skauge, M. G. Aarra, and M. T. Tweheyo, "A New Polymer Application for North Sea Reservoirs," *SPE Reservoir Evaluation & Engineering,* vol. 12(03), 2009, pp. 427-432.

[5] M. A. E. Kokubun, F. A. Radu, E. Keilegavlen, K. Kumar, and K. Spildo, "Transport of Polymer Particles in Oil–Water Flow in Porous Media: Enhancing Oil Recovery," *Transport in Porous Media,* vol. 126(2), 2018, pp. 501-519.

[6] T. Skauge, K. Spildo, and A. Skauge, "Nano-sized Particles For EOR," in *SPE Improved Oil Recovery Symposium*, Tulsa, Oklahoma, USA, 2010, pp. 10. https://doi.org/10.2118/129933-MS.
[7] Z. Hu, S. M. Azmi, G. Raza, P. W. J. Glover, and D. Wen, "Nanoparticle-assisted water-flooding in Berea sandstones," *Energy & Fuels,* vol. 30(4), 2016, pp. 2791-2804.
[8] S. Ayatollahi and M. M. Zerafat, "Nanotechnology-Assisted EOR Techniques: New Solutions to Old Challenges," in *SPE International Oilfield Nanotechnology Conference and Exhibition*, Noordwijk, The Netherlands, 2012, pp. 15. https://doi.org/10.2118/157094-MS.
[9] M. V. Bennetzen and K. Mogensen, "Novel Applications of Nanoparticles for Future Enhanced Oil Recovery," in *International Petroleum Technology Conference*, Kuala Lumpur, Malaysia, 2014, pp. 14. https://doi.org/10.2523/IPTC-17857-MS.
[10] M. I. Youssif, R. M. El-Maghraby, S. M. Saleh, and A. Elgibaly, "Silica nanofluid flooding for enhanced oil recovery in sandstone rocks," *Egyptian Journal of Petroleum,* vol. 27(1), 2018, pp. 105-110.
[11] M. Zallaghi, R. Kharrat, and A. Hashemi, "Improving the microscopic sweep efficiency of water flooding using silica nanoparticles," *Journal of Petroleum Exploration and Production Technology,* vol. 8(1), 2018, pp. 259-269.
[12] L. Hendraningrat, S. Li, and O. Torsater, "Effect of Some Parameters Influencing Enhanced Oil Recovery Process using Silica Nanoparticles: An Experimental Investigation," in *SPE Reservoir Characterization and Simulation Conference and Exhibition*, Abu Dhabi, UAE, 2013, pp. 10. https://doi.org/10.2118/165955-MS.
[13] B. A. Suleimanov, F. S. Ismailov, and E. F. Veliyev, "Nanofluid for enhanced oil recovery," *Journal of Petroleum Science and Engineering,* vol. 78(2), 2011, pp. 431-437.
[14] K. Xu, D. Agrawal, and Q. Darugar, "Hydrophilic Nanoparticle-Based Enhanced Oil Recovery: Microfluidic Investigations on Mechanisms," *Energy & Fuels,* vol. 32(11), 2018, pp. 11243-11252.
[15] N. A. Ogolo, O. A. Olafuyi, and M. O. Onyekonwu, "Enhanced Oil Recovery Using Nanoparticles," in *SPE Saudi Arabia Section Technical Symposium and Exhibition*, Al-Khobar, Saudi Arabia, 2012, pp. 9. https://doi.org/10.2118/160847-MS.
[16] L. Hendraningrat, S. Li, and O. Torsæter, "A coreflood investigation of nanofluid enhanced oil recovery," *Journal of Petroleum Science and Engineering,* vol. 111(2013, pp. 128-138.
[17] A. Bila, Å. J. Stensen, and O. Torsæter, "Experimental Investigation of Polymer-Coated Silica Nanoparticles for Enhanced Oil Recovery," *Nanomaterials,* vol. 9(6), 2019.
[18] R. Li, P. Jiang, C. Gao, F. Huang, R. Xu, and X. Chen, "Experimental Investigation of Silica-Based Nanofluid Enhanced Oil Recovery: The Effect of Wettability Alteration," *Energy & Fuels,* vol. 31(1), 2017, pp. 188-197.
[19] M. Adil, K. Lee, H. M. Zaid, N. R. A. Latiff, and M. S. Alnarabiji, "Experimental study on electromagnetic-assisted ZnO nanofluid flooding for enhanced oil recovery (EOR)," *PloS one,* vol. 13(2), 2018, pp. e0193518.
[20] T. Zhang, D. Davidson, S. L. Bryant, and C. Huh, "Nanoparticle-Stabilized Emulsions for Applications in Enhanced Oil Recovery," in *SPE Improved Oil Recovery Symposium*, Tulsa, Oklahoma, USA, 2010, pp. 18. https://doi.org/10.2118/129885-MS.
[21] I. Kim, A. J. Worthen, K. P. Johnston, D. A. DiCarlo, and C. Huh, "Size-dependent properties of silica nanoparticles for Pickering stabilization of emulsions and foams," *Journal of Nanoparticle Research,* vol. 18(4), 2016, pp. 82.
[22] K. R. Aurand, G. S. Dahle, and O. Torsæter, "Comparison of oil recovery for six nanofluids in Berea sandstone cores," in International Symposium of the SCA, France, 2014, vol., no., pp. 1-12.
[23] A. I. El-Diasty and A. M. Aly, "Understanding the Mechanism of Nanoparticles Applications in Enhanced Oil Recovery," in *SPE North Africa Technical Conference and Exhibition*, Cairo, Egypt, 2015, pp. 19. https://doi.org/10.2118/175806-MS.
[24] A. Bila, J. Å. Stensen, and O. Torsæter, "Experimental Evaluation of Oil Recovery Mechanisms Using a Variety of Surface-Modified Silica Nanoparticles in the Injection Water," in *SPE Norway One Day Seminar*, Bergen, Norway, 2019, pp. 19. https://doi.org/10.2118/195638-MS.
[25] H. ShamsiJazeyi, C. A. Miller, M. S. Wong, J. M. Tour, and R. Verduzco, "Polymer-coated nanoparticles for enhanced oil recovery," *Journal of Applied Polymer Science,* vol. 131(15), 2014.
[26] A. O. Gbadamosi, R. Junin, M. A. Manan, N. Yekeen, A. Agi, and J. O. Oseh, "Recent advances and prospects in polymeric nanofluids application for enhanced oil recovery," *Journal of Industrial and Engineering Chemistry,* vol. 66(2018, pp. 1-19.
[27] N. K. Maurya and A. Mandal, "Studies on behavior of suspension of silica nanoparticle in aqueous polyacrylamide solution for application in enhanced oil recovery," *Petroleum Science and Technology,* vol. 34(5), 2016, pp. 429-436.
[28] R. Ponnapati, O. Karazincir, E. Dao, R. Ng, K. K. Mohanty, and R. Krishnamoorti, "Polymer-functionalized nanoparticles for improving waterflood sweep efficiency: Characterization and transport properties," *Industrial & Engineering Chemistry Research,* vol. 50(23), 2011, pp. 13030-13036.
[29] A. Behzadi and A. Mohammadi, "Environmentally responsive surface-modified silica nanoparticles for enhanced oil recovery," *Journal of Nanoparticle Research,* vol. 18(9), 2016, pp. 266.
[30] S. K. Choi, H. A. Son, H. T. Kim, and J. W. Kim, "Nanofluid enhanced oil recovery using hydrophobically associative zwitterionic polymer-

coated silica nanoparticles," *Energy & Fuels,* vol. 31(8), 2017, pp. 7777-7782.

[31] W. G. Anderson, "Wettability Literature Survey Part 5: The Effects of Wettability on Relative Permeability," *Journal of Petroleum Technology,* vol. 39(11), 1987, pp. 1453-1468.

[32] R. S. Seright, A. Campbell, P. Mozley, and P. Han, "Stability of Partially Hydrolyzed Polyacrylamides at Elevated Temperatures in the Absence of Divalent Cations," *SPE Journal,* vol. 15(02), 2010, pp. 341-348.

[33] C. A. McPhee and K. G. Arthur, "Relative permeability measurements: an inter-laboratory comparison," in, vol., no.: Society of Petroleum Engineers, pp.

[34] K. H. Hosseinzade and O. Torsæter, *Injected pore volume on lab and field scales* (9th International Conference on Porous Media & Annual Meeting). Rotterdam, Netherlands: International Society for Porous Media, Interpore, (2017).

[35] A. Habibi, M. Ahmadi, P. Pourafshary, s. Ayatollahi, and Y. Al-Wahaibi, "Reduction of Fines Migration by Nanofluids Injection: An Experimental Study," *SPE Journal,* vol. 18(02), 2012, pp. 309-318.

[36] P. Bacchin, Q. Derekx, D. Veyret, K. Glucina, and P. Moulin, "Clogging of microporous channels networks: role of connectivity and tortuosity," *Microfluidics and Nanofluidics,* vol. 17(1), 2014, pp. 85-96.

[37] K. J. Humphry, B. M. J. M. Suijkerbuijk, H. A. van der Linde, S. G. J. Pieterse, and S. K. Masalmeh, "Impact of Wettability on Residual Oil Saturation and Capillary Desaturation Curves," *Petrophysics,* vol. 55(04), 2014, pp. 313-318.

[38] J. J. Sheng, "Status of surfactant EOR technology," *Petroleum,* vol. 1(2), 2015, pp. 97-105.

[39] C. Dai *et al.*, "Spontaneous Imbibition Investigation of Self-Dispersing Silica Nanofluids for Enhanced Oil Recovery in Low-Permeability Cores," *Energy & Fuels,* vol. 31(3), 2017, pp. 2663-2668.

[40] X. Zhang, M. R. Servos, and J. Liu, "Ultrahigh Nanoparticle Stability against Salt, pH, and Solvent with Retained Surface Accessibility via Depletion Stabilization," *Journal of the American Chemical Society,* vol. 134(24), 2012, pp. 9910-9913.

[41] W. Kuang, S. Saraji, and M. Piri, "A systematic experimental investigation on the synergistic effects of aqueous nanofluids on interfacial properties and their implications for enhanced oil recovery," *Fuel,* vol. 220(2018, pp. 849-870.

[42] V. N. Paunov, B. P. Binks, and N. P. Ashby, "Adsorption of charged colloid particles to charged liquid surfaces," *Langmuir,* vol. 18(18), 2002, pp. 6946-6955.

[43] M. Mohammed and T. Babadagli, "Wettability alteration: A comprehensive review of materials/methods and testing the selected ones on heavy-oil containing oil-wet systems," *Advances in colloid and interface science,* vol. 220(2015, pp. 54-77.

[44] D. T. Wasan and A. D. Nikolov, "Spreading of nanofluids on solids," *Nature,* vol. 423(6936), 2003, pp. 156-159.

[45] S. Li and O. Torsaeter, "The Impact of Nanoparticles Adsorption and Transport on Wettability Alteration of Intermediate Wet Berea Sandstone," in *SPE Middle East Unconventional Resources Conference and Exhibition*, Muscat, Oman, 2015, pp. 14. https://doi.org/10.2118/172943-MS.

[46] D. A. DiCarlo, B. Aminzadeh, M. Roberts, D. H. Chung, S. L. Bryant, and C. Huh, "Mobility control through spontaneous formation of nanoparticle stabilized emulsions," *Geophysical Research Letters,* vol. 38(24), 2011.

[47] E. Rodriguez Pin, M. Roberts, H. Yu, C. Huh, and S. L. Bryant, "Enhanced Migration of Surface-Treated Nanoparticles in Sedimentary Rocks," in *SPE Annual Technical Conference and Exhibition*, New Orleans, Louisiana, 2009, pp. 21. https://doi.org/10.2118/124418-MS.

 https://doi.org/10.1051/e3sconf/202014604001

Permeability Prediction using multivariant structural regression

Matthew Andrew[1,*]

[1]Carl Zeiss Microscopy, 4385 Hopyard Rd., Pleasanton, CA, 94588, USA

Abstract. A novel method for permeability prediction is presented using multivariant structural regression. A machine learning based model is trained using a large number (2,190, extrapolated to 219,000) of synthetic datasets constructed using a variety of object-based techniques. Permeability, calculated on each of these networks using traditional digital rock approaches, was used as a target function for a multivariant description of the pore network structure, created from the statistics of a discrete description of grains, pores and throats, generated through image analysis. A regression model was created using an Extra-Trees method with an error of <4% on the target set. This model was then validated using a composite series of data created both from proprietary datasets of carbonate and sandstone samples and open source data available from the Digital Rocks Portal (www.digitalrocksporta.org) with a Root Mean Square Fractional Error of <25%. Such an approach has wide applicability to problems of heterogeneity and scale in pore scale analysis of porous media, particularly as it has the potential of being applicable on 2D as well as 3D data.

1 Introduction

Flow and transport in porous media are fundamentally rooted at the scale of the tiny tortuous pore pathways through which the flow takes place. As such, pore scale investigation is now a widely adopted tool across a range of disciplines associated with the examination of flow and transport. Such an approach has applications ranging from understanding the flow properties of porous ceramics (used as a catalytic substrate for vehicle emission reduction) [1], [2], examining battery electrolyte exchange [3], [4], characterizing geological formations for the purpose of understanding groundwater flow [5], carbon capture and storage [6]–[8] and (in its most industrially applied application) oil and gas recovery [9]–[12]. One of the most useful results from a holistic pore-scale characterization is the prediction of effective medium properties, such as permeability [13], relative permeability [14], capillary pressure curves [15], and effective acoustic and electrical properties [16], [17]. Traditional approaches to the prediction of effective properties from porous media (e.g. permeability, diffusivity or effective conductivity) focus on coupling the 3D structural imaging of porous media with the full physical simulation of the partial differential equations governing the property of interest [13], [18]. Once the domain and physics have been defined (with appropriate boundary conditions), the physics then converges over multiple iterations. Such an approach is contrasted to more traditional (legacy) approaches for effective property estimation, such as Kozeny-Carman or Kuwabara [19], [20] techniques which feed a relatively limited and difficult to measure set of structural properties into quasi-analytical models to make flow estimations. These are typically inaccurate (e.g. [21]), may require significant (and arbitrary) correction factors and are challenging to effectively apply to real problems [22]. For example, different variants of the Kozeny-Carman equation require the knowledge of characteristic particle size, tortuosity or effective channel diameter – something typically ambiguous to measure in a real system of arbitrary pore network type.

The past 10 years have seen an explosion in the availability of open source machine learning and computer vision tools, enabling both in-depth multivariant classification and analysis of porous media structure [23], [24] and powerful new regression techniques, facilitating the usage of such multivariant statistics for effective prediction [25]. The application of these tools and techniques to flow and transport in porous media has been limited. In this study these techniques are used to present a new technique for prediction of Stokes flow (viscous dominated) permeability from pore-scale images of rocks using multivariant statistical regression coupled with extensive synthetic geometry generation and pore-scale computational simulation. Pore structures were parameterized using 34 distinct statistical parameters, derived from specific discretized structural properties. This multivariant model is validated on an extensive open source digital rock dataset, comprising carbonate and sandstone pore geometries, giving an average prediction error of around 25% for predicted permeability values ranging over 6 orders of magnitude (10μD to 10D).

2 Materials and Methods

One of the biggest challenges in the application of machine learning based computer vision techniques to

* Corresponding author: matthew.andrew@zeiss.com

digital rock analysis is that of data. Pore networks are complex, and the range of tools available for characterizing them is broad, so any multivariant characterization and prediction will require many datasets (potentially many thousands or millions) to effectively train while avoiding model overfitting, a difficult feat when each 3D image may take several hours to acquire. To overcome this obstacle, synthetic pore network generation techniques were used to generate a more extensive training set than would be available for real (imaged) structures. Early digital rock studies used synthetic pore space generation extensively to examine simple rock systems at the pore scale. While such an approach has been largely supplanted by direct imaging, synthetic techniques do present specific advantages, especially when examining mechanisms behind various processes while controlling the amount of heterogeneity [26], [27], or when trying to develop an holistic understanding of systems which are extremely challenging to characterize experimentally, such as the organic hosted pore networks present in unconventional shale reservoirs [28]. Synthetic pore spaces can either be constructed physically, usually by glass beads or etchings in glass (e.g. [29], [30]) or numerically using a pore space generator, using stochastic or object-based techniques, subject to with various constraints (e.g. [31]–[33]).

Two thousand, one hundred and ninety (2,190) synthetic intergranular pore networks were generated by modelling grains as convex polyhedral [34], randomly placed within the pore space under a range of packing conditions and statistical descriptions. Polyhedra were defined by picking vertices around a bounding ellipse, defined by 3 axes (L_1, L_2 & L_3, such that $L_1 < L_2 < L_3$). All geometries were given a nominal voxel size of 1μm – a single nominal voxel size was used as the resulting permeability simulation results were then scaled across multiple nominal voxel sizes using the Hagen-Poisseuille equation as described below.

To allow for a range of granular aspect ratios the axis lengths for each grain were picked from a normal statistical distribution, each defined by an average and standard deviation (μ_1, μ_2, μ_3 and σ_1, σ_2, σ_3). For simplicity the same distribution was used for the two shortest axes ($\mu_1 = \mu_2$ and $\sigma_1 = \sigma_2$). The ratio $L_3/L_{1,2}$ (defining average grain aspect ratio) was randomly defined at the beginning, being a uniformly distributed random variable between 1 and 3. Similarly target porosity (φ) was a uniformly distributed random variable in the range $0.05 < \varphi < 0.7$, as were $\mu_{1,2}$ (in the range 15-60 voxels). σ_1 and σ_2 were determined using a coupled random variable in the range 5-35 voxels, and σ_3 an independently determined variable in the same range. Multiple geometric packing strategies were used, including (in order of increasing computational expense) allowing grains to overlap [28], removing grain overlap and the explicit physical modelling of gravitational settling. The entire range of target porosities could be effectively modelled when allowing grain overlap; however when using the other techniques (which achieved higher grain packing densities than when allowing overlap) once the maximum packing density was reached grain networks were eroded and dilated to extend the networks over a wider array of porosities. Computationally cheaper techniques allowed for more complex statistical descriptions (along with larger total geometries) to be used, including multi size component grain distributions. A set of various different synthetic images are shown in figure 1.

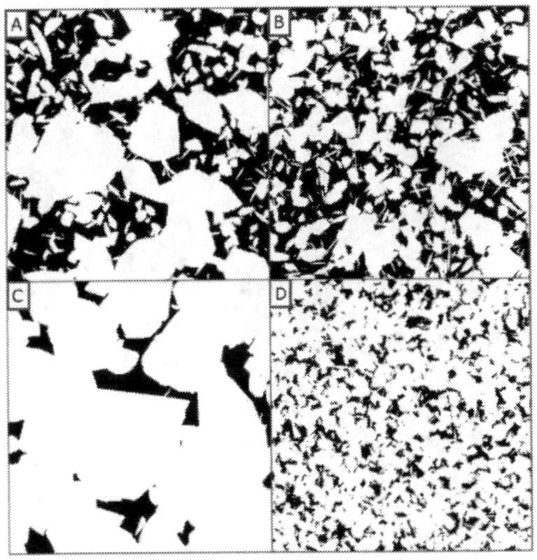

Figure 1 - A) A dual grain size system, with one grain population having a largest axis of 25μm and one population having a largest axis of 43μm. B) A dual grain size system, with one grain population having a largest axis of 22μm and one population having a largest axis of 35μm. C) A dual grain size system, with one grain population having a largest axis of 56μm and one population having a largest axis of 86μm. D) A single grain size system with a largest axis length of 19μm.

Structural statistics were calculated on a range of 2D slices through each 3D volume. It was decided to make measurements in 2D rather than on the full 3D network as it was more computationally tractable (and trivially parallelizable) and allows for the same model to be applied on 3D networks (acquired with XRM and FIB-SEM) and 2D structures (such as those characterized using light or electron microscopy). The application of these techniques to 2D systems is an interesting target of future work, but beyond the scope of this paper. Inherent biases and errors associated with 2D rather than 3D measurements are inherently accounted for in the relative weightings of different measurements during statistical multivariant regression. Grains and pores were then separated using a watershed algorithm operating across an Euclidian distance map [35], [36], with object seeds filtered using the "h-maxima" algorithm (figure 2). While such a technique has its limitations when dealing with complex network geometries [37], relatively simple networks can be effectively characterized [38]. Perhaps more important was that a perfect granular separation was not required, just that any object separation errors were consistent between training and validation series of data. For example, differential errors in grain separation for systems with large grains (relative to the voxel size) relative to systems with smaller grains (relative to the voxel size) are implicitly accounted for by the large array of different grain sizes in the training set. The same errors will be present in both the training and the final (imaged) datasets, and therefore the impact of any errors on prediction will be minimized.

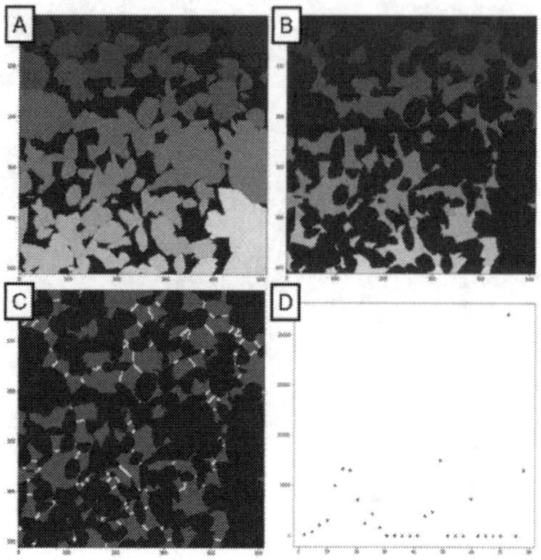

Figure 2 - Feature vector extraction. A) Grain separation. B) Pore separation. C) Throat identification (throats shown in yellow). D) Grain size distribution, showing a multimodal population distribution.

Once grains and pores were separated, population statistics were computed on features from these different populations, including grain, pore and pore throat sizes, shape averages, distributions, and network connectivity. This created a large (34 component) feature vector set for each network. The range of structural feature vectors (as well as their relative importance is shown in figure 3. The relative feature importance is defined as fraction of the decision tree associated with the permeability prediction decision tree [39], or the number of nodes in the decision graph associated with that feature. The 21 most important features had feature importances larger than 10% of the most important feature vector in the set (the standard deviation in the inscribed throat radius). Such a broad distribution in importances is an indication of both the complexity of the problem and the power of such a multivariant approach – a large number of different structural metrics are require to achieve a high quality prediction. An extensive evaluation of specific feature impacts in such a complex multivariant model, as well as an examination of the impact of changing numbers of input parameters and parameter selection on model performance would be an interesting target for future work. The requirement of the use of a large number of network characteristics is unsurprising, particularly given the requirement to make predictions using only features extracted in 2D. The goal for this approach is ultimately to generate a model which can be applied on 2D data (for example light or electron microscopy), and a greater exploration of this is the target for future work. It is likely that, should features be extracted in 3D, a more accurate prediction could be made, potentially with a smaller number of extractable features, but this model would not have the potential of being (ultimately) transferred to 2D analytical techniques.

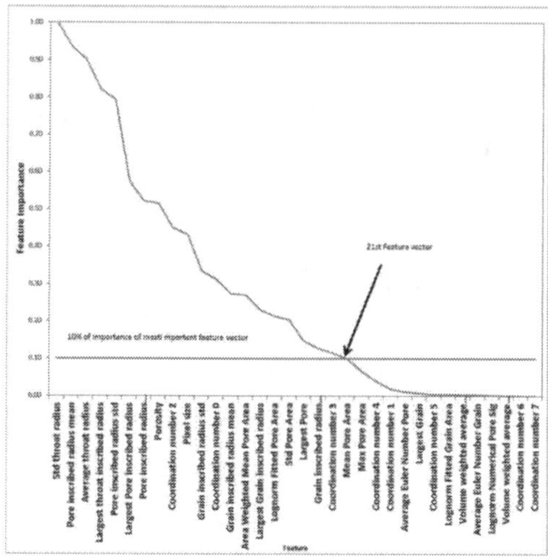

Figure 3 - distribution of feature vector importances.

It is interesting that the most important parameter (the parameter which is responsible for the greatest number of nodes in the forest of decision trees) is that of the standard deviation in inscribed pore throat radii. This corresponds to whether or not the porous medium is homogenous (has a uniform fluid velocity distribution) or heterogeneous (has a more channelized fluid flow distribution). As the flow rate is non-linearly sensitive to changes in throat size (equation 1, below) it is not surprising that the distribution in throat size has a strong control on flow rate. Similarly it is unsurprising that the second and third most important features are the average inscribed pore and pore throat radii respectively.

Flow was then simulated on each of these synthetic volumes using traditional digital rock techniques [40], giving the permeability for each network. Each of the synthetic networks had a nominal voxel size of 1μm; however, a much broader range of pore network statistics (with associated permeabilities) could be generated by examining analytical solutions to the Navier-Stokes equation within the viscous dominated regime (such as the Hagen-Poiseuille equation, equation 1)

$$\frac{\Delta P}{\Delta L} = \frac{8\mu}{r^2} v \qquad (1)$$

where $\Delta P/\Delta L$ is the pressure gradient, μ is the viscosity, r the pipe radius and v the (linear) fluid average velocity. For any (defined) structure, a Stokes flow solution to the Navier-Stokes equation will predictably scale as the square of the spatial length scale. As such any change in nominal voxel size (for a topologically and topographically defined network) should change permeability values similarly. An arbitrarily extended set of network statistics could therefore be created, scaling each (dimensioned) measure appropriately with the voxel size, and the permeability values with its square. An extended set of 219,000 network statistics were created using this technique, with nominal voxel sizes ranging from 10nm to 10um and predicted permeability values varying over 10 orders of magnitude (from from 10^{-25}m^{-2} to 10^{-7}m^{-2}, with a 10-90 percentile range of 1.2×10^{-18}m^{-2} – 9.25×10^{-12}m^{-2}) (figure 4).

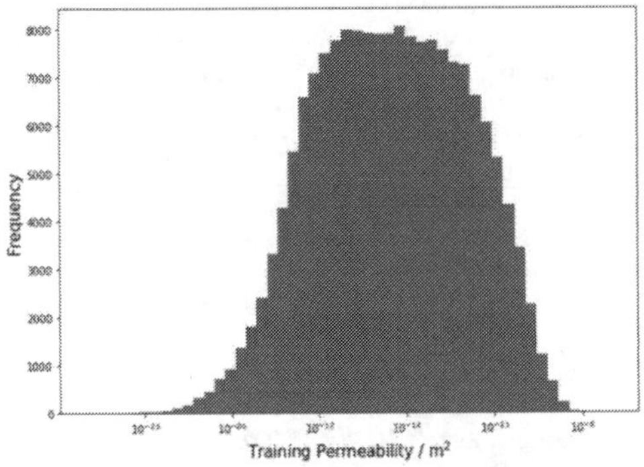

Figure 4 - distribution of training permeability values.

At first inspection it seems like such an approach adds little to the overall training dataset – they are just generated by extrapolation. The key to understanding its utility is that the regression is designed to be performed on real (measured) datasets with real pixel sizes. The same network will have different permeability values if you change its nominal voxel size, and any model must be trained on volumes spanning the entire range of pixel sizes of interest. One way to extend the pixel size range of the training set is to randomly assign pixel sizes, however this will decrease the training dataset density at any specific spatial length scale. Instead Hagen-Poiseuille extrapolation was used as it allows for all the training datasets to be available to the training algorithm at all spatial lengthscales.

3 Results and discussion

This extended dataset of multivariant structural statistics was split into "training" and "testing" sets (with 80% of the data belonging to the training set, and 20% belonging to the test set). The training set was then regressed against the natural logarithm of the permeability using open-source machine learning tools (www.scikit-learn.org) using a randomly seeded Extra-Trees (decision tree based) ensemble algorithm [41]. Total training time on a 40-core workstation was around 12 seconds. A natural logarithmic representation of permeability was used for regression as it allowed prediction errors to be evenly distributed across the entire range of permeability – a linear representation would have concentrated prediction errors strongly towards the largest permeabilities, overfitting these values and underfitting the rest of the dataset. Another way of viewing this is that a logarithmic representation essentially forces the algorithm to minimize fractional (rather than absolute) errors in the regression model. This model was then tested on the test set, finding a Root-Mean-Square-Fractional-Error (RMSFE) of <4% (figure 5).

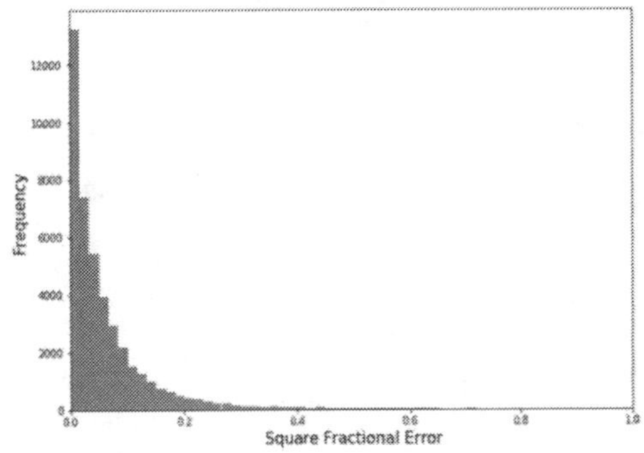

Figure 5 - Root Mean Square Fractional Error (RMSFE) distribution

One major question when applying this model (trained on synthetic networks) to real datasets is whether the synthetic images really represent real imaged pore networks. A composite set of real digital rock images to validate the predictive model for more general use. This composite set of 36 images consisted of a range of open source volumes taken from the Digital Rock Portal (www.digitalrocksportal.org), previously published data, and new nano-CT datasets from micritic carbonate microporosity. This constituted two qualitatively different pore networks (the intergranular pore networks common in sandstones and micritic microporosity common in dual porosity carbonate reservoirs. While there are significant differences in chemistry and scale for the networks, in both cases the grain networks consist of euhedral to sub-hedral sub-angular clasts, so they may have a similar relationship between pore structure and flow properties and have grains which are well represented by convex polyhedra.

The unsegmented datasets were segmented using state-of-the-art machine learning based segmentation [24] (Zen Intellesis). Flow was then simulated on each of the segmented datasets, with predicted permeabilities ranging over 6 orders of magnitude (10µD to 10D). These predictions were then compared with multivariant predictions with a RMSFE of <25% across the entire range (figure 6).

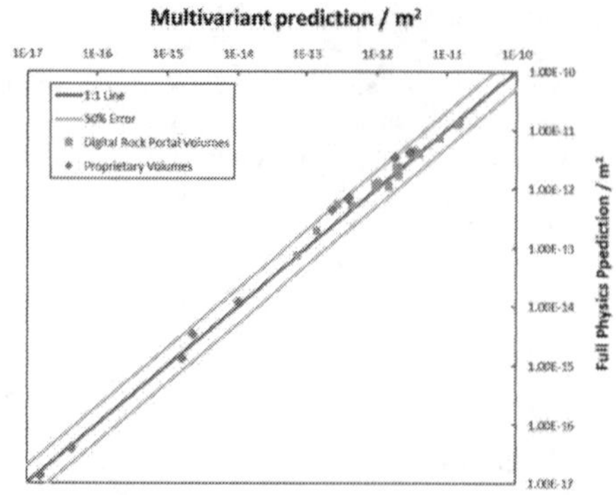

Figure 6 - correlation between simulated permeabilities and permeability predicted using multivariant regression.

Such a result is extremely encouraging – while the prediction error was higher in the (real) validation datasets than the (synthetic) test datasets, the prediction error was still relatively small. Predictions were accurate across multiple pore network types (both intergranular sandstone pore networks and micritic microporosity), and a wide range of length scales and permeability values.

This technique offers multiple advantages over both traditional (full physics) simulations and legacy estimation techniques. First, while analysing and simulating flow on the thousands of training images was computationally expensive, applying it to each imaged dataset was relatively cheap. As such it is significantly faster to apply than traditional techniques, with a large volume prediction taking less than a minute on standard lab computational resources, where a full physics simulation of the same result might take many hours. Perhaps the most important advantage of this technique, however, is it has the potential of greatly expanding the forms of data applicable to quantitative flow prediction, as the features used for both regression and prediction can be extracted in 2D as well as 3D. This allows for effective property prediction from structural analysis of 2D data (e.g. light or electron microscopy), which can be acquired over much larger areas (allowing a much better characterization of structural heterogeneity), can be acquired much faster, and offers qualitatively richer data (giving information about mineralogy, geochemistry or texture not present when using 3D imaging techniques).

A significant drawback to traditional techniques for permeability estimation (like Kozeny-Carman) is the challenging and arbitrary nature of the measurement of their constituent parameters [42]. A significant advantage of multivariant regression-based techniques is that all inherent errors in measurement are included when performing the regression - the minimization of the prediction error implicitly includes the minimization of the impact of any systematic measurement errors. The general approach for the prediction of effective medium properties based on multivariant structural statistics can be extended to other systems through the incorporation of either other synthetic pore network generation routines or re-enforcement learning with real (imaged) data. There is also significant scope for future development in the extraction of structural statistics, including the usage of multi-point statistical characterization [43] or recently developed persistent homology metrics [44]. Future work will focus on the examination of the usage of such large area predictive statistical techniques to validate and inform upscaling techniques, allowing for integration of multiple scales of data in single predictive models.

4 Conclusions

A novel method for permeability prediction using multivariant statistical regression is presented. Over 2,000 synthetic geometries were created by modelling granular media using convex polyhedral. Full physical simulations were performed on all of these datasets. The resulting permeability values were used as a target function for a high dimensional description of the volumes, created from structural statistics of rock grains, pores and throats. The regression model was trained with an error of less than 4% in the test data. Validation of this model was performed on a composite series of 36 independent datasets , taken from both an open source data source (www.digitalrocksportal.org) and proprietary data from carbonate microporosity. These datasets range in pore system from carbonate microporosity to intergranular sandstone pores, with pixel sizes ranging from 10µm to 32nm and permeabilities ranging over 6 orders of magnitude from 10µD to 10D. Predictions in this validation set showed a RMSFE of less than 25% over the entire range.

Differences in error between the validation and test data provide an interesting insight into the validity of the synthetic datasets in training multivariant models – while they do not perfectly represent real (imaged) data, they are close enough to accurately predict data across length scales. Interesting future work could focus on expanding the training dataset to include networks representative of other pore systems, either through an expansion of synthetic generation techniques or through the incorporation of real imaged geometries.

Such an approach to permeability prediction is extremely promising for addressing many problems associated with pore scale analysis of geological materials, particularly those associated with scale and heterogeneity. By reducing the relative computational complexity of simulation larger geometries can be efficiently analysed. Perhaps more exciting is the application of these tools (trained on 2D slices through 3D volumes) to other pore structural analytical technologies operating in 2D. These techniques are able to address much larger linear length scales, thereby enabling for a much more effective assessment of structural heterogeneity. They also allow for the incorporation of analytical information not accessible in 3D.

Acknowledgements

I thank ZEISS microscopy and Math2Market Gmbh. for the equipment and software access required for this work. Many thanks to Dr. Hannah Menke for assistance in the preparation of this manuscript.

References

[1] G. Bruno, A. Efremov, B. Wheaton, I. Bobrikov, V. G. Simkin, and S. Misture, "Micro- and macroscopic thermal expansion of stabilized aluminum titanate," *J. Eur. Ceram. Soc.*, 2010.

[2] G. Bruno *et al.*, "On the stress-free lattice expansion of porous cordierite," *Acta Mater.*, 2010.

[3] P. R. Shearing, L. E. Howard, P. S. Jørgensen, N. P. Brandon, and S. J. Harris, "Characterization of the 3-dimensional microstructure of a graphite negative electrode from a Li-ion battery," *Electrochem. commun.*, 2010.

[4] D. Kehrwald, P. R. Shearing, N. P. Brandon, P. K. Sinha, and S. J. Harris, "Local Tortuosity Inhomogeneities in a Lithium Battery Composite Electrode," *J. Electrochem. Soc.*, 2011.

[5] R. Dann, M. Turner, M. Close, and M. Knackstedt, "Multi-scale characterisation of coastal sand aquifer media for contaminant transport using X-ray computed tomography," *Environ. Earth Sci.*, 2011.

[6] H. P. Menke, M. G. Andrew, M. J. Blunt, and B. Bijeljic, "Reservoir condition imaging of reactive transport in heterogeneous carbonates using fast synchrotron tomography - Effect of initial pore structure and flow conditions," *Chem. Geol.*, vol. 428, pp. 15–26, 2016.

[7] M. Andrew, B. Bijeljic, and M. J. Blunt, "Pore-by-pore capillary pressure measurements using X-ray microtomography at reservoir conditions: Curvature, snap-off, and remobilization of residual CO2," *Water Resour. Res.*, vol. 50, no. 11, pp. 8760–8774, 2014.

[8] S. Iglauer, S. Favretto, G. Spinelli, G. Schena, and M. J. Blunt, "X-ray tomography measurements of power-law cluster size distributions for the nonwetting phase in sandstones," *Phys. Rev. E - Stat. Nonlinear, Soft Matter Phys.*, vol. 82, pp. 10–12, 2010.

[9] M. J. Blunt *et al.*, "Pore-scale imaging and modelling," *Adv. Water Resour.*, vol. 51, pp. 197–216, 2013.

[10] a. V. Ryazanov, K. S. Sorbie, and M. I. J. van Dijke, "Structure of residual oil as a function of wettability using pore-network modelling," *Adv. Water Resour.*, vol. 63, pp. 11–21, 2014.

[11] J. Schmatz, J. L. Urai, S. Berg, and H. Ott, "Nano-scale imaging of pore-scale fluid-fluid-solid contacts in sandstone," *Geophys. Res. Lett.*, p. n/a-n/a, 2015.

[12] D. Silin, L. Tomutsa, S. M. Benson, and T. W. Patzek, "Microtomography and Pore-Scale Modeling of Two-Phase Fluid Distribution," *Transp. Porous Media*, vol. 86, pp. 495–515, 2011.

[13] H. Andrä *et al.*, "Digital rock physics benchmarks-part II: Computing effective properties," *Comput. Geosci.*, vol. 50, pp. 33–43, 2013.

[14] T. Ramstad, N. Idowu, C. Nardi, and P. E. Øren, "Relative Permeability Calculations from Two-Phase Flow Simulations Directly on Digital Images of Porous Rocks," *Transp. Porous Media*, 2012.

[15] I. Shikhov and C. H. Arns, "Evaluation of Capillary Pressure Methods via Digital Rock Simulations," *Transp. Porous Media*, 2015.

[16] M. A. Knackstedt, S. Latham, M. Madadi, A. Sheppard, T. Varslot, and C. Arns, "Digital rock physics: 3D imaging of core material and correlations to acoustic and flow properties," *Lead. Edge*, 2009.

[17] Y. Keehm, T. Mukerji, and A. Nur, "Computational rock physics at the pore scale: Transport properties and diagenesis in realistic pore geometries," *Lead. Edge*, 2002.

[18] H. Andrä *et al.*, "Digital rock physics benchmarks-Part I: Imaging and segmentation," *Comput. Geosci.*, vol. 50, pp. 25–32, 2013.

[19] P. C. Carman, "Fluid flow through granular beds," *Trans. Inst. Chem. Eng.*, vol. 15, pp. 150–166, 1937.

[20] S. Kuwabara, "The Forces experienced by Randomly Distributed Parallel Circular Cylinders or Spheres in a Viscous Flow at Small Reynolds Numbers," *J. Phys. Soc. Japan*, 1959.

[21] S. Mauran, L. Rigaud, and O. Coudevylle, "Application of the carman-kotenzy correlation to a high-porosity and anisotropic consolidated medium: The compressed expanded natural graphite," *Transp. Porous Media*, 2001.

[22] P. N. Singh and W. W. Wallender, "Effects of Adsorbed Water Layer in Predicting Saturated Hydraulic Conductivity for Clays with Kozeny–Carman Equation," *J. Geotech. Geoenvironmental Eng.*, 2008.

[23] L. Mosser, O. Dubrule, and M. J. Blunt, "Reconstruction of three-dimensional porous media using generative adversarial neural networks," *Phys. Rev. E*, 2017.

[24] M. Andrew, "A quantified study of segmentation techniques on synthetic geological XRM and FIB-SEM images," *Comput. Geosci.*, vol. 22, no. 6, pp. 1503–1512, 2018.

[25] O. Sudakov, E. Burnaev, and D. Koroteev, "Driving digital rock towards machine learning: Predicting permeability with gradient boosting and deep neural networks," *Comput. Geosci.*, vol. 127, pp. 91–98, 2019.

[26] J. Bear, "Dynamics of Fluids in Porous Media," *Soil Sci.*, 2006.

[27] P. S. Ringrose, a. W. Martinius, and J. Alvestad, "Multiscale geological reservoir modelling in practice," *Geol. Soc. London, Spec. Publ.*, vol. 309, no. 1, pp. 123–134, 2008.

[28] M. Andrew, "Comparing organic-hosted and intergranular pore networks: topography and topology in grains, gaps and bubbles," *Geol. Soc. London, Spec. Publ.*, vol. 484, pp. SP484--4, 2018.

[29] T. Phenrat, H. J. Kim, F. Fagerlund, T. Illangasekare, R. D. Tilton, and G. V. Lowry, "Particle size distribution, concentration, and magnetic attraction affect transport of polymer-modified Fe0 nanoparticles in sand columns," *Environ. Sci. Technol.*, 2009.

[30] F. De Smedt and P. J. Wierenga, "Solute Transfer Through Columns of Glass Beads," *Water Resour. Res.*, 1984.

[31] P. E. Øren and S. Bakke, "Process based reconstruction of sandstones and prediction of transport properties," *Transp. Porous Media*, vol. 46, no. 2–3, pp. 311–343, 2002.

[32] S. Bakke and P.-E. Øren, "3-D Pore-Scale Modelling of Sandstones and Flow Simulations in the Pore Networks," *SPE J.*, vol. 2, no. 02, pp. 136–149, 1997.

[33] T. Lerdahl, P. E. Oren, and S. Bakke, "A predictive network model for three-phase flow in porous media," *SPE/DOE Improv. Oil Recover. Symp.*, 2000.

[34] P. A. Cundall, "Formulation of a three-dimensional distinct element model-Part I. A scheme to detect and represent contacts in a system composed of many polyhedral blocks," *Int. J. Rock Mech. Min. Sci.*, 1988.

[35] A. Rabbani, S. Jamshidi, and S. Salehi, "An automated simple algorithm for realistic pore network extraction from micro-tomography images," *J. Pet. Sci. Eng.*, vol. 123, pp. 164–171, 2014.

[36] M. G. Andrew, H. P. Menke, M. J. Blunt, and B. Bijeljic, "The Imaging of Dynamic Multiphase Fluid Flow Using Synchrotron-Based X-ray Microtomography at Reservoir Conditions," *Transp. Porous Media*, vol. 110, pp. 1–24, 2015.

[37] A. Q. Raeini, B. Bijeljic, and M. J. Blunt, "Generalized network modeling: Network extraction as a coarse-scale discretization of the void space of porous media," *Phys. Rev. E*, vol. 96, no. 1, 2017.

[38] H. Dong and M. J. Blunt, "Pore-network extraction from micro-computerized-tomography images," *Phys. Rev. E - Stat. Nonlinear, Soft Matter Phys.*, vol. 80, no. 3, 2009.

[39] Gilles Louppe, "Understanding Random Forests: From Theory to Practice," 2014.

[40] S. Linden, A. Wiegmann, and H. Hagen, "The LIR space partitioning system applied to the Stokes equations," *Graph. Models*, vol. 82, pp. 58–66, 2015.

[41] P. Geurts, D. Ernst, and L. Wehenkel, "Extremely randomized trees," *Mach. Learn.*, 2006.

[42] H. P. Menke, Y. Gao, S. Linden, and M. G. Andrew, "Using nano-XRM and high-contrast imaging to inform micro-porosity permeability during Stokes-Brinkman single and two-phase flow simulations on micro-CT images," *Adv. Water Resour.*

[43] H. Okabe and M. J. Blunt, "Prediction of permeability for porous media reconstructed using multiple-point statistics," *Phys. Rev. E - Stat. Physics, Plasmas, Fluids, Relat. Interdiscip. Top.*, 2004.

[44] A. L. Herring, V. Robins, and A. P. Sheppard, "Topological Persistence for Relating Microstructure and Capillary Fluid Trapping in Sandstones," *Water Resour. Res.*, 2019.

A fast FFT method for 3D pore-scale rock-typing of heterogeneous rock samples via Minkowski functionals and hydraulic attributes

Han Jiang[1] and *Christoph H.* Arns[1,*]

[1]UNSW, School of Minerals & Energy Resources Engineering, Sydney 2052 NSW, Australia

Abstract. The integration of numerical simulation and physical measurements, e.g. digital and conventional core analysis, requires the consideration of significant sample sizes when heterogeneous core samples are considered. In such case a hierarchical upscaling of properties may be achieved through a workflow of partitioning the sample into homogeneous regions followed by characterization of these homogeneous regions and upscaling of properties. Examples of such heterogeneities are e.g. fine laminations in core samples or different micro-porosity types as consequence of source rock components and diagenesis. In this work we utilize regional measures based on the Minkowski functionals as well as local saturation information derived through a morphological capillary drainage transform as a basis for such a classification/partitioning. An important consideration is the size of the measurement elements utilized, which could be considerable in the case of larger heterogeneities; in such case the calculation of the regional measures can be computationally very expensive. Here we introduce an FFT approach to calculate these measures locally, utilizing their additivity. The algorithms are compared against direct summation techniques and shift-overlap approaches for a selection of different averaging supports to illustrate their speed and practical applicability. We consider a range of artificial Boolean models to illustrate the effect of including hydraulic information on the resulting classifications scheme. This allows the determination of bias, since for these model systems local classes are known ab-initio. The classification framework is tested by comparing to the known initial micro-structure distribution and relative bias quantified in terms of choice of averaging elements (size and shape). Importantly, depending on the actual morphological transition between micro-type partitions, partitions including hydraulic attributes differ from pure morphological partitions with applications to electrofacies and hydraulic unit definitions.

1 Introduction

Rock type classification is an important topic in reservoir characterization and spans a large range of scales from seismic interpretation (meter scale [1]) to well logging (down to centimeter scale [2]) and further down in scale to core analysis. It forms part of an upscaling strategy involving the correlation of classes of different type, purpose, or origin; e.g. lithofacies vs. electrofacies (distinguishable classes based on well-logs). However, reservoir heterogeneity extends to even smaller scales.

The rapid development of (micro-) X-ray CT techniques exposes rock heterogeneity down to the order of microns, at the pore-scale, in 3D. While the calculation of petrophysical properties and correlations has been demonstrated for a range of properties including e.g. electrical conductivity, elastic moduli, mercury intrusion capillary pressure, and permeability [3]-[7], this is for homogeneous sandstones of limited clay content. For more complex carbonates cross-variograms between porosity and permeability and multiple imaging length scales [8] may be utilized. Often one may however want to utilize a hierarchical upscaling approach based on core-scale rock-types to integrate several length scales and multiple physical properties, including for thinly laminated sandstones. While petrophysical rock-types may include such laminations and thus anisotropy caused, describing fluid flow leads to a homogenization problem which can be approached at the Darcy scale [9], or numerically through partial homogenization [10] by solving the Brinkmann equation. Either way, a classification technique for scale separation is required.

Classification techniques targeting petrophysical interpretation are not new. Archie developed a method of skeleton classification with a few petrophysical parameters in [11]. Skalinski and his colleagues used pore throat distributions as the main parameter for pore type classification with mercury injection pressure measurement data of many rock samples [12]. Characteristics of fractures and vugs of carbonate rocks were extracted from micro-CT images in [13]. Dernaika et al. [14] calculated porosity and mineralogy of core samples from advanced dual energy XCT imaging and linked to plug data after upscaling.

Schmitt and colleagues considered pore shapes (length, width, thickness, orientation) in reservoir rock from 3-D X-ray micro-CT images [15], limiting themselves to large pores due to the higher resolution

* Corresponding author: c.arns@unsw.edu.au

requirements for deriving these measures. Alternately, one may characterize microstructure utilizing the additive morphological measures volume, surface area, mean and total curvature, namely the Minkowski functionals [16], which can be evaluated easily on a segmented tomogram [17], [18] and have been used in the past as sensitive descriptors of morphology [17]-[22]. The strong correlation of physical properties of Boolean models to morphological properties, the Minkowski functionals [23],[24], motivates the use in rock-typing approaches. This was done in [25] for a thinly laminated sandstone using a 1D sliding measuring window approach and in 3D using non-overlapping regions of the image. A high-resolution field of morphological regional measure could not be computed due to computational cost. In [26] a support shift technique was utilized for the computations, which however still does not scale to large supports, as we require in this work. We define as support of a measurement a small (regional) volume within the sample, and over which an average is taken. A larger support is desirable to reduce the standard deviation of such a measure.

Another relationship often used in rock-typing approaches is given by mercury intrusion capillary pressure (MICP) curves [27]. In terms of micro-CT approaches this type of information can be gained either by direct imaging approaches [28], or by modeling the MICP process numerically [29],[30]. Capillary pressure heterogeneity was also inferred from regional saturation distributions in [31], while [32] considered non-wetting phase cluster distributions.

The resulting question we target in this work is whether the inclusion of hydraulic information – taken to be local fluid saturation – changes rock-types defined on the basis of the Minkowski functionals alone as basis for petrophysical rock-types due to their strong relationship to physical properties. We utilize the decomposition of the Minkowski functionals into histograms over vertex configurations to implement (fast) calculations of the Minkowski measures over large support via Fast Fourier Transforms (FFTs). The technique is compared against more direct implementation schemes on a set of artificial microstructures and applied to the classification of laminated Boolean composites. Classifications based on the Minkowski measures alone are compared with ones including a saturation attribute and resulting bias is discussed.

2 Methodology

In this section we review the basic techniques utilized in this work. We introduce the Minkowski functionals and measures as a full set of additive measures of 3D space, regional measures based on these definitions over a finite support, as well as resulting curvature density fields and algorithms to calculate them. Finally, for demonstration of the algorithms we introduce a set of Boolean composites featuring forming layered media.

2.1 Minkowski functionals

In this work we follow the notation of [26]. Consider a body Y in $\mathbb{R}^3$ with sufficiently smooth surface ∂Y. Then the additive geometrical characteristics are volume $V(Y)$, surface area $S(Y)$, integral of mean curvature $M(Y)$ and integral of total curvature $K(Y)$ with surface integrals

$$\begin{aligned} S(Y) &= \int_{\partial Y} ds\,, \\ M(Y) &= \int_{\partial Y} \frac{1}{2}\Big(\frac{1}{r_1(s)} + \frac{1}{r_1(s)}\Big)\, ds\,, \\ K(Y) &= \int_{\partial Y} \frac{1}{r_1(s)\, r_2(s)}\, ds\,. \end{aligned} \tag{1}$$

Here $r_1(s)$ and $r_2(s)$ are the maximum and minimum curvature radii in s. The above characteristics are related to the intrinsic volumes via

$$\begin{aligned} V_0(Y) &= \frac{1}{4\pi} K(Y)\,, \\ V_1(Y) &= \frac{1}{\pi} M(Y)\,, \\ V_2(Y) &= \frac{1}{2} S(Y)\,, \\ V_3(Y) &= V(Y)\,. \end{aligned} \tag{2}$$

On the segmented tomogram the intrinsic volumes can be determined by evaluating the histogram of the individual vertex configurations. Excellent illustrations of this process have been given in [17], [33]. Here we consider the grain phase as the phase of interest; all other phases are merged for this calculation and the characteristic function of the structure takes the value of 1 for the granular phase and 0 otherwise. A vertex made up of eight voxels can then be in $2^8 = 256$ different configurations, where each configuration has a certain contribution to volume, surface area, mean, and total curvature. Lookup tables for these contributions has been given in [17], [33]. We use the 8-neighbourhood for 2D connectivity and 26-neighbourhood for 3D connectivity as given in [17]. For a recent review of the applications of integral geometry in porous media see [19].

2.2 Regional curvature measures

To define a measure over a finite volume we consider the average over regions of spherical and ellipsoidal shape. For a given position of this support in the actual tomographic images we then evaluate all vertices of the discrete tomogram, where the centers of the vertices are within the measurement window.

Consequently, the regional measures can then be derived from the local configuration histogram over the given spatial support. Evaluating the regional measures for a regular grid of support vectors will generate a set of morphological fields $V(\boldsymbol{r})$, $S(\boldsymbol{r})$, $M(\boldsymbol{r})$, $K(\boldsymbol{r})$. We report all measures as curvature densities, e.g. normalized by the averaging volume.

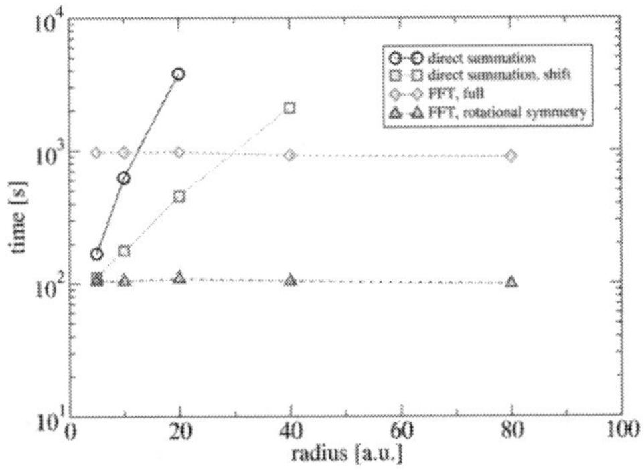

Fig. 1. Computational times for the Minkowski measure field generation for a 400^3 sub-volume of a segmented Bentheimer sandstone sample as function of radius of a spherical support (radius in voxel, single processor).

2.2.1 Curvature density fields by direct summation

The most direct approach for calculating the curvature density fields simply places the spatial support at a position-of-interest, evaluates the configuration histogram, and stores the result at the respective position, looping over the curvature measure grid. For dense strides of the support this leads to significant overlap of support volumes for adjacent evaluations; thus, large numbers of vertices are evaluated multiple times for large measurement support, e.g. large radii of a spherical support. This naïve approach therefore does not scale well with increasing support size, as it scales with support volume.

2.2.2 Curvature density fields by support-shift algorithms

A somewhat more efficient way recognizes this partial support overlap and only evaluates vertices anew which are added to the support volume after a shift on the lattice, e.g. stride along a lattice direction with a fixed step size; the same number of vertices will be removed and added from the histogram for each support shift when shifting in the same direction. This approach scales significantly better, and essentially scales with the surface "area" of the support for small strides; it takes direct advantage of the additivity of the Minkowski measures.

2.2.3 Curvature density fields by FFT methods

The additivity of the curvature measures allows a much more efficient approach to deriving regional measures. This approach utilizes a convolution of the support with the vertex configuration field, thus deriving the curvature density fields by a series of Fast Fourier Transforms (FFTs) and adding up the curvature contributions to each regional measure according to the lookup table of [17]. In this case the size of the support has essentially no influence on the speed of the algorithm. Very large support volumes and arbitrary support shapes can be utilized this way.

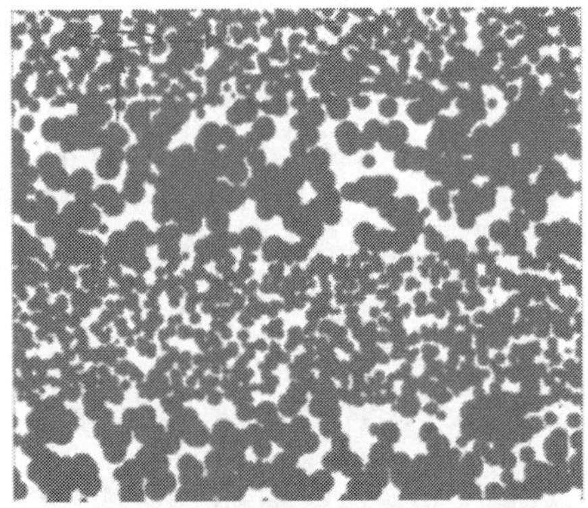

[a]

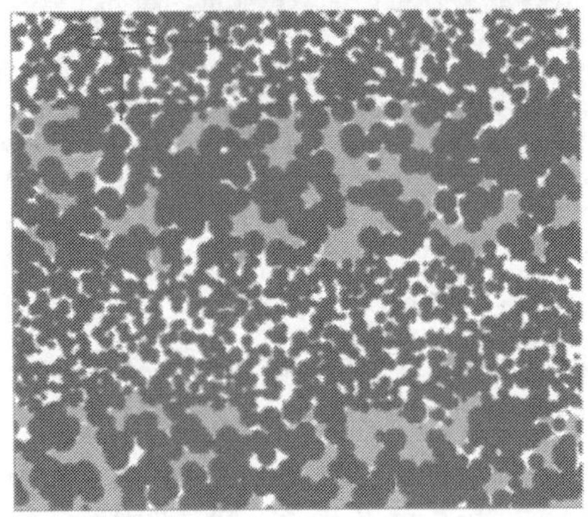

[b]

Fig. 2. 2D slice through a layered Boolean model with two generating sphere processes, one for each layer. [a] generated microstructure, [b] microstructure partially saturated by a non-wetting phase with invasion radius $r_i = 6.6$. Sphere radii are $r_1 = 13$ and $r_2 = 26$. The microstructure has dimensions of 1200 x 1000 x 800 voxels after removal of boundary regions in arbitrary units. Blue denotes the invading non-wetting phase.

A comparison of the relative performance of the algorithms is given in Fig. 1 for a sub-volume of a segmented tomogram of Bentheimer sandstone. Clearly, using large averaging volumes becomes very costly with the non-FFT based algorithms applied here, while the FFT approach scales well, e.g. is independent of averaging volume. A significant speedup of the FFT method can be achieved by utilizing the rotational symmetry of the possible vertex configurations [17]. For the FFT method we consider both spherical and ellipsoidal measure supports. In practice the size and shape of the measurement support may be chosen through structural analysis, utilizing variograms [26], and considering that the resulting rock-types should be compact, e.g. do not have too many holes. Due to additivity and the availability of parallel FFT libraries for various architectures the proposed technique easily scales to largest datasets.

2.3 Boolean model

We generate layered synthetic micro-structures by controlling the generating grains of a Boolean (or Poisson particle) process according to an initial (known) layered distribution of two rock-types: after choosing a **Table. 1**

Structural details of dry sample in Figure 2. The generating shape for both layers is spheres.

Formation	Thickness (voxel)	Porosity	Particle Radius (voxel)
Layer 1	300	0.242	13
Layer 2	300	0.244	26

random location to place a particle the particle shape and size is decided by the corresponding pre-defined rock-type: Similar microstructure modeling approaches were presented in [10], [25], [30] for a more general 3D case, which was not required here for the demonstration of the technique. Placing a spherical particle of radius r at a random location results in a decrease of porosity. Particles can overlap into adjoining rock-types and the particle placement process is stopped when both rock-types achieve their target porosity. A 2D slice through a 3D realization of a Boolean layered system with two different sphere sizes of $r_1 = 13$ voxels and $r_2 = 26$ voxels is given in Fig. 2a. The structure parameters are summarized in Table 1.

2.4 Simulated non-wetting fluid invasion

We model the invasion of the synthetic rock with a non-wetting fluid by utilizing the capillary drainage transform [7], [24] with a fixed saturation chosen to provide additional contrast between the different rock-types. This is achieved by selecting an invasion radius below the critical radius of percolation for the larger-scale porosity, but above the critical radius of percolation for the smaller-scale porosity. A 2D slice through the saturated medium at $S_w = 0.38$, corresponding to an invasion radius of $r_i = 6.6$ is given in Fig. 2b. The critical radii for the two sphere composites making up the layers are $r_{c1} = 4$ and $r_{c2} = 9$. Note that the original samples were 1600 x 1200 x 1600 voxels. We selected the inner 1200 x 1000 x 800 region for further analysis in order to avoid boundary effects resulting from simulated capillary pressure fluid distributions.

2.5 Rock-type classification

We carry out the classification into different rock-types on the regional curvature measures and saturation field. This is achieved by defining the centroids of a Gaussian Mixture Model (GMM) in a supervised way by selecting the central zones of the layers as can be seen visually. The GMM centroids have cross-variance matrices of size 4 x 4 each (curvature measures only) or 5 x 5 (including saturation), which together with the respective means are derived upfront from statistics of unambiguous regions.

3 Results

The application of an ellipsoidal measurement support with half-axes of a=120, b=30, c=120 and resulting curvature and saturation fields is given in Fig. 3. The averaging support size and shape was chosen to avoid generating holes in the resulting classification and respecting the evident layering. The corresponding histograms of the regional measures are given in Fig. 4. Recall that the porosity of the two rock-types is essentially the same. Consequently, there is no real contrast in the solid volume measure (Fig. 4a). The different color scales (Fig. 3b) are a consequence of different grain sizes, the larger structure showing the more extreme values. We observe a clear separation of rock-types in the surface area density as one would expect (Fig. 3c, Fig. 4b), with small spheres corresponding to the larger surface area contributions. The 2D slice illustrates the strong layering detected by surface area, while the histogram illustrates a good separation with a moderate area one could interpret as overlap region. An equally sharp contrast in layers is visible in Fig. 3d, corresponding to mean curvature. The smaller values (larger in magnitude) correspond to the small spheres. The corresponding histogram (Fig. 4c) shows an excellent separation with only a small area between peaks. Consider now the total curvature measure (Fig. 3e, Fig. 4d). The smaller sphere regions show a stronger connected behavior expressed by the more negative total curvature measure. Compared to this, the larger sphere regions exhibit a negative peak suggesting connectivity of both phases, but with some values becoming positive. Even larger measurement supports may lead to an even better separation of layers for this measure. Finally, the saturation field offers an approach to generate contrast between different rock-types by capillary pressure, here in a static sense; we consider this a hydraulic attribute for rock-typing. The measure is potentially attractive for low resolution image classification as such contrast can be generated either by simulation or by actual differential tomographic imaging (dry & wet imaging). From the 2D cross-section of the saturation map we observe that for the given invasion radius the rock-type with smaller features remains largely fully saturated with wetting fluid, with some exceptions close to the boundary, making the interface between rock-types less consistent (Fig. 3f). The corresponding histogram shows a huge spike for $S_w = 0$ (truncated), and a somewhat distributed peak with a significant transition zone otherwise.

The distribution of the curvature and saturation measures is further summarized in Fig. 5 through a 1D profile. We emphasize that the procedures introduced here are general for 3D; however, given the special nature of the laminated samples it is logical to also provide a corresponding 1D log.

Consider now the resulting rock-type classifications of the laminated microstructure for the case of the ellipsoidal measurement support of a=c=120 voxel and b=30 voxel via the GMM procedure. We illustrate the sharpness of the boundary both by taking 2D slices perpendicular to the bedding direction (Fig. 6) and by illustrating the resulting classification as 1D log (Fig. 7). For the case of using the Minkowski measures only as basis for the classification the recovered interface between the rock-types is sharp and coincides well with the original interface. Adding water saturation to the classification basis results in a somewhat smoother

https://doi.org/10.1051/e3sconf/202014604002

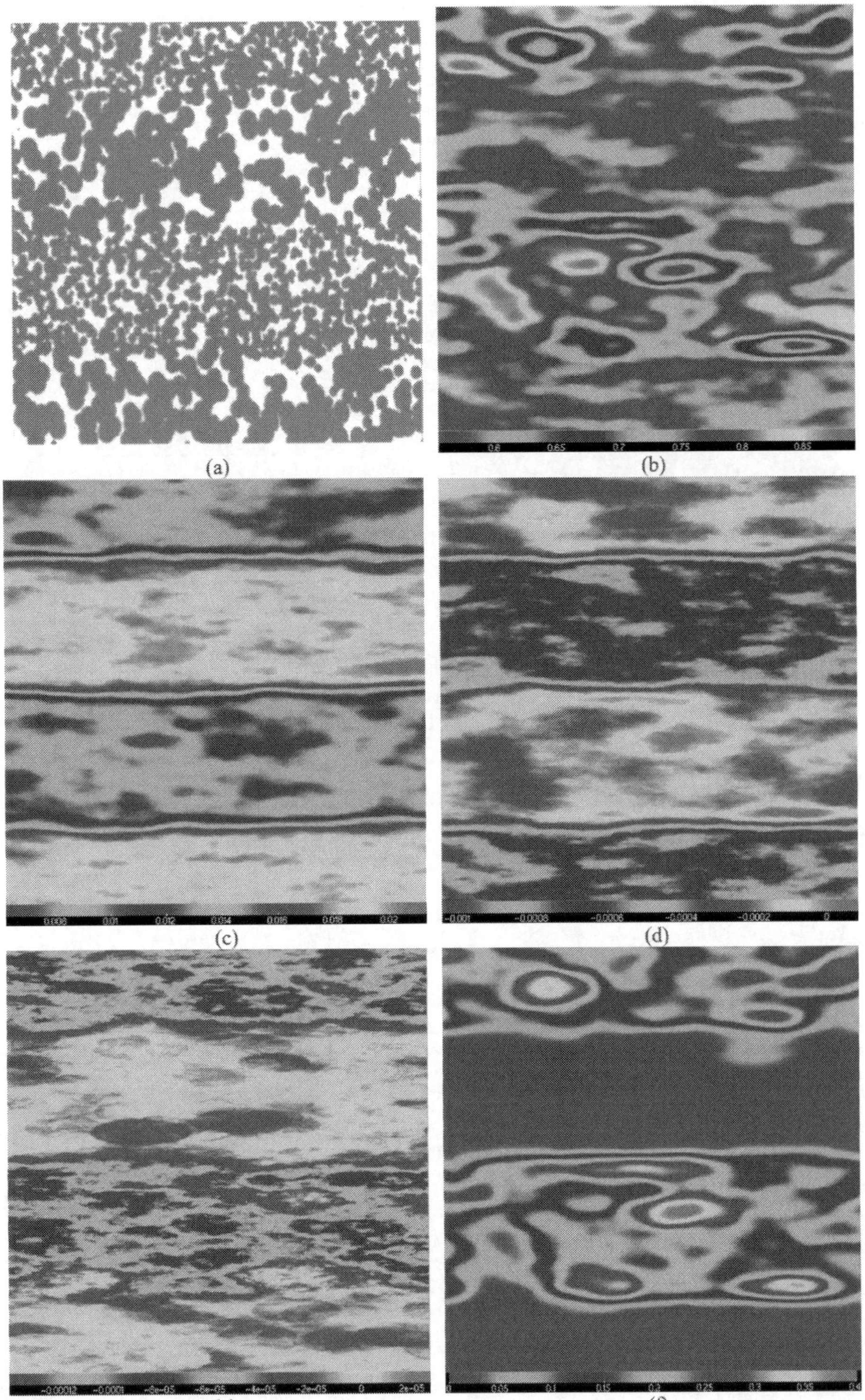

Fig. 3. 2D-cuts through fields of the regional Minkowski measures and hydraulic property. The averaging support is of ellipsoid shape with half-axes of a=120, b=30 and c=120 voxels. (a) original sample, (b) solid volume solid, (c) surface area, (d) integral of mean curvature, (e) integral of total curvature, and (f) fluid saturation over total volume. All volumes are 960 x 940 x 560 voxels corresponding to all locations where the full size of the measuring window is contained within the sample (see also Fig. 2). Scales are given at the bottom of the figures; for distribution of the values in (b-f) see Fig. 4.

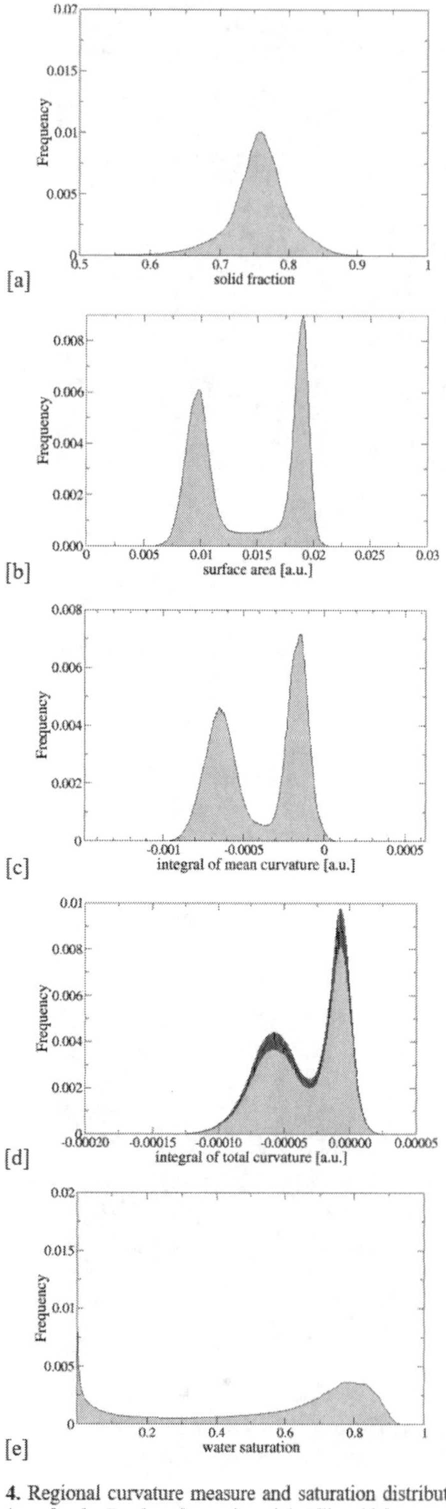

Fig. 4. Regional curvature measure and saturation distribution functions for the Boolean layered model. Ellipsoidal averaging element with a=c=120 voxels and b=30 voxels. [a] Solid volume, [b] surface area, [c] mean curvature, [d] total curvature, [e] defending phase saturation (point at zero saturation of 0.16 is truncated, e.g. of the scale).

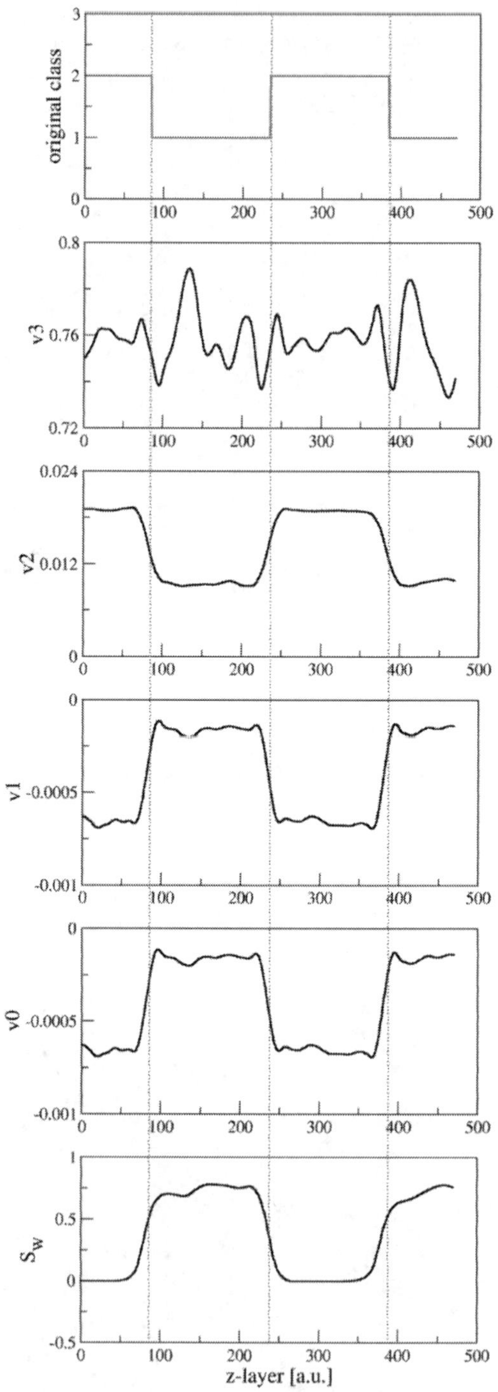

Fig. 5. Profile of the curvature and saturation measures along the y-dimension, e.g. averages over slices parallel to the layering. Ellipsoidal averaging element with a=c=120 voxels and b=30 voxels. The red lines indicate the actual rock types and their boundaries. For a cut through the corresponding 3D fields of regional measures in color see Fig. 3b-f. All measures are given in lattice units.

https://doi.org/10.1051/e3sconf/202014604002

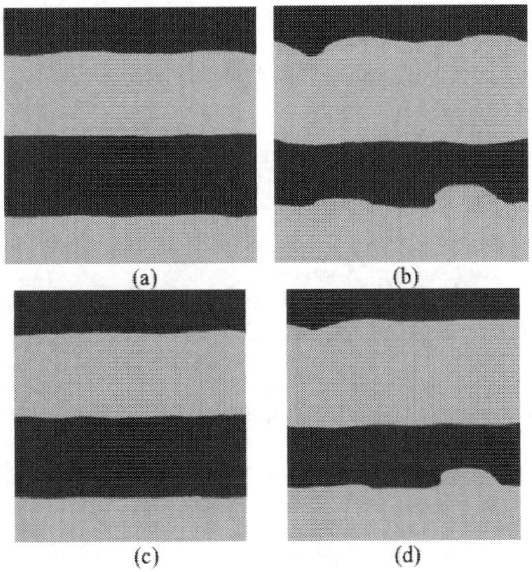

Fig. 6. 2D slices through the recovered layer systems from morphological descriptors and fluid saturation with different measuring window size. *Left:* classification based on Minkowski descriptors only. *Right:* combination of Minkowski descriptors and fluid saturation. (a,b) ellipsoidal averaging elements with a=c=120 and b=30 voxels, (c,d) a=c=160 voxels and b=40 voxels.

transition of the rock-types; furthermore, an apparent bias in the position of the boundary is introduced. We attribute this to an internal invasion boundary condition, e.g. a pore-scale saturation transition zone. This interpretation is supported by Fig. 6b and Fig. 6d, showing local misclassifications likely due to local heterogeneities in structure, allowing invasion of the boundary at lower capillary pressure. This local variation in saturation is also evident in Fig. 7.

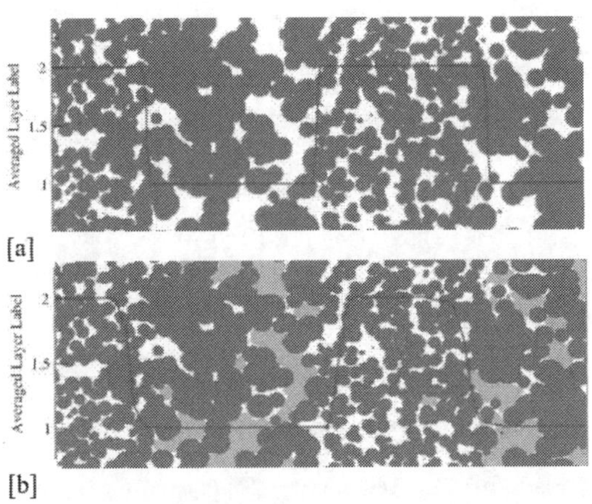

Fig. 7. 2D slices through the recovered 3D layer type distribution along Y dimension, overlapping original sample. The curve in (a) is recovered from the four Minkowski descriptors while (b) is recovered from Minkowski descriptors and fluid saturation. The blue color depicts the non-wetting phase saturation calculated by a MICP simulation.

We further quantify the relative boundary bias by scanning through each x-z column along the y-axis, calculating both the mean and standard deviation of the interface location for the case two different averaging element sizes and the Minkowski measures only versus including water saturation as additional hydraulic discriminator. The respective measures are reported in Table 2. The curvature measures alone recover the thickness of all layers almost exactly for both averaging ellipsoidal supports (b = 30 and b = 40, a = c = 4b). There is little variation in the location of the interface as well, e.g. no significant undulations are introduced, which is visible in the small standard deviation. Adding water saturation to the classification basis increases standard deviation significantly and introduces considerable bias towards the larger-scale rock-type.

Table 2. Boundary information of different descriptors under various window sizes.

b			Layer 1	Layer 2	Layer 3	Layer 4
30	Original		170	300	300	170
	MFs	Mean	168	298	300	173
		Std	4	6	6	10
	MFs + Sw	Mean	132	367	225	216
		Std	7	9	15	17
40	Original		160	300	300	160
	MFs	Mean	155	299	293	172
		Std	15	29	29	72
	MFs + Sw	Mean	122	374	220	204
		Std	16	33	29	63

4 Conclusions

In this work we introduced a fast FFT based algorithm for the calculation of regional curvature fields in 3D for the case of large measurement supports. The algorithm was demonstrated on a layered Boolean microstructure mimicking a laminated sandstone. It provides linear scaling with sample size and does not depend on the spatial support size. This allows the use of large spatial supports for rock-typing purposes utilizing the Minkowski functionals.

We further illustrated the classification bias introduced by including hydraulic information into the classification basis. For the given model rock, the strongest discriminator in terms of interface sharpness was regional mean curvature, followed by regional surface area. Including hydraulic information into the classification basis resulted in a smoother boundary transition which overestimated the relative fraction of the larger scale porosity.

CHA thanks the Australian Research Council (ARC) for funding through grant DP160104995 and through an ARC Future Fellowship (FT120100216). HJ and CHA thank the Chinese Scholarship Council (CSC) for their support. This research was undertaken with the assistance of resources and services from the National Computational Infrastructure (NCI), which is supported by the Australian Government. We thank reviewers Ryan Armstrong and Steffen Berg for a careful review, which further improved the quality of this manuscript.

References

1. W. Xu, T. Tran, R. Srivastava, A. G. Journel, et al., Srivastava, R. M. "Reservoir characterization with probability field simulation," in *Proc. SPE Annu. Tech. Conf. Exhib,* Society of Petroleum Engineers, 1992.
2. M. Pirrone, A. Battigelli, L. Ruvo, et al., "Lithofacies Classification of Thin-Layered Turbidite Reservoirs Through the Integration of Core Data and Dielectric-Dispersion Log Measurements," *SPE Reservoir Eval. Eng.*, vol. 19, no. 2, pp. 226-238, Apr. 2016, Art. no. SPE-170748-PA.
3. C. H. Arns, M. A. Knackstedt, M. V. Pinczewski, and W. Lindquist, "Accurate estimation of transport properties from microtomographic images," *Geophysical Res. Lett.*, vol. 28, no. 17, pp. 3361-3364, Sep. 2001.
4. C. H. Arns, M. A. Knackstedt, W. V. Pinczewski, and E. J. Garboczi, "Computation of linear elastic properties from microtomographic images: Methodology and agreement between theory and experiment," *Geophys.*, vol. 67, no. 5, pp. 1396-1405, Sep. 2002.
5. C. H. Arns, M. A. Knackstedt, W. V. Pinczewski, and N. S. Martys, "Virtual permeametry on microtomographic images," *J. Petroleum Sci. Eng.*, vol. 45 no. 1-2, pp. 41-46, Nov. 2004.
6. J. Fredrich, A. DiGiovanni, and D. Noble, "Predicting macroscopic transport properties using microscopic image data." *J. Geophysical Res.: Solid Earth*, vol. 111, no. B3, Mar. 2006.
7. C. H. Arns, M. A. Knackstedt, and N. S. Martys, "Cross-property correlations and permeability estimation in sandstone," *Phys. Rev. E*, vol. 72, no. 4, pp. 046304, Oct. 2005.
8. A. Dehghan Khalili, J.-Y. Arns, F. Hussain, Y. Cinar, W.V. Pinczewski, and C.H. Arns, "Permeability upscaling for carbonates from the pore-scale using multi-scale Xray-CT images," *SPE Reservoir Eval. Eng.*, vol. 16, no. 4, pp. 353, Mar. 2012.
9. A. Fadili, and R. Ababou, "Dual homogenization of immiscible steady two - phase flows in random porous media," *Water Resour. Res.*, vol. 40, no. 1, pp. 1-18, Jan. 2004.
10. S. Norouzi Apourvari and C. H. Arns, "Image-based relative permeability upscaling from the pore scale," *Adv. Water Resour.*, vol. 95, pp.161-175, Sep. 2016.
11. G. E. Archie, "Classification of carbonate reservoir rocks and petrophysical considerations," *AAPG Bull.*, vol. 36, no. 2, pp. 278-298, Feb. 1952.
12. M. Skalinski, J. Kenter, S. Jenkins, et al., "Rock type definition and pore type classification of a carbonate platform, Tengiz Field, Republic of Kazakhstan," in *Proc. SPWLA 50th Annu. Logging Symp.*, Society of Petrophysicists and Well-Log Analysts, 2009, SPWLA-2009-62185.
13. B. Li, X. Tan, F. Wang, P. Lian, W. Gao, and Y. Li, "Fracture and vug characterization and carbonate rock type automatic classification using X-ray CT images," *J. Petroleum Sci. Eng.*, vol. 153, pp. 88-96, May. 2017.
14. M. Dernaika, Y. N. Uddin, S. Koronfol, O. Al Jallad, G. Sinclair, Y. Hanamura, K. Horaguchi, et al., "Multi-scale rock analysis for improved characterization of complex carbonates," in *Proc. SPE Reservoir Characterisation Simul. Conf. Exhib.,* Society of Petroleum Engineers, 2015, SPE-175598.
15. M. Schmitt, M. Halisch, C. M¨uller, and C. P. Fernandes, "Classification and quantification of pore shapes in sandstone reservoir rocks with 3-D X-ray micro-computed tomography," *Solid Earth,* vol. 7, no. 1, pp. 285-300, Feb. 2016.
16. K. R. Mecke, T. Buchert, and H. Wagner, "Robust morphological measures for large-scale structure in the universe," *arXiv preprint astro-ph/9312028*, Dec. 1993.
17. C. H. Arns, M. A. Knackstedt, W. V. Pinczewski, and K. R. Mecke, "Euler-Poincaré characteristics of classes of disordered media," *Phys. Rev. E*, vol. 63, no. 3, pp. 031112, Feb. 2001.
18. C. Arns, M. Knackstedt, and K. Mecke, "3D structural analysis: sensitivity of Minkowski functionals," *J. Microsc.*, vol. 240, no. 3, pp. 181-196, Dec. 2010.
19. R. T. Armstrong, J. E. McClure, V. Robins, et al., "Porous Media Characterization Using Minkowski Functionals: Theories, Applications and Future Directions," *Transport in Porous Media,* pp. 1-31, Nov. 2018.
20. J. Schmalzing and K. M. G´orski, "Minkowski functionals used in the morphological analysis of cosmic microwave background anisotropy maps," *Monthly Notices Roy. Astronomical Soc.*, vol. 297, no. 2, pp. 355-365, Jun. 1998.
21. C. H Arns, MA Knackstedt, KR Mecke, "Characterisation of irregular spatial structures by parallel sets and integral geometric measures," *Colloids and Surfaces A: Physicochemical and Engineering Aspects,* vol. 241, no. 1-3, pp. 351-372, Jul. 2004.
22. S. Winitzki and A. Kosowsky, "Minkowski functional description of microwave background Gaussianity," *New Astron.*, vol. 3, no. 2, pp. 75-99, Mar. 1998.
23. C. H. Arns, M. A. Knackstedt, and K. R. Mecke, "Reconstructing complex materials via effective grain shapes," *Phys. Rev. Lett.* Vol. 91, no. 21, pp. 215506 Nov. 2003, doi: 10.1103/PhysRevLett.91.215506.
24. C. H. Arns, M. A. Knackstedt, and K. R. Mecke, "Boolean reconstructions of complex materials: Integral geometric approach," *Phys. Rev. E*, vol. 80, no. 5, pp. 051303, Nov. 2009, doi: 10.1103/PhysRevE.80.051303.

25. N. I. Ismail, "Rock-typing using the complete set of additive morphological descriptors, " Ph.D. dissertation, School of Minerals and Energy Resour. Eng., Univ. New South Wales, Sydney, NSW, Australia, 2014.
26. C. H. Arns, J. Mecke, K. R. Mecke, and D. Stoyan, "Second-order analysis by variograms for curvature measures of two-phase structures," *Europ. Phys. J. B: Cond. Matt.*, vol. 47, no. 3, pp. 397-409, Oct. 2005.
27. J. A. Rushing, K. E. Newsham, and T. A. Blasingame, "Rock typing: Keys to understanding productivity in tight gas sands," in *Proc. 2008 SPE Unconventional Reservoirs Conf.*, Keystone, Colorado, Jan. 2008.
28. M. Turner, L. Knuefing, C. Arns, A. Sakellariou, T. Senden, A. Sheppard, R. Sok, A. Limaye, W. V. Pinczewski,and M. Knackstedt, "Three-dimensional imaging of multiphase flow in porous media," *Physica A: Statistical mechanics and its applications*, vol. 339, no. 1-2, pp. 166-172, Aug, 2004.
29. M. Hilpert and C. T. Miller, "Pore-morphology-based simulation of drainage in totally wetting porous media," *Adv. Water Resour.*, vol. 24, no. 3-4, pp. 243-255, Feb. 2001.
30. I. Shikhov and C. H. Arns, "Evaluation of capillary pressure methods via digital rock simulations," *Transport in Porous Media*, vol. 107, no. 2, pp. 623-640, Mar. 2015.
31. R. Pini, S. C. Krevor, & S. M. Benson, "Capillary pressure and heterogeneity for the CO2/water system in sandstone rocks at reservoir conditions," *Adv. Water Resour.*, vol. 38, pp. 48-59, Mar. 2012.
32. A. Georgiadis, S. Berg, A. Makurat, G. Maitland, and H. Ott, "Pore-scale micro-computed-tomography imaging: Nonwetting-phase cluster-size distribution during drainage and imbibition," *Phys. Rev. E*, vol. 88, no. 3, pp. 033002, Sep. 2013, doi: https://doi.org/10.1103/PhysRevE.88.033002
33. S. T. Hyde, I. S. Barnes, B. Ninham, "Curvature energy of surfactant interfaces confined to the plaquettes of a cubic lattice," *Langmuir*, vol. 6, no. 6, pp. 1055-1062, Jun. 1990.

E3S Web of Conferences **146**, 05001 (2020) https://doi.org/10.1051/e3sconf/202014605001
SCA 2019

Towards Relative Permeability Measurements in Tight Gas Formations

Denis Dzhafarov[1], and *Benjamin* Nicot[1*]

[1]Total SA, 64018 Avenue Larribau PAU cedex, France

Abstract. Relative permeability is a concept used to convey the reduction in flow capability due to the presence of multiple fluids. Relative permeability governs the multiphase flow, therefore it has a significant importance in understanding the reservoir behavior. These parameters are routinely measured on conventional rocks, however their measurement becomes quite challenging for low permeability rocks such as tight gas formations.
This study demonstrates a methodology for relative permeability measurements on tight gas samples. The gas permeability has been measured by the Step Decay method and two different techniques have been used to vary the saturations: steady state flooding and vapor desorption.
Series of steady-state gas/water simultaneous injection have been performed on a tight gas sample. After stabilization at each injection ratio, NMR T2, NMR Saturation profile and low pressure Step Decay gas permeability have been measured. In parallel, progressive desaturation by vapor desorption technique has been performed on twin plugs. After stabilization at each relative humidity level the NMR T2 and Step Decay gas permeability have been measured in order to compare and validate the two approaches.
The techniques were used to gain insight into the tight gas two phase relative permeability of extremely low petrophysical properties (K<100 nD, phi < 5 pu) of tight gas samples of Pyrophyllite outcrop.
The two methods show quite good agreement. Both methods demonstrate significant permeability degradation at water saturation higher than irreducible. NMR T2 measurements for both methods indicates bimodal T2-distributions, and desaturation first occurs on low T2 signal (small pores).
Comparison of humidity drying and steady-state desaturation technique has shown a 12-18 su difference between critical water saturation (Swc) measured in gas/water steady-state injection and irreducible saturation (Swirr) measured by vapor desorption.

1 Introduction

In tight-gas sandstone, the productivity of a well is sometimes quite different from that of a nearby well. Wells also can be very sensitive to small amounts of water, whether from an aquifer associated with the reservoir, from hydraulic fracturing, or from other completion operations. In tight rocks, in case of small porosity and very small permeability, a significant influence water saturation on gas phase permeability is expected [1].

Relative permeability governs the multiphase flow; therefore, it has a significant importance in understanding the reservoir behavior. These parameters are routinely measured on conventional rocks, however their measurement becomes quite challenging for low permeability rocks such as tight gas formations.

Tight gas rocks generally have various initial water saturation values. Due to tight nature of the rock, a significant transition zone above gas water contact exists. In this case, presence of water can have a big impact on the effective permeability and, therefore, on the wells productivity. Thus, a proper measurement of effective permeability over different partial water saturations is an important subject for the reservoir characterization and production forecast.

1.1 Literature review and Issues of Kr tight gas

Liquid distribution in a tight gas reservoir controls the flow of gas more critically than in conventional systems [2]. Earlier experimental studies [3, 4] on tight gas sandstone show that the relative permeability to gas decreases dramatically as water saturation increases. While this implies that measurements relating to two-phase flow in tight sands should be performed under carefully controlled conditions of stress and water saturation, such elaborate experiments are difficult and time-consuming because of the extremely small flow rates involved.

It was indicated [3, 5] that in low-permeability reservoir irreducible water saturation and critical water saturation can be dramatically different. "Permeability jail" concept in low-permeability reservoir was introduced, there is a

* Corresponding author: benjamin.nicot@total.com

broad range of water saturations in which neither gas nor water can flow.

Convention holds that connate water has little effect on oil or gas permeability because it occupies the smaller pores [1].

In the recent study [2], by use of laboratory scale experiments and modelling, it was shown that capillary driven transport is an important mechanism that helps to redistribute the water within the tight gas rock samples. Strong effect of water dissipation due to capillary suction (during shut-in period) was demonstrated.

Effect of water saturation on the effective permeability to gas has been the subject of numerous experiments [6,4,3]. The common laboratory practice is to use evaporation to desaturate samples, effective gas permeability is measured by either the steady-state or the pulse technique at various levels of water saturation established by evaporation.

However, one can argue that the nature of water distribution resulting from evaporation has certain disadvantages of heterogeneity of water distribution and possible salt precipitation of brine.

2 Samples and Methods

2.1 Our methods

In this study, we demonstrate a methodology for relative permeability measurements on tight gas samples, with particular attention to the monitoring (quality control) of desaturation process, using NMR T2 and NMR saturation profiles measurements. Two different techniques were used to vary the saturations: steady state flooding and vapor desorption. The gas-phase effective permeability at certain water saturation has been measured by the Step Decay method [7].

Steady-state gas/water simultaneous injection has been performed at elevated pressure drop across the sample in order to achieve faster stabilization. After stabilization at each injection ratio, NMR T2, NMR Saturation profile has been measured to control the water saturation homogeneity.

In parallel, progressive desaturation by vapor desorption (humidity drying) technique has been performed on twin plugs. The experimental technique implies the samples saturation equilibration in a series of relative humidity levels. Saturation is then determined gravimetrically and by NMR T2 volume measurements.

After stabilization at each relative humidity level, NMR T2 and Step Decay gas permeability have been measured in order to compare and validate two approaches.

The Step Decay method is an advanced technique allowing the determination of absolute permeability of tight permeability samples (<0.1 mD) that is not possible or very long in time with standard methods under steady-state flow using Darcy's equation The principle of the Pulse Decay method consists in applying a pulse of pressure on one face of the plug and recording the differential pressure ΔP calculated from measurements taken at both extremes of the plug. The Step Decay method is based on the application of several consequently increasing pressure pulses [7]. The principal schema of Step Decay experimental set up is shown on Figure 1.

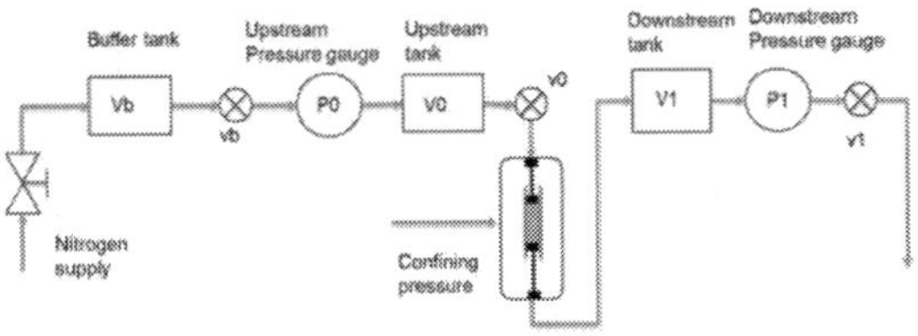

Fig. 1. Schematic of Step Decay experimental device

Step-Decay allows the simultaneous determination of intrinsic permeability (Kl), Klinkenberg coefficient (b) and porosity in one test only, speeding up the laboratory measurement.

Currently, the Step Decay method is used for absolute permeability determination (gas monophasic condition) of permeability of tight rock /gas shale.

Step Decay permeability measurements performed at different gas/water saturations provides gas relative permeability data. In this case, the procedure of experiment is complicated by several steps of sample (de)saturation and have been performed using the following two methods:

- Steady-state gas/water simultaneous injection
- Vapor Desorption (humidity drying)

2.2 Steady-state gas/water simultaneous injection

The experiment concerns the progressive desaturation of the rock sample by the steady-state technique

In the steady state test, a sequence of fixed ratios of water/gas fluids is injected through the core sample until saturation and pressure equilibrium is established at each ratio.

Since the permeability involved is extremely low (<100 nD,), extremely low flow rates are required (< 5 μL/min). To ensure such injection requirements an ultra high pressure pump has been employed.

 https://doi.org/10.1051/e3sconf/202014605001

Humid nitrogen gas and brine are injected to displace gradually the water from the initially fully saturated tight rock sample at constant temperature. Prior to each fractional flow stage, the sample is fully saturated every time by water flooding.

Injection of gas is conducted at constant pressure drop across the core. The water injection is conducted at imposed flow rate on one side of the core (injection). Pressures at the inlet and outlet and the gas flow rate at the outlet of the core sample are monitored with time. The pressure differential to gas and water is assumed to be equal to the measured total pressure drop. Figure 2 shows a schematic of the experimental setup.

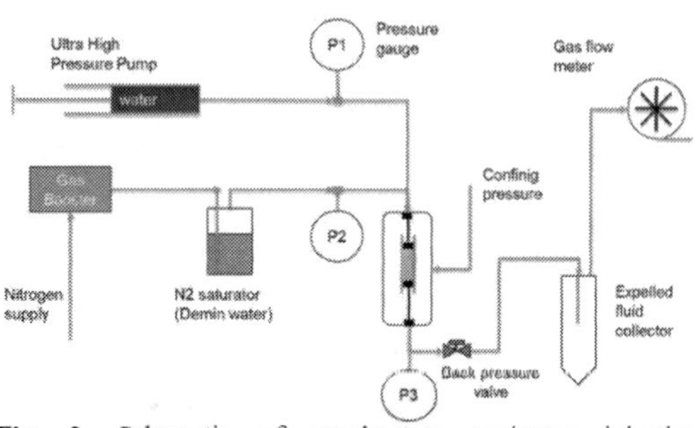

Fig. 2. Schematic of steady-state gas/water injection experimental set up

Each time when stabilization is achieved, injection is stopped and equipment is progressively depressurized. The sample is taken out from the cell and water saturation is determined by gravimetric measurements. NMR T2 and NMR Saturation profiles are also measured to ensure water saturation uniform distribution.

Capillary end effects are a particular problem in gas floods and the potential presence, influence and impact of end effects should be carefully considered when designing a gas displacement relative permeability test program [8]. Steady state water displacing experiments are not generally recommended unless one is aware of the limited saturation range expected from these analyses and of the potential saturation inaccuracies because of disequilibria effects [9]. Thus, the steady-state set up in this case is used only for saturation variation and the gas phase effective permeability is measured by Step Decay method.

2.3 Vapor Desorption (humidity drying)

The fundamental principal of vapor desorption method is the observation that the vapor pressure (Pv) above a liquid's curved surface is a function of the liquid surface curvature. Since the capillary pressure (Pc) is a function of the liquid surface curvature as well, both capillary and vapor pressure are the functions of saturation (Sw). Pc, therefore, can be calculated from Pv by controlling the relative humidity in the rock pores [10,11].

Initially fully saturated samples are placed in a sealed environment with relative humidity control (Humidity drying)

The test sequence represents a drainage type test (water is displeased by gas) where the wetting phase saturation Sw (water) is reduced as the humidity of the chamber is reduced. In practical manner, water is removed sequentially from the largest pores down through the smallest pores. Figure 3 schematically illustrates how the water vapor condenses within the pores of different size at different RH conditions.

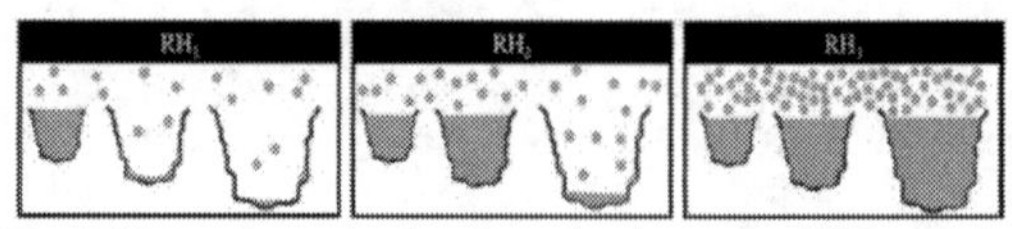

Fig. 3. Vapor desorption/condensation in pores of different size at different relative humidity environments RH1<RH2<RH3, after [11]

An electronically controlled humidity chamber is used. It allows investigating the relative humidity from 10 % to 95 % at room temperature. At each humidity level the mass stabilization is achieved and NMR T2 and Step Decay permeability are measured. Water saturation is determined by gravimetric techniques.

The main advantage of this method is the ability to achieve very high capillary pressures (the case for tight gas samples). The salinity increase or probable salt crystallization are the main problems of vapor desorption method that can potentially influence the Pc curve and permeability measurements especially at low saturations. The stabilization time may be several weeks/months.

2.4 Samples description and preparation

Identical plugs of Pyrophyllite quarry have been used to perform the test. Pyrophyllite is a homogeneous quarry rock sourced from the United Sates. The samples have preferential water wet wettability. They are mostly composed of clay and consequently has a low permeability (10-200 nD).

A Mercury Injection Capillary Pressure (MICP) and scanning-electron-microscope (SEM) analysis have been carried out on the fragments of the rock (Figures 4 and 5). In Figure 4, the pore throat distribution obtained by MICP shows a continuous distribution of pores approximately in the range between 2 and 30 nm. SEM images analysis shows that two types of pores coexist in this rock: interconnected system of small elongated pores and big pores weakly connected to the main pore system.

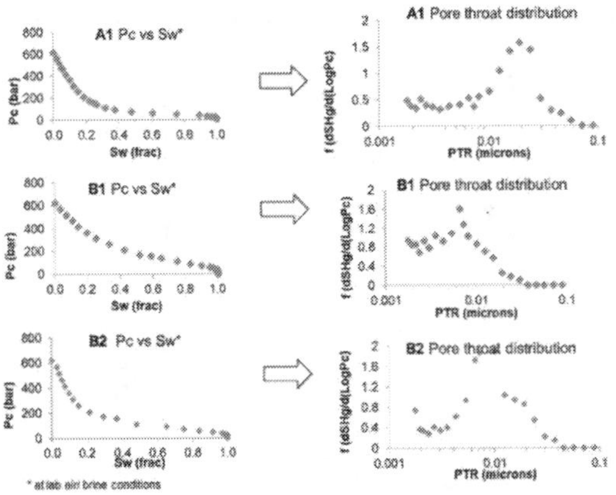

Fig. 4. Capillary pressure (Pc) and Pore throat size distribution (PSD) of Pyrophyllite quarry rock

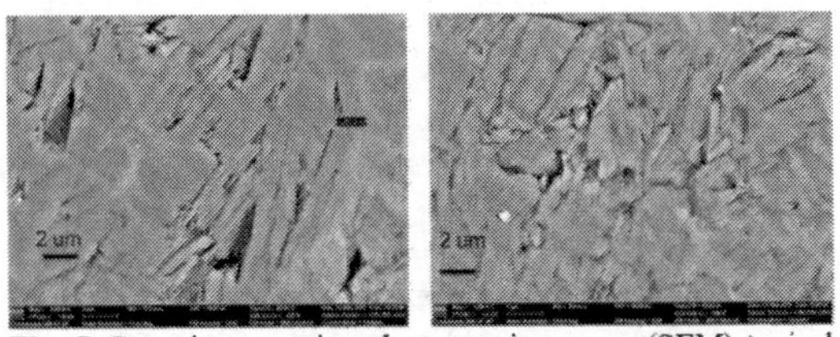

Fig. 5. Scanning scanning-electron-microscope (SEM) typical images of Pyrophyllite quarry rocks

Before starting the experiments the Step Decay permeability (Kl) of each plug were measured at initial "as received" conditions. In this case, the significance of the "as received" condition is that the samples are at the same saturation conditions as in the quarry (atmospheric water) and do not contain hydrocarbons.

Next, the plugs were subjected to Soxhlet cleaning with isopropanol and gentle oven drying at 60 °C until mass stabilization. Then plugs were fully saturated by 70 g/l NaCl water brine at 500 bar during two weeks, after having created a vacuum over a period of four days.

Total of five pyrophyllite plugs with diameter 23 mm and height 25 mm have been prepared:

- 1 plug A1 (Permeability-230 nD and Porosity-5.6 pu) has been used for Steady-State flooding experiment
- 4 plugs B1/B2/B3/B4 (Permeability- 80 – 350 nD and Porosity 4.6 - 6.6 pu) have been used for Vapor desorption experiment.

3 Results

3.1 Steady-state desaturation experiment results

The experiment started by only water injection to fully water saturated sample in order to determine brine permeability.

Before each fractional flow stage the sample was fully re-saturated to 100 % Sw. Every fractional flow stage starts by one phase water injection cycle. As soon as the water stabilization is achieved, the gas at higher pressure is applied, while water is continuously injected at the same rate.

Four different gas /water injection ratios have been applied, they are listed in Table 1.

At the final stage, only gas has been injected. For this stage the sample has been again fully re-saturated (100% Sw) and the only gas injection has been applied until pressure and gas rate stabilization.

Table 1. Main parameters of gas/water steady-state experiment (sample A1).

TEST	Water rate, ml/hr	Water Pres, bars	Gas Pres, bars	NMR Vol, ml	Sw, su
Only water	0.10	100	-	0.59	100.0
1	0.30	260	310	0.52	87.5
2	0.20	205	310	0.45	75.8
3	0.13	120	310	0.40	68.3
4	0.03	53.6	310	0.34	57.3
Only Gas	-	-	150	0.34	56.8

Back pressure for all tests has been set to 32 bars. Net overburden stress has been set to 100 bars.

Critical water saturation (Swc) has been defined at -57 su. High Swc value is expected due to tight nature of the rock (high capillary pressure)

NMR Saturation profile has been performed after each injection ratio:

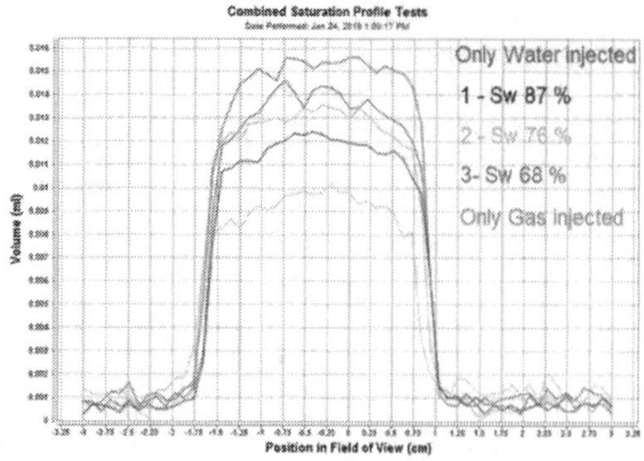

Fig. 6. NMR Saturation profile. Measurements performed after each injection ration at ambient conditions.

Based on Figure 6, saturation profiles appear relatively flat. NMR Saturation profiles do not have any evidence of pronounced capillary end effects. Such saturation behavior could be explained by the combinations of following phenomenon that potentially could influence the saturation distribution and reduce the capillary end effect:

- Water capillary suction. There is around 10 min time lag between the experiment and NMR saturation measurements, so the water is redistributed by capillarity
- Drying effect
- Effect of gas expansion/compression

The results of Step-Decay gas phase effective permeability measurements over different partial saturation are presented in Figure 7.

It is assumed that use of low pressure step decay method is not influenced by the capillary end effects, as it does not require the water flooding during the experiment.

Water phase effective permeability has been calculated assuming steady-state water flow (see table above). Water phase effective permeability should be treated with caution since it may be biased by saturation disequilibria effects [9]

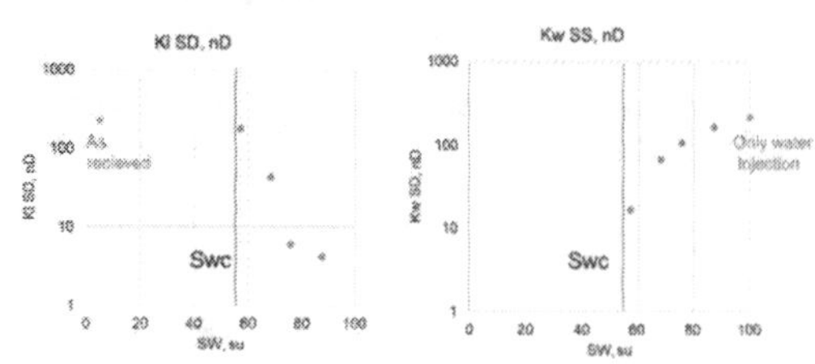

Fig. 7. Gas phase effective permeability measurements by Step-Decay technique. Steady-state gas/water flooding was used to vary the saturation. Water phase effective permeability have been measured by steady-state gas/water simultaneous injection.

Step-Decay gas phase permeability measurements revealed insignificant difference (by factor 1.2-1.5) between the permeability in "as received" conditions and at critical water saturation (Swc). At the same time significant permeability degradation was observed at water saturation being higher than critical. The water phase effective permeability did not show significant decay over Sw decrease (about a decade).

NMR T2 distribution has been measured at each injection ratio. Combined T2 plot is shown on Figure 8.

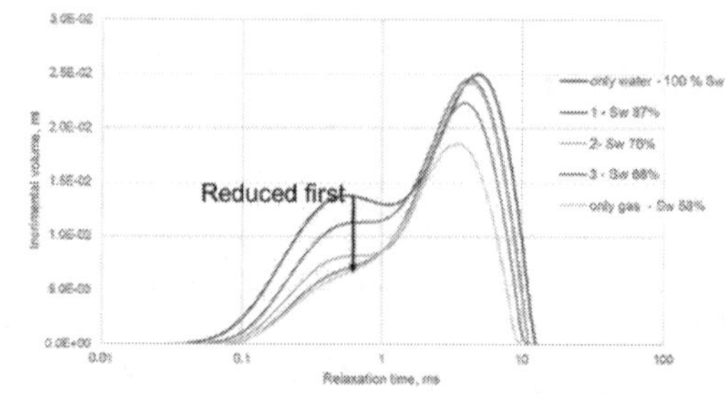

Fig. 8. NMR T2 distribution for different water saturations during steady-state gas/water simultaneous injection

One notes bimodal T2-distributions with higher intensity on second pic. As it can be seen from Figure 8, the desaturation first occurs on low T2 signal. Initially, first decrease of low T2 signal was not expected, as it implies that water desaturation occurs first in low porosity. This could be explained by the assumptions that gas initially flows the path of interconnected system of small elongated pores and desaturates bigger pores at the later stages.

3.2 Vapor desorption results

Low permeability sample B1 (76.4 nD) and high permeability samples B2, B3, B4 (230-350 nD) have been used to perform the vapor desorption test.

Initially, fully saturated samples were subjected to four different relative humidity (RH) levels: 90%, 85%, 80%, 75%. After stabilization at each relative humidity level the NMR T2 and Step Decay gas permeability have been measured in order to compare the two approaches. Step-decay permeability measurements results are shown on Figure 9.

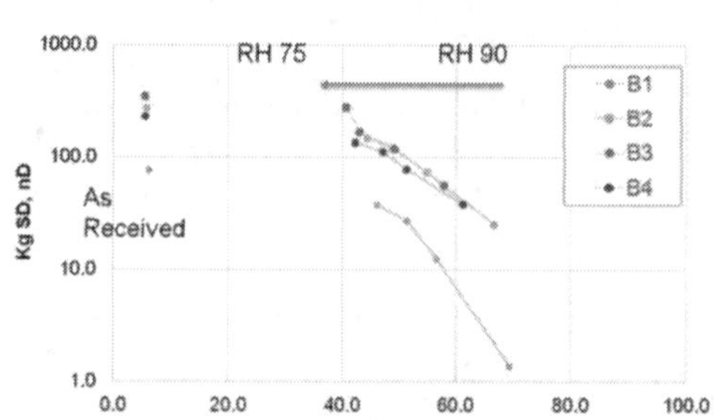

Fig. 9. Gas phase effective permeability measurements by Step-Decay technique. Humidity drying was used to vary the saturation

 https://doi.org/10.1051/e3sconf/202014605001

Effective gas permeability decay was observed with saturation increase, but less pronounced compared to steady-state desaturation (see Figure 7), since the gas effective permeability high water saturation was not measured (the maximum measured water saturation is limited to RH 90%).

At Relative Humidity of 75 % saturation levels ranged from 40 su (for B3 sample) to 46 su (for B1 sample). According to the Kelvin equation (Newsham et al 2003), RH 75 % corresponds to approximately 400 bars of capillary pressure in laboratory air/ brine conditions. So, it is assumed that at this pressure irreducible saturation (Swirr) is achieved. Swirr values appear systematically smaller by 12-18 su compared to Swc measured by steady-state desaturation (see Figure 7).

NMR T2 distribution at each humidity level is shown on Figure 10.

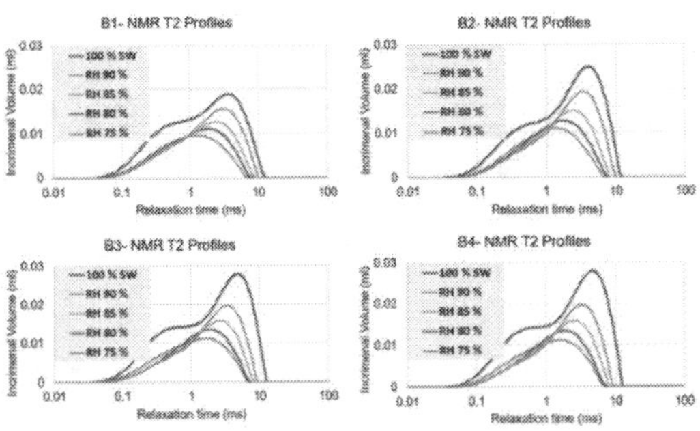

Fig. 10. NMR T2 distribution for different water saturations during vapor desorption desaturation process

NMR T2 distribution during vapor desorption process (Figure 10) confirms that desaturation first occurs on low T2 signal, which was previously observed at steady-state gas/water desaturation experiment (Figure 8).

4 Discussion

This study aims at developing new method of tight gas relative permeability measurement and gain insights into the tight flooding specific features. Steady-state regime has been achieved at elevated pressure drop and low injection volumes (see table 1). In this case the direct determination of effective gas phase permeability from experiment is complicated by numerous effects: gas compression/expansion effects, end scale effect, gas viscous instability, etc. On top of that, it is extremely difficult to monitor the downstream fluids production. Thus, the steady-state gas water set-up was used only for the saturation variation. Effective permeability at known saturation was measured by low pressure Step Decay technique. In such way it helps us to avoid many problems related to steady-state relative permeability determination. However, the sample is exposed to the atmosphere which can lead to different undesired influences such as drying and capillary suction. In this case, it is important to control the homogeneity of water saturation. We use NMR Saturation profile measurements to prove the saturation homogeneity.

NMR T2 Monitoring at each saturation level was used to perform analysis of fluids distribution within the pore medium. Despite the fact that the gas is non-wetting phase and it initially flows the path of least resistance through the largest pores, we observe first decrease of low T2 signal at high saturation. Detailed analysis and explanation of this observation is yet to be done.

Comparison of humidity drying and steady-state desaturation technique has shown the difference in critical water saturation (Swc) measured by steady-state gas/water injection and irreducible saturation (Swirr) measured by vapor desorption is 12-18 su. This difference appears mainly in high T2 signal (big pores). Sharp permeability decrease (about a decade) with saturation increase has been observed for both desaturation methods.

5 Conclusion

In this study, we demonstrate a methodology for relative permeability measurements on tight gas samples of extremely low petrophysical properties (K<100 nD) of Pyrophyllite outcrop. We pay attention to the monitoring (quality control) of desaturation process, using NMR T2 and NMR saturation profiles measurements. The gas permeability has been measured by the Step Decay method, and two different techniques have been used to vary the saturations: steady state flooding and vapor desorption. The following conclusion can be made:

- Step-Decay gas phase permeability measurements revealed insignificant difference (by factor 1.2-1.5) between the permeability in "as received" conditions and at critical water saturation (Swc). At the same time significant permeability degradation was observed at water saturation being higher than critical (about a decade).

- NMR T2 measurements indicate bimodal T2-distributions and that desaturation first occurs on low T2 signal (small pores). First decrease of low T2 signal was not expected and further detailed analysis of this phenomenon is required

- Comparison of humidity drying and steady-state desaturation technique has shown the difference in critical water saturation (Swc) measured in steady-state and irreducible saturation (Swirr) measured by vapor desorption is 12-18 su. This difference appears mainly in high T2 signal (big pores).

References

1. S. Motealleh, S. L. Bryant, "Quantitative Mechanism for Permeability Reduction by Small Water Saturation in Tight-Gas Sandstones," *SPE Journal*, vol. 14, no. 2, pp. 252-258, June 2009, doi:10.2118/107950-PA
2. D. H. Le, H. N. Hoang, J. Mahadevan, "Gas Recovery From Tight Sands: Impact of Capillarity," *SPE Journal*, vol. 17, no. 4, pp. 981-991, December 2012, doi:10.2118/119585-PA.
3. K.W. Shanley, R.M. Cluff, J.W. Robinson, "Factors controlling prolific gas production from low-permeability sandstone reservoirs: Implications for resource assessment, prospect development, and risk analysis," *AAPG Bull*, vol. 88, pp.1083-1121, 2004, doi:10.1306/03250403051
4. P. Chowdlah, "Influence of Water-Desaturation Technique and Stress on Laboratory Measurement of Hydraulic Properties of Tight Sandstones" *SPE Formation Evaluation*, vol. 3, no. 4, pp. 679-685, December 1988, doi:10.2118/15210-PA
5. R. M. Cluff, A. P. Byrnes, "Relative Permeability In Tight Gas Sandstone Reservoirs - The "Permeability Jail" Model," presented at the SPWLA 51st Annual Logging Symposium, Perth, Australia, June19-23, 2010, Paper SPWLA-2010-58470
6. J. S. Ward, N. R. Morrow, "Capillary Pressures and Gas Relative Permeabilities of Low-Permeability Sandstone," *SPE Formation Evaluation*, vol. 2, no.3, pp 345-356, September 1987, doi:10.2118/13882-PA
7. D. Lasseux, Y. Jannot, S. Profice, M. Mallet, G. Hamon, "The 'Step Decay': A New Transient Method for the Simultaneous Determination of Intrinsic Permeability, Klinkenberg Coefficient and Porosity on Very Tight Rocks," presented at the International Symposium of the Society of Core Analysts, Aberdeen, Scotland, 27-30 August, 2012, Paper SCA2012-25
8. C. McPhee, J. Reed, I. Zubizarreta, "Relative Permeability," in *Core Analysis: A Best Practice Guide*, vol. 64, 1st ed., Amsterdam, Netherlands: Elsevier B.V., 2015, pp. 561-562
9. A. Cense, J. Reed, P. Egermann, "SCAL for gas reservoirs: a contribution for Better experiments", presented at the International Symposium of the Society of Core Analysts, Snowmass, Colorado, USA, 21-26 August, 2016, Paper SCA-2016-23
10. K. E. Newsham, J. A. Rushing, P. M. Lasswell, "Use of Vapor Desorption Data to Characterize High Capillary Pressures in a Basin-Centered Gas Accumulation with Ultra-Low Connate Water Saturations," presented at SPE Annual Technical Conference and Exhibition, Denver, Colorado, 5-8 October, 2003, Paper SPE-84596-MS
11. A. Zolfaghari, H. Dehghanpour, "Pore Size Distribution from Water Adsorption Isotherm," presented at SPE Annual Technical Conference and Exhibition, Houston, Texas, USA, 28-30 September, 2015, Paper SPE-175155-MS

Storing CO_2 as solid hydrate in shallow aquifers: Electrical resistivity measurements in hydrate-bearing sandstone

Jarand Gauteplass*, *Stian* Almenningen, and *Geir* Ersland

Department of Physics and Technology, University of Bergen, 5007 Bergen, Norway

Abstract. A recent proposed carbon dioxide (CO_2) storage scheme suggests solid CO_2 hydrate formation at the base of the hydrate stability zone to facilitate safe, long-term storage of anthropogenic CO_2. These high-density hydrate structures consist of individual CO_2 molecules confined in cages of hydrogen-bonded water molecules. Solid-state storage of CO_2 in shallow aquifers can improve the storage capacity greatly compared to supercritical CO_2 stored at greater depths. Moreover, impermeable hydrate layers directly above a liquid CO_2 plume will significantly retain unwanted migration of CO_2 toward the seabed. Thus, a structural trap accompanied by hydrate layers in a zone of favorable kinetics are likely to mitigate the overall risk of CO_2 leakage from the storage site. Geophysical monitoring of the CO_2 storage site includes electrical resistivity measurements that relies on empirical data to obtain saturation values. We have estimated the saturation exponent in Archie's equation, $n \approx 2.1$ (harmonic mean) for CO_2 and brine saturated pore network, and for hydrate-bearing seal ($S_H < 0.4$), during the process of storing liquid CO_2 in Bentheimer sandstone core samples. Our findings support efficient trapping of CO_2 by sedimentary hydrate formation and show a robust agreement between saturation values derived from PVT data and from modifying Archie's equation.

1 Introduction

Carbon capture and storage (CCS) technologies are expected to play a substantial role in the transformation of the energy sector toward reduced emissions of anthropogenic greenhouse gases [1]. Captured CO_2 is typically injected and stored in a supercritical state [2] in aquifers and depleted reservoirs at great depths. Once injected, CO_2 is retained in the sediments by physicochemical processes including structural trapping, capillary trapping, fluid dissolution, and mineral reactions. The contribution from each trapping process varies greatly with time [3]. Structural and capillary trapping are highly relevant from the onset of injection, while fluid dissolution and mineral reaction are believed to have a significant impact at a later stage.

More recently, an additional trapping mechanism suggests liquid CO_2 stored and contained by an upper CO_2 hydrate layer located at the base of the gas hydrate stability zone (GHSZ). This self-sealing hydrate layer makes an artificial cap rock that can prevent upward migration of CO_2 [4,5]. Cooler storage conditions enhances the CO_2 storage capacity due to increased CO_2 density, increased mobility control (important if long inter-well distance), and increased CO_2 solubility in water compared to storage of supercritical CO_2.

Experimental work has verified that CO_2 hydrate can form at the base of the GHSZ and reduce the CO_2 diffusion rate significantly in unconsolidated media [6]. Furthermore, high-density storage of CO_2 hydrate in silica sand has been demonstrated [7], as well as CO_2 hydrate acting as permeability barriers and successfully sealing off the pore space [8,9]. CO_2 immobilization by hydrate formation was directly visualized using MRI and micromodels [10]. A substantial GHSZ is ideal to make sure escaped liquid CO_2 is immobilized and converted to solid hydrate before reaching the seabed, and thus extending the hydrate sealing layer. Predicted thickness of the GHSZ for offshore Western Europe is nearly 0.5 km of the upper sediments [11], showing great potential for liquid CO_2 storage at shallow depths.

Resistivity measurements are routinely used to determine presence of sedimentary hydrates both in the field and in the laboratory. However, relevant empirical data are needed for saturation quantifications. These can be obtained and calibrated based on controlled laboratorial experiments. For a medium with uniform cross-section transmitting a uniform flow of electric current, resistivity is found from:

$$R_t = Z\frac{A}{L}\cos(\theta) \tag{1}$$

where R_t is the bulk resistivity, Z is the measured impedance, A is the cross-sectional area of the sample, L is the length of the sample, and θ is the phase angle. The Formation Factor (F) relates empirically to porosity through [12]:

$$F = \frac{R_0}{R_w} = a\phi^{-m} \tag{2}$$

* Corresponding author: Jarand.Gauteplass@uib.no

where R_0 is the resistivity of a fully brine-saturated sample, R_w is the resistivity of the brine, ϕ is the porosity of the sample, m is the cementation exponent and a is the tortuosity factor. R_w is calculated using a standard conversion [13]:

$$R_w = 0.0123 + \frac{3647.5}{[C]^{0.955}} \quad (3)$$

where C is the ion content of brine. The R_w value is corrected for temperature variations by [14]:

$$R_w(2) = R_w(1) * \frac{(T_1 + 21.5)}{(T_2 + 21.5)} \quad (4)$$

where T_1 is ambient temperature and T_2 is sample temperature. The Resistivity Index (RI) that applies to sediments partially saturated with a non-conductive material such as oil, gas, or hydrate, is defined as:

$$RI = \frac{R_t}{R_0} = S_w^{-n} \quad (5)$$

where R_t is the measured bulk resistivity, S_w is the brine saturation and n is the saturation exponent.

Hydrate growth is accompanied by an effective reduction of the pore space as well as a salinity increase of the remaining brine that is not converted to solid hydrate. Both of these processes results in a continuous change in R_0 as hydrate grows, and a dynamic R_0* needs to be implemented in eq. 5. This R_0* is calculated from eq. 2 by adjusting R_w and ϕ as hydrate grows. R_w is found from eq. 3-4 by keeping track of the salinity increase during hydrate growth from PVT data. PVT data is also used to monitor the hydrate saturation during hydrate growth, and ϕ_{eff} is then found from the following relation:

$$\phi_{eff} = \phi(1 - S_H) \quad (6)$$

where S_H is the hydrate saturation. The cementation exponent m is calculated by eq. 2 when the sample (with known porosity) is completely filled with brine. This m is then assumed constant as hydrate grows in the pore space [15]. The tortuosity factor a is set to 1 to ensure that $R_w = R_0$ in the limiting case where $\phi \rightarrow 1$.

Finally, the saturation exponent n is derived during hydrate growth by a modified version of eq. 5:

$$\log\left(\frac{R_t}{R_0 *}\right) = -n \log S_w \quad (7)$$

The n is found as the slope when plotting the left side of eq. 7 as a function of $-\log S_w$.

The majority of hydrate resistivity studies presented are related to CH_4 hydrate in the context of mapping and production of natural gas through various dissociation processes [15-18]. To the best of the authors' knowledge, this paper presents the first reported resistivity measurements on sedimentary CO_2 hydrate. We provide the saturation exponent n during CO_2 injection into brine-filled cores and subsequent CO_2 hydrate formation. Saturation values derived from resistivity measurements are compared with PVT derived saturations to investigate the applicability of using resistivity measurements to monitor the evolving CO_2 hydrate seal in subsurface carbon storage.

2 Materials and methods

Homogenous and quartz-dominated Bentheimer sandstone (95.5% quartz, 2.0% kaolinite, 1.7% K-feldspar, 0.8% other [19]) was used in this study. Average porosity and absolute permeability were measured to 0.22 and 1.1 D, respectively. Twin samples, all with diameter of 5 cm and length 15 cm, were cleaned, dried at 70 °C for 24 hours, and fully saturated with brine (3.5 or 5.0 weight% NaCl) under vacuum. The brine-saturated cores were positioned in a core holder containing a rubber sleeve, fixed upstream end-piece, and floating downstream end-piece (see Fig. 1). A nitrogen-supported back-pressure regulator was connected downstream and an effluent sample collector measured brine production. A refrigerated circulator supplied the system with cooling fluid. Precise high-pressure pumps regulated overburden and pore pressure. The laboratory setup allowed pressure differences and bulk resistivity (Hewlett-Packard LCR-meter; 1 kHz, two-electrode setup) across the core sample to be logged.

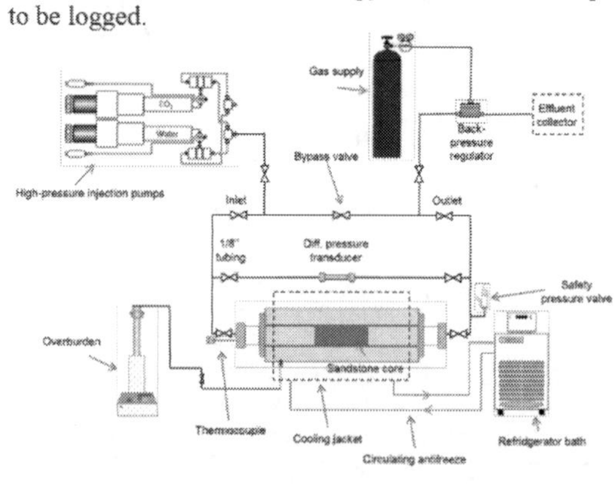

Fig. 1. Coreflooding laboratory setup including sandstone core sample, core holder, cooling system, pressure and temperature measurements, back-pressure regulator, and high-pressure pumps to regulate pore pressure and overburden. Modified from [20].

The pore space was pressurized with brine to 7.0 MPa, while the confinement pressure was set to 10.0 MPa. The core was then flooded with brine (μ=1.07 cP) over a range of injection rates and absolute permeability was calculated. The waterflood was followed by liquid CO_2 (μ=0.07 cP) injection at constant volumetric flow rate (0.5, 5 or 10 cm^3/min) to achieve a mixture of water and CO_2 in the pore space mirroring CO_2 invasion into an aquifer.

Two different hydrate growth conditions were designed in the laboratory: i) hydrate formation at constant pressure (CO_2 pressurized from both core ends, bypass valve open) and ii) flow-induced hydrate formation during CO_2 injection with constant volumetric flow rate. Onset of hydrate formation within the pore space was determined from the increase in resistivity, temperature, and differential pressure.

3 Results and discussion

3.1 CO_2 – brine system (outside GHSZ)

The CO_2 injection rate into a porous media affects the displacement efficiency and fluid saturations within the pore network. Achieving substantial CO_2 storage capacity in a multiple well scenario relies on an efficient displacement process from injector to producer(s). Though supercritical CO_2 can recover about the same amount of water at core-scale, the more optimum mobility ratio between liquid CO_2 and water is likely to be important at long interwell distances.

Fig 2 shows the rate dependency on macroscopic sweep efficiency, which govern the initial fluid distribution before hydrate formation. Three experiments with CO_2 injection rate of 0.5 cm^3/min (circles), 5 cm^3/min (diamonds) and 10 cm^3/min (triangles) were conducted outside of the GHSZ (20 °C) at 7.0 MPa. Injection rate (capillary number) affects both overall S_w and pore-level fluid distribution. A linear production profile is valid before CO_2 breakthrough (BT), supplemented with water production measurements. Delayed CO_2 BT (indicated with broken vertical lines) for 5 and 10 cm^3/min implies improved sweep compared to the low injection rate (0.5 cm^3/min). However, doubling the injection rate from 5 to 10 cm^3/min had insignificant effect on the displacement process, reaching a plateau where approximately 50% of the brine remained after injecting several pore volumes (PV) of liquid CO_2. Though the experiments were designed to minimize capillary end effects by increasing the core length and using relatively high flow rates, the lowest rate (0.5 cm^3/min) experiment is prone to a more heterogeneous saturation profile due to reduced displacement efficiency. At breakthrough, saturation fractions in the pore space were S_w= 0.77 and S_{CO2}= 0.23 (0.5 cm^3/min), S_w= 0.59 and S_{CO2}= 0.41 (5 cm^3/min), and S_w= 0.57 and S_{CO2}= 0.43 (10 cm^3/min).

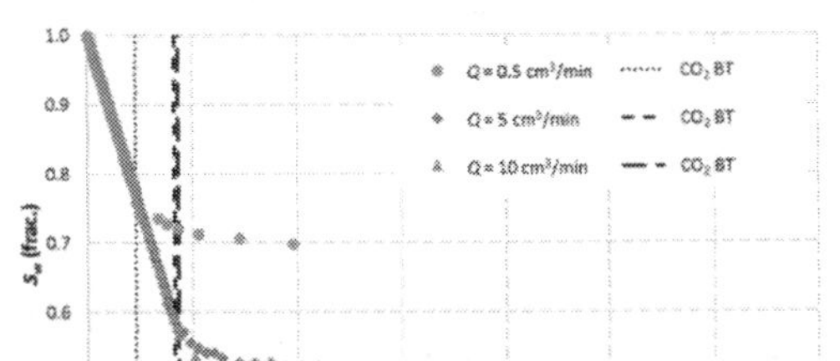

Fig 2. Drainage of initial water by liquid CO_2 in Bentheimer core samples at different injection rates; 0.5 cm^3/min, 5 cm^3/min, and 10 cm^3/min.

The bulk resistivity increased with increasing CO_2 saturation because electrically conductive brine was replaced by insulating CO_2 in the pore space. Fig 3 presents a logarithmic cross plot of water saturation (S_w) and resistivity index (RI) during CO_2 injection into the cores. The saturation exponent n was found as the slope of the curves, with emphasis on the first saturation point (S_w = 1) and the last saturation points (after CO_2 BT in the cores) when determining the slope of the curves (solid filled markers). The use of Archie's equation is not applicable until CO_2 has reached the end of the core, achieving a predominantly uniform two-phase saturation profile throughout the entire core length. This is illustrated by the data points (no fill) obtained before CO_2 BT, which deviate from the linear trend lines in Fig. 3.

The saturation exponent n increased with decreasing CO_2 flow rate, and was 1.9 for 10 cm^3/min, 2.1 for 5 cm^3/min, and 2.3 for 0.5 cm^3/min. This again reflects the different macroscopic sweep efficiencies that were achieved for the different flow rates, and highlights that n is sensitive to the displacement history. The n values identified in our CO_2-brine systems corroborate with n values reported for similar conditions [21,22]. CO_2 is a highly reactive fluid that can influence resistivity measurements through i) dissolution and dissociation where new ions are provided to the solution, and ii) contribution of surface conductivity – even in clay free rocks [21,23]. Both processes lead to increased electrolytic conductivity and may thus overestimate the water saturation if not accounted for. These effects are negligible if the water is highly saline [24]. In the next section, we will compare the water saturation derived from Archie's using the estimated n values with measured PVT data, to find if these effects are relevant to our systems with seawater salinity or higher.

3.2 CO_2 hydrate – brine system (within GHSZ)

To simulate the conditions of shallow CO_2 storage sites in offshore Western Europe, a temperature and pressure regime of 4 °C and 7.0 MPa pore pressure were chosen. This puts the system well within the GHSZ for CO_2 hydrate. Two hydrate growth scenarios were tested; i) static hydrate formation at constant pressure and ii) flow-induced hydrate formation under continuous CO_2 injection. Both approaches resulted in formation of CO_2 hydrate and subsequently immobilization of the injected CO_2 over a range of thermodynamic conditions.

Fig 4 shows a logarithmic cross plot of water saturation (S_w) and resistivity index (RI) during CO_2 hydrate formation in the pore space. The saturation exponent n is derived from the slope of the best-fit linear model to all measured data points during CO_2 hydrate formation. Final hydrate saturation depended on initial displacement of water by CO_2, and the more efficient displacement, the more hydrates were formed. Improved CO_2 sweep led to higher number of interfaces that acted as potential nucleation sites, and reduced water shielding in the samples. Increasing the initial CO_2 flow rate from 0.5 cm^3/min to 10 cm^3/min, increased the final hydrate saturation by almost a factor of 2.

 https://doi.org/10.1051/e3sconf/202014605002

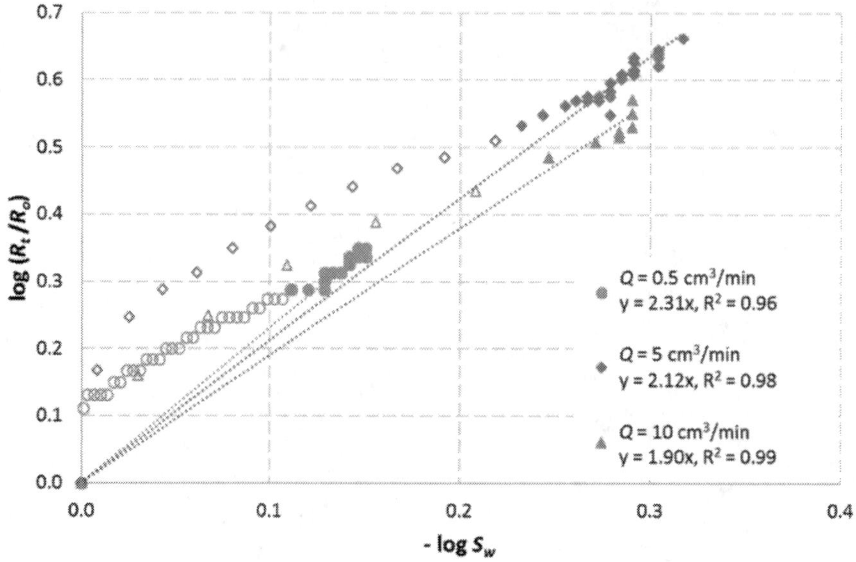

Fig 3. Logarithmic cross plot of RI and S_w during CO_2 injection in brine-filled Bentheimer core samples at injection rates: 0.5, 5 and 10 cm^3/min at experimental conditions of 7.0 MPa and 20 °C. Data point markers with no fill (non-uniform saturation profile) were excluded from the linear trend lines. The saturation exponent n ranged between 1.9 – 2.3.

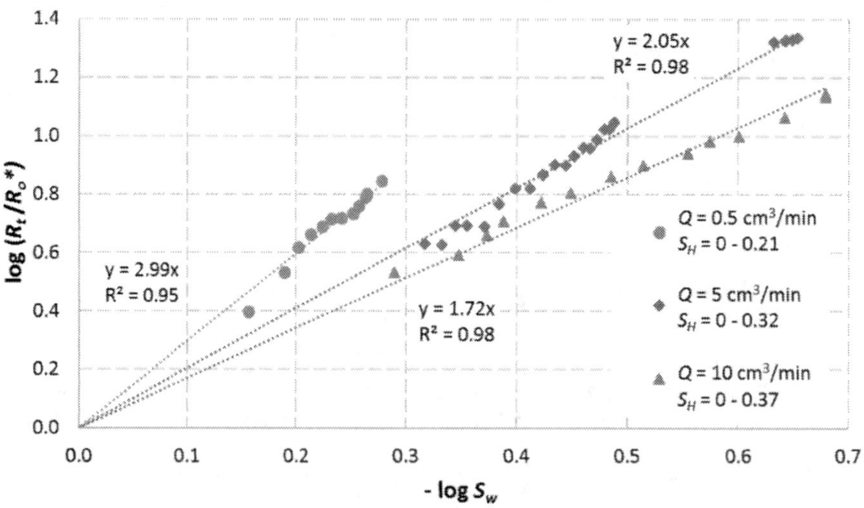

Fig 4. Logarithmic cross plot of RI and S_w during CO_2 hydrate formation in Bentheimer core samples from a range of initial brine saturations caused by varying CO_2 injection rates: 0.5, 5 and 10 cm^3/min at 7.0 MPa and 4 °C. Depending on the saturation profiles, the saturation exponent n ranged from 1.7 to 3.0.

When hydrates, water and CO_2 were present simultaneously in the pore space, n corresponded to 1.7 for $S_H = 0.37$ ($S_w = 0.21$), 2.1 for $S_H = 0.32$ ($S_w = 0.22$), and 3.0 for $S_H = 0.21$ ($S_w = 0.53$). Compared to the CO_2-brine system, hydrate formation changed the n value for the 0.5 cm^3/min drainage experiment (least uniform saturation distribution) from 2.3 to 3.0, for 10 cm^3/min from 1.9 to 1.7, while for 5 cm^3/min n remained unchanged (2.1). In the case of very limited hydrate formation ($S_H = 0.21$), low bulk resistivity measurements can be ascribed to substantial connectivity and increased ion content of the remaining free water. The obtained n values are nonetheless in good agreement with recent studies for natural gas hydrate in coarse-grained reservoirs [15], and for glass bead specimen [25].

The next three figures show a direct comparison of saturation values derived from Archie's and from measured PVT data.

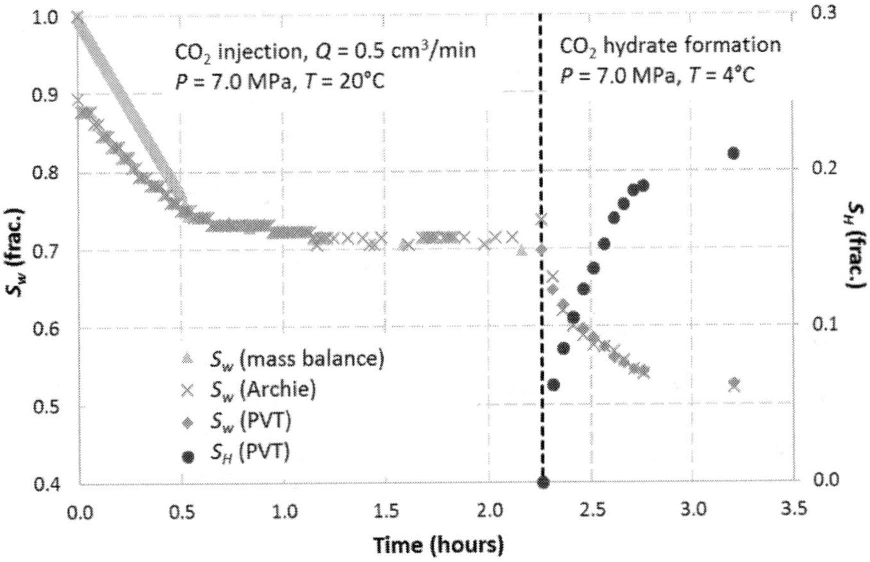

Fig 5. Comparison of saturation calculations from Archie and measured PVT data for injection rate 0.5 cm³/min during initial displacement of water by CO_2 (7.0 MPa and 20 °C) and the following CO_2 hydrate formation (7.0 MPa and 4 °C). Obtained saturation values correlates very well after CO_2 breakthrough as a uniform saturation profile is established, and throughout CO_2 hydrate formation.

In Fig 5, saturation profiles during the initial displacement and the following hydrate nucleation and growth are displayed for flow rate 0.5 cm³/min. The aforementioned unsteady-state regime occurring before CO_2 breakthrough, leads to severe deviation between the Archie saturation and correct linear displacement (mass balance) due to temporarily non-uniform saturation profiles. Once CO_2 breaks through at the outlet end of the sample, the saturation values from Archie's match actual PVT values very well at the plateau ($S_w \approx 0.7$). The onset of hydrate formation is indicated with a vertical line (broken). At this point, Archie's overestimate the water saturation somewhat compared to actual measurements. This apparent increase in water saturation is likely due to a short drop in resistivity linked to hydrate nucleation as reported in the literature [8,16,26]. Another possibility is the aforementioned CO_2 effects that may overestimate the water saturation, although the effects are most likely inhibited by the saline brine present. For the following hydrate growth process there is a very good agreement between the two water saturation profiles.

In Fig 6, drainage of water by CO_2 at injection rate of 5 cm³/min and subsequent hydrate formation is displayed. Again, we observed a deviation in saturation profiles before CO_2 BT, and a good agreement after the CO_2 front reached the outlet end of the sample. The consistency continues from the onset of hydrate formation until hydrates occupy approximately 15% of the pore space. At this point the hydrate formation rate decreased substantially and the saturation profiles temporarily plateaued (for 0.2 hours). This period of hampered hydrate growth is not captured using Archie's saturation calculations, thus underestimating the water saturation here. Accelerated hydrate formation followed next and this "normalization" caused the end-point saturation values from PVT data and resistivity measurements to match once again.

Fig 7 shows saturation profiles during the initial displacement, and the following hydrate nucleation and growth for flow rate of 10 cm³/min. The remaining water saturation in the core after CO_2 breakthrough was almost identical to the 5 cm³/min experiment. There is a good agreement between the water saturation profiles after this point including the whole hydrate formation period in Fig 7.

CO_2 dissociation effects are highly sensitive to salinity. The 3.5 weight% NaCl solution used in Fig 5-7, belongs in a "high-salinity regime" where the conductivity was actually reduced by up to 15% due to reduced ion mobility [27]. This CO_2 dissolution effect, if not accounted for, will underestimate the water saturation derived from resistivity measurements. At the time-scale investigated in our study, no consistent impact of CO_2 dissociation on resistivity measurements was observed. Modifying Archie's equation by accounting for reduced effective porosity and increased salinity of the remaining water for each time step [16], resulted in resistivity saturation values agreeing very well with obtained PVT measurements.

 https://doi.org/10.1051/e3sconf/202014605002

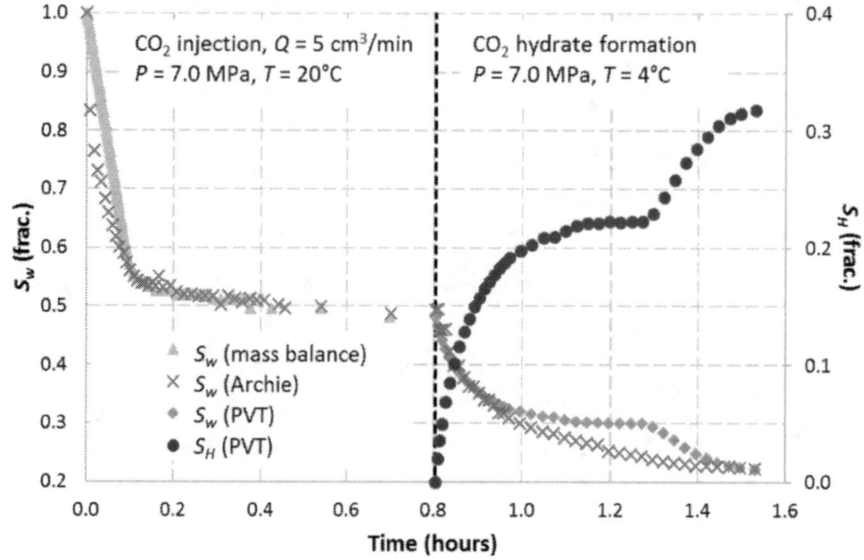

Fig 6. Comparison of saturation values obtained from Archie and measured PVT data for injection rate 5 cm³/min during drainage (7.0 MPa and 20 °C) and during the following CO_2 hydrate formation (7.0 MPa and 4 °C).

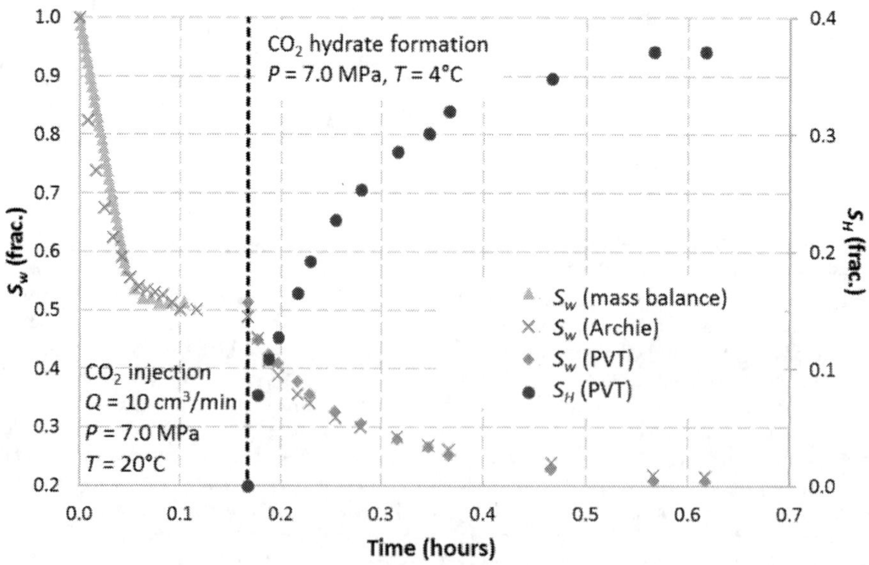

Fig 7. Comparison of saturation calculations from Archie and measured PVT data for injection rate 10 cm³/min during initial displacement of water by CO_2 (7.0 MPa and 20 °C) and the subsequent CO_2 hydrate formation (7.0 MPa and 4 °C).

In addition to the constant pressure experiments, a series of flow-induced CO_2 hydrate formation experiments were tested for various thermodynamic conditions (within the GHSZ). CO_2 was injected into fully brine-saturated core samples at 7.0 MPa pore pressure and aquifer temperature of 4 °C or 6 °C. In Fig 8, resistivity profiles for different CO_2 flow rates, and salinity and temperature regimes are compared as a function of time. Here, increased flow rate (from 0.5 to 5 cm^3/min) accelerated hydrate formation and subsequent CO_2 trapping and immobilization. However, in terms of pore volumes (PV) CO_2 injected, we observed no effect of injection rate on hydrate induction time. The initial displacement of brine by liquid CO_2 increased the bulk resistivity from approximately 5 Ωm to 10 Ωm in all four corefloods. Two experiments were flooded with CO_2 at a constant rate of 5 cm^3/min at 7.0 MPa and 4 °C, where one core contained 3.5 weight% NaCl (red curve) and the other 5 weight% NaCl (yellow) – to demonstrate the effect of salinity increase on hydrate formation. Furthermore, two experiments were flooded with CO_2 at a constant rate of 0.5 cm^3/min at 7.0 MPa and salinity of 3.5 weight% NaCl, one experiment at 4 °C (blue) and the other at 6 °C (light blue) - to demonstrate the effect of temperature increase.

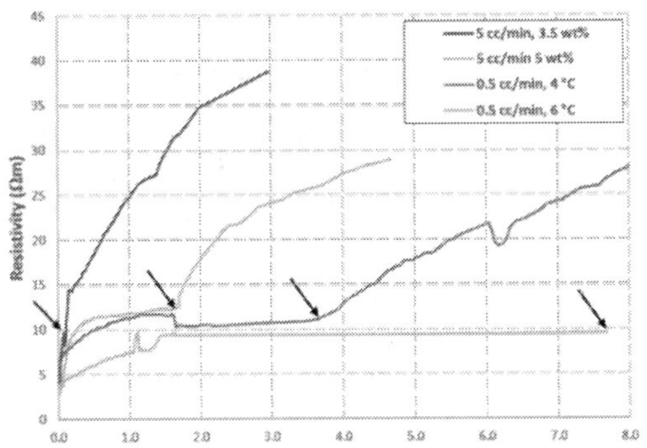

Fig 8. Resistivity profiles for various temperature and salinity conditions. Arrows indicate hydrate nucleation detected by a combination of pressure, resistivity, and temperature readings. Increase in salinity/temp caused a delayed CO_2 hydrate seal formation during continuous flow experiments.

The 5 cm^3/min constant rate experiment at lowest salinity (Fig 8 – red curve) started forming solid hydrates in the pore space shortly after CO_2 breakthrough (nucleation indicated with black arrows). By increasing the brine salinity from 3.5 to 5 weight% (yellow curve), we observed a prolonged induction time of approx. 1.5 hours (factor 9 increase) from flow-induced hydrate formation. When injecting CO_2 at 0.5 cm^3/min at 3.5 weight%, the effect of increasing the sandstone temperature from 4 °C to 6 °C was a factor 2 increase in induction time from 3.8 hours (blue) to 7.6 hours (light blue – resistivity data beyond this point is missing, however point of hydrate nucleation was identified from pressure and temperature logs).

The flow-induced hydrate induction time was evidently sensitive to salinity and temperature variations, and must be taken into consideration when screening for potential carbon storage sites. All four experiments led to solid CO_2 hydrate formation eventually. The steady increasing resistivity profiles after nucleation demonstrated hydrate growth in the pore network and decreased effective porosity and permeability. All corefloods experienced significant differential pressure build-up across the samples after hydrate formation, effectively stopping the CO_2 production at the outlet. These observations suggest that the injected CO_2 phase is made discontinuous by pore-spanning hydrate layers acting as permeability barriers, and thus successfully obstruct viscous CO_2 flow in the core sample for the time investigated.

Conclusions

Electrical resistivity measurements providing fluid saturations relevant for CO_2 hydrate storage, resulted in the following key experimental observations:

For two-phase CO_2-brine systems, the saturation exponent n ranged from 1.9 – 2.3 (harmonic mean $n \approx 2.1$) depending on the CO_2 injection rate used during the drainage process. Because the saturation exponent is sensitive to the saturation profile along the core length, it is not recommended to rely on saturation values derived from resistivity measurements using a 2-electrode setup in non-uniform fluid distribution processes.

During CO_2 hydrate formation, the saturation exponent n ranged from 1.7 – 3.0 (harmonic mean $n \approx 2.1$) depending on the initial distribution of brine, which resulted in different final CO_2 hydrate saturations. The estimated values of n may be used to map the brine saturation S_w and the CO_2 hydrate saturation ($S_H = 1 - S_w$) in excess water conditions, and are in good agreement with previously measured n values during methane hydrate growth. Resistivity measurements are increasingly important for $S_H < 0.4$, as acoustic methods currently cannot obtain sufficient velocity contrasts in zones of low hydrate saturation.

CO_2 hydrates effectively blocked the CO_2 flow path and sealed off the sandstone pore network during flow-induced hydrate formation for different injection rates and thermodynamic conditions. Moderate increase in brine salinity or aquifer temperature resulted in significantly prolonged induction time before CO_2 hydrate formed under constant flow rate. Once stored, unwanted CO_2 re-mobilization/migration was obstructed by formation of sedimentary hydrate layers. This observed mechanism could act as an additional safety factor against leakage from geological stored CO_2 located below the gas hydrate stability zone.

The authors gratefully acknowledge financial support from Equinor and the Research Council of Norway. One of the authors is supported by CLIMIT grant number 255490.

Nomenclature

a = tortuosity constant
A = area, L^2, m^2
C = ion content of brine, ppm
F = formation factor
L = length, L, m
m = cementation exponent
n = Archie's saturation exponent
RI = resistivity index
R_o = resistivity of fully water saturated sample, $(mL^3)/(tq^2)$, Ωm
R_t = resistivity of sample, $(mL^3)/(tq^2)$, Ωm
R_w = resistivity of brine, $(mL^3)/(tq^2)$, Ωm
S_{CO2} = saturation of CO_2, fraction
S_H = saturation of hydrate, fraction
S_w = saturation of water, fraction
S_{wi} = initial saturation of water, fraction
T = temperature, T, °C
Z = impedance, $(mL^2)/(tq^2)$, Ω
ΔP = differential pressure, $m/(Lt^2)$, bar
Θ = phase angle, °
φ = porosity, fraction
φ_{eff} = effective porosity, fraction
μ = viscosity, cP

References

1. International Energy Agency (IEA/OECD), "20 years of Carbon Capture and Storage - Accelerating future deployment." Review report (2016)
2. O. Eiken, P. Ringrose, C. Hermanrud, B. Nazarian, T. A. Torp and L. Høier, "Lessons learned from 14 years of CCS operations: Sleipner, In Salah and Snøhvit." Energy Procedia **4**, 5541-5548 (2011)
3. S. Benson and D. R. Cole, "CO2 Sequestration in Deep Sedimentary Formations." *Elements* **4**, 325-331 (2008)
4. H. Koide, M. Takahashi, H. Tsukamoto and Y. Shindo, "Self-trapping mechanisms of carbon dioxide in the aquifer disposal." *Energy Conversion and Management* **36**, 505-508 (1995)
5. K.Z. House, D.P. Schrag, C.F. Harvey and K.S. Lackner, "Permanent carbon dioxide storage in deep-sea sediments." *Proc Natl Acad Sci USA* **103**, 12291-12295 (2006)
6. B. Tohidi, J. Yang, M. Salehabadi, R. Anderson, and A. Chapoy, "CO2 hydrates could provide secondary safety factor in subsurface sequestration of CO2." *Environ. Sci. Technol.* **44**, 1509-1514 (2010)
7. M. Massah, D. Sun, H. Sharifi and P. Englezos, "Demonstration of gas-hydrate assisted carbon dioxide storage through horizontal injection in lab-scale reservoir." *Journal of Chemical Thermodynamics* **117**, 106-112 (2018)
8. J. Gauteplass, S. Almenningen, G. Ersland and T. Barth, "Hydrate seal formation during laboratory CO2 injection in a cold aquifer." *International Journal of Greenhouse Gas Control* **78**, 21-26 (2018)
9. J. Gauteplass, S. Almenningen, G. Ersland, T. Barth, J. Yang and A. Chapoy, "Multiscale investigation of CO2 hydrate self-sealing potential for carbon geo-sequestration." *Chemical Engineering Journal* **381**, 3122646 (2020)
10. S. Almenningen, J. Gauteplass, P. Fotland, G. L. Aastveit, T. Barth and G. Ersland, "Visualization of hydrate formation during CO2 storage in water-saturated sandstone." *International Journal of Greenhouse Gas Control* **79**, 272-278 (2018)
11. C. A. Rochelle, A. P. Camps, D. Long, A. Milodowski, K. Bateman, D. Gunn, P. Jackson, M. A. Lovell and J. Rees, "Can CO2 hydrate assist in the underground storage of carbon dioxide?" *Geological Society* **319**, 171-183 (2009)
12. G. E. Archie, "The Electrical Resistivity Log as an Aid in Determining Some Reservoir Characteristics." *AIME* **146**, 54-62 (1942)
13. Baker-Hughes. *Introduction to Wireline Log Analysis* (1992)
14. J. J. Arps, "The Effect of Temperature on the Density and Electrical Resistivity of Sodium Chloride Solutions." *Journal of Petroleum Technology* **5**, 17-20 (1953)
15. A. E. Cook and W. F. Waite, "Archie's Saturation Exponent for Natural Gas Hydrate in Coarse-Grained Reservoirs." *Journal of Geophysical Research: Solid Earth* **123**, 2069-2089 (2018)
16. K. A. Birkedal, G. Ersland, L. P. Hauge, A. Graue, K. Hester, J. Stevens and J. Howard, "Electrical resistivity measurements of CH4 hydrate-bearing sandstone during formation." *7th International Conference on Gas Hydrates* (2011)
17. A. E. Cook, B. I. Anderson, J. Rasmus, K. Sun, Q. Li, T. S. Collett and D. S. Goldberg, "Electrical anisotropy of gas hydrate-bearing sand reservoirs in the Gulf of Mexico." *Marine and Petroleum Geology* **34** (2012)
18. Y. F. Sun, and D. Goldberg, "Dielectric method of high-resolution gas hydrate estimation." *Geophysical Research Letters* **32** (2005)
19. T. Ramstad and H. Rueslåtten, "Pore scale numerical analysis for geological sequestration of CO_2." Technical report, 1-63 (2013)
20. C. Hågenvik, "CO2 Injection in Hydrate Bearing Sandstone with Excess Water." University of Bergen, MSc thesis (2013)
21. D. Bosch, J. Ledo, P. Queralt, F. Bellmunt, L. Luquot and P. Gouze, "Core-scale electrical resistivity tomography (ERT) monitoring of CO2–brine mixture in Fontainebleau sandstone." *Journal of Applied Geophysics* **130**, 23-36 (2016)
22. M. Han, S. Youssef, E. Rosenberg, M. Fleury and P. Levitz, "Deviation from Archie's law in partially saturated porous media: Wetting film versus disconnectedness of the conducting phase." *Physical Review E* **79** (2009)
23. J. H. Börner, V. Herdegen, J.-U. Repke and K. Spitzer, "The impact of CO2 on the electrical properties of water bearing porous media – laboratory experiments with respect to carbon capture and storage." *Geophysical Prospecting* **61**, 446-460 (2013)
24. M. Fleury and H. Deschamps, "Electrical Conductivity and Viscosity of Aqueous NaCl Solutions with Dissolved CO2." *Journal of Chemical & Engineering Data* **53**, 2505-2509 (2008)

https://doi.org/10.1051/e3sconf/202014605002

25. E. Spangenberg and J. Kulenkampff., "Influence of methane hydrate content on electrical sediment properties." *Geophysical Research Letters* **33** (2006)
26. Y. Liu, W. Zhang, Y. Liu, S. Ren, "Experimental characterization and modelling of acoustic and electrical resistance in hydrate bearing sediments." *6th International Conference on Gas Hydrates* (2008)
27. J. H. Börner, K. Spitzer, J.-U. Repke and V. Herdegen, "The electrical conductivity of CO2-bearing pore waters at elevated pressure and temperature: a laboratory study and its implications in CO2 storage monitoring and leakage detection." *Geophysical Journal International* **203**, 1072-1084 (2015)

Local Capillary Pressure Estimation Based on Curvature of the Fluid Interface – Validation with Two-Phase Direct Numerical Simulations

Takashi Akai[1,*], *Branko* Bijeljic[1], and *Martin* Blunt[1]

[1]*Imperial College London, Department of Earth Science and Engineering, London, SW7 2AZ, United Kingdom*

Abstract. With the advancement of high-resolution three-dimensional X-ray imaging, it is now possible to directly calculate the curvature of the interface of two phases extracted from segmented CT images during two-phase flow experiments to derive capillary pressure. However, there is an inherent difficulty of this image-based curvature measurement: the use of voxelized image data for the calculation of curvature can cause significant errors. To address this, we first perform two-phase direct numerical simulations to obtain the oil and water phase distribution, the exact location of the interface, and local fluid pressure. We then investigate a method to compute curvature on the oil/water interface. The interface is defined in two ways. In one case the simulated interface which has a sub-resolution smoothness is used, while the other is a smoothed interface which is extracted from synthetic segmented data based on the simulated phase distribution. Computed mean curvature on these surfaces are compared with that obtained from the fluid pressure computed directly in the simulation. We discuss the accuracy of image-based curvature measurements for the calculation of capillary pressure and propose the best way to extract an accurate curvature measurement, quantifying the likely uncertainties.

1 Introduction

Capillary pressure, P_c, is a pressure discontinuity across the interface between oil and water, defined as $P_c = P_o - P_w$, where P_o and P_w are the pressures of oil and water phase, respectively. Traditionally, capillary pressure for oil/water systems has been measured in a laboratory using the porous plate method in which the pressure of each phase is measured using two external pressure transducers. Based on the Young-Laplace equation, capillary pressure locally is defined as:

$$P_c = 2\sigma\kappa_m, \qquad (1)$$

where σ is the interfacial tension between two phases and κ_m is the mean curvature of the interface.

With the advancement of high-resolution three-dimensional X-ray imaging, it is now possible to directly measure the curvature of the interface extracted from segmented CT images during two-phase flow experiments to derive capillary pressure. Armstrong et al. [1] demonstrated this approach using synchrotron-based tomographic datasets of oil/water drainage and imbibition cycles on a bead pack structure [2]. They compared the capillary pressure obtained from curvature measurements with that obtained from pressure transducers. Fairly good agreement was obtained for imbibition, whereas the curvature measurement showed a systematically lower value than that obtained from the transducers for drainage cycles. Later, using the same dataset, Li et al. [3] presented that their proposed curvature measurement method improved the agreement with the transducer based capillary pressure. Using a similar curvature measurement method, Herring et al. [4] estimated the capillary pressure for a range of curvature between 0 and 0.225 voxel^{-1} based on their air/water drainage and imbibition experiments on a Bentheimer sandstone. However, there is an inherent difficulty of this image-based curvature measurement: the use of voxelized image data for the calculation of curvature can cause significant errors, resulting in a wide range of measured values, with some negative curvature values, which are not expected in a water-wet system. Hence, it is not clear how the distribution of measured curvature values represents the true range of local capillary pressure.

We investigate the accuracy of curvature measurement on the basis of pore-by-pore comparison using direct numerical simulations of two-phase flow. The color-gradient lattice Boltzmann method is employed to generate an oil and water phase distribution in pore structures of a bead pack and Bentheimer sandstone. From the simulated phase distribution, synthetic segmented data is generated, then curvature computation on the interface extracted from this segmented data is performed by employing several smoothing methods. These curvature values are then compared with those obtained from the simulated local fluid pressure. We discuss the accuracy of image-based curvature measurements for the calculation of capillary pressure and suggest the best method with associated errors for capillary pressure estimation.

* Corresponding author: t.akai17@imperial.ac.uk

2 Methods

2.1 The color-gradient lattice Boltzmann method

The color-gradient LB model proposed by Halliday et al. [5] was used. Our LB model was constructed with a 3D19Q lattice model which consists of a set of 19 discrete lattice velocity vectors, $\mathbf{e}_i$, in three-dimensional space. We defined particle distributions of two immiscible fluids, labeled red and blue, as f_i^r and f_i^b, respectively. The fluid density, ρ_r and ρ_b, and velocity, $\mathbf{u}$, at position $\mathbf{x}$ and time t are obtained by:

$$\rho^k = \sum_i f_i^k(\mathbf{x},t), \tag{2}$$

$$\rho\mathbf{u} = \sum_i f_i \mathbf{e}_i, \tag{3}$$

where f_i is the total particle distribution given by $f_i = f_i^r + f_i^b$; ρ is the total fluid density given by $\rho=\rho_r+\rho_b$. The lattice Boltzmann equation for the total particle distribution is written as:

$$f_i(\mathbf{x}+\mathbf{e}_i\delta t, t+\delta t) = f_i(\mathbf{x},t) + \Omega_i(\mathbf{x},t) + \phi_i, \tag{4}$$

where t denotes the lattice time step which was set to unity and Ω_i and ϕ_i are the collision operator and the body force term, respectively. For the collision operator, we used the Multiple-relaxation-time (MRT) collision operator [6] expressed as:

$$\Omega_i = -\left(\mathbf{M}^{-1}\mathbf{S}\mathbf{M}\right)_{i,j}\left(f_i - f_i^{eq}\right), \tag{5}$$

where $\mathbf{M}$ and $\mathbf{S}$ are the transformation matrix and the diagonal relaxation matrix, respectively. f_i^{eq} is the equilibrium distribution function which is obtained by a second order Taylor expansion of the Maxwell-Boltzmann distribution with respect to the local fluid velocity. The location of the interface was tracked using a color function ρ^N defined by:

$$\rho^N(\mathbf{x},t) = \frac{\rho_r(\mathbf{x},t) - \rho_b(\mathbf{x},t)}{\rho_r(\mathbf{x},t) + \rho_b(\mathbf{x},t)}, \quad -1 \le \rho^N \le 1. \tag{6}$$

Using the color function, the interfacial tension between two fluids was computed as a spatially varying body force, $\mathbf{F}$, based on the continuum surface force (CSF) model [7] given by:

$$\mathbf{F} = -\frac{1}{2}\sigma\kappa\nabla\rho^N, \tag{7}$$

where σ is the interfacial tension and κ is the curvature of the interface. This spatially varying body force $\mathbf{F}$ was incorporated into the lattice Boltzmann equation through the body force term ϕ_i. For the MRT model, this was performed by transforming the forcing term proposed by Guo et al. [8] using the scheme presented in Yu and Fan [9]. After the application of the interfacial tension ($\mathbf{F}$) to the particle distributions, the recoloring algorithm proposed by Latva-Kokko and Rothman [10] applied to these distributions to ensure phase segregation and maintain the interface. This results in a slightly diffusive interface whose thickness is about 2 to 3 lattice units. Further details of our LB model can be found in Akai et al. [11, 12]. The only difference is that we used the MRT collision operator, while the single-relaxation-time (SRT) collision operator [13] was used in Akai et al. [11].

At solid-fluid boundary lattice nodes, a full-way bounce back boundary condition was implemented to achieve a non-slip boundary condition. In addition, the wettability of solid surface was modeled by specifying contact angles using the wetting boundary condition presented in Akai et al. [11]. This wetting boundary condition accurately assigns contact angles for 3D arbitrary geometries with smaller spurious currents compared to the widely-used fictitious density boundary condition [11].

2.2 Curvature computation on voxelized images

There are mainly two approaches to compute curvature from voxelized data such as micro-CT images. One approach calculates curvatures from the gradient of 3D float data (e.g., raw or processed grayscale data), while the other approach estimates curvature through the fitting of a quadratic form locally to a surface extracted from voxelized, segmented data [14]. We used the latter approach as this has been presented in previous studies [14, 1, 15, 4, 3, 16].

We started with three-phase segmented label data (oil, water and solid) obtained from raw grayscale CT images. Using the marching cubes algorithm [17], the oil/water interface was extracted from the label data. Since this surface had a staircase shape, it had to be smoothed before computing curvature. In this study, we compare three smoothing methods: Constrained Gaussian smoothing (CGS), Laplacian smoothing (LPS) and boundary preserving Gaussian smoothing (BPGS). CGS is applied when the surface is extracted using the marching cubes algorithm. A Gasussian kernel filter with different kernel sizes is applied to label data, then isosurface is extracted. In this process, the constraint to preserve the location of an original label is imposed.

LPS [18] and BPGS [19] are applied after the extraction of the surface with the marching cubes algorithm. The extracted surface with a staircase shape is modeled as a triangulated surface. Then, the vertices of the triangle elements are moved with these smoothing methods. In LPS, the position of a vertex is moved to the average position of its neighboring vertices. BPGS also moves the position of a vertex based on the position of its neighboring vertices. A scale factor which determines the degree of the movement in one iteration is defined. Two consecutive smoothing steps with a positive and negative scaling factor are performed in one iteration. This smoothing produces a surface which preserves the original boundary without shrinkage [19]. The degree of the smoothing in LPS and BPGS is controlled by the number of iterations. In this study, CGS and BPGS were performed with commercial image processing software, Avizo, while LPS was performed with Paraview.

After the generation of a smoothed triangulated surface, the elemental triangles were fitted by a quadratic form:

$$ax^2 + by^2 + cz^2 + 2exy + 2fyz + 2gzx + 2lx + 2my + 2nz + d = 0. \tag{8}$$

Then, the principal curvature values and directions of principal curvature were obtained from the eigenvalues and eigenvectors of the fitted quadratic form in Eq. 8 [1]. Avizo was used to perform this computation. Users can choose the number of neighboring triangles to be used for the fitting at the center of a triangle. We used a fixed value of 4 neighbors in the following analyses.

3 Results

In section 3.1, we provide validation of our direct numerical simulations for different grid resolutions. Next, in section 3.2, using a single oil droplet test case, we investigate the accuracy of the curvature computation method described in section 2.2. Here, a smoothed interface which was extracted from synthetic segmented data based on the simulated phase distribution is used to compute curvature, then the resultant values are compared with those obtained from the simulated fluid pressure. Finally, in section 3.3, the same approach is applied to the simulated phase distributions in two realistically complex pore structures of a bead pack and Bentheimer sandstone.

3.1 Validation of the two-phase lattice Boltzmann model

To validate our two-phase lattice Boltzmann model and investigate its accuracy as a function of grid resolutions, the oil/water interface in a corner of triangular pore space was simulated. A 2D pore structure with an isosceles triangle shape as shown in Fig. 1 was used. Here, the length of L_x and L_y were set to 70 µm. R is the radius of curvature of the interface; θ is the contact angle; β is the half angle of the corner of the triangle, which is given by: $\tan\beta = L_x/2L_y$. This pore structure was modeled with 4 grid sizes of Δ=1.0, 2.0, 3.5 and 5.0 µm.

The identical density and viscosity of the water and oil phases were set to 1,000 kg/m^3 and 1 mPa, respectively. The interfacial tension and contact angle were set to 18 mN/m and 45°, respectively. Initially, the lower part of the pore structure was filled with oil to a specified oil saturation, while the other part at the top corner was filled with water. Then, simulations were performed for 50,000 time steps until they reached equilibrium.

In this pore geometry, the radius of curvature, R, can be analytically derived based on the geometrical relationship, which is given by:

$$R^2 = \frac{S_w L_x^2}{4\tan\beta}\left(\frac{\cos^2\alpha\cos\beta}{\sin\beta} + \sin\alpha\cos\alpha - \left(\frac{\pi}{2} - \alpha\right)\right)^{-1}$$

$$\alpha = \theta + \beta, \tag{9}$$

where S_w is the water saturation and the other parameters are shown in Fig 1. This analytically derived radius of curvature was used to compare with the simulated radius of curvature obtained using the input contact angle of θ=45°. Based on the simulated fluid configurations, the exact location of the interface corresponding to ρ^N=0 (see Eq. 6) was extracted as shown in Fig. 2. Then, the radius of the curvature of this interface was obtained by fitting a circle to the interface, since in 2D in equilibrium the analytical shape of the interface is a part of a circle. Table 1 shows the comparison between the radius of curvature obtained from the analytical solution and that obtained with a circle fit to the simulated interface for different grid sizes. The relative error in the radius of curvature is less than 3% for these grid resolutions. This suggests that good agreement in capillary pressure is also obtained since the capillary pressure is directly linked to the radius of curvature through Eq. 1.

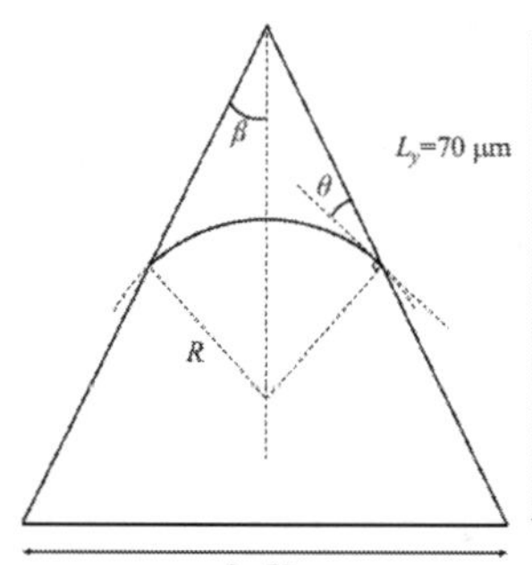

Fig. 1. A schematic of the isosceles triangle pore used for the simulations.

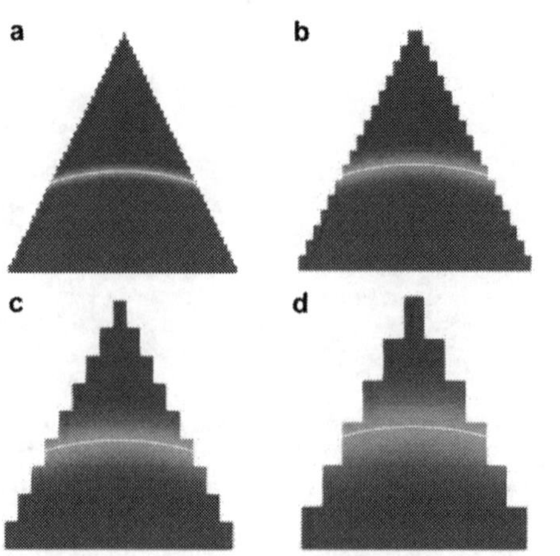

Fig. 2. Simulated fluid configurations for grid sizes of (a) Δ=1.0 µm, (b) Δ =2.0 µm, (c) Δ =3.5 µm and (d) Δ =5.0 µm. Here, oil and water are shown in red and blue, while the exact location of the oil/water interface is shown by the white line.

Table 1. Comparison between the radius of curvature obtained from the analytical solution and that obtained with a circle fit to the simulated interface for different grid sizes.

grid size	S_w	analytical R	simulated R	%Diff.
μm	%	μm	μm	%
1.0	35.9%	70.3	70.9	0.9%
2.0	35.6%	69.9	68.2	-2.4%
3.5	36.0%	70.3	68.3	-3.0%
5.0	32.7%	67.0	65.9	-1.7%

3.2 Curvature of a single oil droplet in triangular pore space

To investigate the curvature computation algorithm described in section 2.2, a simple 3D test case was performed. A cylindrical pore structure with the length of 171.5 μm with an isosceles triangular cross-section as in the previous test case was used. This pore structure was modeled with a grid size of 3.5 μm. The same fluid properties and contact angle as in the previous section were used. Initially, oil was placed in the central region of the pore space as shown in Fig. 3a. Then, simulations were performed for 50,000 time steps until they formed a single oil droplet after equilibrium (Fig. 3b).

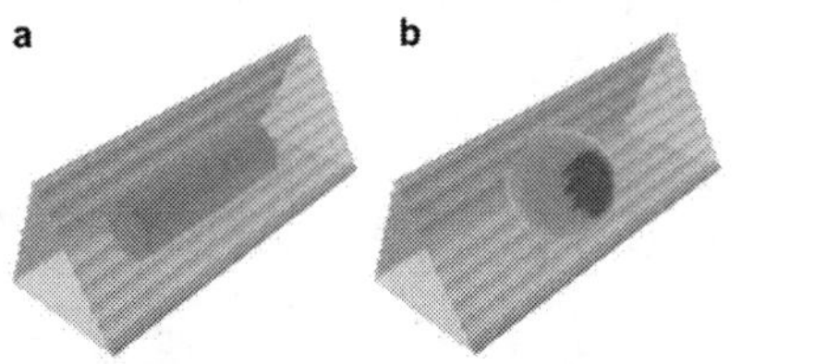

Fig. 3. An oil droplet in a 3D triangular pore space. (a) Initial condition. (b) Equilibrium condition. Here, solid is shown in transparent green and the oil phase is shown in red.

Two types of the oil/water interface were prepared based on the simulation results: a "simulated interface" and a "smoothed interface". The simulated interface was obtained by extracting the contour line of the color function $\rho^N=0$. This surface originally had a sub-resolution smoothness. The other surface, the smoothed interface, was obtained from synthetic voxelized label data. The simulated distribution of the color function was segmented into label data with the threshold of $\rho^N=0$, then the oil/water interface was extracted using the marching cubes algorithm. This surface has a staircase shape due to the shape of a voxelized grid system. Therefore, it needs to be smoothed before computing curvature. As discussed in section 2.2, three smoothing methods were applied: Constrained Gaussian smoothing (CGS), Laplacian smoothing (LPS) and boundary preserving Gaussian smoothing (BPGS). The parameters used for the smoothing are summarized in Table 2.

Based on the simulated fluid pressure, the capillary pressure was obtained using:

Table 2. Parameters used for Constrained Gaussian smoothing (CGS), Laplacian smoothing (LPS) and boundary preserving Gaussian smoothing (BPGS).

smoothing method	parameters	smoothing level: level 1	level 2	level 3
CGS	kernel size	1	3	9
LPS	number of iterations	200	600	1800
BPGS	number of iterations	50	150	450

$$P_c^{sim} = P_o^{avg} - P_w^{avg} = \frac{1}{N_o}\sum_{\rho^N>0.9} P - \frac{1}{N_w}\sum_{\rho^N<-0.9} P \quad (10)$$

where P_o^{avg} and P_w^{avg} are the average fluid pressure in the oil and water phases, respectively; N_o and N_w are the number of grid blocks having $\rho^N>0.9$ and $\rho^N<-0.9$, respectively; P is the simulated fluid pressure in each grid block. Here, to exclude the interface region, we computed an average pressure for each phase for $|\rho^N|>0.9$. Then, P_c^{sim} was converted to the simulated mean curvature using Eq. 1. We refer to this mean curvature as the fluid pressure derived mean curvature, κ_m^P.

Fig. 4 shows the oil/water interface after smoothing, colored by computed mean curvature. For all of the smoothed surfaces, we see a variation in the computed mean curvature, although the correct mean curvature should give a uniform value as shown in the curvature computed on the simulated interface, Fig. 4(j), because the oil droplet is in capillary equilibrium. Furthermore, we see errors around the edge of the interface (close to the three-phase contact lines) for all of the smoothed interfaces, Fig. 4(a)~(i). Therefore, we decided to discard the data points whose distance from solid surface is fewer than 3 voxels. This 3 voxels distance cutoff was used for all of the following analyses. Table 3 summarizes the average and standard deviation of the computed mean curvature and the relative difference to the fluid pressure derived mean curvature, κ_m^P =0.133 voxel^{-1} for the smoothing level 2 as shown in Fig. 4(b), (e), (h). We note that Li et al. [3] proposed that in addition to the distance cutoff, taking an average weighted by the distance from the solid surface further improves the accuracy of curvature measurement using synthetic test cases of the interface in a cylindrical structure with a circular-cross section; however, taking a distance-weighted average did not improve the estimation of curvature in our cases as we can conclude from the overestimated curvature values observed in the central region of the oil droplet in Fig. 4(a) ~ (i).

In fact, the optimum smoothing method and its level of smoothing is dependent on the shape and the size of an object. Therefore, here, we only provide the qualitative features of the three smoothing methods. CGS preserves the shape of an object, but significant smoothing cannot be applied even with increasing the kernel size, because the resultant surface is constrained by the original voxel data. As a result, CGS tends to give a wider range of variation. LPS can apply significant smoothing by

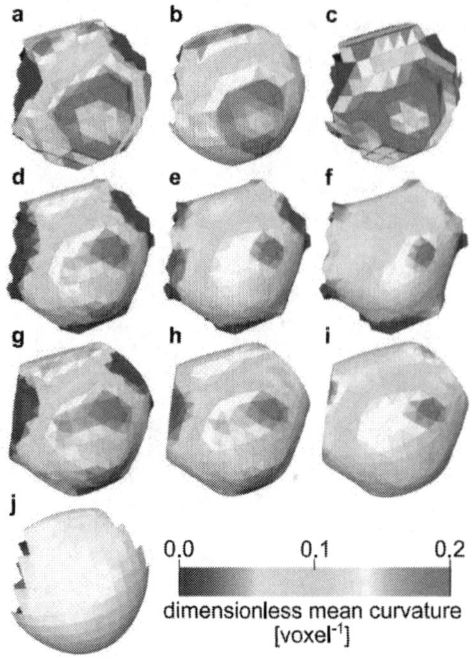

Fig. 4. The oil/water interface after the application of the several smoothing methods with different smoothing levels. CGS with a kernel size of (a) 1 voxel, (b) 3 voxels and (c) 9 voxels. LPS with (d) 200 iterations, (e) 600 iterations and (f) 1800 iterations. BPGS with (g) 50 iterations, (h) 150 iterations and (i) 450 iterations. (j) The simulated interface. Here, the surface is colored by the value of mean curvature. The correct surface should present uniform mean curvature as seen in the simulated interface since the droplet is in capillary equilibrium for which the capillary pressure is uniform.

increasing the number of iterations; however, this could change the shape of the interface as shown in a top part of the interface in Fig. 4(f) where we see the bending of the interface towards the opposite direction. This is caused by the propagation of errors in the three-phase contact line through many iterations. However, combining with the distance cutoff, this erroneous part can be effectively removed while preserving a well smoothed surface in the middle of the interface. BPGS gives the similar results to LPS, while better keeping the shape of an object compared to LPS.

For the simulated interface which was directly obtained from the simulated color function without any smoothing, the computed curvature showed good agreement with the fluid pressure derived curvature with a smaller standard deviation (Table 3). Moreover, since this surface did not show the errors close to the three-phase contact line, the average and standard deviation without the distance cutoff also gave similar values, i.e., the average of 0.132 voxel^{-1} and standard deviation of 3.99×10^{-3} voxel^{-1}.

Table 3. Voxel based interfaces with the smoothing level 2 (see Table 2) and a distance cutoff of 3 voxels. Here, the relative difference was computed to the fluid pressure derived mean curvature of $\kappa_m{}^P$=0.133 voxel^{-1}.

surface	computed curvature avg. [voxel^{-1}]	computed curvature std. [voxel^{-1}]	relative difference [%]
simulated interface	0.134	3.23×10^{-3}	0.7%
smoothed with CGS	0.149	3.23×10^{-2}	12.1%
smoothed with LPS	0.132	2.40×10^{-2}	-0.1%
smoothed with BPGS	0.139	2.68×10^{-2}	5.0%

3.3 Local capillary pressure estimation for complex porous media

Two realistically complex pore structures were used: a synthetic bead pack pore structure and micro-CT images of Bentheimer sandstone. For these structures, simulation domains were composed of 256×256×256 voxels with a grid size of 3.56 μm as shown in Fig 5. For the analysis of the simulation results, the pore structures were divided into pore regions using the separate pore algorithm in Avizo, resulting in 402 and 272 pore regions for the bead pack and Bentheimer sandstone, respectively. The mean pore radius which accounts for 50% of the pore volume was 37 μm and 30 μm for the beadpack and Bentheimer sandstone, respectively.

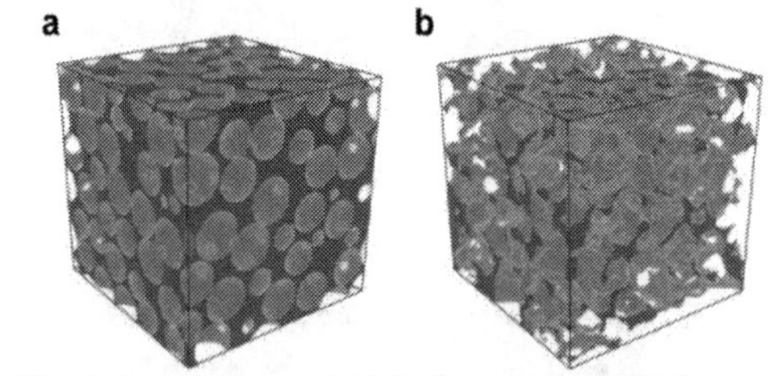

Fig. 5. Pore structures of (a) the bead pack and (b) the Bentheimer sandstone.

The identical density and viscosity of water and oil phase were set to 1,000 kg/m^3 and 1 mPa, respectively. The interfacial tension and contact angle were set to 25 mN/m and 45°, respectively. All 6 faces of the cubic simulation domain were covered with solid voxels, i.e., a no-slip boundary condition was applied to all 6 faces. Initially, 50% of oil and 50% of water were randomly placed in pore voxels, then the simulations were performed with no external force for 250,000 time steps (corresponding to 0.1 seconds) until they reached equilibrium conditions.

Fig. 6 shows the simulated phase distribution at the equilibrium condition. Similar to the analysis presented in the previous section, the simulated interface and the smoothed interface were prepared from the simulation results. Curvature computation was performed on these surfaces. Figs. 7 and 8 show the histogram of the computed curvature for the bead pack and Bentheimer sandstone, respectively. The fluid pressure derived mean

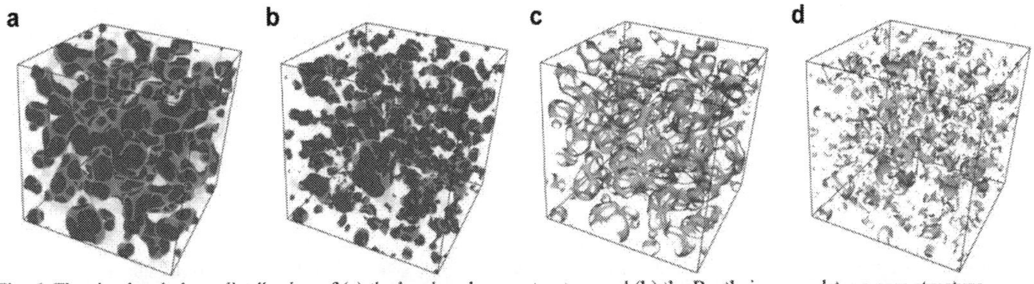

Fig. 6. The simulated phase distribution of (a) the bead pack pore structure and (b) the Bentheimer sandstone pore structure. Here, only oil is shown in red, while water and solid are transparent. The extracted interface based on the simulation results for (c) the bead pack and (d) the Bentheimer sandstone. This was performed extracting the contour line of ρ^N=0.

curvature, $\kappa_m{}^P$, obtained with Eq. 10 is also shown by the vertical line. In both figures, the histograms obtained from the smoothed interface without and with the distance cutoff are shown in (a) and (b), while those obtained from the simulated interface without and with the distance cutoff are shown in (c) and (d). The average and standard deviation of these distributions are summarized in Table 4 for the bead pack and Table 5 for the Bentheimer sandstone.

Here, we discuss the following two key observations. First, the computed curvatures obtained with the smoothed interface show the wide range of the distribution when the cutoff is not applied, while the curvatures obtained with the simulated interface show a much narrower range. For the histograms without the cutoff, the difference between the distributions obtained with the smoothed interfaces and the simulated interface suggests that there are many erroneous values in curvature computation on the smoothed surfaces. The histograms obtained with CGS appeared to have the most similar distribution to that obtained with the simulated interface. However, as shown in Table 4 and 5, their average values are much higher than that obtained from the fluid pressure and simulated interface because of the long tails of their distribution toward values higher than 0.3, which are not shown in Figs. 7 and 8. Second, after the application of the distance cutoff, all the histograms obtained with the smoothed interface become similar to that obtained with the simulated interface. However, in both the smoothed and the simulated interface, the data points of high curvature values have been lost. This is because the distance cutoff removes the data points of not only the edges of the interface but also the entire parts of the interface in small pores, which tend to give a high curvature value. Although a high curvature value is not captured when the distance cutoff is applied, the smoothed interface seems to provide a good estimate of local mean curvature values of the interface. This will be

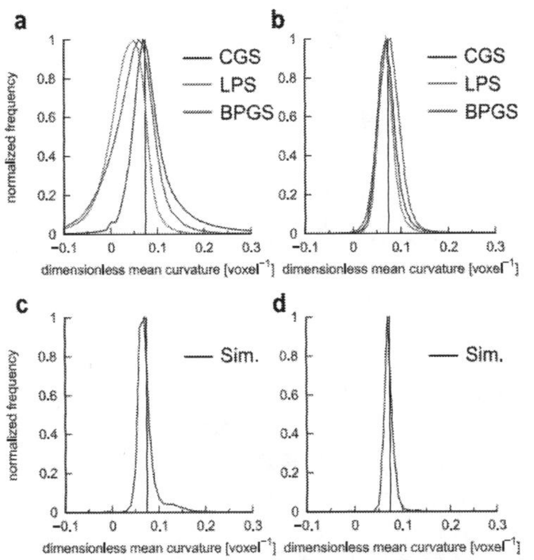

Fig. 7. The histogram of computed mean curvature for the bead pack. The histograms computed on the smoothed interface without and with the distance cutoff (a and b). The histograms computed on the simulated interface without and with the distance cutoff (c and d). The fluid pressure derived mean curvature obtained with Eq. 10 is also shown by the vertical line.

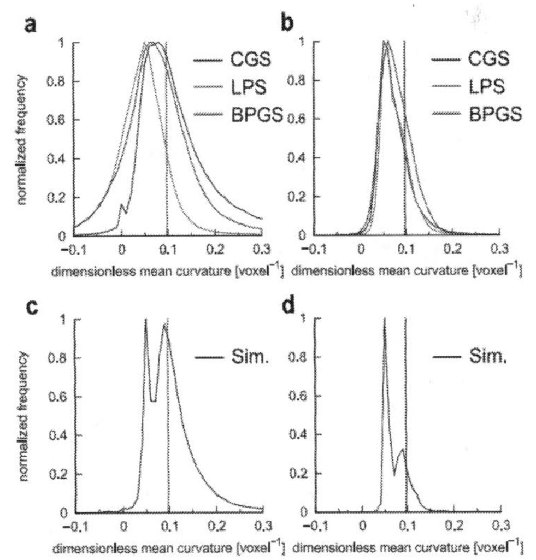

Fig. 8. The histogram of computed mean curvature for the Bentheimer sandstone. The histograms computed on the smoothed interface without and with the distance cutoff (a and b). The histograms computed on the simulated interface without and with the distance cutoff (c and d). The fluid pressure derived mean curvature obtained with Eq. 10 is also shown by the vertical line.

Table 4. The average and standard deviation of the distribution of computed mean curvature for the bead pack shown in Fig. 7.

	computed curvature avg. [voxel^{-1}]	computed curvature std. [voxel^{-1}]
simulated fluid pressure	7.40×10^{-2}	N/A
without the cutoff		
simulated interface	7.36×10^{-2}	4.25×10^{-2}
smoothed with CGS	11.5×10^{-2}	16.1×10^{-2}
smoothed with LPS	3.06×10^{-2}	5.63×10^{-2}
smoothed with BPGS	4.54×10^{-2}	6.87×10^{-2}
with the cutoff		
simulated interface	7.12×10^{-2}	1.22×10^{-2}
smoothed with CGS	8.02×10^{-2}	8.43×10^{-2}
smoothed with LPS	7.01×10^{-2}	2.03×10^{-2}
smoothed with BPGS	7.62×10^{-2}	2.54×10^{-2}

Table 5. The average and standard deviation of the distribution of computed mean curvature for Bentheimer sandstone shown in Fig. 8.

	computed curvature avg. [voxel^{-1}]	computed curvature std. [voxel^{-1}]
simulated fluid pressure	9.73×10^{-2}	N/A
without the cutoff		
simulated interface	11.1×10^{-2}	8.20×10^{-2}
smoothed with CGS	19.3×10^{-2}	27.0×10^{-2}
smoothed with LPS	4.24×10^{-2}	7.64×10^{-2}
smoothed with BPGS	8.51×10^{-2}	10.8×10^{-2}
with the cutoff		
simulated interface	7.17×10^{-2}	2.51×10^{-2}
smoothed with CGS	8.16×10^{-2}	8.48×10^{-2}
smoothed with LPS	7.45×10^{-2}	3.12×10^{-2}
smoothed with BPGS	7.81×10^{-2}	3.60×10^{-2}

further discussed in the following analysis. Among the three smoothing methods, when the cutoff is applied, LPS gave the closest average value of the mean curvature to that obtained from the simulated interface with the smallest standard deviation.

To further investigate the accuracy of curvature computation on the interface, the comparison was made on a pore-by-pore basis. First, we computed local capillary pressure applying Eq. 10 for each pore region. This local capillary pressure was converted to the mean curvature for each pore region using Eq. 1. Next, we also obtained the mean curvature for each pore region by taking the average of the computed curvature values on the interfaces in that pore region. Figs. 9 and 10 show the comparison between the dimensionless mean curvature obtained from the simulated fluid pressure and that obtained from the computed curvature on the interface for the bead pack and Bentheimer sandstone, respectively. In both, the curvature computed on the simulated interfaces gave consistent values with those obtained from the fluid pressure. This means that when the interface is reasonably smooth, the curvature computation gives an accurate estimation of local capillary pressure. For the curvature computed on the smoothed interface, CGS shows highly overestimated curvature values for some pore regions, while LPS and BPGS estimate curvature within a range of about ±20% difference from the curvatures obtained with the fluid pressure for the mean curvature smaller than

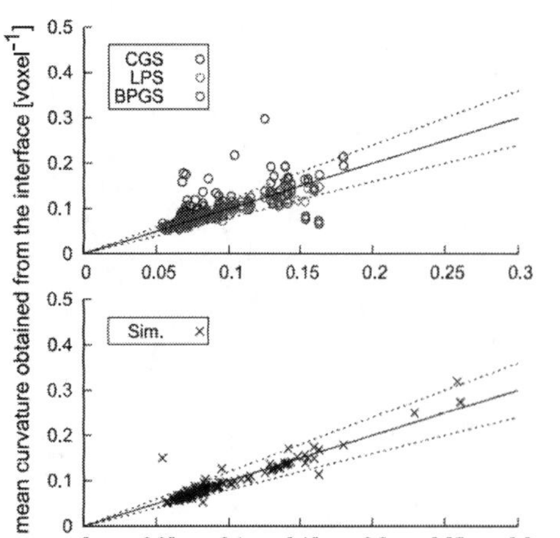

Fig. 9. Comparison of local mean curvature for the bead pack on pore-by-pore basis. X-axis shows the dimensionless mean curvature obtained from the simulated fluid pressure, while the Y -axis shows that computed on the interface. A unit slope indicating perfect agreement in these values is shown by a black solid line, while, ±20% difference is shown by black dotted lines.

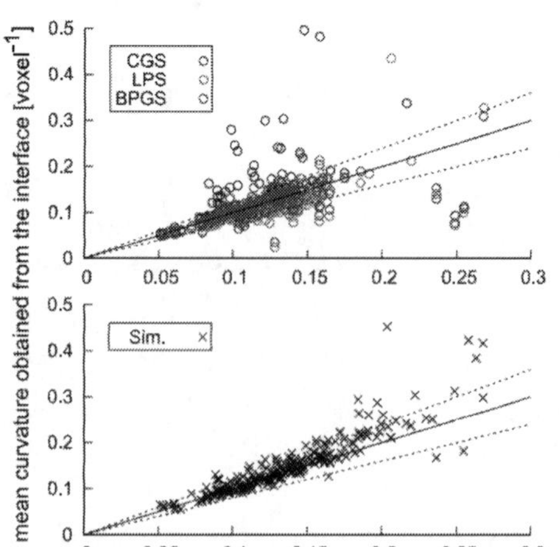

Fig. 10. Comparison of local mean curvature for the Bentheimer sandstone on pore-by-pore basis. X-axis shows the dimensionless mean curvature obtained from the simulated fluid pressure, while the Y -axis shows that computed on the interface. A unit slope indicating perfect agreement in these values is shown by a black solid line, while, ±20% difference is shown by black dotted lines.

about 0.15 $voxel^{-1}$. Therefore, with LPS or BPGS incorporating the distance cutoff of 3 voxels, local capillary pressure can be measured with a ±20% error up to 0.15 $voxel^{-1}$ at this resolution.

4 Conclusions

Direct numerical simulations were conducted to obtain oil and water phase distribution, the exact location of the oil/water interface and local fluid pressure. From the simulation results, two types of the oil/water interface were generated. One is a simulated interface which has a sub-resolution smoothness, while the other is a smoothed interface. The smoothed interfaces were generated by applying three types of smoothing methods to a staircase shape surface extracted from synthetic segmented data based on the simulated phase distribution. Then, curvature computation was performed on these surfaces. The resultant mean curvature value from these surfaces was compared with that obtained from the simulated fluid pressure. This was performed to find the best way to extract accurate curvature measurement, and to quantify the likely uncertainties.

First, we tested our approach using a single oil droplet in triangular pore space. Computed curvature values on the simulated interface showed a narrow distribution whose average value was consistent with that obtained from the fluid pressure. This means that when the surface is sufficiently smooth, the fitting of a quadratic form equation to the surface properly computes local curvatures. On the other hand, computed curvature values on the smoothed interface showed a wide distribution with erroneous values close to the three-phase contact lines. Therefore, we discarded values whose distance from solid surface was fewer than 3 voxels. After the application of the distance cutoff, the average of computed mean curvature values became closer to that obtained from the fluid pressure. However, their standard deviation was on the order of 10^{-2} $voxel^{-1}$ against the mean curvature value of 0.133 $voxel^{-1}$, which was 10 times higher than that obtained from the simulated interface. These deviated values were caused by error, since an oil droplet in capillary equilibrium should give a uniform value of mean curvature.

Next, realistically complex pore structures of a bead pack and Bentheimer sandstone were used to simulate oil droplets in capillary equilibrium. When the distance cutoff was not applied, curvature computed on the simulated interface gave an average value of mean curvature consistent with that obtained from the fluid pressure for both the bead pack and Bentheimer sandstone. However, when the distance cutoff was applied, the average value became lower than that from the fluid pressure. This is because the distance cutoff removes the data points of not only the edges of the interface but also the entire parts of the interface in small pores, which tend to give a high curvature value. For the smoothed interface, the distributions obtained without the cutoff were quite different from that obtained from the simulated interface. Many negative values of computed mean curvatures suggest significant errors in these cases. After the application of the distance cutoff, the distribution became closer to that obtained with the simulated interface.

Finally, we compared computed mean curvature on a pore-by-pore basis. As opposed to the test case performed on the single oil droplet, oil droplets with different mean curvature values corresponding to different pore sizes can be obtained in these simulations. Therefore, we need to investigate whether the range of computed mean curvature is caused by error or it reflects a variation in local capillary pressure. Good agreement between mean curvature obtained from the fluid pressure and that computed from the simulated surface for each pore region suggested that the range of distribution observed in the simulated surface properly captures variation in local capillary pressure. For the smoothed interface with the cutoff, LPS and BPGS estimated curvature within a range of about ±20% difference from the curvatures obtained with the fluid pressure for mean curvatures smaller than about 0.15 $voxel^{-1}$, while CGS showed highly overestimated curvature values for some pore regions.

In conclusion, the application of the distance cutoff is necessary to remove the errors close to the three-phase contact line. Among the three tested smoothing methods, LPS appeared to be the best method since it gave the closest average of mean curvature values to that obtained from the simulated interface with the smallest standard deviation. In addition, it gave good estimates of the local mean curvature for each pore with a ±20% error up to 0.15 $voxel^{-1}$ at this resolution.

Currently, we are using the similar approach to quantify the accuracy of curvature computation for drainage and imbibition events on the bead pack and Bentheimer sandstone. This will be discussed in a future publication.

Acknowledgements

The authors gratefully thank Japan Oil, Gas and Metals National Corporation (JOGMEC)

References

1. R. T. Armstrong, M. L. Porter, and D. Wildenschild, "Linking pore-scale interfacial curvature to column-scale capillary pressure," Adv. Water Resour., vol. 46, no. 6, pp. 55–62, Sep. 2012, doi: 10.1016/j.advwatres.2012.05.009.
2. M. L. Porter, D. Wildenschild, G. Grant, and J. I. Gerhard, "Measurement and prediction of the relationship between capillary pressure, saturation, and interfacial area in a NAPL-water-glass bead system," Water Resour. Res., vol. 46, no. 8, pp. 1–10, Aug. 2010, doi: 10.1029/2009WR007786.
3. T. Li, S. Schlüter, M. I. Dragila, and D. Wildenschild, "An improved method for estimating capillary pressure from 3D microtomography images and its application to

the study of disconnected nonwetting phase," Adv. Water Resour., vol. 114, pp. 249–260, 2018, doi: 10.1016/j.advwatres.2018.02.012.

4. A. L. Herring, J. Middleton, R. Walsh, A. Kingston, and A. Sheppard, "Flow rate impacts on capillary pressure and interface curvature of connected and disconnected fluid phases during multiphase flow in sandstone," Adv. Water Resour., vol. 107, pp. 460–469, 2017, doi: 10.1016/j.advwatres.2017.05.011.
5. I. Halliday, A. P. Hollis, and C. M. Care, "Lattice Boltzmann algorithm for continuum multicomponent flow," Phys. Rev. E, vol. 76, no. 2, p. 026708, Aug. 2007, doi: 10.1103/PhysRevE.76.026708.
6. D. D'Humieres, I. Ginzburg, M. Krafczyk, P. Lallemand, and L.-S. Luo, "Multiple-relaxation-time lattice Boltzmann models in three dimensions," Philos. Trans. R. Soc. A Math. Phys. Eng. Sci., vol. 360, no. 1792, pp. 437–451, Mar. 2002, doi: 10.1098/rsta.2001.0955.
7. J. Brackbill, D. Kothe, and C. Zemach, "A continuum method for modeling surface tension," J. Comput. Phys., vol. 100, no. 2, pp. 335–354, Jun. 1992, doi: 10.1016/0021-9991(92)90240-Y.
8. Z. Guo, C. Zheng, and B. Shi, "Discrete lattice effects on the forcing term in the lattice Boltzmann method," Phys. Rev. E, vol. 65, no. 4, p. 046308, Apr. 2002, doi: 10.1103/PhysRevE.65.046308.
9. Z. Yu and L.-S. Fan, "Multirelaxation-time interaction-potential-based lattice Boltzmann model for two-phase flow," Phys. Rev. E, vol. 82, no. 4, p. 046708, Oct. 2010, doi: 10.1103/PhysRevE.82.046708.
10. M. Latva-Kokko and D. H. Rothman, "Diffusion properties of gradient-based lattice Boltzmann models of immiscible fluids," Phys. Rev. E, vol. 71, no. 5, p. 056702, May 2005, doi: 10.1103/PhysRevE.71.056702.
11. T. Akai, B. Bijeljic, and M. J. Blunt, "Wetting boundary condition for the color-gradient lattice Boltzmann method: Validation with analytical and experimental data," Adv. Water Resour., vol. 116, no. March, pp. 56–66, Jun. 2018, doi: 10.1016/j.advwatres.2018.03.014.
12. T. Akai, A. M. Alhammadi, M. J. Blunt, and B. Bijeljic, "Modeling Oil Recovery in Mixed-Wet Rocks: Pore-Scale Comparison Between Experiment and Simulation," Transp. Porous Media, vol. 127, no. 2, pp. 393–414, Mar. 2019, doi: 10.1007/s11242-018-1198-8.
13. Y. H. Qian, D. D'Humières, and P. Lallemand, "Lattice BGK models for Navier-Stokes equation," EPL (Europhysics Lett., vol. 17, no. 6, p. 479, 1992.
14. R. T. Armstrong, C. H. Pentland, S. Berg, J. Hummel, D. Lichau, and L. Bernard, "Estimation of curvature from micro-CT liquid-liquid displacement studies with pore scale resolution," Int. Symp. Soc. Core Anal., vol. SCA2012-55, no. August, p. 6, 2012, doi: 10.1109/CVPR.1994.323794.
15. M. Andrew, B. Bijeljic, and M. J. Blunt, "Pore-by-pore capillary pressure measurements using X-ray microtomography at reservoir conditions: Curvature, snap-off, and remobilization of residual CO 2," Water Resour. Res., vol. 50, no. 11, pp. 8760–8774, Nov. 2014, doi: 10.1002/2014WR015970.
16. Q. Lin, B. Bijeljic, R. Pini, M. J. Blunt, and S. Krevor, "Imaging and Measurement of Pore-Scale Interfacial Curvature to Determine Capillary Pressure Simultaneously With Relative Permeability," Water Resour. Res., vol. 54, no. 9, pp. 7046–7060, Sep. 2018, doi: 10.1029/2018WR023214.
17. W. E. Lorensen and H. E. Cline, "Marching cubes: A high resolution 3D surface construction algorithm," in Proceedings of the 14th annual conference on Computer graphics and interactive techniques - SIGGRAPH '87, 1987, pp. 163–169, doi: 10.1145/37401.37422.
18. L. R. Herrmann, "Laplacian-isoparametric grid generation scheme," J. Eng. Mech. Div., vol. 102, no. 5, pp. 749–907, 1976.
19. G. Taubin, "Curve and surface smoothing without shrinkage," in Proceedings of IEEE International Conference on Computer Vision, 1995, pp. 852–857, doi: 10.1109/ICCV.1995.466848.

Inverted bucket centrifugation with fluorinated oils and its applications to T_2 cut-offs

Ben Anger[1,*], *Stefan* Hertel[2], *Keith* Love[3], *Michael* Ehiwario[3], *Matthias* Appel[1]

[1]Shell Global Solutions International B.V., Shell Technology Centre Amsterdam, Amsterdam, Netherlands
[2]Shell International Exploration and Production Inc., Shell Technology Center Houston, Houston, TX, USA
[3]Shell Exploration & Production Company, Shell Woodcreek, Houston, TX, USA

Abstract. Hydrocarbon reservoirs with a large column height as well as tight gas rocks require a large range of capillary pressures to describe the saturation of fluids present in these formations. While mercury injection capillary pressure (MICP) can achieve high equivalent capillary pressures, the tests are destructive to the core plugs. Centrifuge techniques have gained in popularity since they are faster than the porous plate technique, but they are limited in the achievable pressure range. Here, we propose the use of fluorinated oils to extend the achievable capillary pressure of the air-brine centrifuge technique by a factor of two. We use Fluorinert FC-70 in an inverted bucket configuration which doubles the radius of rotation and keeps the density contrast comparable to an air-brine system. Furthermore, we show the application to NMR T_2 cut-off determination as a function of capillary pressure. Since Fluorinert does not contain any hydrogen, there is no signal overlapping with the brine in the core plugs. Furthermore, in the inverted bucket configuration, the outlet face of the plug is not in contact with a drainage surface so that the Hassler-Brunner boundary condition of Pc = 0 is satisfied. Additionally, the method allows the storage under a liquid Fluorinert phase, which prevents evaporation and significantly extends the available time for NMR measurements at low water saturations.

1 Introduction

The interactions between wetting and non-wetting fluids in a porous rock are controlled in large part by capillary forces [1, 2, 3, 4]. These forces determine fluid distributions and control the ease with which pore fluids can be recovered during production. The relationship between fluid saturations and capillary pressure is represented in capillary pressure curves. Lab measurements of these curves provide information on important petrophysical properties such as irreducible water saturation and residual oil saturation as well as insight into the initial hydrocarbon saturation in a reservoir. Reliable estimations of petroleum reserves depend on a complete understanding of the capillary pressure curves for a given reservoir.

Capillary pressure is defined as the pressure difference between the wetting and non-wetting phases as a function of the saturation of the wetting-phase in a porous medium. This pressure difference arises thanks to the interfacial tension that exists at the boundary between two immiscible fluids. Together with gravity, the capillary pressure in a porous medium controls both the distribution of fluid saturations as well as the ease with which fluids can move within the rock. As such, the capillary pressure distribution within a reservoir has a strong influence on the producible hydrocarbon volume.

Several methods exist to measure capillary pressure curves in porous rock through the displacement of the wetting fluid phase with a non-wetting phase [5]. Mercury intrusion capillary pressure (MICP) is a fast method for determining saturation as a function of capillary pressure, but rock samples are destroyed during MICP measurements. Steady-state displacement against a saturated porous plate provides an accurate measurement of saturations, but since the time required for equilibration at each capillary pressure can be on the order of several weeks, steady-state measurements are extremely time consuming.

A third common technique for the determination of capillary pressure curves is the use of a high-speed centrifuge to displace one reservoir fluid with another [6]. This method is non-destructive and is significantly faster than steady-state methods. However, accurate interpretation of high-speed centrifuge data is dependent on a series of assumptions that can lead to uncertainty [7]. Efforts have been made to address these inherent uncertainties using novel non-wetting fluids [8] or the use of magnetic resonance imaging. Green *et al.* [9],

* Corresponding author: benjamin.anger@shell.com

https://doi.org/10.1051/e3sconf/202014603006

combined the saturation gradient inherent in centrifuge measurements with one-dimensional NMR images to extract capillary pressure data from a sample that had been exposed to only one centrifuge speed [9, 10].

Proton magnetic resonance provides a powerful and non-destructive method for probing the plug saturation and fluid distributions within a rock sample. The combination of NMR methods and the high-speed centrifuge techniques allows for the determination of bound and free fluid volumes through the measurement of a quantity called the T_2 cut-off [11]. As illustrated in **Figure *1***, the T_2 cut-off is the boundary in T_2 space between movable and bound fluids at a given capillary pressure. NMR data from sample desaturated at a given capillary pressure is compared to the fully saturated NMR data to determine the T_2 cut-off.

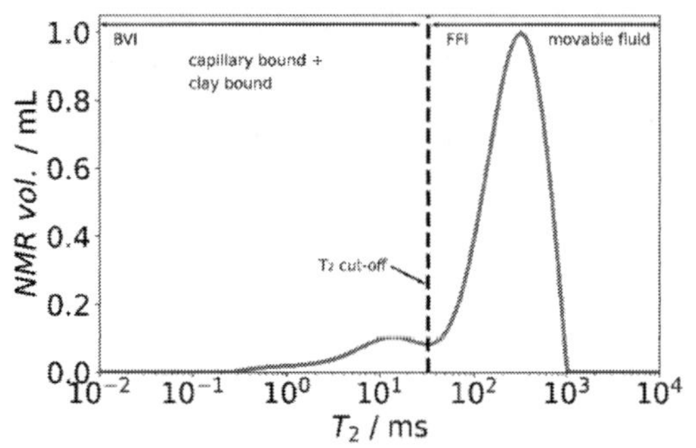

Figure 1. Fluid distribution and T_2 cut-off.

Generally, a single cut-off value is used as an input for NMR logging curves, which allows for separation of NMR T_2 distributions into the movable fluid fraction and the bound fluid fraction. However, it is not always clear which capillary pressure should be used to determine the irreducible water saturation and hence the T_2 cut-off value. It would therefore be desirable to develop a method to rapidly determine cut-off values at a range of capillary pressures.

In this work we present a method for rapid determination of movable and bound fluid volumes as a function of capillary pressure through a combination of NMR relaxometry and centrifuge primary drainage experiments in which formation brine is displaced by a dense fluorinated oil (Fluorinert FC-70). Because the non-wetting FC-70 is denser than the displaced brine, an inverted bucket configuration can be used in the centrifuge. The combination of comparable density contrast and the larger distance of the sample from the axis of rotation allows for desaturation at twice the capillary pressures compared to standard centrifuge methods. Further, in the inverted bucket configuration, the outlet face of the plug is not in contact with a drainage surface so the Hassler-Brunner boundary condition of P_c=0 is satisfied.

NMR T_2 measurements were made on a group of samples after multiple centrifuge speeds to produce T_2 cut-offs as a function of capillary pressure. This work showed that for this sample set, the commonly used cut-off of 33 ms should be replaced with a cut-off value of approximately 5 ms at the maximum capillary pressure achieved in the reservoir.

2 Experimental methods

2.1 Samples

The samples used in this work were sourced from a sandstone formation. All plugs were 2.54 cm in diameter. They were cleaned according to API RP 40 protocols, including UV fluorescence of solvent effluent and silver nitrate testing for salt content and then pressure saturated with 90 kppm NaCl equivalent synthetic formation brine. Klinkenberg-corrected permeability, helium porosity and MICP measurements were performed on twin plugs in a vendor lab. MICP saturations were calculated based on the normalization of the volume of Hg injected to the He porosity. Closure corrections and Hill-Shirley-Klein corrections were applied to correct for surface roughness and the presence of clay-bound water respectively. Sample details can be found in **Table *1***.

Table 1. Sample details, d = 2.54 cm for all plugs.

Sample	Klinkenberg Perm. (mD)	He porosity (frac.)	Length (cm)
C04	1.82	0.148	1.765
C05	2.43	0.178	1.602
C06	2.62	0.170	1.816
C07	2.96	0.174	2.034
C08	3.02	0.160	1.583
C09	3.26	0.158	1.836

2.2 Desaturation techniques

2.2.1 Steady-state displacement

Several methods were tested for the displacement of brine with a fluorinated oil (Fluorinert FC-70.) Standard NMR 2 MHz relaxometers are insensitive to FC-70, so an experiment in which brine is displaced by a fluorinated oil provides maximum contrast. At any given capillary pressure, the only fluid present after the displacement that is visible in the NMR measurements is the brine remaining in the plug.

Steady-state displacement methods against a brine-saturated porous plate produces a uniform saturation distribution throughout the sample. While the results of these experiments are very robust, this type of desaturation can be extremely time consuming due to the very low flow rates typically associated with steady-state displacement. Our initial attempts were performed in hydrostatic Hassler-type core holders. These measurements were complicated by the fact that we were required to unload and reload each sample between each injection pressure in order to measure the brine saturation with NMR. During our initial attempts at this method, we determined that extraneous brine entered the samples during the process of unloading and loading as a result of the removal of the confining stress. This unavoidable ingress of brine added an unacceptable amount of uncertainty to the NMR brine volumes.

We then successfully performed a steady-state desaturation against a saturated porous plate in an NMR pressure cell. Since the core flood was performed in the NMR instrument, there was no need to remove the sample between subsequent desaturation steps. This effort led to generally satisfactory results, but several complications led us to abandon these efforts. The presence of the brine-saturated porous plate made precise determination of the saturation values at each injection pressure difficult since signal from the porous plate and core plug were combined in the NMR response. Separation of the two signal sources through subtraction is possible, but can lead to a significant decrease in signal to noise and an increase in uncertainty. Further, covering the full range of desired capillary pressures required more than two months for a single sample. The long measurement time combined with the fact that the NMR spectrometer was unavailable for other work during the experiment led us to seek alternative solutions.

2.2.2 Centrifuge desaturation

In order to achieve the desired range of capillary pressures on a realistic number of samples in an acceptable time frame, we turned to high-speed centrifuge displacement of the synthetic formation brine with FC-70. Two centrifuges were used in this work. The first was a Beckman Coulter Optima L-80XP Ultracentrifuge, and the second was a Beckman Coulter Optima L-90K. For all samples the inverted bucket orientation was used (see **Figure 2**), with the inlet face at a radius of 16.66 cm from the axis of rotation. Samples were centrifuged in sets of three or six simultaneously for 4-7 days per capillary pressure at rotation speeds of 1000, 1500, 2250, 3200, 3900, 5515, 10000 and 15000 RPM.

There are two main benefits to the use of the inverted bucket orientation. The distance of the plug from the axis of rotation is roughly double that of the radius in conventional buckets leading to a factor of two increase in the obtainable P_c values. Further, no end piece is required with inverted buckets. Since the outlet face of the plug is not in contact with another surface, the Hassler-Brunner boundary condition of P_c=0 at the outlet face is satisfied.

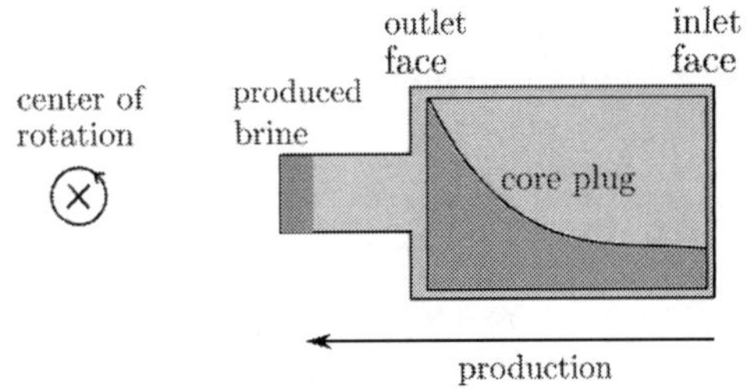

Figure 2. Inverted bucket orientation. In the plug, the remaining brine is blue and the FC-70 is yellow.

In the centrifuge, synthetic formation brine (ρ = 1.057 g/cm^3, η = 0.99 cP) was displaced by FC-70 (ρ = 1.93 g/cm^3, η = 24 cP). The change in weight of each plug was measured after each centrifuge speed to provide an independent measurement of the change in saturation at each capillary pressure. At all times during the series of measurements, the brine-saturated plugs were submerged in FC-70. This is important because the produced brine was never in contact with the plugs at any time, preventing brine from being imbibed into the samples and introducing uncertainty.

In order to correlate the FC-70/ brine capillary pressure dependent data with the Hg/air from the MICP measurements, we measured the contact angle of FC-70 in brine and the IFT to be 140° and 55.47 dynes/cm respectively. Each centrifuge speed produced a range of Hg/air equivalent capillary pressures dependent on plug length. A list of pressure ranges can be found in **Table 2**.

2.4 Nuclear magnetic resonance

2.4.1 Bulk NMR

Bulk T_2 NMR measurements were performed on each sample after each centrifuge speed. Relaxation distributions were obtained via CPMG measurements at 2.3 MHz in Oxford Instruments GeoSpec2 relaxometers with TE = 400 μs (chosen to match common logging tool values), 12,500 echoes and a recycle delay of 7.5 s. The NMR response of the samples were compared to a calibration sample of known water volume to calculate brine saturations. All NMR experiments were performed with the plugs submerged in a clean bath of NMR-silent FC-70. This provided the opportunity for extended signal averaging without the worry of the loss of brine through evaporation – an important consideration since the total brine volume present in plugs after high rotation speeds was on the order of 0.5 cm^3 or less.

2.4.2 1D NMR imaging

A uniform saturation distribution is not expected as a result of centrifuge desaturation experiments. To quantify this, one-dimensional saturation profiles were acquired at 12.7 MHz on an Oxford Instruments spectrometer using a simple frequency encode one-dimension profiling sequence.

3 Results and discussion

All samples were centrifuged at the range of speeds described earlier. An example of the results for a single plug at all centrifuge speeds can be found in **Figure 3**. This plot shows a continued decrease in the total brine saturation as irreducible brine saturation is obtained at all capillary pressures. Further, the main peak of the T_2 distributions shifts to shorter times as brine in larger pores and more loosely bound brine is expelled from the sample.

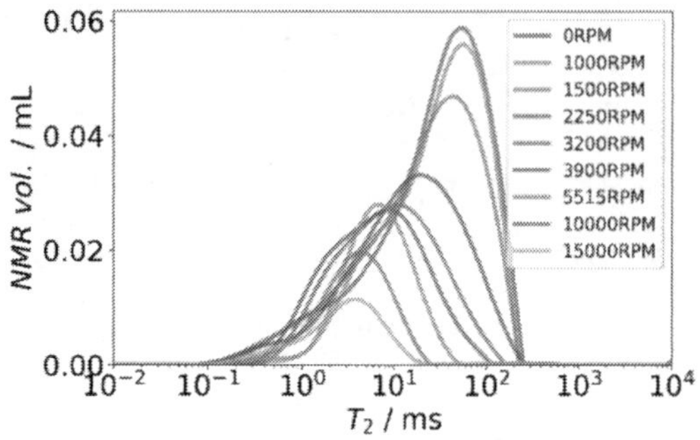

Figure 3. T_2 spectra for all capillary pressures for sample C09.

Since the synthetic brine was displaced by NMR-silent FC-70, the calibrated NMR volume after each rotation speed corresponds to the remaining water volume in the plug at a given capillary pressure and all movable brine has been displaced by Fluorinert. From this brine volume, a water saturation S_w can be calculated for each plug as a function of capillary pressure. A comparison of NMR saturations to vendor-supplied MICP saturation data is shown in **Figure 4**. The Fluorinert-brine capillary pressures were averaged over the core plug and converted to the mercury/air capillary pressure through the well-known conversion

$$P_{Hga} = P_{Fb}\frac{\sigma_{Hga}\cos(\theta_{\mathrm{Hga}})}{\sigma_{Fb}\cos(\theta_{Fb})} \quad (1)$$

where σ_{Hga} and σ_{Fb} are are the interfacial tensions for mercury/air and FC-70 brine and θ_{Hga} and θFb are the contact angles for mercury/air and Fc-70/brine.

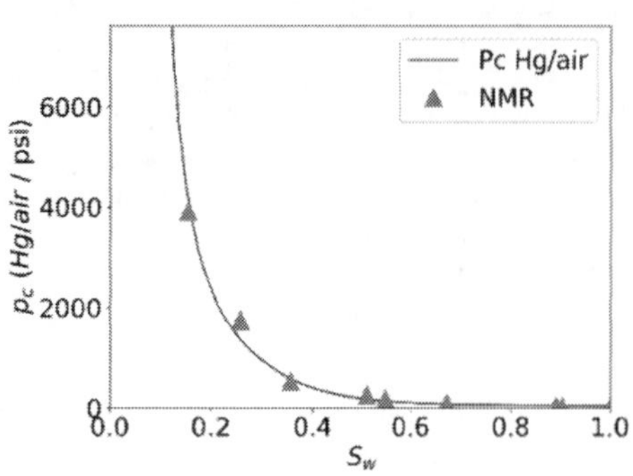

Figure 4. Comparison of NMR and MICP saturations for sample C09.

There is excellent agreement between the MICP and NMR data over the entire range of capillary pressures.

As seen in the comparison of NMR brine volumes to brine volumes calculated from the change in plug mass after centrifuging in **Figure 5**, there was good agreement between the two measurement methods.

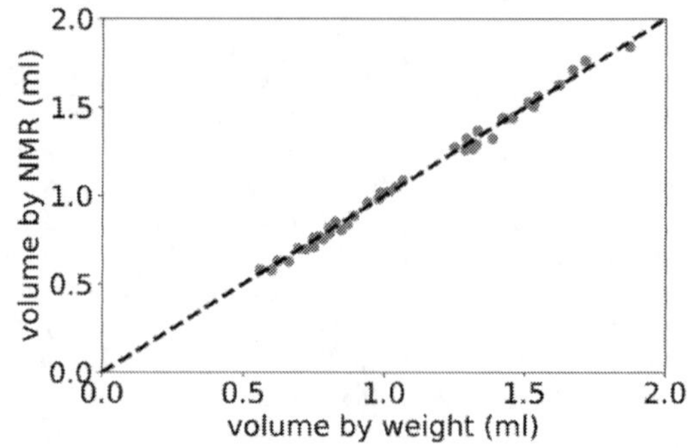

Figure 5. Comparison of NMR brine volume to calculated brine volume from change in sample mass.

A representative one-dimensional NMR saturation profile is shown in **Figure 6**. As expected S_w approaches 100% at the outlet face.

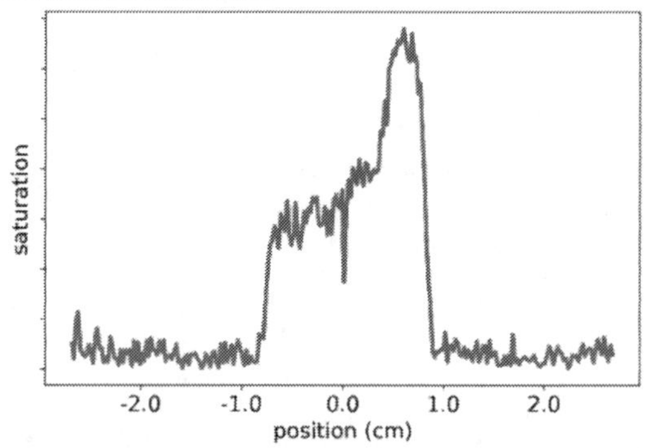

Figure 6. Saturation profile for sample C09 after centrifuging. Brine saturation approaches 100% at the outlet face on the right side of the plot.

As observed in the 1D NMR images, brine saturation in the plug after centrifugation is not uniform. If the Hassler-Brunner boundary condition of P_c=0 at the outlet phase is assumed, the variation in capillary pressure as a function of position along the plug is given by:

$$P_c(r) = \frac{\Delta\rho\omega^2}{2}(R_1^2 - r^2) \quad (2)$$

where $\Delta\rho$ is the density contrast between FC-70 and the expelled brine, ω is the angular velocity of the centrifuge and R_1 is the distance of the inlet face from the axis of rotation. Given this distribution of capillary pressures and the one-dimensional profiles observed in several samples, an integrated average of the distribution described in equation 1 was calculated for each sample. The average Pc value is given by:

 https://doi.org/10.1051/e3sconf/202014603006

$$\bar{P_c} = \frac{\Delta\rho\omega^2}{2(R_2 - R_1)^2}\int_{R_1}^{R_2}(R_1^2 - r^2)dr \quad (3)$$

where R_2 is the distance of the outlet face from the axis of rotation. A range of Pc values were calculated for each centrifuge speed and each sample used in this work. These values can be found in **Table 2**.

Table 2. Equivalent FC-70/brine and Hg/air P_c ranges for all centrifuge rotation rates.

Rotation Speed (RPM)	P_c (Fc-70/brine, psi)	P_c (Hg/air, psi)
1000	3- 5	15-19
1500	7-10	34-43
2250	18-22	77-97
3200	36-45	155-196
3900	54-68	231-291
5515	108-134	461-581
10000	348-442	1515-1910
15000	786-995	3410-4297

As shown in **Figure 7**, the T_2 cut-off at a given capillary pressure can be determined by comparing the cumulative NMR signal at S_w=100% to the cumulative NMR signal from the desaturated sample. The T_2 cut-off is the value at which a horizontal line drawn through the final cumulative volume of the desaturated sample intersects with the cumulative volume of the fully saturated sample.

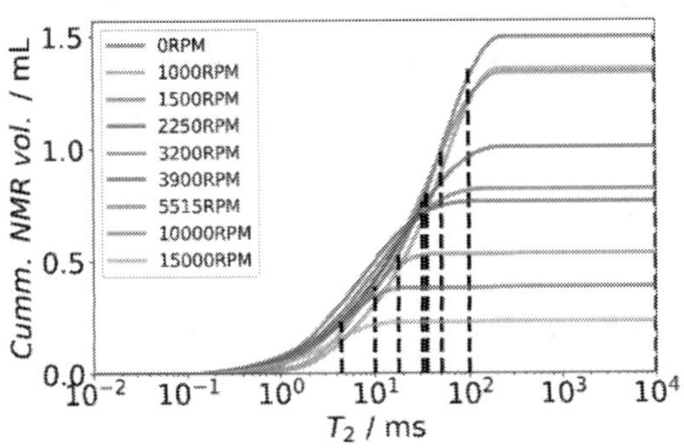

Figure 7. T_2 cut-off determination for sample C09.

Pressure dependent cut-off values were extracted for all samples. The results for sample C09 can be found in **Figure 8**. The commonly used sandstone cut-off of 33 ms is indicated by a dashed vertical line. While there is good agreement with the C09 data at a Hg/air equivalent of about 300 psi, it is clear that for higher capillary pressures the cut-off value between movable and bound fluid is much lower. For all samples in this work, the asymptotic T_{2irr} value was 4-5 ms. In this case, use of the standard cut-off value would lead to large errors in the estimates of movable fluid volumes.

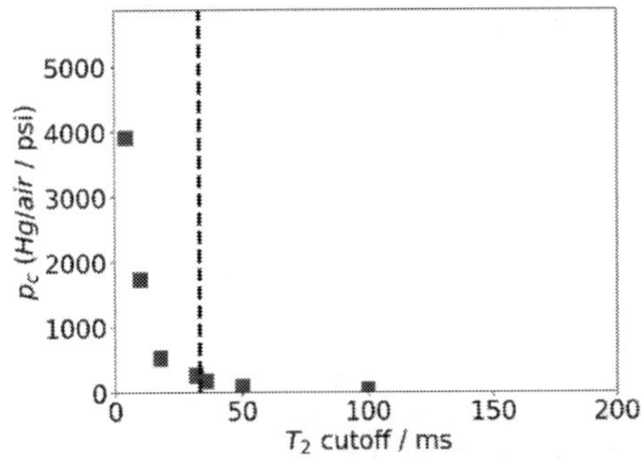

Figure 8. T_2 cut-off as function of Hg-air capillary pressure for sample C09. The dotted line is the standard 33 ms cut-off.

A modified Brooks-Corey model was applied to the data to extract asymptotic T_2 cut-off values for each sample. The standard Brooks-Corey *[12]* model relates capillary pressure to brine saturation as:

$$P_c = P_e\left(\frac{S_{w100} - S_{wi}}{S_w - S_{wi}}\right)^n \quad (4)$$

where P_e is the capillary entry pressure, S_{wi} is the irreducible water saturation and S_{w100} (typically given as 1) is the maximum saturation. Following the recently published work of Al-Harbi *et al. [13]*, this relationship can be recast in terms of relaxation times as:

$$P_c = P_e\left(\frac{T_{2\max} - T_{2irr}}{T_{2cut} - T_{2irr}}\right)^n \quad (5)$$

where S_{w100} is replaced by T_{2max} (longest T_2 value in the fully T_2 distribution of the saturated sample), S_w is replaced by T_{2cut} (T_2 cut-off), and S_{wi} is replaced by T_{2irr} (T_2 cut-off at irreducible saturation). T_{2max} was between 300-400 ms for all samples.

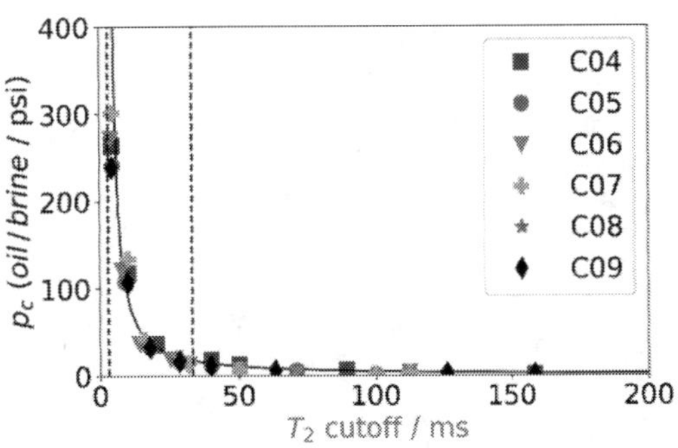

Figure 9. Brooks-Corey T_2 cut-off model applied to samples C04-C09 with T_{2irr} = 3.3 ms, P_e = 1 psi and n = 1.1. Dashed lines are drawn at T_{2irr} and at 33 ms.

The application of the modified Brooks-Corey model to a selection of samples can be found in **Figure 9**. For all samples in this set, an asymptotic cut-off value in the range of 4-5 ms was found to provide a more accurate measure of the division between mobile and immobile pore fluids at irreducible water saturation at capillary pressures corresponding to those found in the reservoir from which the samples were obtained.

5 Conclusions

In this work we presented a novel method for quantifying movable fluid volumes as a function of capillary pressure through a combination of NMR relaxometry and primary drainage experiments in a centrifuge in which the formation brine is displaced by Fluorinert FC-70. The combination of a large density contrast between the wetting and non-wetting fluid phases and the increased distance of the sample from the axis of rotation allows for significantly larger capillary pressures compared to standard air-brine displacement measurements. This method allows for the storage and measurement of samples submerged in Fluorinert, which prevents evaporation and significantly extends the available time for NMR measurements at low saturations.

The samples measured in this work showed asymptotic cut-off values in the range of 4-5 ms, significantly shorter than the commonly used cut-off value of 33 ms. The determination of a more accurate division between mobile and immobile fluid allows for more accurate estimates of movable fluids in place.

This work is directly applicable to reservoirs with large transition zones. In this specific case, it will be useful to produce a model in which the T_2 cut-off depends on capillary pressure and hence on the height above free water level. In a future paper, we will present the application of this method to the determination of a saturation model in which the T_2 cut-off varies as a function of the height above free water level.

Work is also currently underway to incorporate spatially resolved relaxation measurements of plugs after centrifuging using the 1D NMR profiling method developed by Hertel *et al.* [14]. This technique provides a snapshot of fluid distributions within a centrifuged plug on short time scale compared to standard 1D MRI methods. With this method, a significantly smaller number of centrifuge speeds will be needed to determine cut-off values at a full range of relevant capillary pressures.

The authors would like to thank Shell management for the permission to use the samples, and Kirk Smith, Aaron Dhaemers and Vivian Morah for their technical assistance.

References

1. J. Schön, Physical Properties of Rocks, Amsterdam: Elsevier, 2015.
2. E. Thomas, "Capillary Pressure Tutorial: Part 1 - It's a Jungle in Here," *Petrophysics,* vol. 59, no. 4, pp. 421-427, 2018.
3. E. Thomas, "Capillary Pressure Tutorial: Part 2 - The Path Out of the Jungle," *Petrophysics,* vol. 59, no. 5, pp. 557-563, 2018.
4. E. Thomas, "Capillary Pressure Tutorial: Part 3 - The Jungle Gives Us Many Things," *Petrophysics,* vol. 59, no. 6, pp. 739-747, 2018.
5. C. McPhee, J. Reed and I. Zubizarreta, Core Analysis: A Best Practice Guide, Amsterdam: Elsevier, 2015.
6. G. Hassler and E. Brunner, "Measurement of capillary pressures in small core samples," *Trans. of AIME,* vol. 160, pp. 114-123, 1945.
7. D. O'Meara, G. Hirasaki and J. Rohan, "Centrifuge Measurements of Capillary Pressure: Part 1-Outflow Boundary Condition," *SPE Reservoir Engineering,* vol. 7, no. 1, pp. 133-142, 1992.
8. C. Straley and G. Leu, "Low-field NMR profiles for verification of oil and water saturations in cores," *Proc. Soc. Core Anal.,* pp. SCA2006-31, 2006.
9. D. Green, J. Gardner, B. Balcom, M. McAloon and J. Cano-Barrita, "Comparison Study of Capillary Pressure Curves Obtained Using Traditional Centrifuge and Magnetic Resonance Imaging Techniques," in *SPE Symposium on Improved Oil Recovery,* Tulsa, OK, 2008.
10. Q. Chen and B. Balcom, "Measurement of rock-core capillary pressure curves using a single-speed centrifuge and one-dimensional magnetic-resonance imaging," *J. Chem. Phys.,* vol. 122, p. 214720, 2005.
11. A. Timur, "Effective Porosity and Permeability of Sandstones Investigated Through Nuclear Magnetic Principles," *The Log Analyst,* vol. 10, no. 1, 1969.
12. R. Brooks and A. Corey, "Hydraulic Properties of Porous Media," *Hydrology Papers,* no. 3, 1964.
13. A. Al-Harbi, G. Hursan, H. Kwak and J. Gao, "A New NMR-based Height Saturation Model of a Low Permeability Carbonate Reservoir," *Saudi Aramco Journal of Technology,* pp. 49-59, 2018.
14. S. Hertel, D. de Kort, I. Bush, A. Sederman, L. Gladden, B. Anger, H. de Jong and M. Appel, "Fast spatially-resolved T2 measurements with constant-gradient CPMG," *Magnetic Resonance Imaging,* vol. 56, pp. 70-76, 2019.

Digital core repository coupled with machine learning as a tool to classify and assess petrophysical rock properties

Vanessa Hébert[1*], *Thierry* Porcher[1], *Valentin* Planes[1], *Marie* Léger[1], *Anna* Alperovich[2], *Bastian* Goldluecke[2], *Olivier* Rodriguez[1] and *Souhail* Youssef[3]

[1]Voxaya, Cap Oméga, rond-point Benjamin Franklin, CS 39521, 34960 Montpellier, France

[2]University of Konstanz, Universitätsstraße 10, 78464 Konstanz, Germany

[3]IFP Energies Nouvelles, 1-4 Avenue du Bois Préau, 92852 Rueil-Malmaison, France

Abstract. To make efficient use of image-based rock physics workflow, it is necessary to optimize different criteria, among which: quantity, representativeness, size and resolution. Advances in artificial intelligence give insights of databases potential. Deep learning methods not only enable to classify rock images, but could also help to estimate their petrophysical properties. In this study we prepare a set of thousands high-resolution 3D images captured in a set of four reservoir rock samples as a base for learning and training. The Voxilon software computes numerical petrophysical analysis. We identify different descriptors directly from 3D images used as inputs. We use convolutional neural network modelling with supervised training using TensorFlow framework. Using approximately fifteen thousand 2D images to drive the classification network, the test on thousand unseen images shows any error of rock type misclassification. The porosity trend provides good fit between digital benchmark datasets and machine learning tests. In a few minutes, database screening classifies carbonates and sandstones images and associates the porosity values and distribution. This work aims at conveying the potential of deep learning method in reservoir characterization to petroleum research, to illustrate how a smart image-based rock physics database at industrial scale can swiftly give access to rock properties.

Introduction

Digital rock analysis recently becomes an important part of the laboratory services in oil and gas industry. Numerical properties accelerate and improve the understanding of the reservoir behavior.

Computed tomography scans from rocks associated with Digital Rock Physics (DRP) analysis receives considerable and extended use in the oil and gas laboratory services. Computed micro-tomography (MCT) scans of rocks record multiple structural information such as the texture and the rock fabric. A 3D image of a rock sample gives access to the actual representation of the mineral phase and the pore space. Rock images with voxel size down to the micrometer resolution allow to extract the topology of the pore space. Once segmented in two (usually pore and solid) or more phases of interest, the MCT images can be used to simulate porous media physical properties such as fluids transport, electrical and geo-mechanical properties as well as processes like enhanced oil recovery simulation [1, 2]. Ongoing improvements in MCT systems and image analysis software quickly provide increasing amounts of data. Properties computed from rock images accelerate and improve the understanding of the reservoir behavior.

In DRP analysis, key elements are the voxel size and the representative elementary volume (REV) determined during the acquisition phase. A too low image resolution can restrict the pore space reconstruction, causing bad pore space topology determination and rock properties estimation as demonstrated in [3]. Al-Raoush [4] and Papadopoulos reveals that the minimal REV for particle size distribution and coordination number is larger than the minimal REV for porosity. This technical restriction results in a permanent search of increasing resolution and volume size of the acquired images. At the same time the constant progress of acquisition technique leads to more and more image acquisition generating dozen terabytes of data needing to be processed afterward. As a consequence, standard image processing techniques become more and more limited and need to be automated to work with an efficient workflow.

Recent advances in high performance computing and machine learning (ML) will probably lead to new and more efficient computations. At the present time, research carried out on deep learning produces various frameworks easily accessible that particularly democratize its uses. Deep neural network provides excellent results for X-ray computed tomography (Wang [5], Würfl *et al.* [6]). Convolution neural network (CNN) is an important deep learning architecture. It can extract the image features automatically and has high classification accuracy. CNNs have achieved a wide range of applications such as plant classification, face

* Corresponding author: vanessa.hebert@voxaya.com

recognition, handwritten Chinese character recognition and so on Mikia *et al.* [7] and Lopes *et al.* [8].

Naranjo Leon *et al.* [9] provided a permeability-porosity relationship for each rock type, allowing to complement the reservoir characterization in the un-cored wells. Chen and Zeng [10] demonstrated the capability of machine learning in rock facies classification from wireline log scalar attributes improving by feature augmentation. Compared to conventional image segmentation methods, machine-learning segmentation could come closer to the ground truth for determining the porosity from noisy MCT images (Berg *et al.* [11]). Karimpouli and Tahmasebi [12] revealed that a CNN algorithm improves the accuracy of a segmentation comparing with a multiphase thresholding segmentation. Their network also produced valid results for unseen images with a categorical accuracy of about 96%. Araya-Polo *et al.* [13] used a deep learning architecture to instantaneously predict permeability of clastic rocks from high resolution Scanning Electron Microscopy images. Sudakov *et al.* [14] validated a 3D CNN method for predicting permeability of digital Berea sandstone volume subsets.

In this study we pursue tests on the CNN potential for DRP predictions. We provide some proofs of concept and discuss how they could be designed to be integrated in a more global usual analysis workflow. The typical size of a MCT image, a few dozen gigabytes, is huge in comparison to hardware capacity (some gigabytes for GPU) and time consuming. That is why the CNN is relevant to emphasize relevant features with a reasonable time and standard computing resources.

The goal of this study is thus to explore and evaluate the contribution of machine learning as a tool to evaluate geological and petrophysical properties directly from grayscale MCT scan images. The purpose is to deploy an automated workflow directly after the reconstruction of 3D rock images.

The paper is organized as follow: we first introduce the basics of CNN algorithms and the selected image database used for this study. Second, we present the classification of reservoir sedimentary rock types from grayscale MCT images using an adapted version of the Inception-V3 network, named RockClass model. Then an optimized regression-CNN network, named RegPhi model, is used to estimate the total porosity from MCT scan images without segmentation. The porosity is evaluated thanks to an AutoEncoder model realizing an automated segmentation from the grayscale images and associating the total porosity. Finally, the results are discussed.

Workflow description

Machine learning concepts

Convolutional neural networks (CNNs) are a powerful tool widely used in various computer vision problems, like image classification [15], object detection [16], segmentation [17] and image enhancement. As convolutional networks assume that pixels that are close to one another are semantically related, they seem a good candidate to extract physical properties from MCT images.

During the last decade, many network architectures (types of layers, size of layers, interconnexion of layers) have been experimented. In this paper, we use 3 types of networks.

The first one is a customization of the Inception-V3 network as presented in [18] developed by Google to solve the ImageNet Large Scale Visual Recognition Challenge [19]. This network has been trained on a very large dataset (1.2 million 2D 299 x 299 images) and is able to classify 1,000 kinds of images. The output of the network is a vector of probabilities for the image to belong to a specific class.

Then we use a network proposed by Sudakov *et al.* [14] to estimate the porosity from a MCT image. This network has fewer layers than Inception-V3 but is able to accept 3D 100 x 100 x 100 voxel images. Instead of providing a classification, this network is trained to fit a function (porosity in our case). The output of the network is a scalar.

Finally, an AutoEncoder (or encoder-decoder) network is used to produce a pixel-wise output. The key idea is to extract a relatively small set of features (bottleneck) from the input image data and then decode those features into the desired output.
Although this kind of CNNs is able to model complicated phenomena due to a large number of parameters, it is still hard to predict the behavior of neural network on "unfamiliar" test examples. There are various techniques to overcome this difficulty: for instance, semi-supervised learning where a mathematical model is incorporated into CNN architecture or loss function as demonstrated in [20 – 22].

Dataset

The petroleum reservoir rocks are widely composed of two main lithological classes, sandstones and carbonates [23]. The original database is made of a series of reservoir analogue samples including 2 carbonates from Estaillades [24, 25] and Savonnieres [26-28] formations, and 2 sandstones from Fontainebleau [29, 25] and Berea [30-33] formations. For each sample, the rock type and the rock formation are notified in the generated database which will provide the predicted class of rock type.

See below Table 1 for a description and illustrations of the image collection.

The core plugs are imaged in 3D with a voxel size of 3 to 4 µm. Having the same voxel size ensures a consistent learning of the patterns. This voxel size proposed by most of micro-CT scanners is commonly used for imaging rock porous media. 3-µm voxel size enables the internal fine structures to be imaged accurately, though this is the limit between macropore and micropore for the Estaillades sample. Although it remains a challenge to accurately predict porosity from subresolved imaged porous media, we choose to disregard the porosity resolution influence (Saxena *et al.* [34]). We use only 2-phase segmentation which totally

 https://doi.org/10.1051/e3sconf/202014601003

ignores the subresolved porosity. The 2-phase segmentation handed by an expert user reflects the most common arbitrary segmentation and its known sensibility.

Sub-resolution porosity involves low contrast and blurred limits between the different phases, which can partly cause a dispersion of the results. These segmented images are taken as references for the training network predicting the porosity. The goal is to try to predict an estimation based on particular subjective expert appreciation, which may differ from the ground truth porosity. For monomineral sample, it is possible to take into account grey levels reported to microporosity thanks to a mean grey segmentation method [25].

Table 1: List of the selected rock images, a collection based on sandstones and carbonates, associated to the CT image information and to the lab and computed basic measurements.

Rock type & formation	Lab measurements	Cross-section of the CT image, CT image information and Property estimations	
Carbonate Estaillades	Porosity: 25% Permeability: 273 mD	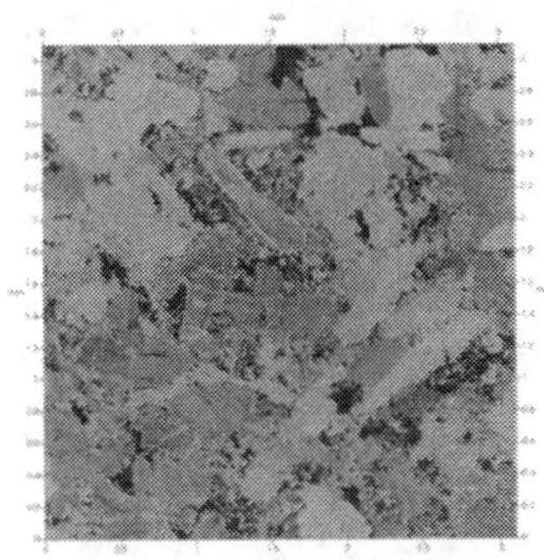	Voxel size: 3.1 µm Image size: 1000 x 1000 x 1000 voxels Porosity: 15% Permeability: 475 mD
Carbonate Savonnieres	Porosity: 22% Permeability: 115 mD	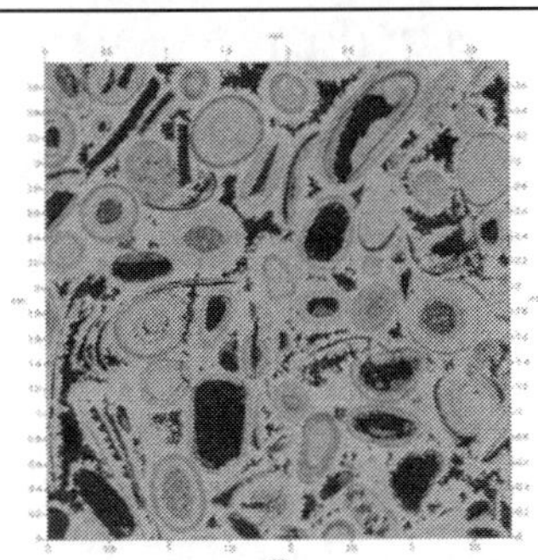	Voxel size: 3.8 µm Image size: 1000 x 1000 x 1000 voxels Porosity: 22% Permeability: 50 mD
Sandstone Berea	Porosity: 20% Permeability: 500 mD [35]	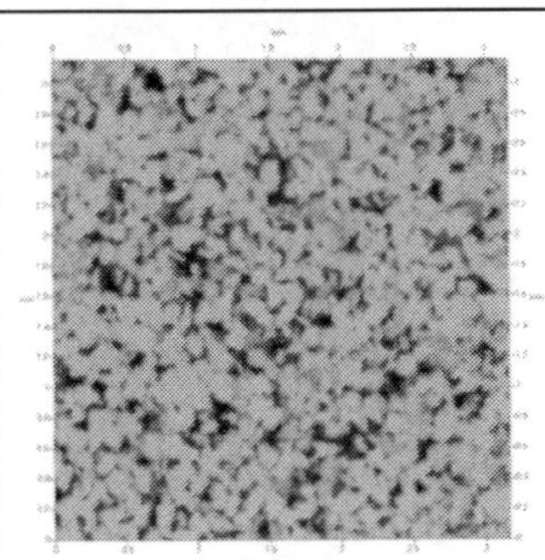	Voxel size: 3.2 µm Image size: 1000 x 1000 x 1000 voxels Porosity: 21% Permeability: 620 mD

Sandstone Fontainebleau	Porosity: 12% Permeability: 320 mD	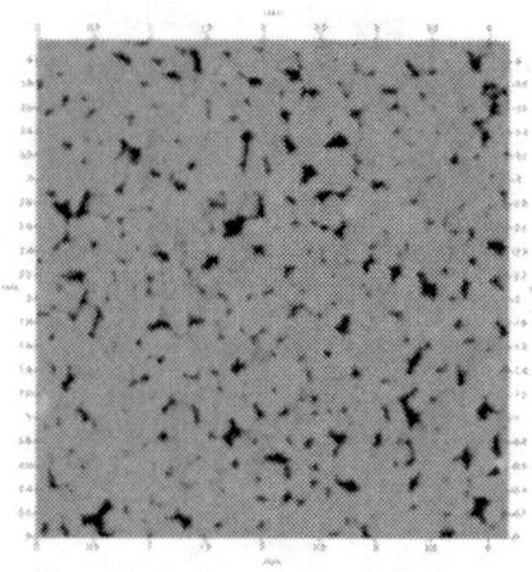	Voxel size: 3.2 µm Image size: 1000 x 1000 x 1000 voxels Porosity: 12% Permeability: 380 mD

Voxaya's software Voxilon [36] is used to extract a dataset of 18 images of size up to 1000 x 1000 x 1000 voxels from those 4 digital plugs. An expert user generates the 3D binary segmented images by selecting the threshold values. The total porosity and absolute permeability are computed from each segmented image. From these large blocks, small 100 x 100 x 100 voxels non-overlapping images are extracted using Python scripting tools [36], leading to a database with more than 36,000 images (18,000 CT and 18,000 segmented). For each segmented image, related pore volume fraction, related permeability and tortuosity are computed, the last two properties not being presented in this paper. For each study, part of the database is used for training and some data is always kept aside for evaluation (training assessment) and testing (model assessment).

Rock classification

Convolution neural network (CNN) can automatically extract image features and presents high classify accuracy.

The goal is to prove that the rock types and formations classification can be realized using existing technologies for image classification.

Network architecture

Here we choose to use the pre-trained Inception version 3 (Inception-V3) model specialized in 2D image classification by feature extraction, provided by the TensorFlow framework. This network is quite optimized and contains 48 layers (see Figure 1 below). A specialized algorithm for training is designed to overcome the limitations of usual algorithms.

Transfer Learning Concept

All the power of the inception approach lives in the Transfer Learning concept. The inception network model is a complex network, so training the model directly from the beginning would cost at least a few days. However, using the method of Transfer Learning, the parameters of the first layers are kept unchanged and only the last layer from the network is adapted to the use case and trained. The last layer is a Softmax classifier, a mathematical function which outputs a probability distribution [37]. The porosity for a patch 100 x 100 x 100 is individually estimated. To evaluate the porosity for the whole rock sample one could compute porosity for every patch and average the result. It is replaced by a layer with as many neurons as there are classes to choose from. The network has to be retrained to update the weights of this layer using a back propagation algorithm and the cross entropy loss function. In our case, we use 4 classes: Berea, Fontainebleau, Estaillades and Savonnieres.

Training

As input data, we use 4,000 grayscale 2D slices (1,000 per class) resized to be compatible with the network (initially 1000 x 1000 resized to 299 x 299) and rescaled between 0 and 1.0. Training duration is about 6 minutes on a 32 cores Intel Xeon(R) CPU E5-2667 v3 @ 3.20GHz system.

Results

For testing, 1,200 slices (300 slices per class) from images were analyzed and the computation took 64 seconds. Eventually the data for testing reveals a final accuracy of 100% for the selected rock typing classification. Recognition between the 2 lithological rock types (sandstone and carbonate) is 100% and recognition of the 4 rock formations is also 100%.

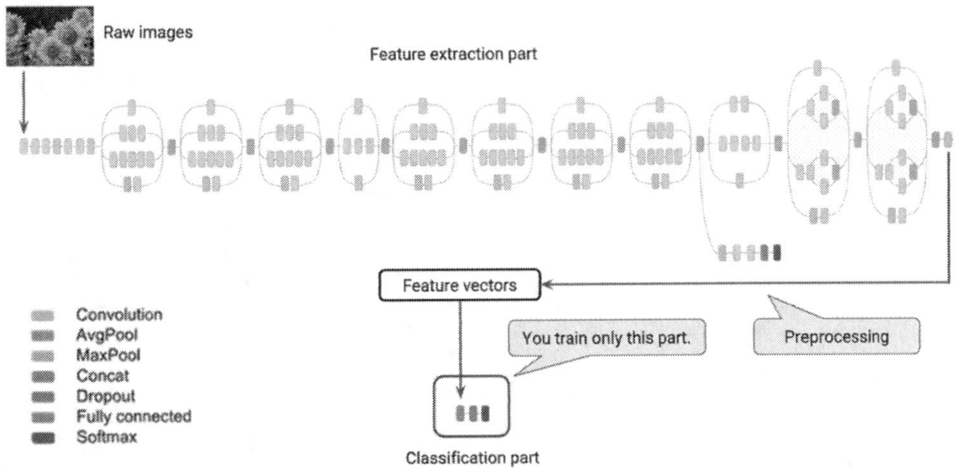

Figure 1: Inception V3 architecture (source: Google codelabs).

Porosity estimation

Porosity from CT scan without segmentation

In this paper we try to predict porosity from grayscale 3D images. Araya *et al.* [13] already used grayscale SEM image as input for DL-based permeability predictions avoiding the segmentation, the highest sensitive step linked to the resolution, and speeding-up processing workflows.

Following that work, we focus on the Berea sandstone sample for training and prediction imaged in 3D. For the network architecture we take inspiration from Sudakov *et al.* [14]. Using this network, the authors try to predict permeability by Linear Regression and CNN from segmented images. Our model RegPhi is implemented using the Keras framework [38] and is described in Table 2.

Training

In our study we use non-overlapping sub-blocks (100 x 100 x 100 voxels) from 2 Berea CT images. We have 8,000 of these sub-blocks, 3,500 being used for training while keeping 600 blocks for validation.

Table 2: The RegPhi Network Architecture.

Layer type	Parameters
3D Convolution	[filters=32, kernel_size=(5, 5, 5), strides=(2,2,2), padding='valid', activation='relu')]
3D Convolution	[filters=32, kernel_size=(5, 5, 5), strides=(2,2,2), padding='valid', activation='relu')]
3D Max Pooling	[pool_size=(2, 2, 2)]
3D Convolution	[filters=32, kernel_size=(3, 3, 3), padding='valid', activation='relu']
3D Convolution	[filters=32, kernel_size=(3, 3, 3), padding='valid', activation='relu']
3D Max Pooling	[filters=32, kernel_size=(3, 3, 3), padding='valid', activation='relu']
Dense	[units=128, activation='relu']
Dense	[units=64, activation='relu']
Flatten	
Dense	[units=1]

Results

Figure 2 shows the distribution of the relative error on porosity prediction compared to the reference.

$$Error = 100 * | \varphi_{RegPhi} - \varphi_{Reference} | / \varphi_{Reference}$$

These predictions are realized on a set of 1,000 sub-blocks of 100 x 100 x 100 voxels from grayscale Berea sandstone images without any segmentation or improvement process. We use a REV-independent approach in order to provide prediction on large amount of images. Here we can see encouraging results about the porosity prediction, the median of the series is under 15% despite a mean relative error of 18% (see Figure 3).

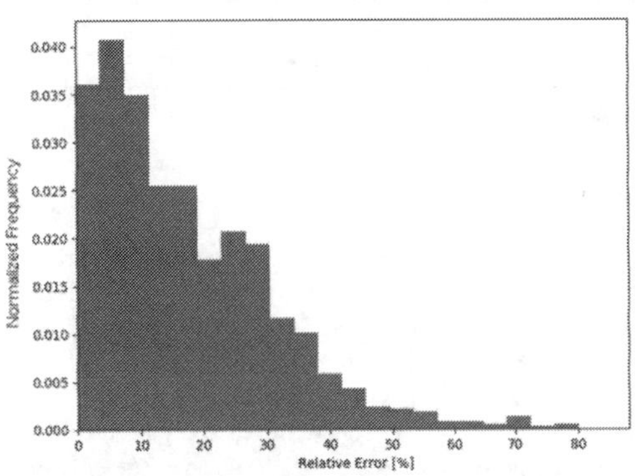

Figure 2: Distribution of absolute error for porosity prediction in percentage from the same Berea sandstone image.

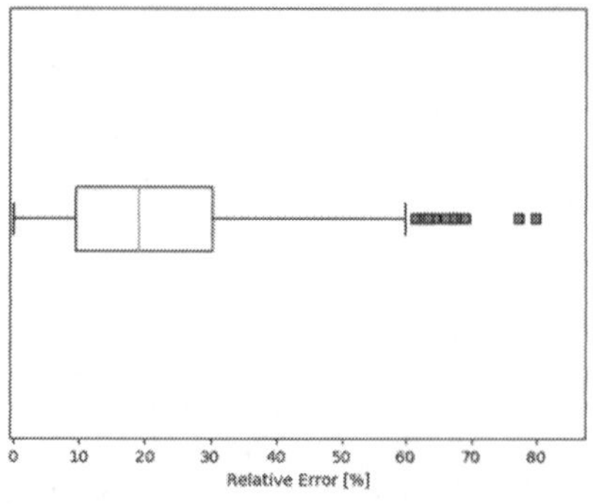

Figure 3: Corresponding box plot of absolute error for porosity prediction in percentage from the same Berea sandstone image.

Porosity from automated segmentation

Network architecture

In this work, we propose an AutoEncoder network that segments pores in 3D rock images (see Figure 4 for a detailed description of the network architecture).

The idea is to compute the high-level features from the input 3D volume that already contain information about porosity in the training example and then decode those features into the segmentation mask, where every value corresponds to the probability of the voxel being a pore. To make sure that the bottleneck layer captures information about porosity, we predict it as an additional pathway in the decoder. We believe that this type of guidance helps the network to cope well with the segmentation task since it already takes the porosity into account while decoding.

Theoretically speaking, we should have trained network with the loss function for porosity and without. Nevertheless we could in general argue that some guidance for the bottleneck is shown to be a promising technique in deep learning. For instance, variational autoencoders are good example how this type of guidance improves the robustness of classification problem (Pu *et al.* [39]).

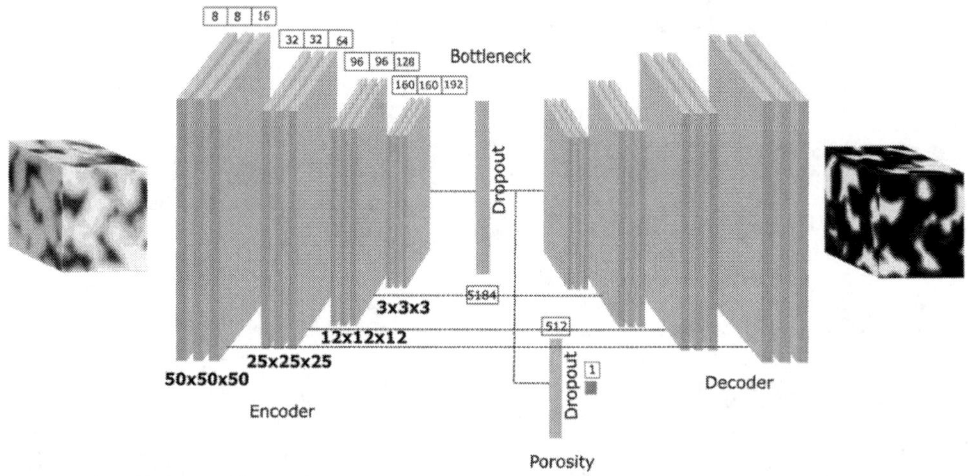

Figure 4: Network architecture with 12 convolutional layers, bottleneck, and 12 corresponding upsampling layers. The layers are organized in 4 blocks with 3 layers each. On top of the encoder part, we show the number of features after each layer. On the bottom of the encoder, we show the tensor size. With dashed lines, we illustrate skip connections [40], where encoder features are copied to the corresponding decoder layers. This helps to bring small details that were lost after downscaling. The decoder part has two pathways, the first one recovers the segmentation mask and the second one computes the porosity.

Computational framework

The input data of the network is a set of grayscale CT image volumes of dimension 100 x 100 x 100 voxels. Larger images are segmented into these patches, so that our network can deal with input data of any shape.

The basic ingredients for the encoders and decoders are residual blocks. To decrease resolution, we employ strided convolutions with stride 2, so that our network is fully convolutional. The choice of this architecture is well supported in the literature [41, 42] as it allows to speed up the training process and deals with "vanishing gradients", which is a common problem in deep learning. The residual blocks have a very simple structure and allow direct pass-through of the batch normalized input, see Figure 4.

In the encoder pathway, four groups of three residual blocks are chained together, 12 blocks in total. In the first three groups, every third block reduces the patch resolution via strided convolution while increasing feature depth, with the overall goal of gradually reducing dimensionality. Others keep spatial resolution the same while increasing the number of features. In the last group, both blocks decrease spatial resolution such that the output shape is 3 x 3 x 3 x 192, see Figure 4. The feature output at the representation level is concatenated. This is the final output of the encoder, and the bottleneck of the network.

After passing the bottleneck, the low-dimensional representation is decoded again by a chain of residual layers. The latent variables enter two decoder pathways. One is the path that predicts the porosity and another one outputs the segmentation mask. Decoder pathways that lead to the segmented 3D volume use transposed convolutions to exactly revert the encoder on the corresponding level. However, the only link between them is through the latent representation and skip connections, see Figure 5. To prevent overfitting, we apply dropout layers on the bottleneck with 0.2 probability of each node to be discarded.

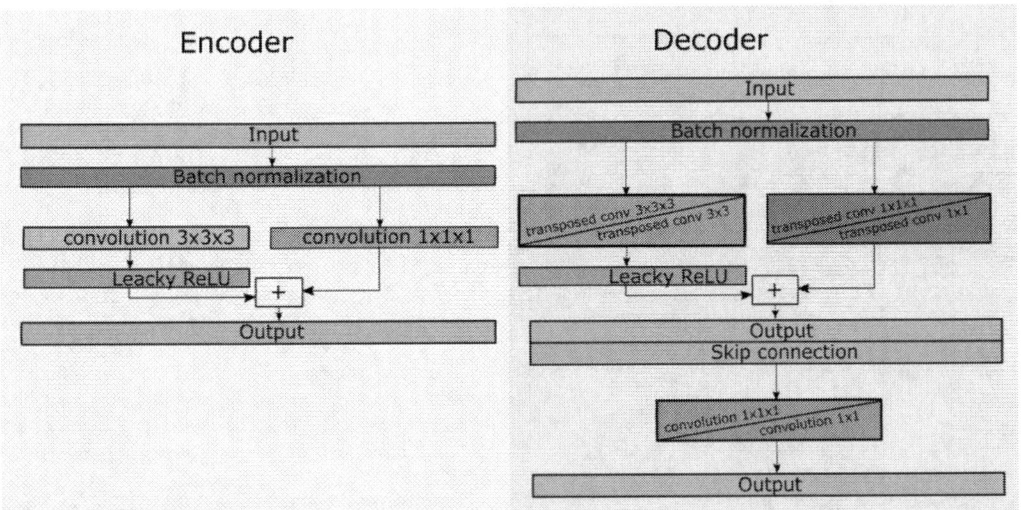

Figure 5: Single residual block of the encoder and decoder networks. After batch normalization, a first path leads through a (possibly strided) convolution or upsampling layer and a leaky ReLU. A second path either keeps the input or passes it through a strided (transposed) convolution in case it needs to be resampled. Both paths are added together to produce the final output. In the decoder block we skip connect the corresponding encoder features to the output of the decoder. The idea is that it is much easier for such blocks to learn the identity transformation, or perform only small modifications to the input [42], which helps the encoder-decoder paths to gradually add details.

Loss function

In the classical supervised machine learning used to train the model, we need training data together with a label or target. By observing the random variable X and its label Y in the training set, the supervised learning tries to fit the model f to the training samples $(x_0, y_0), \dots, (x_n, y_n)$.

Test data has similar structure as training data, but the network never sees it.

The training process can be considered as learning the mapping f to predict label y^* from the new instance (x^*, y^*) of the test data via $f(x^*) = y^*$.

A typical cost function for image classification and segmentation tasks is cross-entropy with logits.

Let $p(x_i) = f(x_i)$ be the probability that point x_i is a pore. Thus $p(x_i) \in [0,1]$ with $p(x_i)$ close to 0 if the probability is low. The cross-entropy between $p(x)$ and y is given by:

$$E_{cross-entrop}\ (y, p) = -\sum_{i \in \Omega} y_i \log(p(x_i)) \quad (1)$$

with Ω being a set of all points in the training example.

In order to predict the porosity from the bottleneck, we use the standard mean squared error (MSE):

$$E_{MSE} = ||\phi_y - \phi_e||^2 \quad (2)$$

where ϕ_y denotes the target porosity and ϕ_e is the predicted porosity computed at the bottleneck level.

Finally, to link the predicted porosity ϕ_e and final output we compute $\phi_{out} = \sum_{i \in \Omega} p(x_i)/|\Omega|$ and introduce another loss function:

$$E_{model} = ||\phi_{out} - \phi_e||^2 \quad (3)$$

that ensures that the porosity computed from the segmentation mask equals the predicted porosity obtained from the bottleneck features.

Final loss function is the sum of the previously defined loss functions (1), (2) and (3):

$$E = E_{cross-entrop} + E_{MSE} + E_{model} \quad (4)$$

Training the CNN model

From the training data, we leave aside 10% for a validation set whereas the rest is used for training. Several images are also completely held back and used only for testing. We implement the network using TensorFlow in Python3, and train it on an Intel Xeon(R) CPU E5-2667 v3 @ 3.20GHz system with one Nvidia Quadro M6000. Weights are initialized using the same strategy as for residual networks [40]. Stochastic optimization using the Adam optimizer [43] took roughly two days, after which loss remained stable. We train with batch size 10 and learning rate 1-e4. Reconstruction of a single pathway during evaluation requires about 0.5 - 1.5 seconds for the 3D volume of size 100 x 100 x 100. The complete segmentation of a 1000 x 1000 x 1000 voxels image takes about 15 - 20 minutes.

Results

Figures 6 and 7 show the source CT image and the resulting segmentations using Hysteresis segmentation and the AutoEncoder network. Globally, AutoEncoder segmentations tend to be sharper than the binary thresholding but also thinner (one voxel difference on the boundaries). Thus, porosity computed on the AutoEncoder images will always be smaller than the reference obtained by Hysteresis segmentation.

CT image	Reference	AutoEncoder segmentation
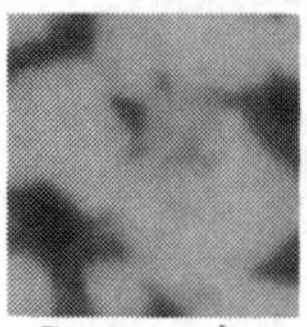 Cross-section	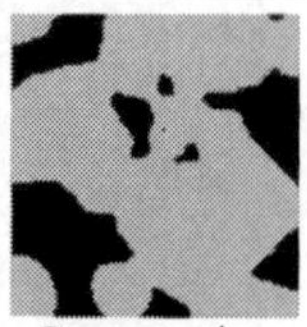 Cross-section segmented image	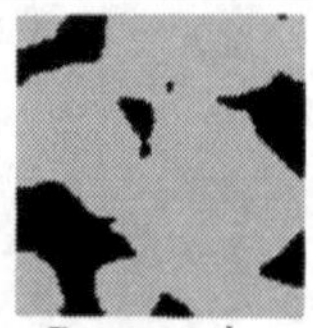 Cross-section segmented image
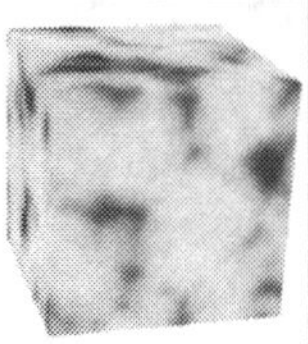 3D volume	 3D volume of pore space	 3D volume of pore space
	Computed porosity volume: 20%	Computed porosity volume: 15%

Figure 6: Comparison between Hysteresis segmentation (Reference) and AutoEncoder model methods: original CT image in grayscale, segmented image with black pixels for pore and greyish for solid.

A) 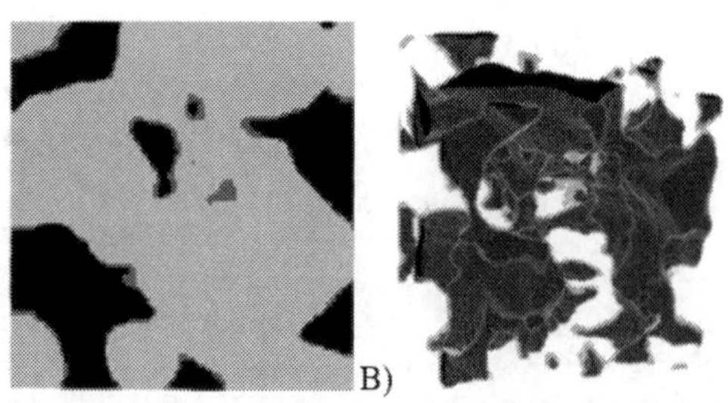 B)

Figure 7: Cross-section (A) and 3D volume (B) images resulted from the comparison of segmentation methods: Hysteresis segmentation and AutoEncoder model. The black voxels are joint porosity of both segmentations and dark grey voxels are supplementary pore voxels in the Hysteresis segmentation.

Discussion and perspectives

Computing resources

The most time-consuming task is the training part of the networks. GPU runs (using one Nvidia Quadro M6000 graphics card) are 15 faster than CPU runs (on a 32 cores Intel Xeon workstation).

Rock classification

In this study, 4 classes of lithological rock formations are distinguished with a highly successful recognition rate.

This work will be extended to numerous classes of reservoir rock facies. All authors [4, 10, 12, 14] referring to machine learning mentioned the importance of the quality of the dataset and required a massive amount of data. This database generation with various available reservoir rock formations imaged at several voxel sizes will be a crucial step.

Rock classification using Inception-V3 (RockClass) proves to be a very promising tool. It can be used for pre-calibrating geological-oriented workflow in digital rock analysis proposing a suitable image processing and analysis workflow. The classification rate could be used as a quality control indicator for validating the image quality before any numerical computations. Then a rock class recognition is interesting to offer a fast estimation of the numerical properties already processed from similar rock class.

To further improve digital rock classification, additional physical attributes as mineralogy and pore fabric could probably enhance RockClass capability in detailing geological rock types. For example, pore fabric is directly related to the hydraulic radius size distribution, which is the key parameter for the permeability computation.

Porosity estimation

The porosity is computed on 1,000 sub-blocks of 100 x 100 x 100 voxels by 3 porosity modules: Hysteresis segmentation, RegPhi an AutoEncoder networks.

The porosities estimated from Hysteresis segmentation are used as reference values. They are computed from segmented images obtained by binary thresholding. Porosity estimation obtained by Hysteresis segmentation is benchmarked and validated in several studies with diverse partners (IFPEN, Geosciences Montpellier, [2]).

Figure 8 illustrates the porosity distribution obtained by these 3 methods on 1,000 juxtaposed blocks of size 100 x 100 x 100 voxels belonging to the same 1000 x 1000 x 1000 voxels image.

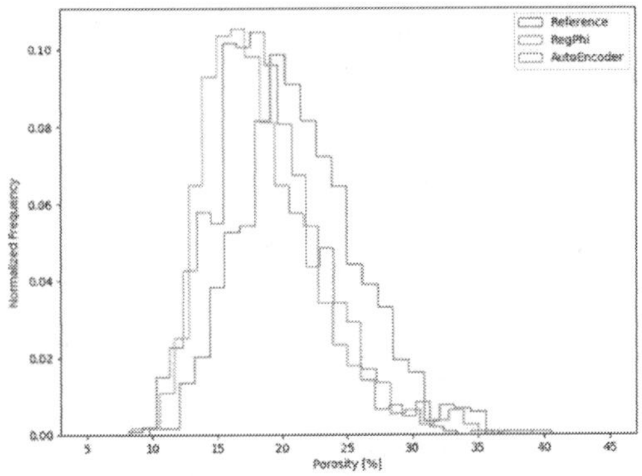

Figure 8: Porosity distribution computed by image analysis (reference), RegPhi and AutoEncoder models for Berea Sandstone sub-blocks.

The reference porosity values indicate that the porosity ranges from 9 to 36% with a mean value of 21.5%. This large distribution is mostly due to variation of the pore-solid ratio along the series of small REV of 100 x 100 x 100 voxels with a high voxel size. The RegPhi and AutoEncoder results present a wider porosity distribution, respectively, from 8 to 41% with a mean value of 18.6% and from 8 to 37% with a mean value of 18.8%. The AutoEncoder model is considered being a more reliable network for predicting porosity values close to the values obtained by conventional image analysis.

RegPhi model allows to estimate the porosity without segmentation whereas AutoEncoder generates a segmented image to assess the porosity. Both networks seem to predict similar porosity distributions and from 2.7% (AutoEncoder) to 2.9% (RegPhi) under evaluate the porosity compared to the reference one. This promising result validates improbable capabilities of both models to evaluate porosity directly from grayscale images. An intensified training is required to enhance their performance.

RegPhi and AutoEncoder networks are both trained on these similar subs-volumes of Berea grayscale image. This similarity of results demonstrates how efficient and accurate segmented images are generated by the AutoEncoder model in this study. From these automated segmented images, series of numerical petrophysical properties could be swiftly computed. A complete digital petrophysical workflow can thus be proposed, here for Berea sandstone.

Conclusion

In this study we aim at demonstrating the potential of neural network algorithms to make digital rock analysis simple. We work on implementing an automated workflow from MCT images to digital petrophysical measurements without operator intervention.

The numerical petrophysics workflow is based on a 3-step approach. First, a rock classification from grayscale MCT images using RockClass model is investigated in the two main lithological rock reservoirs types, carbonates and sandstones. Then RegPhi model is used to predict the total porosity from grayscale MCT images.

An AutoEncoder network allows to automatically generate the pore space segmented image and then evaluate the total porosity.

The results of this study are not limited to 4 reservoir rock classes and porosity estimation, but the proposed workflow can be adapted for any CT rock types in order to deliver a complete series of numerical analysis.

This work was supported by the ERC Starting Grant ``Light Field Imaging and Analysis" (LIA 336978, FP7-2014).

References

[1] P. Gouze and L. Luquot, "X-ray microtomography characterization of porosity, permeability and reactive surface changes during dissolution", J. of Contaminant Hydrology, 120-121, pp. 45–55, doi: 10.1016/j.jconhyd.2010.07.004, (2011)

[2] L. Luquot, V. Hebert and O. Rodriguez, "Calculating structural and geometrical parameters by laboratory measurements and X-ray microtomography: a comparative study applied to a limestone sample before and after a dissolution experiment", Solid Earth, 7, pp. 441-456, doi: 10.5194/se-7-441-2016 (2016)

[3] S. Youssef, E. Rosenberg, N. Gland, S. Bekri and O. Vizika, "Quantitative 3D characterisation of the pore space of real rocks: Improved μ-CT resolution and pore extraction methodology", Int. Symp. Soc. Core Analysts, Calgary, Canada, Sept. 10-12, Paper SCA2007-17 (2007)

[4] R. Al-Raoush and A. Papadopoulos, "Representative elementary volume analysis of porous media using X-ray computed tomography", Powder Technology, 200, pp. 69-77, doi: 10.1016/j.powtec.2010.02.011 (2010)

[5] G. Wang, "A Perspective on deep imaging", IEEE Access, **4**, 8914-8924, Nov. 3, doi: 10.1109/ACCESS.2016.2624938 (2016)

[6] T. Würfl et al., "Deep learning computed tomography: Learning projection-domain weights from image domain in limited angle problems", IEEE Trans. Med. Imaging, 37, pp. 1454-1463 (2018)

[7] Y. Mikia et al., "Tooth labelling in cone-beam CT using deep convolutional neural network for forensic identification", Computers in Biology and Medicine, 80, pp. 24-29 (2017)

[8] A. T. Lopes, E. de Aguiar, A. F. de Souza, T. Oliviera-Santos, "Facial expression recognition with convolutional neural networks: Coping with few data and the training sample order", Pattern Recognition, 61, pp. 610-628 (2017)

[9] M. A. Naranjo Leon et al., "Rock typing mapping methodology based on indexed and probabilistic self-organized map in Shushufindi field", SPE Latin American and Caribbean Petroleum Engineering Conf., Rio de Janeiro, Brazil, Nov 18-20, Paper SPE-177086-MS, doi: 10.2118/177086-MS (2015)

[10] J. Chen and Y. Zeng, "Application of machine learning in rock facies classification with physics-motivated feature augmentation", Machine Learning, Geophysics, arXiv: 1808.09856 (2018)

[11] S. Berg, N. Saxena, M. Shaik and C. Pradhan, "Generation of ground truth images to validate micro-CT image-processing pipelines", The Leading Edge, Advancements in image processing, 37, pp. 412-420 (2018)

[12] S. Karimpouli and P. Tahmasebi, "Image-based velocity estimation of rock using convolutional neural networks", Neural Networks, 111, pp. 89-97 (2019)

[13] M. Araya-Polo, F. O. Alpak, S. Hunter, R. Hofmann and N. Saxena, "Deep learning-driven pore-scale simulation for permeability estimation", ECMOR XVI, Barcelona, Spain, Dec. 3 (2018)

[14] O. Sudakov, E. Burnae and D. Koroteev, "Driving digital rock towards machine learning: Predicting permeability with gradient boosting and deep neural networks", Computer and Geosciences, 127, pp. 91-98 (2019)

[15] A. Krizhevsky, I. Sutskever and G. E. Hintonn, "Imagenet classification with deep convolutional neural networks", Advances in neural information processing systems, 25 (2012)

[16] Z. Zhao, P. Zheng, S. Xu and X. Wu., "Object detection with deep learning: A review", IEEE Transactions on Neural Networks and Learning systems, Apr. 19, arXiv: 1807.05511v2 (2018)

[17] Y. Guo, Y. Liu, T. Georgiou and M. S. Lew, "A review of semantic segmentation using deep neural networks", Int. J. of Multimedia Information Retrieval, 7, pp. 87-93 (2018)

[18] C. Szegedy, V. Vanhoucke, S. Ioffe, J. Shlens and Z. Wojna, "Rethinking the inception architecture for computer vision", IEEE Conf. on Computer Vision and Pattern Recognition, pp. 2818-2826 (2016)

[19] ImageNet. http://www.image-net.org

[20] A. Alperovich, O. Johannsen, M. Strecke and B. Goldluecke, "Light field intrinsics with a deep encoder-decoder network", poster, Proc. CVPR (2018)

[21] W. Shi et al., "Real-time single image and video super resolution using an efficient sub-pixel convolutional neural network", Proc. CVPR, pp. 1874–1883 (2016)

[22] A. Alperovich, O. Johannsen and B. Goldluecke, "Intrinsic light field decomposition and disparity estimation with a deep encoder-decoder network", IEEE European Signal Processing Conf. (2018)

[23] S. N. Ehrenberg and P. H. Nadeau, "Sandstone vs. carbonate petroleum reservoirs: A global perspective on porosity-depth and porosity-permeability relationships", AAPG Bulletin, 89, pp. 435-445 (2005)

[24] J. Dautriat, N. Gland, A. Dimanov and J. Raphanel, "Hydromechanical behavior of heterogeneous carbonate rock under proportional triaxial loadings", J. of Geophysical Research, 116, B01205 (2011)

[25] S. Youssef et al., "High resolution μ-CT combined to numerical models to assess electrical properties of bimodal carbonates", Int. Symp. Soc. Core Analysts, Abu Dhabi, UAE, Paper SCA2008-37 (2008)

[26] T. Bultreys, W. De Boever, L. Van Hoorebeke and V. Cnudde, "A multi-scale image-based pore network modeling approach to simulate two-phase flow in heterogeneous rocks", Int. Symp. Soc. Core Analysts, St. John's Newfoundland and Labrador, Canada, Paper SCA2015-027 (2015)

[27] H. Derluyn, J. Dewanckele, M. N. Boone, V. Cnudde, D. Derome and J. Carmeliet, "Crystallization of hydrated and anhydrous salts in porous limestone resolved by synchrotron X-ray microtomography", Nuclear Instruments and Methods in Physics Research, DOI: 5 10.1016/j.nimb.2013.08.065 (2014)

[28] S. Roels, J. Elsen, J. Carmeliet and H. Hens, "Characterization of pore structure by combining mercury porosimetry and micrography", Materials and Structures, 34, pp. 76-82 (2001)

[29] T. Bourbié and B. Zinszner, "Hydraulic and acoustic properties as a function of porosity in Fontainebleau sandstone", J. of Geophysical Research, 90, pp. 1524-1532 (1985)

[30] A. Sinnokrot, H. J. Ramey and S. Marsden, "Effect of temperature level upon capillary pressure curves", Soc. Petroleum Engineering, 11, pp. 13-22 (1971)

[31] P.L. Churcher, P. R. French, J. C. Shaw and L. L. Schramm, "Rock properties of Berea sandstone, Baker dolomite, and Indiana limestone", SPE Int. Symp. Oilfield Chemistry, Anaheim, California, Paper SPE-21044-MS, Feb. 20-22 (1991)

[32] P.-E. Oren and S. Bakke, "Reconstruction of Berea sandstone and pore-scale modeling of wettability effects", J. Petroleum Science and Engineering, 39, pp. 177-199 (2003)

[33] S. Youssef, D. Bauer, Y. Peysson and O. Vozoka, "Investigation of pore structure impact on mobilization of trapped oil by surfactant injection", Int. Symp. Soc. Core Analysts, Avignon, France, Sept. 11-18, Paper SCA2014-64 (2014)

[34] N. Saxena, R. Hofmann, F.O. Alpak, J. Dietderich, S. Hunter and R.J. Day-Stirrat, "Effect of image segmentation & voxel size on micro-CT computed effective transport & elastic properties", Marine and Petroleum Geology, 86, pp. 972-990 (2017)

[35] S. Cyprien et al., "The impact of sub-resolution porosity of X-ray microtomography images on the permeability", Transport in Porous media, 113, pp. 227-243 (2016)

[36] O. Rodriguez, T. Porcher, V. Planes, G. Mecuson and R. Bouvier, "Non destructive testing of CMC engine internal parts from X-ray tomographic images", 9th Int. Conf. Industrial Computed Tomography, Padova, Italy, Feb. 13-15 (2019)

[37] A. Krizhevsky, I. Sutskever and G. E. Hinton, "ImageNet classification with deep convolutional neural networks", Advances in Neural Information Processing Systems, 25, pp. 1097-1105 (2012)

[38] F. Chollet et al., https://keras.io (2015)

[39] Y. Pu et al., "Adversarial symmetric variational autoencoder", Advances in Neural Information Processing Systems, 30, pp. 4330-4339 (2017)

[40] X.-J. Mao, C. Shen and Y.-B. Yang, "Image restoration using very deep convolutional encoder-decoder networks with symmetric skip connections", Proc. Neural Information Processing Systems, arXiv: 1603.09056 (2016)

[41] K. He, X. Zhang, S. Ren and J. Sun, "Deep residual learning for image recognition", Proc. Int. Conf. Computer Vision and Pattern Recognition, arXiv: 1512.03385 (2016)

[42] J. Springenberg, A. Dosovitskiy, T. Brox and M. Riedmiller, "Striving for simplicity: The all convolutional net", Proc. 3rd Int. Conf. Learning Representations, workshop track (2015)

[43] D. Kingma and J. Ba, "Adam a method for stochastic optimization", Proc. 3rd Int. Conf. Learning Representations (2015)

Using Capillary Condensation and Evaporation Isotherms to Investigate Confined Fluid Phase Behavior in Shales

Elizabeth Barsotti[1,*], *Evan* Lowry[1], *Mohammad* Piri[1], and *Jin-Hong* Chen[2]

[1]Center of Innovation for Flow through Porous Media, Petroleum Engineering Department, University of Wyoming, Laramie, Wyoming, USA
[2]Aramco Services Company, Aramco Research Center-Houston, Houston, Texas, USA

Abstract. The abundance of nanopores (pores with diameters between 2 and 100 nm) in shale and ultra-tight reservoirs precludes the use of common pressure-volume-temperature (PVT) analyses on reservoir fluids. The small sizes of the pores cause capillary condensation, which is a nanoconfinement-induced gas-to-liquid phase change, that can occur at pressures more than 50% below the corresponding bulk phase change of the fluid due to strong fluid-pore wall interactions. We quantify this phenomenon by measuring propane isotherms both in a synthetic nanoporous medium and a core from a shale gas reservoir. Comparison of our results in the two porous media indicates the occurrence of capillary condensation in shale rock. At the same time, we observe capillary condensation hysteresis for shale, in which the density of the fluid is significantly lighter during desorption than adsorption. This indicates structural changes to the rock matrix caused by the phase behavior of the confined fluid. We use scanning electron microscopy to corroborate our findings. These results have significant implications for determining the PVT properties, porosity, and permeability of shale and ultra-tight formations for use in reservoir modeling and production estimations.

1 Introduction

Current methods of shale reservoir evaluation do not account for (i) capillary condensation, a confinement-induced gas-to-liquid phase change, or (ii) continuous pore filling, a sharp increase in the density of supercritical fluids, that occur in nanopores. [1] Even though it has been shown that accounting for nanoconfinement-induced phase phenomena such as capillary condensation could increase reserves estimates by 600%, [2] the current prevailing opinion only considers the existence of an adsorbed, or physisorbed, phase on the pore walls and free gas in the pore bodies. [3]

The primary reason for this is misinterpretation of isotherms measured in shale. Throughout the literature, isotherms measured for both hydrocarbons and gases, such as CO_2, do not exhibit the distinct capillary condensation step defined in the IUPAC Type IV isotherm [4] or the continuous filling step that has recently been elucidated by Tan et al. [5] and Barsotti et al. [6] Instead, they appear similar to IUPAC Type I and II isotherms, [7], [8] which are attributed to mono-layer and multi-layer adsorption, respectively. [4] Therefore, they are given the same interpretation even though the IUPAC isotherms rely on fundamental assumptions about the adsorbate and adsorbent that are not true of shale rock. The Type I isotherm assumes a microporous material, while the Type II isotherm assumes a macroporous or non-porous surface. [4] Both of these cases are far removed from the wide pore size distributions, complex wettabilities, unpredictable pore geometries, and complex fluid compositions in shale. Even so, the ease with which longstanding, simplistic adsorption theories, such as Langmuir and BET, can be fit to shale adsorption isotherms has led to their widespread incorporation into shale characterization [9], [3], [10], [11] and reservoir modeling. [12]-[14]

However, we prove here that despite the fact that the shapes of shale isotherms are often similar to IUPAC isotherms, the underlying physics are generally not the same. By creating controlled pore size distributions with a synthetic nanoporous medium called MCM-41 and comparing the isotherms measured in them to those measured in shale, we show that the broad pore size distribution of shale can alter the appearance of capillary condensation and continuous pore filling such that they manifest much differently from the traditional IUPAC Type IV isotherm. We further demonstrate this with desorption measurements where hysteresis belies the underlying phase change. Closer scrutiny of the hysteresis loop in combination with scanning electron microscopy not only confirms the occurrence of capillary condensation but also provides evidence that just as the nanopores can alter the phase behavior of fluids, so can the nanoconfinement-induced phase behavior of the fluids impart irreversible changes to the rock, including fractures. Our observations of capillary condensation and continuous pore filling cast significant doubt upon the

* Corresponding author: ebarsott@uwyo.edu

ability of simple adsorption theories to accurately model phase behavior in shale.

2 Materials and Methods

2.1. Isotherm Measurements

All isotherms were measured using a novel gravimetric apparatus. [15], [16] Unlike other systems available commercially and in the literature, our apparatus is compatible with macro-scale cores enclosed in core holders. [15] All core holders employed in this study were ¾" in internal diameter and 7" long and comprised a titanium body with stainless steel endcaps. During the isotherm measurements, the core holders were suspended from Mettler Toledeo XPE505C mass comparators inside of an environmental chamber (Thermotron). The environmental chamber uses air as the thermostating fluid such that experiments between 5.4 and 16.1°C were characterized by a temperature control of +/- 0.1°C. In other words, the experimental fluid resided solely in the core holder, while the temperature and pressure of the thermostatic air were constant throughout all experiments. Therefore, no correction for buoyant force had to be made, [16] underscoring another unique feature of our apparatus.

By injecting or retracting the experimental fluid into or out of the core holders, its pressure could be changed, such that the change in mass and pressure at constant temperature could be used to plot an isotherm. Equilibrium was defined as the time at which the pressure of the fluid became constant as read by a Rosemount pressure transducer (Emerson). [16] At equilibrium, pressure and mass data points were generated by averaging data collected at a rate of 1/sec over a timespan of 1 minute. [16] Statistical analyses of these time series showed the standard error for mass and pressure to be less than 0.008 g and 0.008 psi, respectively, for all measurements. Because these errors are negligible, no error bars are included in the isotherm plots. We refer interested readers to our previous publications for more comprehensive discussions of the operating principle, repeatability, and accuracy of the apparatus. [15]-[17]

For adsorbents, both synthetic silica nanopores (MCM-41) and shale cores from a gas reservoir in the Middle East were used. The shale used here is from the same core that was used in our previous work. [18]

MCM-41 is a synthetic nanoporous silica popular for use in capillary condensation experiments because its singular pore size distribution and cylindrical pores simplify data analysis. Three different pore sizes of MCM-41 (Glantreo, Ltd.) were characterized using Non-Local Density Functional Theory analysis of nitrogen isotherms measured at 77K. [6] The characteristic pore diameters of the samples were 2.90 nm, 4.19 nm, and 8.08 nm. [6] Henceforth, the small, medium, and large pore sizes of MCM-41 will be referred to as MCM-41-S, MCM-41-M, and MCM-41-L to be consistent with our previous work. [6] To create controlled pore size distributions, known amounts of the different adsorbents were mixed together. In one core holder, 8.2946 grams of MCM-41-S were packed. [6] In a second core holder, 4.96016 grams of MCM-41-S were mixed with 3.51795 grams of MCM-41-M. In a third core holder, 5.02250 grams of MCM-41-S were mixed with 2.31973 grams of MCM-41-L. The three different core holders allowed us to determine how large and small differences in pore size affect the shape of an isotherm.

As received shale cores were crushed by hand following the same procedure in our previous work. [18] Pieces of the crushed rock were randomly selected for characterization using a Helios 650 (ThermoFisher Scientific) focussed ion beam scanning electron microscope (FIB-SEM). Energy dispersive X-ray scattering revealed the mineralogy to comprise kerogen and calcite with minor amounts of quartz, pyrite, and illite. [18] High resolution imaging using a Through-the-Lens detector in downhole vision mode showed the kerogen to be highly nanoporous, as shown in Figure 1. Approximately 67 grams of the shale rock were packed into a titanium core holder for the isotherm measurement.

Propane (99%, AirGas) was used as an adsorbate for the measurements in both types of porous media.

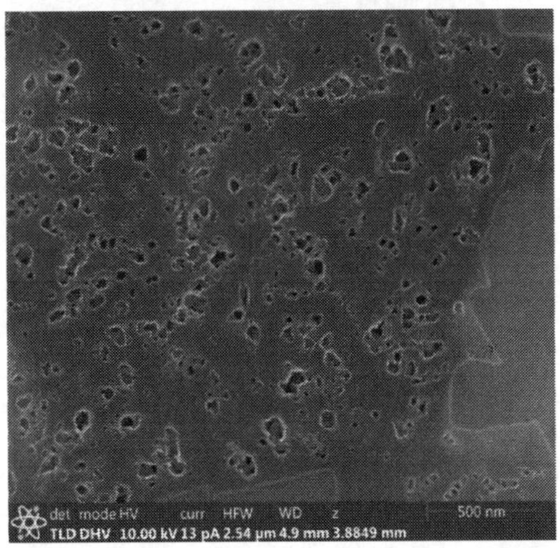

Figure 1. Scanning electron micrograph of kerogen nanopores in the shale rock taken at a voltage of 10kV and a current of 13 pA with an instrument resolution of 0.8 nm.

2.1. SEM Imaging

One piece of the crushed shale was randomly selected for FIB-SEM imaging to determine whether the fluid phase behavior in the rock affected the rock properties. The sample was mounted on a 45-degree, pre-tilted aluminium stub (Ted Pella) using carbon tape and silver paint (Ted Pella). Then, a 20 µm by 20 µm cross-section of the sample was polished with the FIB-SEM's gallium ion beam. Subsequently, a 1 µm diameter, 2 µm deep circular hole was milled into the kerogen in the rock using the ion beam. Imaging of the surface was done prior to and after exposure to propane. To expose the sample to propane, it was removed from the sample stub, and placed into a pressure cell. The pressure cell was first vacuumed out using a Welch vacuum pump to remove any air and then pressurized to 6.9 bar with propane at 20°C. After approximately 1.5 hrs, the pressure of the cell was reduced to atmospheric and the rock sample was removed, re-mounted on the 45-degree, pre-tilted stub, and imaged again. The circular hole was to determine whether any permanent changes to the kerogen structure, such as irreversible swelling, had occurred.

3 Results and Discussion

3.1. Isotherms in MCM-41

The isotherms measured in all three of the MCM-41 packs are shown in Figure 2. Isotherms were measured at 5.4°C. Note that the bulk saturation pressure for propane at this temperature is 5.5766 bar, as reported by NIST. [19] In Figure 2, the bulk saturation pressure is measured as the rightmost abrupt increase in amount adsorbed as indicated in the plots. In all three cases, the measured bulk saturation pressure fell within 2% (percent difference) of that reported by NIST, affirming the accuracy of our measurements.

In all three isotherms, the initial increase in the amount adsorbed is propane adsorption on the pore walls. The downward concavity indicates that the propane is a wetting fluid [4] to the MCM-41.

At higher pressures capillary condensation or continuous pore filling occurs. For the pure MCM-41-S, only a single step in the isotherm is observed, indicating filling of the pores with a dense supercritical phase. Of the three pore sizes only MCM-41-L was sufficiently large to exhibit the subcritical phase transition, capillary condensation. [5] Therefore, in the mixture of MCM-41-S and MCM-41-L, the two distinct steps in the isotherm indicate continuous pore filling for MCM-41-S and capillary condensation in MCM-41-L. When the pore sizes were closer together in the case of the mixture of MCM-41-S and MCM-41-M, the continuous pore filling in both blended together into a single step.

This has significant implications for shale rock, where the pore size distribution is broad but consecutive sizes are close together. It is often assumed that, because a distinct capillary condensation or continuous pore filling step cannot be observed in shale isotherms, neither of the nanoconfined phase phenomena is occurring. However, if we extend our understanding from the MCM-41 experiments, the absence of a step in the isotherm does not indicate the absence of capillary condensation or continuous pore filling; rather it indicates the absence of a distinct, singular pore size. In other words, phase phenomena occur gradually in the different pore sizes giving rise to a gently sloping isotherm that may be easily mistaken as an IUPAC Type I or II isotherm.

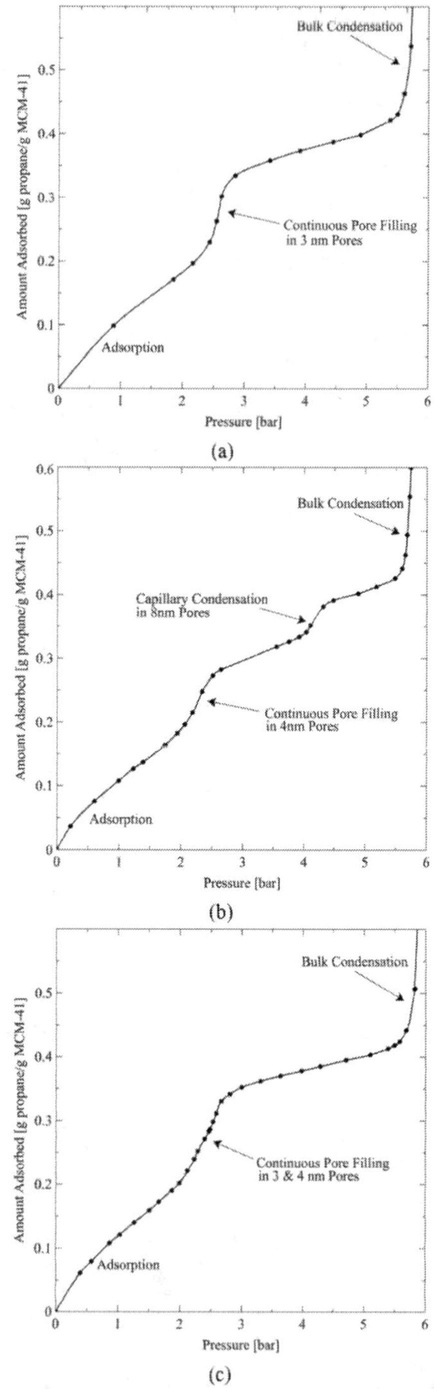

Figure 2. Propane adsorption isotherms measured at 5.4°C in the core holders containing (a) MCM-41-S, (b) MCM-41-S and MCM-41-L, and (c) MCM-41-S and MCM-41-M. Figure (a) adapted with permission from Barsotti et al. [6] Copyright 2018 American Chemical Society.

Likewise, the presence of both a sub-critical and supercritical phase in the nanopores, may not only further reduce the relevance of current modeling efforts but also make accurate modeling more difficult as many equations of state are not well-equipped to work with supercritical fluids.

3.1. Isotherms in Shale Rock

This behavior is evident in Figure 3, where no apparent capillary condensation or continuous pore filling step is observed during adsorption, and the isotherm does, indeed, appear similar to the IUPAC Type I and II isotherms. However, the occurrence of capillary condensation in the shale sample can be verified by also measuring desorption. Upon desorption, hysteresis is observed. Hysteresis at relatively high pressures is commonly attributed to capillary condensation. [4] Therefore, because we observe hysteresis occurring at pressures as high as 90% of the bulk saturation pressure, which is approximately 7.5 bar in Figure 3, we interpret it as indicating the occurrence of capillary condensation. This interpretation is consistent with other studies in the literature where any hysteresis occurring above 50% of the bulk saturation pressure was defined as a key indicator of capillary condensation. [20] Therefore, we show that capillary condensation can occur in shale rock, even when it is not immediately apparent from the adsorption part of the isotherm.

Interestingly, the hysteresis loop for shale also appears to differ from those types classified by IUPAC. Rather than desorption occurring at larger amounts adsorbed than adsorption, [4] we observe it to occur at smaller amounts. This indicates that the fluid in the pore space is less dense during desorption than during adsorption. This can be interpreted to mean that the pore size is larger during desorption than adsorption. Thus, it may be possible not only that the shale nanopores are affecting the phase behavior of the fluid but also that the fluid can affect the morphology of the pores. In addition to our observation of hysteresis at high pressures, our observation of hysteresis at low relative pressure supports this, as it has been found in the literature to indicate adsorbent deformation. [4]

Further evidence for deformation is given by the dips observed in the isotherm at intermediate pressures. In isotherms measured for coal, these dips are often interpreted to indicate swelling of the adsorbent. [21]

To prove this, a polished cross section of the shale was visualized both before and after contact with propane at 0 °C and ~6.9 bar. Both images are presented in Figure 4. After the propane capillary condensation, large fractures were observed in the rock. Our initial hypothesis was that permanent swelling of the kerogen induced the fractures. But comparison of the 1μm diameter holes milled into the kerogen before and after propane exposure shows no change to the diameter or shape of the hole. We, therefore, hypothesize that reversible kerogen swelling causes fracturing. In essence, the kerogen swells during adsorption. During desorption, as the kerogen returns to its initial configuration, differences in the elasticities of the inorganic (i.e., calcite, quartz, and pyrite) and organic minerals result in fractures. Kerogen swelling has been observed previously during capillary condensation, [18], [22] but to the best of our knowledge, this is the first observation of fracturing during capillary condensation.

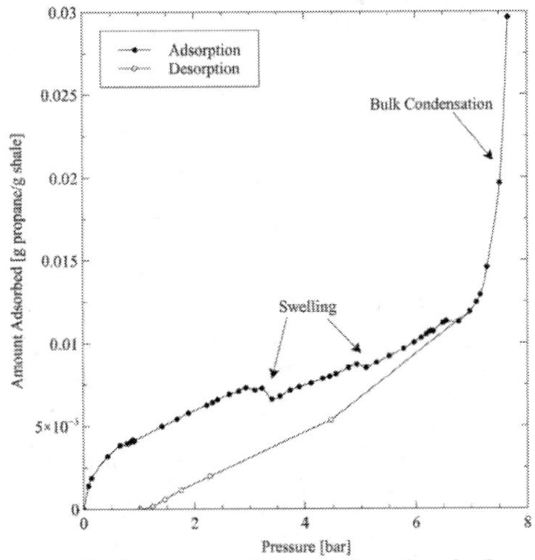

Figure 3. Propane capillary condensation isotherms measured in shale rock at 16.1°C. Note that the bulk saturation pressure for propane at 16.1°C is 7.5379 bar as reported by NIST. [19] Therefore, the percent difference between our measurement and theirs is less than 1%, also affirming the accuracy of this measurement.

4 Conclusions

By measuring isotherms both in controlled pore size distributions of MCM-41 and shale rock, we show that complexities of the adsorbent, such as a wide pore size distribution, can cause the shape of the isotherm to appear significantly different from ideal cases presented in the literature for simple porous media. This discrepancy can lead to the misinterpretation that capillary condensation or continuous pore filling do not occur in shale. However, our observations of capillary condensation occurring gradually at different pressures across the range of pore size both in the MCM-41 and shale disprove this. Observations of hysteresis in the shale further indicate the occurrence of capillary condensation. Indeed, the complex mineralogy of shale may lead to capillary condensation-induced fractures. This indicates that the rock and fluid properties are interrelated and that current methods of PVT and core analysis that do not account for capillary condensation may lead to erroneous estimates of rock and fluid properties. The degree to which capillary condensation affects these phenomena is highly dependent upon the chemical compositions and physical morphologies of the fluid and the rock. We have proven

https://doi.org/10.1051/e3sconf/202014605003

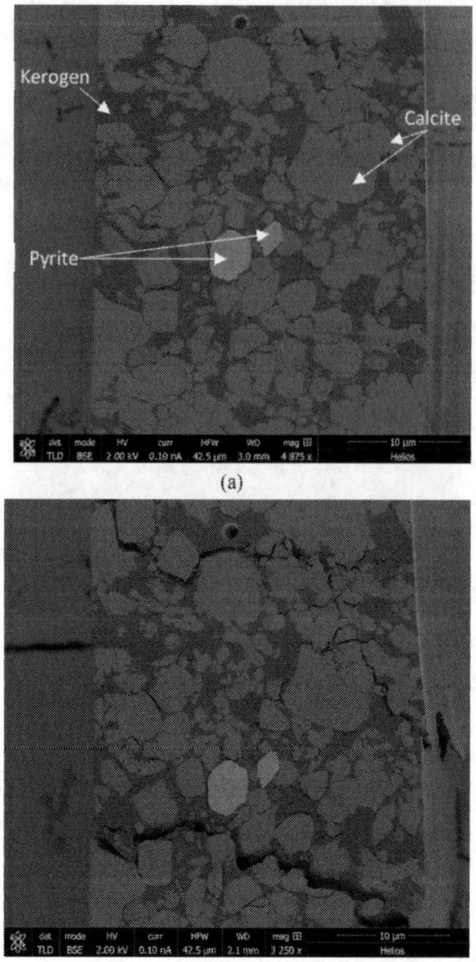

(a)

(b)

Figure 4. SEM images of a polished cross-section of the shale (a) before and (b) after exposure to propane at 20°C and ~6.9bar. The low, medium, and high greyscale intensity minerals are kerogen, calcite, and pyrite, respectively.

this both experimentally [16], [23] and theoretically [24] in MCM-41, where selective adsorption occurs and can affect the phase transition. Future work must include more complex mixtures in different types of reservoir cores to fully characterize this phenomenon.

We gratefully acknowledge the financial support of Saudi Aramco, Hess Corporation, and the School of Energy Resources at the University of Wyoming.

References

[1] E. Barsotti, S. P. Tan, S. Saraji, M. Piri and J.-H. Chen, "A review on capillary condensation in nanoporous media: Implications for hydrocarbon recovery from tight reservoirs," *Fuel*, pp. 344-361, 2016.

[2] J.-H. Chen, A. Mehmani, B. Li, D. Georgi and G. Jin, "Estimation of total hydrocarbon in the presence of capillary condensation for unconventional shale reservoirs," in *SPE Middle East oil and gas show and conference*, Manama, Bahrain, 2013.

[3] J. Richardson and W. Yu, "Calculation of Estimated Ultimate Recovery and Recovery Factors of Shale-Gas Wells Using a Probabilistic Model of Original Gas in Place," *SPE Reservoir Evaluation & Engineering*, pp. 638 - 653, 2018.

[4] M. Thommes, K. Kaneko, A. V. Neimark, J. P. Olivier, F. Rodriguez-Reinoso, J. Rouquerol and K. S. Sing, "Physisorption of gases, with special reference to the evaluation of surface area and pore size distribution (IUPAC Technical Report)," International Union of Pure and Applied Sciences & De Gruyter, 2015.

[5] S. P. Tan, E. Barsotti and M. Piri, "Criticality of Fluids Confined in Nanopores Based on the Liquid Spinodal," *Under Review*, 2019.

[6] E. Barsotti, S. P. Tan, M. Piri and J.-H. Chen, "Phenomenological Study of Confined Criticality: Insights from the Capillary Condensation of Propane, n-Butane, and n-Pentane in Nanopores," *Langmuir*, pp. 4473-4483, 2018.

[7] S. Duan, M. Gu, X. Du and X. Xian, "Adsorption Equilibrium of CO2 and CH4 and Their Mixture on Sichuan Basin Shale," *Energy & Fuels*, p. 2248–2256, 2016.

[8] S. Rani, B. K. Prusty and S. K. Pal, "Methane adsorption and pore characterization of Indian shale samples," *Journal of Unconventional Oil and Gas Resources*, pp. 1-10, 2015.

[9] W. Yu, K. Sepehrnoori and T. W. Patzek, "Modeling Gas Adsorption in Marcellus Shale With Langmuir and BET Isotherms," *SPE Journal*, pp. 589 - 600, 2016.

[10] S. Rani, E. Padmanabhan and B. K. Prusty, "Review of gas adsorption in shales for enhanced methane recovery and CO2 storage," *Journal of Petroleum Science and Engineering*, pp. 634-643, 2019.

[11] M. Eliebid, M. Mahmoud, S. Elkatatny, M. Abouelresh and R. Shawabkeh, "Adsorption Role in Shale Gas Recovery and the Feasibility of CO2 in Shale Enhanced Gas Recovery: A Study on Shale Gas from Saudi Arabia," in *SPE Kuwait Oil & Gas Show and Conference*, Kuwait City, Kuwait, 2017.

[12] V. Nguyen-Le and H. Shin, "Development of reservoir economic indicator for Barnett Shale gas potential evaluation based on the reservoir and hydraulic fracturing parameters," *Journal of Natural Gas Science and Engineering,* pp. 159-167, 2019.

[13] S. Yang, K. Wu, J. Xu, J. Li and Z. Chen, "Roles of multicomponent adsorption and geomechanics in the development of an Eagle Ford shale condensate reservoir," *Fuel,* pp. 710-718, 2019.

[14] H. Belyadi and M. Smith, "A Fast-Paced Workflow for Well Spacing and Completions Design Optimization in Unconventional Reservoirs," in *SPE/AAPG Eastern Regional Meeting*, Pittsburgh, Pennsylvania, USA, 7-11 October 2018.

[15] E. Barsotti, S. Saraji and M. Piri, "Nanocondensation Apparatus". United States of America Patent US15/588,094, 2017.

[16] E. Barsotti, S. Saraji, S. P. Tan and M. Piri, "Capillary Condensation of Binary and Ternary Mixtures of n-Pentane–Isopentane–CO2 in Nanopores: An Experimental Study on the Effects of Composition and Equilibrium," *Langmuir,* p. 1967–1980, 2018.

[17] E. Barsotti, S. P. Tan, M. Piri and J.-H. Chen, "Capillary-Condensation Hysteresis in Naturally-Occurring Nanoporous Media," *Under Review,* 2019.

[18] E. Barsotti, S. P. Tan, M. Piri and J.-H. Chen, "Capillary-Condensation Hysteresis in Naturally-Occurring Nanoporous Media," *Under Review,* 2019.

[19] E. Lemmon, M. McLinden and D. Friend, *NIST Standard Reference Database Number 69,* P. Linstrom and W. Mallard, Eds.

[20] H. Bu, Y. Ju, J. Tan, G. Wang and X. Li, "Fractal characteristics of pores in non-marine shales from the Huainan coalfield, eastern China," *Journal of Natural Gas Science and Engineering,* vol. 24, pp. 166-177, 2015.

[21] R. Pini, S. Ottiger, G. Storti and M. Mazzotti, "Role of Adsorption and Swelling on the Dynamics of Gas Injection in Coal," *Journal of Geophysical Research Solid Earth,* vol. 114, no. 84, 2009.

[22] M. Pathak, H. Kweon, M. Deo and H. Huang, "Kerogen Swelling and Confinement: Its implication on Fluid Thermodynamic Properties in Shales," *Scientific Reports,* vol. 7, 2017.

[23] E. Barsotti, M. Piri, J.-H. Chen and S. Althaus, "Solution Gas Drive in Tight Oil Reservoirs: New Insights from Capillary Evaporation Experiments," in *SPE/AAPG/SEG Unconventional Resources Technology Conference*, Houston, Texas, USA, 2018.

[24] S. P. Tan, E. Barsotti and M. Piri, "Application of material balance for the phase transition of fluid mixtures confined in nanopores," *Fluid Phase Equilibria,* 2019.

Methane Isotherms and Magnetic Resonance Imaging in Shales

M. J. Dick[1,*], *D.* Heagle[2], *D.* Veselinovic[1] and *D.* Green[1]

[1]Green Imaging Technologies, Fredericton, NB, Canada
[2]Natural Resources Canada, Canmet Energy – Ottawa, ON, Canada

Abstract. Adsorption isotherms of light hydrocarbons on reservoir rocks are key data used to quantify the total gas content in reservoirs and isotherms are now being used to improve our understanding of the processes affecting subsurface gas flow associated with gas injection from Enhanced Oil Recovery techniques. This project combined elements of the traditional pressure-volume gas adsorption isotherm technique and an NMR-based adsorption isotherm approach to determine the adsorption isotherms of light hydrocarbons on to tight rocks from oil and gas reservoirs. The new approach allows isotherms to be derived from NMR data. First, a T_2 distribution of the gas is determined over a range of gas pressures. Next, the volume of pore gas is estimated using the pore volume of the rock and the Van der Waals gas equation. The adsorbed gas content is then calculated by subtracting pore gas content from the total gas content. This is repeated for a range of gas pressures to determine the adsorption isotherm. This project used the NMR method described above and measured the gas pressure decay in the NMR cell. This combined approach includes the advantages of the NMR method but it also produces a pressure-time curve that can be used to identify when equilibrium is attained in low permeability rocks and can be used to compare adsorption kinetics of different gases. The advantages of our approach are that 1) the samples remain intact and the measurements provide information on the pore size distribution; 2) analyses can be carried out at reservoir pressures; 3) isotherms can be measured for any gas containing hydrogen atoms; and 4) the results can be used to examine the processes controlling gas flow through the rock. Future work to develop this technique will improve our quantification of the amount of pore gas in the cell, which will improve our partitioning between adsorbed gas and pore gas as well as allow for an improved analysis of the pressure response of the sample after degassing.

1 Introduction

Adsorption of gases to rocks is an important process for evaluating natural gas resource potential, coalbed methane recovery, as well as secondary and tertiary enhanced oil recovery techniques. State-of-the-art models are used to evaluate resource potential and enhanced oil recovery but these models require inputs including gas transport processes and adsorption isotherms to help describe the mobility and storage of gasses on the rocks. Gas transport through low-permeability rock, also referred to as tight rock, is a significant factor for evaluating how much of a resource is recoverable and how effective gas injection into a tight reservoir might be. Gas transport by diffusion is a slow process while transport by advection, or Darcy flow, may be orders of magnitude greater. Assessing the contribution of each component to the gas flux requires sophisticated experiments at a variety of pressures and/or multiple gases and for the most part does not lend itself to routine analyses [1].

Adsorption isotherms for rocks commonly report data on the volume of gas contained within the pore space, referred to here as pore gas, and the gas adsorbed on to the surface of the pore walls referred to here as adsorbed gas. Our use of the term adsorbed gas is equivalent to the term excess gas that has been used in some other adsorption isotherm studies [2].

The quantity of pore gas and adsorbed gas is traditionally measured volumetrically using a system of two cells, a sample cell and a reference cell, which are separated by a valve. A crushed rock sample is placed in the sample cell and the reference cell is filled with a gas at a known pressure. The valve is opened and the gas from the reference cell expands into the sample cell and the pressures in the two cells equalize. The procedure is performed first with a non-adsorbing gas, such as helium, to determine the void volume, or free gas content, in the rock sample via the difference in temperature and pressure before and after opening of the valve between the two cells. Then the system is evacuated at below-atmospheric pressures to remove the gas, and the reference cell is filled with an adsorbing gas. The valve is opened again and the adsorbing gas expands into the sample cell, it will fill the void volume and adsorb on the sample. The amount of adsorption can be calculated by subtracting the free gas content. The expansion of the gases is repeated at numerous pressures to generate isotherms. The disadvantages of this method include 1) the sample is crushed for the measurement which alters the structure of the sample; 2) no information is acquired on the size of the pores occupied by the gas; and 3) many analyses do not examine the processes controlling gas flow through the rock.

Low Pressure Gas Adsorption (LPGA) is another technique used to determine adsorption isotherms, pore size distributions and surface area. This technique also uses dry powdered samples that are evacuated of all gas and cooled so that the equilibrium gases, typically N_2 or CO_2, can adsorb to the surface of the sample. After the absorbing gas is added to the sample the change in pressure over time is analyzed to determine the adsorption isotherm as well as the pore size distribution and surface area or the sample. There are disadvantages with this method that impact analysis of low permeability rock including drying and powdering the samples that may change the sample characteristics. Also, the isotherms are carried out at low pressures (< 20 psi) which prevents an analysis of the isotherms at reservoir conditions and the number of gas adsorption sites is related to the gas pressure [2] so conducting the isotherms at lower pressures may underestimate the adsorption capacity of a reservoir or source rock. Finally, the isotherms are not determined using methane or any of the other light hydrocarbon gases, so the interaction between the hydrocarbon gas and organics in the rocks may not be addressed with this method.

Measuring gas in shales is challenging with NMR due to the low hydrogen index of gas and the small pore volume of shales, both of which result in low NMR signal. However, with recent advances in NMR hardware, gas measurements in shales are possible. Low-field NMR can be used to measure gas isotherms with hydrocarbon gases of interest (e.g. methane or ethane), which is important for evaluating gas in place or determining adsorption of hydrocarbon gases during gas injection for Enhanced Oil Recovery (EOR) strategies [3]. The advantages of NMR measurements include 1) the samples remain intact and the measurements provide information on the pore size distribution; 2) analyses can be carried out at reservoir pressures; and 3) isotherms can be measured for any gas containing hydrogen atoms; and 4) the results can be used to examine the processes controlling gas flow through the rock.

Gas adsorption to nanoporous materials and shales have has been previously studied [4,5]. These studies showed it was possible to identify three distributions of methane including 1) methane adsorbed in the micropores, 2) adsorbed methane in fast exchange with the surrounding mesopore network, and 3) adsorbed methane in exchange with the interparticle void space of the adsorbent beds by characteristic relaxation times and pressure dependencies in the relaxation time distributions.

In this study we use low-field NMR to generate methane isotherms by measuring the total gas in the rock and distinguishing the amount of free gas and adsorbed gas in a rock. We calculate the pore gas volume using Van der Waals gas equation to determine the methane adsorption isotherm. We also combine pressure measurements with NMR measurements to carry out an initial assessment of the transport processes controlling gas flow in a shale.

2 Method

The method for determining gas isotherms from NMR data is as follows:

1) Record the T_2 distribution of the dry shale sample (95C for 7 days in vacuum oven) with 0 psi of methane. Shale samples often have NMR signal from organic content present in the rock and should not be included in the gas isotherm analysis. It is assumed that this signal will be invariant with methane pressure and should be subtracted from all subsequent T_2 distributions at higher methane pressures.

2) Equilibrate the sample at a range of methane pressures (i.e. 500, 1000, 1500 psi etc.). At each pressure record the T_2 distributions as a function of time. This is necessary as it will take time for the methane to fully penetrate all the pores of the sample. Continue to record the distributions until the observed NMR signal has stabilized.

3) Subtract the background (0 psi methane) echo train from each of the echo trains measured for the various pressures (500, 1000, 1500 psi etc.) to account for the T_2 distribution of the rock and any liquids or gases present in the sample before the experiment began. Retrieve the observed methane signal from each background subtracted distribution by summing the area under each distribution. This will yield the observed methane content in units of equivalent water volume (ml) as the NMR spectrometer has been calibrated using water.

4) Convert the observed methane content in units of equivalent water volume to gas content (GC as scm^3 of methane/ cm^3 sample) of observed methane using the following equation:

$$GC = \frac{PVT_{STP}}{TP_{STP}H_{index}V_{bulk}} \quad (1)$$

Where P, V and T are the pressure (psi), volume (cm^3) and temperature (307 °K) of methane during the experiment. H_{index} is the hydrogen index of methane. P_{STP} and T_{STP} are standard temperature and pressure (14.7 psi and 292 °K). V_{bulk} is the bulk volume (cm^3) of the core sample.

5) Plot the gas content measurements for each pressure of methane observed as a function of time. This plot should show that the gas content stabilizes over time. After stabilization, retrieve the gas content for each pressure and plot the gas content as a function of pressure. This will yield the total gas isotherm.

6) To retrieve the pore gas isotherm employ the pore volume of the core sample (measured in a separate NMR

https://doi.org/10.1051/e3sconf/202014605004

experiment with a 100% brine saturated sample) and the Van der Waals gas equation and the compressibility of methane were used to calculate the pore gas content (scm³/cm³).

$$Pore\ Gas\ Content = \frac{PnT_{STP}c}{P_{STP}V_{bulk}T} \quad (2)$$

Where P is the pressure (psi) of the cell; n is the porosity (unitless) of the rock; T_{STP} is the standard temperature (292 °K); c is the compressibility (unitless) of methane at 307 °K; P_{STP} is the standard pressure (psi); V_{bulk} is the bulk volume of the core; and T is the temperature (307 °K) of the magnet and cell. This assumes that the pore gas behaves like any standard gas in a container of fixed volume (i.e. an individual pore).

7) To retrieve the adsorbed gas isotherm subtract the pore gas isotherm from the total gas isotherm.

3 Experiment

The method for determining gas isotherms via NMR data was tested in three different experiments on three different shale samples. For the first two experiments, the NMR data was recorded on Oxford Instruments GeoSpec 2-75 rock core analyzer [6] equipped with Oxford Instruments P5 overburden NMR probe [7]. All data was then processed using GIT Systems Advanced software [8].

3.1 Experiment 1

The first experiment determined the gas isotherm for a Barnett shale (BN 1). The Mississippian Barnett Shale is an organic rich, thermally mature source rock for oil and gas produced from Paleozoic reservoir rocks in the Bend arch–Fort Worth Basin area of Texas [9]. Mean TOC for core samples averages between 4% and 5% [9] and a porosity of 6% [10]

Figure 1 shows the experimental setup employed for this experiment. Table 1 summarizes the properties of the sample while Table 2 summarizes the NMR parameters employed.

Core Sample	BN 1	LHSVL	Bakken
Origin	Barnett	Lower Haynesville	Bakken
Core Diameter (cm)	3.81	2.51	2.49
Core Length (cm)	4.79	2.57	1.67
Bulk Volume (mL)	54.61	12.9	8.1
Dry Core Mass (g)	129.16	-	-
Pore Volume (mL)	2.60	1.137	0.19
Porosity (p.u.)	4.76	8.80	2.3

Table 1. Sample Information

Measurement	T_2, BN-1	T_2, LHSVL	T_2, Bakken
Recycle delay (ms)	750	7500	2250
Tau (µs)	50	50	54
Number of Echoes	5000	5000	13889
Filter Width (kHz)	125	125	125
90° Pulse Length (µs)	7.5	9.31	6.63
180° Pulse Length (µs)	15.2	18.88	13.71

Table 2. NMR Parameters

The experiment can be summarized as follows, sample BN 1 was confined hydrostatically with FC-40 (CAS Number 86508-42-1), and NMR-inert fluid, to a pressure of 2500 psi in the overburden NMR probe. The system was examined for gas leaks but no obvious leak was detected. The probe was then inserted into the rock core analyzer. The sample was evacuated with a vacuum pump, then as received or dry T_2 measurements were taken before the introduction of methane. Next, methane was introduced to the sample at 500 psi. T_2 measurements were acquired at 2-minute intervals for the first hour, then at 15 minute intervals for the following 3 hours, and at 60 minute intervals for the remainder of the experiment. The pressure was continuously monitored via Arduino micro-controller [11]. The Arduino then reported the pressure to an internal website and a computer program running on our server logs all the system pressures. In addition, a Raspberry Pi [12] computer monitors the overburden pressure via the same internal website and turns a pump on or off depending if the overburden pressure was too high or too low. This ensured the overburden pressure remained constant throughout the experiment. The T_2 measurements were then employed to retrieve the total gas content present in the rock as a function of time

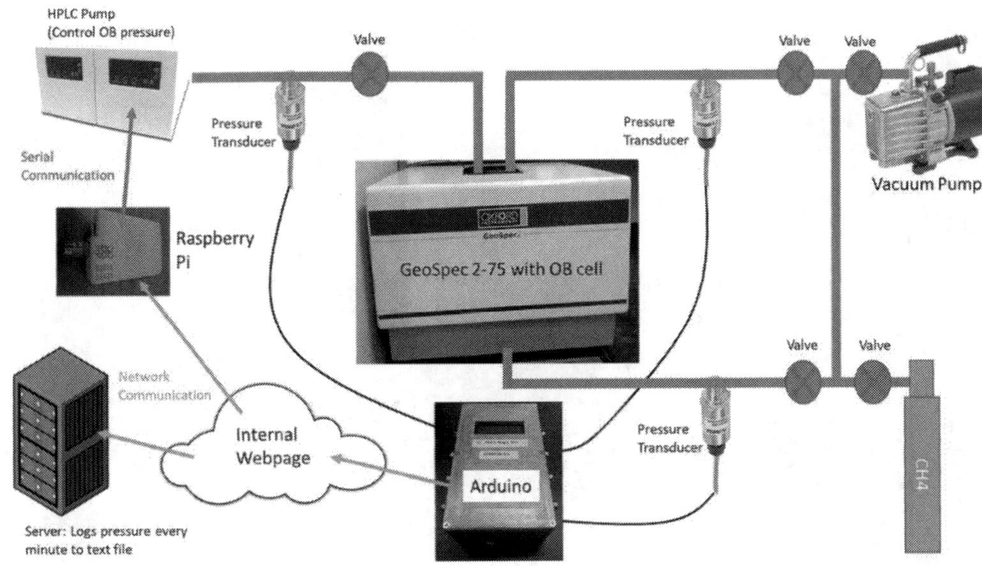

Figure 1 - Experiment 1 Setup

following the procedure outlined in Section 2 of this paper.

3.2 Experiment 2

The second experiment determined the gas isotherm for a Lower Haynesville shale (LHSVL). The Upper Jurassic Haynesville Shale is an organic- and carbonate-rich mudrock deposited in a deep-water environment on the margins of the Gulf of Mexico, with average porosities around 11%. The average TOC of the Haynesville formation is 2.5% [13] with a range of 3% to 14% [14,15].

Table 1 summarizes the properties of the sample while Table 2 summarizes the NMR parameters employed. Sample LHSVL was not a solid core sample as sample BN 1. Instead this sample was made up of three irregularly shaped fragments (Figure 2).

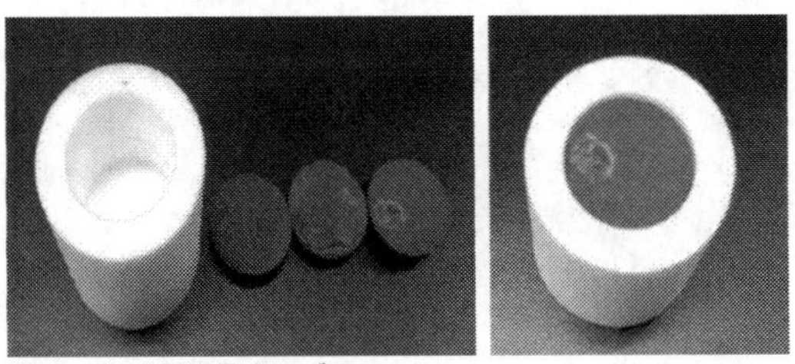

Figure 2: LHSVL Sample

Coring is an issue with most shale samples so the goal of this experiment was to prove that the NMR gas isotherm determination can be employed on samples regardless of shape including crushed and powdered samples. In order to get the fragments into the overburden probe a new Teflon holder with the same outer dimensions as a standard core sample was built. Previous testing with Teflon using a GeoSpec 2-75 rock core analyzer has shown the digital filters remove the fluorine signal, thereby removing the signal from the Teflon core holder from the results. As shown in Figure 2, the samples were then placed into the new holder. The holder and the samples were then placed into the overburden NMR probe and the probe then inserted into the rock core analyzer. As in Experiment 1, the sample was evacuated with a vacuum pump, then as received or dry T_2 measurements were taken before the introduction of methane.

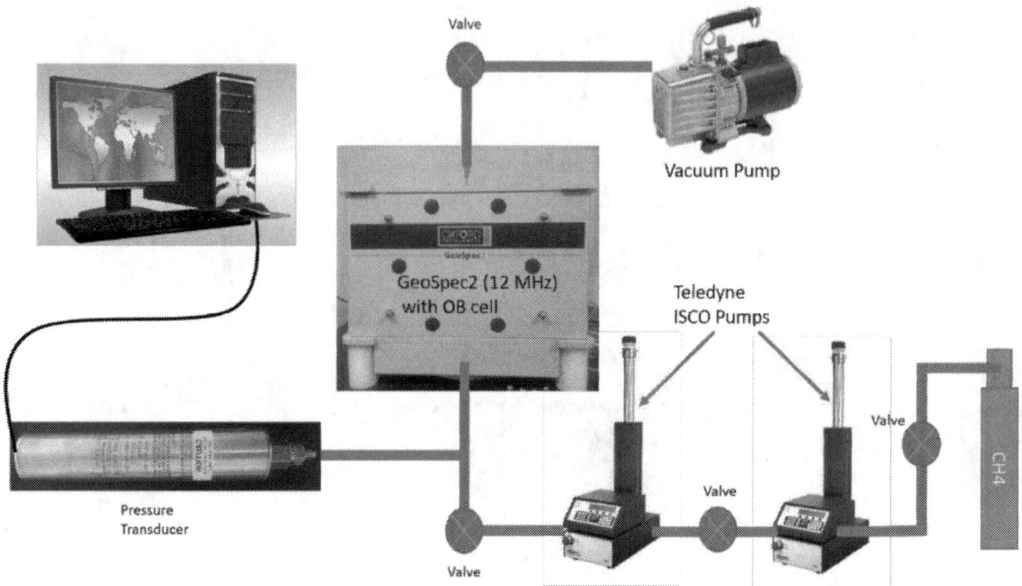

Figure 3: Experiment 3 Setup

No confining fluid was used in this experiment. Instead methane filled the confining volume of the probe. This simplified the experiment as it eliminated the need to use a second pump and an additional fluid as confining fluid. However, as a result of filling the probe with methane, the T_2 distributions recorded at elevated methane pressures had large amounts of signal from free methane. As will be outlined in the Results section of this paper, this added a step to the gas isotherm determination method outlined in the previous section.

3.2 Experiment 3

The experimental design for experiment 3 (Figure 3) was similar to experiment 2 except the core sample was one solid piece of rock and the Teflon core holder was not used. Table 1 shows the rock sample properties for the Bakken core used in this experiment. The Bakken shale sample was collected from the Viewfield area in Saskatchewan, Canada. The Upper and Lower members of the Bakken are black organic-rich shales the Upper Member shale have average TOC of 17.6% with an average porosity of 3% [16].

The methane gas supply was connected to an ISCO 260HL pump and an ISCO 260D pump. The pressure in the overburden cell was controlled using the valves and the constant pressure setting of the ISCO pumps. Independent pressure measurements of the overburden cell were obtained using a Paroscientific 9000K model pressure transducer.

The overburden cell and sample were evacuated using a vacuum pump, then the initial NMR T_2 distribution was obtained using an Oxford Instruments Geospec2 12 MHz rock core analyzer [17]. All data was again processed using GIT Systems Advanced Software. NMR parameters used for all T_2 measurements are shown in Table 2. Adsorption isotherms were determined at the following pressures (psi): 242, 494, 997 and 1593. Adsorption isotherms were carried out by raising the pressure in the cell to the desired pressure, setting the pumps to the constant pressure, and monitored through the course of the experiment. The NMR T_2 distributions were taken minutes apart for the first hour of the experiment and then hours apart as the experiment proceeded for 24 or more hours.

Following measurement of the gas isotherms for the Bakken sample, an experiment was completed examining the desorption isotherm of methane from the shale. This experiment was meant to improve our understanding of the processes controlling methane movement in the rock matrix. After the T_2 data collection was completed at the highest pressure (1593 psi), the overburden cell was allowed to degas over a period of minutes and when the pressure in the cell reached atmospheric pressure, the cell was re-sealed using the valves. The pressure was monitored over the next several days and T_2 measurements were obtained over that time.

4 Results

The left-hand panel of Figure 4 shows three T_2 distributions recorded for sample BN 1. The black trace is the background or as received distributions recorded at 0 psi of methane. The red and blue traces were recorded with a methane pressure of 2000 psi, with the red trace recorded within minutes of methane being introduced to the sample, and the blue trace was recorded hours after the introduction of methane. As expected, the blue trace has more signal than the red trace as methane has more time to penetrate the pores of the sample. If there was free methane surrounding the core then a relatively large peak should be observed near 10^3 ms (see LHSVL and Bakken results below).

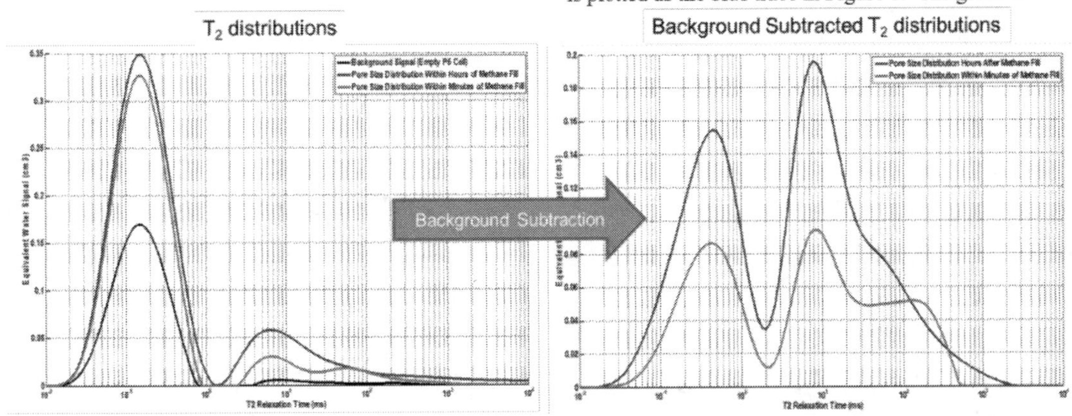

Figure 4: T_2 distributions For BN 1 Sample

The right-hand panel in Figure 4 shows the same T_2-distributions as the right-hand panel, but with the background T_2 distribution (black line in the right-hand panel) subtracted. The bimodal distribution of the subtracted T_2 distributions shows methane occupying small pores ($T_2 < 2$ ms) and methane occupying relatively larger pores ($T_2 > 2$ ms). The area under each of the blue curve and the red curve is the methane content in units of equivalent water volume. These volumes are converted to methane gas content and plotted as a function according to Step 4 of the procedure outlined in Section 2 of this paper.

Figure 5 shows the plot of calculated gas content versus time for sample BN1. The figure shows that for each fill pressure of methane (green 500 psi, black 1000 psi, red 1500 psi, and blue 2000 psi) approximately 4 hours are required for methane to fully penetrate the pores of the sample and the amount of methane to stabilize. After 4 hours the measured gas content remained relatively constant (± 1%). The variability in the gas contents decreases after 5 hours but this is likely due to a reduced sampling frequency.

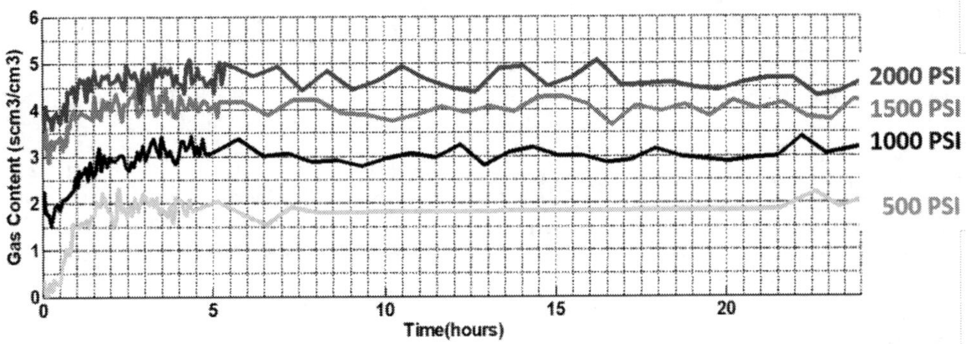

The average amount of methane after four hours was used to calculate the total gas isotherms for each pressure and is plotted as the blue trace in Figure 6. The green trace in Figure 6 is the calculated pore gas isotherm using equation 2 above while the red trace is the adsorbed gas isotherm and is calculated as the difference of the total gas content and the pore gas content. Each were calculated according to the procedure outlined in Section 2 of this paper. The pore gas content increases relatively linearly with pressure and only deviates from linearity at higher pressures where its behaviour begins to deviate from the ideal gas law according of the Van der Waals equation. The adsorbed gas content on the other hand is not linear. The amount of adsorbed gas increases quickly at lower pressures due to the large number of adsorption sites still available but there are a finite number of adsorption sites available so the change in gas content with increasing pressure declines with increasing pressure. Finally, the total gas isotherm is simply the sum of the pore gas isotherm and adsorbed isotherms and has properties similar to each. The adsorbed methane content for BN1 reaches 1.3 scm^3/cm^3 or 0.02 mmol/g of rock at 1500 psi, which is within the reported range of methane adsorption of 0.007 to 0.1 mmol/g [2].

https://doi.org/10.1051/e3sconf/202014605004

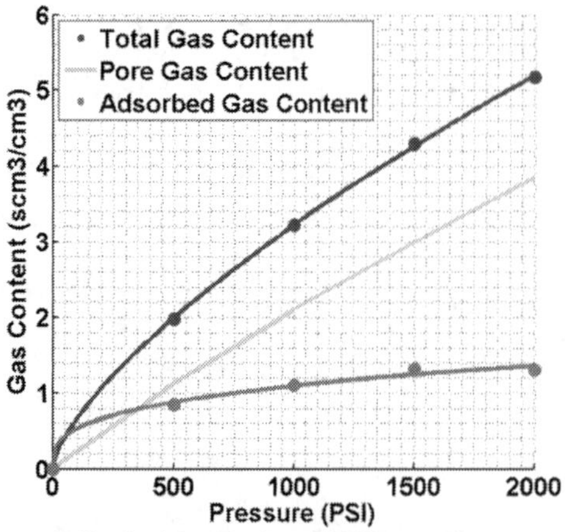

Figure 6: Gas Isotherm For Sample BN1

Figure 7 shows background subtracted T_2 distributions recorded for the fragmented LHSVL sample. The blue trace shows the distribution recorded with a pressure of 500 psi of methane while the red trace shows the distribution recorded at a pressure of 2000 psi. As expected, there is more methane present in the sample at 2000 psi than at 500 psi. What is more interesting about Figure 7, is the large signals present at slow T_2 (i.e. above 100 ms). This signal is too late to be originating in the pores of the shale sample. Typical shales have T_2 distributions with peak values around 0.1 to 10 ms. These late T_2 peaks are instead due to free methane present in the experiment. This was further confirmed in a T1-T2 map in experiment 3, where a large peak appeared at T1 = ~1000 ms, T2 = ~1000 ms and it was assumed to be bulk methane. As mentioned in the previous section, there is significant free methane in the LHSVL experiment because the confining volume of the overburden probe was filled with methane. To calculate the gas isotherm analysis for this sample, the free methane was removed from the T_2 distributions recorded at each pressure. From Figure 7, it is obvious that the removal of signal from free methane can be accomplished by simply applying a cut off to the T_2 distributions. For example, in Figure 7, only the signal below 223 ms (shown by the vertical green line) is included, which removes the free methane signal from the analysis. Once the distributions have the free methane signal removed, the gas isotherm calculation continues according to the method laid out in Section 2 of this paper.

The final gas isotherms generated from the NMR data for LHSVL are shown in Figure 8 and the shape of the curves are similar to what was observed in the BN 1 sample above. The adsorbed methane content for LHSVL reaches 4.0 scm^3/cm^3 or 0.07 mmol/g of rock at 1500 psi, which is close to other reported data 0.09 mmol/g [18].

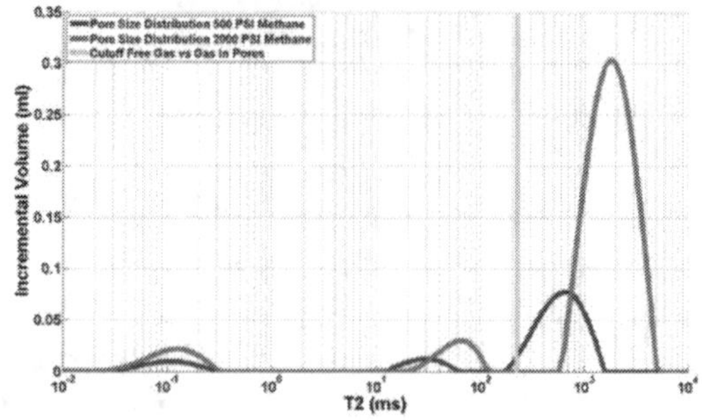

Figure 7: T_2 Distributions for Sample LHSVL

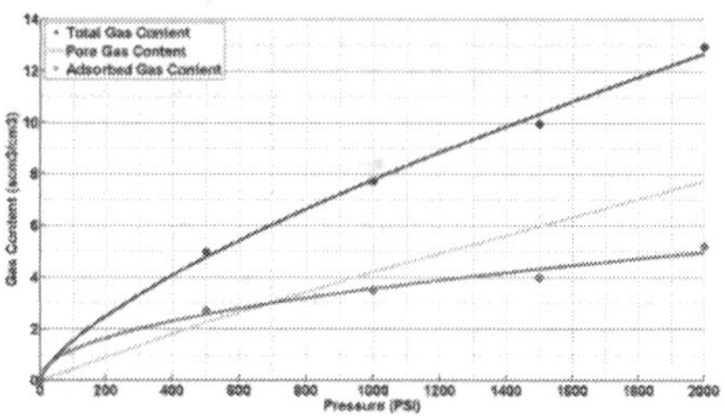

Figure 8: Gas Isotherm For Sample LHSVL

Figure 9 shows a background subtracted T_2 distribution recorded for the Bakken sample at pressure of 1500 psi. T_2 measurements ended before the relaxation was complete because the focus was on characterizing the signal between 0 and 100 ms, the methane associated with the rock. The peak above 200 ms is caused by free methane and was removed to complete the gas isotherm analysis similar to the LHSVL sample described above.

Figure 10 shows the gas isotherm generated for the Bakken core, which is different from the two cores described above. The adsorbed gas content of the Bakken core is greater than the Haynesville core, which is likely due to the higher %TOC in the Bakken core, and TOC preferentially adsorbs hydrocarbon gases. The pore gas content of the Bakken core is significantly lower than the other two cores. The adsorbed gas for the LHSVL sample was approximately 4 scm^3/cm^3 at 1000 psi whereas the pore gas content at 1000 psi for the Bakken sample was approximately 1 scm^3/cm^3. This is consistent with the lower porosity of the Bakken sample compared to the Hayneville sample (Table 1), which limits the volume available for pore gas. The adsorbed methane content for this Bakken core is 0.3 mmol/g of rock at 1500 psi, which is below the reported range of CO_2 adsorption isotherms 1.0 to 3.5 mmol/g [19]. LPGA results for the Bakken were between 1.75 and 3.25 cm^3/g [20], which is less than the adsorption reported here (approximately 7 scm^3/g) but the LPGA analysis was carried out with N_2 at nearly 1500 psi less than the experimental conditions here.

A review of the BN-1, LHSVL and Bakken isotherms shows that the results are consistent with published literature and the distribution of peaks can be used to quantify methane in tight rocks. The results are limited at low pressures because the relatively small amount of methane is insufficient to produce a reliable signal, though the examination of hydrocarbon gases and source or reservoir rocks is typically carried out at higher pressures. Further work is also required to determine the accuracy of the pore gas content using Van der Waals equation as well as exploring employing a Langmuir isotherm to fit to the adsorption isotherm data.

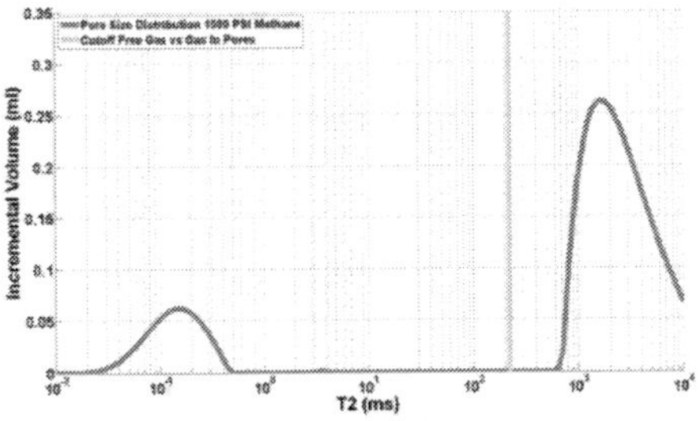

Figure 9: T_2 distribution At 1500 psi for Bakken Sample

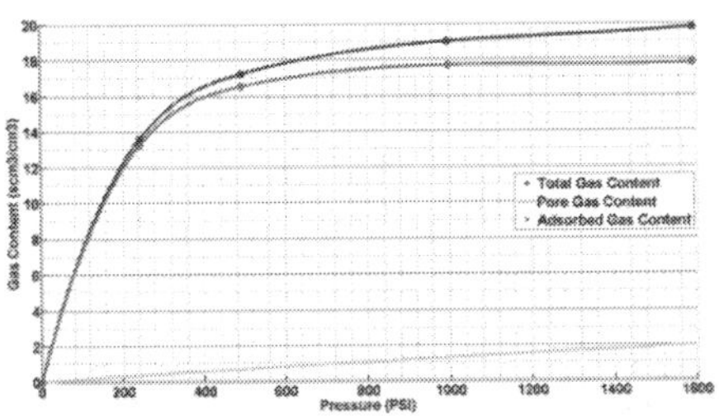

Figure 10: Gas Isotherm for the Bakken Sample

The experiments with the Bakken core continued beyond the measurement of the gas isotherm. T_2 distributions were recorded as the core degassed. Figure 11 shows the final background subtracted T_2 distributions recorded during the degassing stage of the experiment. The upper panel shows the T_2 distributions recorded on the same y-axis scale as the data recorded with a pressure of 1500 psi (Figure 9) where as the lower panel shows the T_2 distributions plotted with a more reduced y-axis scale making it easier to see the peaks. Comparing the upper panel with Figure 9, it is obvious that the amount of gas in the pores (peak near 0.1ms) has been reduced during degassing but gas is still present. This methane signal is most likely now from residual adsorbed gas that is taking time to leave the pore walls.

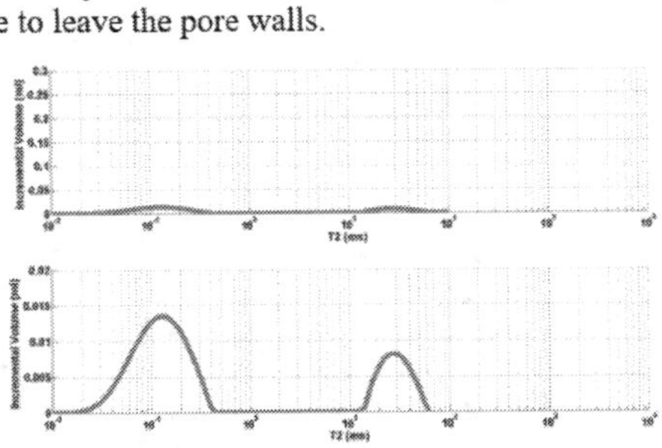

Figure 11: T_2 Distribution Measured For Bakken Sample at the End of Degassing

Further evidence that the methane signal is coming from residual adsorbed gas can be observed by plotting the total gas content determined from the NMR measurements during degassing as a function of time, which is presented in Figure 12. The amount of gas in the core declines to approximately 15 scm^3/cm^3 in the first 12 minutes followed by a gradual decrease in total gas content. The methane is slowly moving out of the core. Once the methane is out of the core it is no longer present in the total gas content calculation as it becomes free methane and is removed from the calculations. The total gas content remains near 15 scm^3/cm^3 after almost two days of degassing. Based on the isotherm (Figure 10), the remaining methane in the core is likely adsorbed to the pore walls but some methane may remain in the pores. .

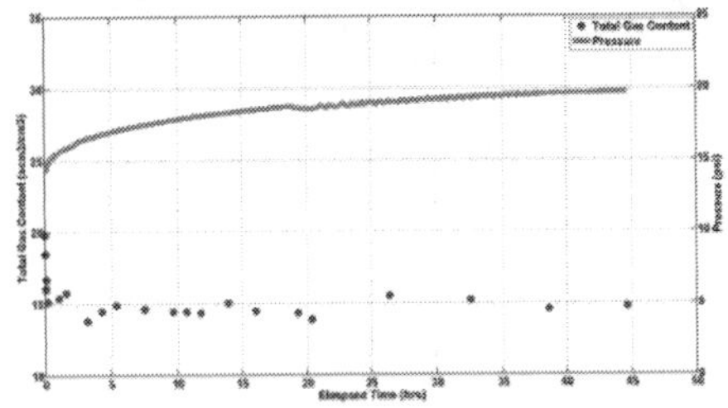

Figure 12: Total Gas Content and Pressure During Degassing for Bakken Sample

Further evidence of this degassing behavior can be interpreted from the experimental pressure that was recorded simultaneously with the total gas content and is also plotted in Figure 12. As expected, the pressure is increasing over time during degassing as the adsorbed gas leaves the pore walls where it does not contribute to measured gas pressure in the overburden cell and migrates to pore gas or free methane resulting in an increase in gas pressure in the cell. After the cell was opened, gas pressure declined to atmospheric pressure (14.3 psi) and increased during the degassing period to 19.7 psi. Future work will determine the volume of the NMR cell so that the declining core gas content can be compared to the rising cell pressure and an improved approach can be used to examine the transport processes controlling methane degassing from the core.

Another interesting result from Figure 11, is the appearance of a second peak between 10 and 100 ms that was not present during the gas isotherm measurement. We believe that this is signal from methane in a fracture within the core. This signal was not present in Figure 9 as the confining pressure was high enough (1500 psi) to close this fracture. Releasing the confining pressure during degassing has caused the fracture to open allowing gas from the smaller pores to migrate into it. In addition, if you look closely at Figure 9 you see a small peak at approximately 2.5 ms. This small peak was present in all the 1500 psi T_2 distributions recorded during the gas isotherm measurement and might be the same fracture as observed in the degassing experiment. At 1500 psi, the fracture would be closed leading to a decrease in the amount of gas in it and a decrease in its T_2 relaxation time.

Figure 13 shows the total gas content of the core as it degases (blue circles) as well as the gas content of the first

T_2 peak (black circles) and the gas content of the second T_2 peak from 10 to 100 ms (grey circles). The total gas content and gas content of the first T_2 peak follow the same pattern. The first T_2 peak represents the majority of the gas in the core, which is corroborated by the larger T_2 peak in Figure 11. The first peak represents the adsorbed gas on the core and the gas in the small pores. The second T_2 peak, contains less gas than the first peak, and increases over time. The higher T_2 value for this peak suggests this peak represents gas filling larger secondary porosity, possibly a fracture that developed in the core during the pressure-depressure cycle and that future degassing phases may need to be carried out at a slower rate to avoid the creation of such features. However, degassing at too slow a rate may miss important information regarding gas transport at early time.

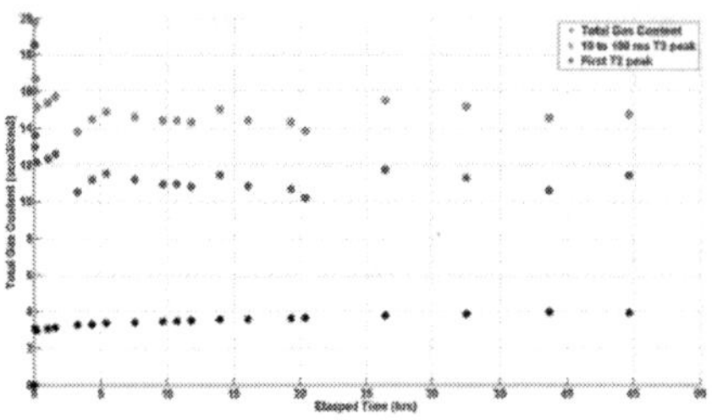

Figure 4: Calculated Total Gas Content For The Core Sample And Calculated Gas Content Both T_2 Peaks Observed.

The change in the total gas content over time shown as blue circles in Figure 13 is the rate that methane adsorbed on to the core and methane in the pores of the core are able to escape the core. The calculated total gas content desorbed from the core was calculated by subtracting the total gas contents from the total gas content of the core at the completion of the 1593 psi adsorption isotherm (19.8 scm^3/cm^3) and are shown in Figure 14 as the experimental data.

The rate of methane loss from the core was examined using the diffusion equation shown by [21].

$$Q(t) = Q_{tot}\sqrt{1 - e^{-BKt}}$$

Where Q(t) is the desorbed gas content (scm^3/cm^3) at time t (s); Q_{tot} is the total desorbed gas content from the beginning of the experiment to the end; $B = 4\pi^2 D/d^2$; and D the diffusion coefficient (m/s^2); d the radius of the sample (m); and K is a unitless correction factor. When the B-value ranges between 6.58 x 10^{-3} to 6.58 x 10^{-6}, which was the case for this work, the K-value is 0.96 [21]. The calculated total degassed contents are shown as blue circles in Figure 14. The calculated gas loss by diffusion matches the pattern of the experimental data suggesting the equation is adequate to examine gas loss from the core. The diffusion coefficient used for this calculation was 2.6 x 10^{-9} m^2/s, which is significantly lower than 2.1 x 10^{-5} m^2/s for methane in air, but greater than the 10^{-12} m^2/s values obtained for coal [21]. The calculated diffusion coefficient suggests that while diffusion may play a role in the rate of methane lost from the core, another process (e.g. advection) may be a significant process. This is not surprising given that at the beginning of the degassing phase the core was at least near equilibrium with 1593 psi and surrounded by atmospheric pressure, which should result in some amount of gas flow out of the core.

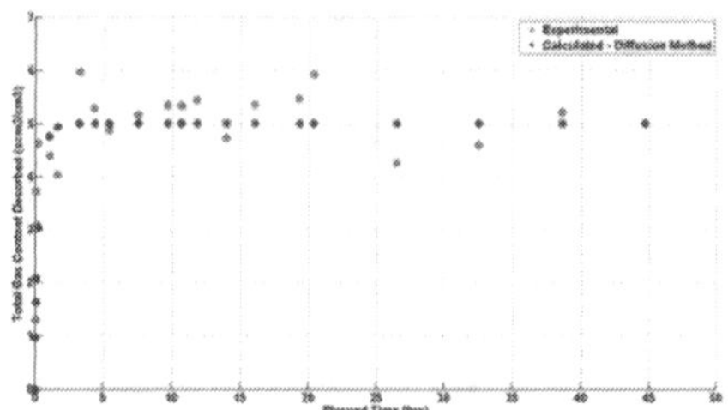

Figure 5: Total Gas Content (scm^3/cm^3) Removed From The Core And Calculated Total Gas Content Removed From Core Using The Diffusion Equation.

References

[1] X. Cui, A.M.M. Bustin, and R.M. Bustin, "Measurements of gas permeability and diffusivity of tight reservoir rocks: different approaches and their applications," *Geofluids*, vol. 9, no. 3, Aug. 2009, doi: 0.1111/j.1468-8123.2009.00244.x.

[2] M. Gasparik, P. Berier, Y. Gensterblum, A. Ghanizadeh, B.M. Kroos and R. Littke, "Geological controls on the methane storage capacity in organic-rich shales," *Int. J. Coal Geol.*, vol. 123, no. 1, pp 34-51, Mar. 2014, 10.1016/j.coal.2013.06.010.

[3] T. Alexander, J. Baihly, C. Boyer, B. Clark, G. Waters, V. Jochen, J. Le Calvez, R. Lewis, C.K. Miller, J. Thaeler, and B.E. Toelle, "Shale Gas Revolution," *Oilfield Rev.*, vol. 23, no. 3, pp. 40-55, Sep. 2011.

[4] C. Horch, S. Schlayer, F. Stallmach, "High-pressure low-field 1H NMR relaxometry in nanoporous materials," *J. of Magn. Reson.*, vol. 240, pp 23-33, Mar. 2014, doi: 1 0.1016/j.jmr.2014.01.002.

[5] Y. Liu, C. Wang, "Determination of the Absolute Adsorption Isotherms of CH_4 on Shale with Low-Field Nuclear Magnetic Resonance," *Energy Fuels*, vol. 32, no. 2, pp. 1406-1415, Jan. 2018, doi: 10.1021 /acs.energyfuels.7b03428.

[6] *Geo-Spec 2-75 Rock Core Analyzer User Manual*, Version 1.8, Oxford Instruments, Tubney Woods, Abingdon, Oxon, UK, 2018.

[7] *P5 Overburden Probe User Manual*, Version 1, Oxford Instruments, Green Imaging Technologies, Oxford Instruments, Tubney Woods, Abingdon, Oxon, UK 2016.

[8] *GIT Systems and LithoMetrix User Manual*, Version 1.9, Green Imaging Technologies, Fredericton, NB, Canada, 2016.

[9] R.M. Pollastro, D.M. Jarvie, R.J. Hill, "Geologic framework of the Mississippian Barnett Shale, Barnett-Paleozoic total petroleum system, Bend

arch–Fort Worth Basin, Texas," *AAPG Bull.*, vol. 91, no. 4, pp. 405-436, doi: 10.1306/10300606008.

[10] K.A. Bowker, "Recent developments of the Barnett Shale play, Fort Worth Basin" *W. Texas Geol. Soc. Bull.*, vol. 42, no. 6, pp 4-11. Art. No. 10126.

[11] Arduino Microcontroler Homepage, www.arduino.cc (accessed Feb. 1, 2018).

[12] Raspberry Pi Computer Homepage, www.raspberrypi.org (accessed Feb. 1, 2018).

[13] N.R. Jabri, "Chemostratigraphy and paleoenvironment of the Haynesville Formation, Harrison County, Texas," M.Sc. thesis, Geol. Dept., Univ. Houston, Arlington, Texas, 2013.

[14] F.P. Wang F.P. and U. Hammes, "Effects of reservoir factors on Haynesville fluid flow and production," *World Oil*, vol. 231, pp. D3-D6.

[15] U. Hammes, H.S. Hamlin, and E.E. Thomas, "Geologic analysis of the Upper Jurassic Haynesville Shale in east Texas and west Louisiana," *AAPG Bull.*, vol. 95, no. 10, pp. 1643-1666, Oct. 2011, doi: 10.1306/02141110128.

[16] L.K. Kreis, A.L. Costa and K.G. Osadetz, "Hydrocarbon potential of Bakken and Torquay formations, southeastern Saskatchewan," in *Saskatchewan and Northern Plains Oil & Gas Symposium* C.F. Gilboy and S.G. Whittaker, Eds. 2006 , pp. 118-137.

[17] *Bench-top Imager Hardware User Manual*, Version 1.1, Oxford Instruments, Tubney Woods, Abingdon, Oxon, UK, 2006, pp. 1-32.

[18] A. Merkel, R. Fink and R. Littke, "The role of pre-adsorbed water on methane sorption capacity of Bossier and Haynesville shales," *International J. Coal Geol.*, vol. 147-148, no.1, pp. 1-8, Aug. 2015.

[19] S. Kumar, "Rock-fluid interaction and phase properties of fluids in nano- and sub-nanopores of shales: sorption-based studies." Ph.D. dissertation, Dept. Petrol. Eng., Colorado School of Mines, 2016.

[20] Y. Liu, B. Shen, Z. Yan, P. Zhao, "Structure characterization and the controlling factors of the Bakken Formation," *Energies*, vol. 11, no. 11, pp. 2879-2894, Oct. 2018, doi: 10.3390/en11112879.

[21] G. Ni, B. Lin, C. Zhai, Q. Li, S. Peng, X. Li, "Kinetic characteristics of coal gas desorption based on the pulsating injection," *Int. J. Min. Sci. Tech.*, vol. 24, no. 5, pp. 631-636, Sep. 2014, doi: 1 0.1016/j.ijmst.2014.03.023.

Core-scale sensitivity study of CO_2 foam injection strategies for mobility control, enhanced oil recovery, and CO_2 storage

Zachary Paul Alcorn[1*], *Sunniva B.* Fredriksen[1], *Mohan* Sharma[2], *Tore* Føyen[1,3], *Connie* Wergeland[1], *Martin A.* Fernø[1], *Arne* Graue[1], *and Geir* Ersland[1]

[1]Dept. of Physics and Technology, University of Bergen, Norway
[2]National IOR Centre of Norway, University of Stavanger, Norway
[3]SINTEF Industry, Trondheim, Norway

Abstract. This paper presents experimental and numerical sensitivity studies to assist injection strategy design for an ongoing CO_2 foam field pilot. The aim is to increase the success of *in-situ* CO_2 foam generation and propagation into the reservoir for CO_2 mobility control, enhanced oil recovery (EOR) and CO_2 storage. Un-steady state *in-situ* CO_2 foam behavior, representative of the near wellbore region, and steady-state foam behavior was evaluated. Multi-cycle surfactant-alternating gas (SAG) provided the highest apparent viscosity foam of 120.2 cP, compared to co-injection (56.0 cP) and single-cycle SAG (18.2 cP) in 100% brine saturated porous media. CO_2 foam EOR corefloods at first-contact miscible (FCM) conditions showed that multi-cycle SAG generated the highest apparent foam viscosity in the presence of refined oil (*n*-Decane). Multi-cycle SAG demonstrated high viscous displacement forces critical in field implementation where gravity effects and reservoir heterogeneities dominate. At multiple-contact miscible (MCM) conditions, no foam was generated with either injection strategy as a result of wettability alteration and foam destabilization in presence of crude oil. In both FCM and MCM corefloods, incremental oil recoveries were on average 30.6% OOIP regardless of injection strategy for CO_2 foam and base cases (i.e. no surfactant). CO_2 diffusion and miscibility dominated oil recovery at the core-scale resulting in high microscopic CO_2 displacement. CO_2 storage potential was 9.0% greater for multi-cycle SAGs compared to co-injections at MCM. A validated core-scale simulation model was used for a sensitivity analysis of grid resolution and foam quality. The model was robust in representing the observed foam behavior and will be extended to use in field scale simulations.

1. Introduction

A major challenge in carbon dioxide (CO_2) enhanced oil recovery (EOR) is poor macroscopic CO_2 sweep efficiency caused by the low viscosity and density of injected CO_2 [1, 2]. These adverse properties can result in viscous fingering and gravity override, greatly hindering oil recovery and sweep efficiency [3, 4]. Reservoir heterogeneity can also cause injected CO_2 to channel through high permeability zones (layers or fractures) resulting in early CO_2 breakthrough, high CO_2 recycling, and low incremental oil recoveries. CO_2 foam injection is a laboratory verified and field-validated technique proven to mitigate poor CO_2 sweep efficiencies through effectively increasing CO_2 viscosity and reducing its relative permeability [5].

Foam is a dispersion of gas (CO_2, in this case) in liquid where the liquid phase is continuous and at least some of the gas phase is discontinuous and separated by lamella [6]. The discontinuous CO_2 phase becomes trapped between lamella [7], effectively reducing its relative permeability. Foam apparent viscosity is used as a measure of foam strength in laboratory experiments, where a stronger foam corresponds to a higher apparent viscosity. Entrapment of CO_2 in foam and CO_2-surfactant emulsification also increases CO_2 apparent viscosity and reduces CO_2 mobility [8, 9]. These combined effects are capable of diverting flow from high permeability, well-swept regions, into less permeable areas with higher oil saturations, thereby increasing macroscopic displacement and oil recovery.

CO_2 foam generation and stability is influenced by oil composition and wettability. The presence of oil can destabilize some foams [10] and it has also been reported that foam can only be generated at strongly water-wet conditions [11]. At strongly water-wet conditions, water-wet films covering the pore surfaces maintain the continuous foam structure [12]. A shift to oil-wet can cause the lamellas to detach from the pore walls and foam may be destabilized. Schramm and Mannhardt (1996) confirmed reduced foam effectiveness at intermediate to oil-wet conditions [13] and Fredriksen et al. (2019) induced surfactant wettability alteration in oil-wet fractures to generate CO_2 foam in matrix below a critical oil saturation [14]. The impacts of wettability and the presence of oil on foam stability are areas under active investigation.

* Corresponding author: zachary.alcorn@uib.no

Foam is generated in porous media by injecting a water (or CO_2) soluble foaming agent (surfactant) with gas, such as CO_2, either simultaneously (co-injection) or in alternating slugs (SAG). The injection strategy must balance *in-situ* foam generation, propagation, and injectivity constraints. In practice, there are two main injection strategies for *in-situ* CO_2 foam generation [15, 16]. The first is simultaneous injection of CO_2 and surfactant solution known as co-injection. In this case, the quality of the foam is determined by the fraction of gas (f_g) and is a function of flow rate [17]. The second injection strategy is surfactant-alternating gas (SAG), where the surfactant solution and CO_2 are injected in alternating slugs and the quality of the foam depends on the slug sizes of CO_2-to-surfactant solution. At laboratory scale, alternating slugs are not typically used due to small fractional flows of gas and the inability to achieve steady-state [18]. At the field scale, operational constraints influence aspects of the injection strategy. For example, co-injection can be difficult to implement due to extremely low injectivity and associated pressure increases, which minimize throughput [19]. In addition, downhole corrosion can occur from carbonic acid during co-injection of CO_2 and surfactant solution. This has led to a majority of field tests using SAG injection for better injectivity control, especially when operating close to the fracture pressure [20, 21, 22].

Few attempts have been made to characterize unsteady-state *in-situ* CO_2 foam behavior. Therefore, this study attempts to begin establishing a knowledge base for investigating core-scale CO_2 foam injection strategies. The aim is to reduce CO_2 mobility, through the generation of foam, in experiments that are representative of the near wellbore region. This work is part of an ongoing field pilot research program utilizing CO_2 foam for mobility control, EOR, and CO_2 storage in the Permian Basin of West Texas. Miscible CO_2 foam injection will be implemented to reduce high producing gas-oil-ratios, CO_2 recycling, and to provide mobility control for increased CO_2 sweep efficiency. The reservoir pressure in the pilot area is 220 bars, which is close to the formation fracture pressure of 269 bars [23]. An injection strategy must be selected that will not fracture the formation, which would cause problems beyond any remediation by the foam. Thus, the primary objective of this work is to evaluate co-injection and a variety of SAG injection strategies for CO_2 foam mobility control, EOR, and CO_2 storage to assist in the design of the field pilot. A secondary objective is to investigate the impacts of miscibility conditions on *in-situ* foam generation and stability. For more detailed information on the field pilot program see Alcorn et al. (2019).

CO_2 foam was evaluated based upon apparent foam viscosity and its impact on oil recovery. In addition, CO_2 storage potential was measured for each injection strategy. Experimentally, foam behavior by co-injection, single-cycle SAG, and multi-cycle SAG were performed in brine saturated systems to generate foam *in-situ* without the presence of oil. CO_2 foam EOR corefloods were then conducted at first-contact miscible (FCM) and multi-contact miscible (MCM) conditions using *n*-Decane and crude oil, respectively, after waterflooding. A local-equilibrium foam model was fitted from foam stability scans, and experimental results from a CO_2 foam EOR coreflood were used to validate a core-scale numerical model to investigate effects of grid resolution, foam quality, and surfactant. The main objective of the numerical sensitivity study was to validate the foam model in representing the observed foam behavior, which is being extended for use in ongoing field-scale simulations.

2. Materials and Fluids

Outcrop limestone core plugs were used as reservoir analogues due to limited reservoir cores available from the field. Core plugs were drilled with a 2-inch diameter from larger slabs, cut, cleaned and dried before being 100% saturated with synthetic Permian Basin brine under vacuum. Values of porosity and pore volumes were calculated based on weight differential before and after fluid saturation. Absolute permeability was measured by injecting brine until a stable differential pressure was obtained for three different flood rates. Permeability measured for the single core plugs varied between 8mD to 73mD and porosities from 19 to 35%.

Synthetic Permian Basin brine was made based upon water analysis from the field, whereas a light North Sea crude oil was used. The North Sea crude oil has an API of 33.6° (calculated from specific gravity), a little above the API gravity of the Permian Basin crude of 31° at standard conditions. Compositions of brine and crude oil are reported elsewhere [14, 24]. The North Sea crude oil is considered MCM with CO_2 at 60°C and 180 bar, with a minimum miscibility pressure (MMP) of 125 bar [25]. Reservoir conditions for the field test are well above MMP for CO_2 and crude oil. To investigate the effects of miscibility conditions on foam generation and stability a refined oil, *n*-Decane ($C_{10}H_{22}$), was also selected for the first set of EOR corefloods to obtain FCM conditions with CO_2.

A non-ionic surfactant (i.e. Huntsman L24-22) was previously selected for the CO_2 foam field pilot from a surfactant screening study that quantified adsorption on reservoir rock with and without CO_2 present [26]. The non-ionic surfactant was also screened for its ability to alter wettability of oil-wet carbonate rocks to weakly oil-wet conditions in favor of foam generation [14]. The non-ionic surfactant was used at a 1.0 wt% concentration in Permian Basin Brine.

3. Procedure

3.1 Coreflood Set-up

The core-scale system was composed of two stacked core plugs providing a total nominal length of 25 cm to generate foam *in-situ*. Cores with similar porosity and absolute permeability were paired and stacked. An overview of experiments and their stacked system properties are presented in Table 1.

Table 1 – Experimental overview and core properties

ID	Experimental overview	Oil phase	Length [cm]	Porosity [%]	K_{abs} [mD]	S_{wi}
D1	Foam Scan: Co-injection	-	7.5 ± 8.8E-03	24.2 ± 0.2	20.5 ± 0.3	1.00 ± 0.01
D2-D3	Foam Scan: Co-injection	-	12.7 ± 3.5E-03	26.2 ± 0.8	31.6 ± 0.2	1.00 ± 0.01
E2	Foam Stability: SAG	-	12.7 ± 2.0E-03	25.4 ± 0.2	42.3 ± 0.1	1.00 ± 0.01
E3-E4	EOR: Co-injection 1.0 ft/day	n-Decane	24.4 ± 2.8E-03	22.3 ± 1.9	15.5 ± 0.2	0.24 ± 7.22E-03
E5-E6	EOR: Co-injection 2.0 ft/day	n-Decane	24.9 ± 2.8E-03	25.8 ± 0.3	21.4 ± 0.3	0.31 ± 7.28E-03
E7-E8	EOR: Single cycle SAG	n-Decane	27.5 ± 2.8E-03	30.7 ± 2.8	24.0 ± 0.0	0.38 ± 7.59E-03
E9-E10	EOR: Multi-cycle SAG	n-Decane	24.8 ± 2.8E-03	25.1 ± 1.5	38.4 ± 0.4	0.29 ± 7.27E-03
E13-E14	Foam Scan/EOR: Co-injection 1.0 ft/day	Crude oil	24.7 ± 2.8E-03	27.5 ± 0.4	31.4 ± 0.5	0.24 ± 7.13E-03
E15-E16	Foam Scan/EOR: Co-injection 1.0 ft/day (base case[1])	Crude oil	27.4 ± 2.8E-03	22.5 ± 2.3	14.3 ± 1.3	0.15 ± 7.04E-03
E17-E18	EOR: Multi-cycle SAG 1.0 ft/day	Crude oil	24.9 ± 2.8E-03	28.4 ± 1.7	31.6 ± 0.2	0.25 ± 7.20E-03
E21-E22	EOR: Multi-cycle WAG 1.0 ft/day (base case[1])	Crude oil	25.0 ± 2.8E-03	25.8 ± 1.8	21.3 ± 0.7	0.24 ± 7.18E-03

[1]Base case without surfactant solution.
[2]Uncertainty calculated as standard deviation of the mean

The cores were mounted in a horizontally oriented Hassler type core holder and experimental conditions were set to 60°C, to avoid crude oil wax precipitation. Pore pressure was set to 180 bars for CO_2 to be MCM with the North Sea crude oil. At these conditions CO_2 is supercritical and will create an emulsified phase (liquid-in-liquid) with the surfactant solution. A differential pressure transducer and two absolute pressure transducers (i.e. one downstream and one upstream) measured pressure response. The standard setup for high-pressure/high-temperature CO_2 foam injection can be found in [27].

In-situ foam generation, strength, and stability is evaluated by its apparent viscosity, which is quantified from flow rate and pressure drop during foam injection [28]. Foam apparent viscosity is calculated by;

$$\mu_{app} = \frac{k\nabla p}{(u_l + u_g)} \tag{1}$$

where, k is the absolute permeability of the porous media, ∇p is the pressure gradient measured, and u_l and u_g are the superficial velocities of liquid and gas, respectively [17].

3.2 Unsteady-State Foam Stability Scan by SAG

Foam generation and stability was investigated for two modes of SAG injection in a foam stability scan. Single-cycle SAG was run at 100% brine saturated conditions prior to multi-cycle SAG on the same core (E2). A waterflood to re-establish initial conditions with close to zero CO_2 saturation was run in between the foam stability scans. For both SAG injections, pre-determined slug sizes were injected for 4.0 pore volumes (PVs) targeting a gas fraction (f_g) of 0.70. For single-cycle SAG, a single slug of surfactant solution was injected (1 PV) before CO_2 was introduced for 3 PVs. For multi-cycle SAG, 12 rapid cycles were run: each cycle consisting of a surfactant slug of 0.11 PVs and a CO_2 slug of 0.22 PVs. Un-steady state apparent foam viscosities were calculated as a function of time (i.e. PVs injected) using Equation 1.

3.3 Steady-State Foam Quality and Rate Scans by Co-injection

Foam generation and stability during co-injection was assessed by foam quality and rate scans. Tests were run in 100% brine saturated cores, where one system was composed of a single core (D1) and the other a stacked system (D2-D3). Scans were also performed at residual oil saturation after CO_2 foam EOR to investigate the effect of residual oil on foam stability (E13-E14), and equivalently for a base case without surfactant (E15-E16). See Table 1 for experimental overview.

Foam quality scans determine the optimal gas fraction (f_g) that will generate the highest apparent viscosity during co-injection. CO_2 fractions were changed from 0.0 to 0.90 for drainage-like co-injection (i.e. increasing CO_2 fraction) at a total superficial injection rate of 1.0 ft/day. Each fraction was injected until steady state pressure drop was achieved before increasing to the next fraction. The apparent foam viscosity was calculated from Equation 1. Rate scans were performed following foam quality scans on the same core(s) to estimate rate-dependency on shear-thinning behavior. Rate scans for co-injection were run by increasing the total injection rate at the optimal CO_2 fraction from quality scans. Starting at a superficial velocity of 1.0 ft/day, the injection rate was increased to 2.0, 3.0, and 4.0 ft/day after reaching steady-state conditions.

3.4 CO_2 Foam EOR

Core plugs were initially 100% brine saturated, stacked, and drained with either *n*-Decane or dead crude oil to irreducible water saturation (S_{wi}) at a constant pressure drop of 2 bar/cm. A waterflood was performed for 1 PV prior to CO_2 foam injection. Injection strategies during CO_2 foam were either co-injection, single-cycle, or multi-cycle SAG. CO_2 foam was injected for no more than 2 PVs, considering volumetric and economic limitations at the field-scale. Produced fluids were measured at ambient conditions as stock tank volumes. Density and compressibility were considered negligible for the specific dead crude oil and refined oil used in these experiments. Saturations were calculated based upon the difference in initial oil in place and oil produced divided the total pore volume.

CO_2 foam EOR was performed at FCM conditions using *n*-Decane and at MCM conditions using North Sea crude oil. At FCM conditions, CO_2 foam co-injection used two injection rates (1.0 ft/day for E3-E4 and 2.0 ft/day for E5-E6) to investigate the shear-thinning behavior of foam

* Corresponding author: zachary.alcorn@uib.no

https://doi.org/10.1051/e3sconf/202014602002

during EOR. Co-injection results were then compared with single cycle (E7-E8) and multi cycle SAG (E9-E10). The most promising (highest apparent foam viscosity and oil recovery) injection strategies were evaluated at MCM conditions, with crude oil, and compared with base case experiments without surfactant solution.

3.5 Core-Scale Model Set-up

Laboratory data from a co-injection experiment (E13-E14) was utilized for a core-scale simulation model. The model was initialized to represent the stacked system during waterflood and co-injection. For the waterflood, simulations were conducted with ECLIPSE 100 Blackoil simulator while the compositional simulator E300 (Schlumberger, 2015.2) was used for co-injection. Experimental data validated the model through matching bottom hole pressure (BHP) and cumulative oil/water production. The validated model was used to conduct a sensitivity analysis on the effect of grid cell size, foam quality, and surfactant on oil recovery and CO_2 mobility reduction. The main objective was to ensure model robustness at representing the observed foam behavior, which is extended to use in ongoing field scale simulations.

The base case model consisted of a rectangular grid with dimensions 1 x 1 x 100 (x, y, z). The length of the model was identical to the experimental core system length of 24.7 cm. The model and individual grid cells were also consistent with the core diameter of 4.8 cm thus, each individual grid cell measured 4.8 x 4.8 cm in the x- and y-directions. The injector was located in the first grid block (inlet), while the producer was located in the last grid block (outlet). The production well was placed on BHP control and the injector was controlled by rate both of which were measured in the laboratory. Relative permeability data for the waterflood was derived from JBN analysis of oil and water displacement during the laboratory experiment [29]. Oil and water densities and viscosities were available from PVT-analysis of the crude oil. The model was initiated with S_{wi} of 0.24 at a system pressure of 182 bars.

The co-injection was initialized at pressure and saturations from the history-matched waterflood. The grid size, orientation, well completions and controls were kept identical, except two injection wells were used to represent the single co-injection well from the experiment (one for CO_2 and one for surfactant solution). A compositional simulation case was generated which contained 14 oil components and 2 water components (water and surfactant). Relative permeability curves were derived from CO_2/brine displacement experiments on similar core material described elsewhere [23]. Despite their influence on foam behavior, capillary pressure effects were not included in this study. This is an area, which merits further investigation.

The injection schedule was identical to the experimental procedure and a foam quality of 70% was targeted using a surfactant solution concentration of 1.0 wt%. The effect of foam was modeled using an empirical local-equilibrium approach where the gas relative permeability in the presence of foam is modified by multiplying the gas relative permeability without foam by a mobility reduction factor (MRF) [30]. The mobility reduction factor is dependent upon water saturation, oil saturation, surfactant concentration and shear rate. A maximum gas mobility reduction (*fmmob*) is also used to set the maximum mobility reduction that can be achieved by foam. These parameters were derived from foam quality and rate scans and fit to the empirical model by curve fitting regression [23, 30].

4. Results and Discussion

Apparent viscosities and incremental oil recoveries were used to evaluate foam generation, stability, and EOR performance during CO_2 foam injection. Several mechanisms contribute to the foam apparent viscosity (and pressure response) including foam generation, trapped gas, and CO_2-surfactant emulsification.

4.1 Unsteady-State Foam Stability by SAG

CO_2 foam stability scans by single-cycle and multi-cycle SAG injections were compared in initially brine saturated systems without oil present (Figure 1). Foam apparent viscosity during single-cycle SAG was 18.2 ± 2.7 cP (dashed curve, Figure 1), and 120.2 ± 0.3 cP for multi-cycle SAG (solid curve, Figure 1). Based upon the increase in apparent viscosity, foam was generated when the first CO_2 slug was injected for both injection modes (Figure 1).

During single-cycle SAG, foam generation occurred within the first PV of CO_2 injected and foam remained stable for the next 2 PV with only a slight dry-out effect towards the end of injection (black dashed curve, Figure 1). Injecting multiple alternating slugs of surfactant solution and CO_2 improved conditions for foam generation and stability. During multi-cycle SAG, surfactant solution was introduced to the system in an imbibition process, which caused a decrease in capillary pressure, likely triggering foam generation. This change in capillary pressure is advantageous for foam generation since the creation of lamella requires exceeding a minimum pressure gradient. Hence, a decrease in capillary pressure during the surfactant slugs improved conditions for foam generation. Increased wetting-phase saturation during surfactant solution slugs also may have mitigated foam dry-out. The growth and propagation of a stable high apparent viscosity foam was clear during multi-cycle SAG (solid curve, Figure 1).

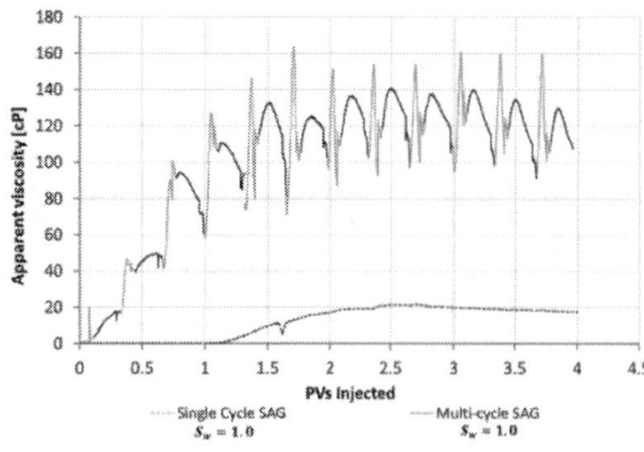

Fig. 1. Foam apparent viscosity as a function of pore volumes injected during single-cycle SAG (dashed curve) and multi-cycle SAG (solid curve). Orange curves represent surfactant solution slugs and the black curves are CO_2 slugs. Single cycle SAG was initiated at $S_w = 1.0$ (no trapped CO_2), whereas multi-cycle SAG injection was initiated with a nominal amount of trapped CO_2 in the core, $S_w \approx 1.0$.

The first CO_2 injection during multi-cycle SAG (from 0.11 to 0.33 PV) is equal to the first CO_2 injection of the single-cycle SAG (from 1 to 1.22 PV), except the amount of surfactant injected (which should benefit single-cycle). We can therefore directly compare the first cycle of the experiments, and evidently, the foam generation is much more immediate and rapid in the multi-cycle experiment, resulting in 15 cP increase in apparent viscosity when 0.22 PV is injected, compared with < 5 cP during the single-cycle. It took 5 cycles to reach peak foam strength of 120 cP during the multi-cycle SAG, where each step had an average increase in apparent viscosity of 25 cP. Extrapolating the observed behavior of the first 0.22 PV of the single-cycle experiment for five consecutive cycles to reach peak apparent viscosity, would result in final apparent viscosity of 25 cP.

Despite efforts to return the core to 100% water saturation, the presence of trapped CO_2 from the previous single-cycle SAG experiment likely reduced CO_2 relative permeability and increased the pressure drop during multi-cycle SAG. This would result in higher foam apparent viscosities. A decrease in effective water permeability was observed for E2 from 42.3 mD to 19.8 mD by capillary trapped CO_2 prior to multi-cycle SAG foam stability. However, during multi-cycle SAG, the step-wise increase in differential pressure and significantly increased apparent viscosity compared to single-cycle SAG suggest that the alternating injection scheme did improve conditions for foam generation and stability.

4.2 Steady-State Foam Quality and Rate Scans by Co-injection

Figure 2 shows co-injection foam quality scans, D1 and D2-D3 (orange and green curves), which have previously been published [14, 31] and are shown here for comparison.

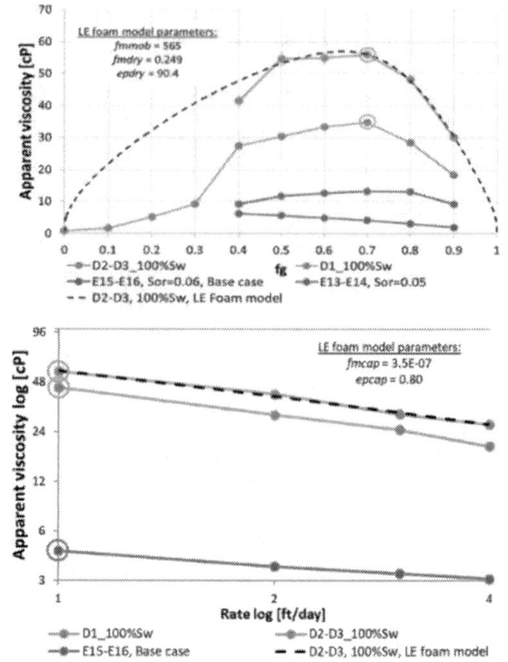

Fig. 2. Foam quality scans $f_g = 0.0$ to 0.90 (top) and foam rate scans at 1.0, 2.0, 3.0 and 4.0 ft/day (bottom) during co-injection of CO_2 and surfactant solution. Foam model parameters derived from the experiments are also shown.

Peak apparent foam viscosities at steady state for $f_g = 0.70$ was 34.9 ± 1.0 cP (D1), and 56.0 ± 1.6 cP (D2-D3), higher than single-cycle SAG (18.2 ± 2.7 cP), but lower than multi-cycle SAG (120.2 ± 0.3 cP) in Figure 1. The foam strengths measured during the two co-injections are relatively similar, considering the heterogeneous core, but discrepancy remains when comparing the three co-injection experiments to the SAGs. It is likely that the injection scheme is not the only explanation for the variation in apparent viscosities.

Foam quality scans with residual oil also investigated the influence of crude oil on foam stability. Experiment E13-E14 in Figure 2 (purple curve) was performed after CO_2 foam EOR at a residual oil saturation (S_{or}) = 0.05 (purple curve, Figure 2). A lower apparent foam viscosity was observed for every f_g compared to foam quality scans conducted with 100% initial water saturation (i.e. strongly water-wet) in Figure 2 (orange and green curves). However, apparent viscosities at fg = 0.70 were 13.20 cP for experiment E13-E14 with surfactant (purple curve, Figure 2) compared to 4.19 cP for the base case (E15-E16, blue curve, Figure 2) at S_{or} = 0.06, without surfactant. Hence, these experiments show that this foam system is capable of generating foam with low apparent viscosities when crude oil is present.

4.3 CO_2 Foam EOR

4.3.1 First-Contact Miscible Conditions

Co-injection and SAG injection strategies were evaluated based upon their apparent viscosity and oil recovery at

FCM conditions (Figure 3). The initial waterfloods recovered between 31.2 % and 47.7% OOIP and a clean water cut was observed for all experiments (left of the vertical black dashed line). The range in waterflood recovery can be attributed to core heterogeneity [32]. Recovery factors and apparent viscosity values are listed in Table 2.

CO_2 foam co-injections were performed at two different injection rates (1.0 ft/day and 2.0 ft/day) to investigate shear-thinning behavior during EOR corefloods (Figure 3). Foam was generated earlier at higher rate (at 1.7 PVs injected, E5-E6, orange dashed curve) compared to the lower co-injection rate (at 2.3 PVs injected, E3-E4, orange solid curve) from dynamic observations of apparent viscosity. The average apparent viscosity, however, was higher at 1.0 ft/day (28.1 cP, E3-E4) than for 2.0 ft/day (18 cP, E5-E6) for the last 0.5 PVs injected (Table 2). This is an indication of the shear-thinning behavior of foam at increasing flow rates. Shear-thinning behavior was also observed in foam rate scans shown in Figure 2 (bottom).

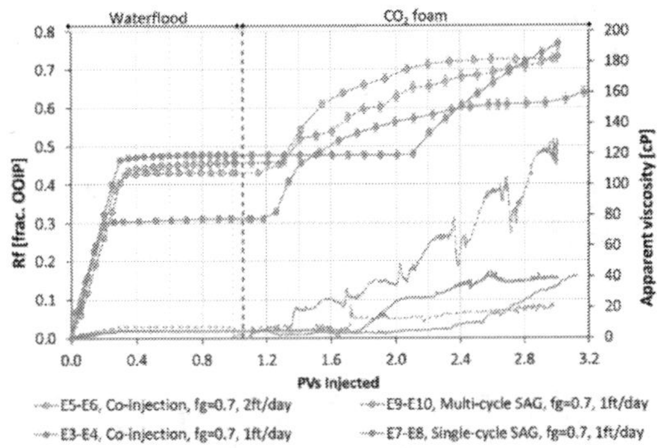

Fig. 3. Recovery factor vs. PVs injected for CO_2 foam EOR corefloods with co-injection and SAG under FCM conditions with *n*-Decane. The orange curves represent CO_2 foam by co-injection at rates of 1.0 ft/day (closed circles) and 2.0 ft/day (open circles), and the green curves are single-cycle SAG (closed circles) and multi-cycle SAG (open circles). The vertical dashed black line separates the waterflood to the left from the CO_2 foam flood to the right. The secondary y-axis shows apparent viscosity.

Table 2. Recovery factors for CO_2 foam EOR corefloods

Core ID	Injection Strategy	Oil phase	$R_{f,\,WF}$ [%OOIP]	$R_{f,\,CO2\,foam}$ [%OOIP]	$R_{f,\,tot}$ [%OOIP]	Apparent visc. CO_2 foam [cP][1]	S_o (after WF)
E3-E4	Co-injection	n-Decane	31.2 ± 0.9	32.7 ± 1.7	63.9 ± 1.4	28.1 ± 7.7	0.48
E5-E6	Co-injection	n-Decane	43.1 ± 0.8	29.6 ± 1.5	72.7 ± 1.2	18.0 ± 1.4	0.39
E7-E8	Single-cycle SAG	n-Decane	47.7 ± 0.8	28.9 ± 1.5	76.6 ± 1.2	37.9 ± 1.1	0.48
E9-E10	Multi-cycle SAG	n-Decane	45.7 ± 1.1	27.6 ± 1.9	73.3 ± 1.6	100.7 ± 14.6	0.46
E13-E14	Co-injection	Crude oil	62.0 ± 0.9	31.0 ± 1.8	93.0 ± 1.3	5.5 ± 0.5	0.29
E15-E16	Co-injection[2]	Crude oil	58.6 ± 1.0	34.9 ± 1.7	93.5 ± 1.4	7.3 ± 0.1	0.35
E17-E18	Multi-cycle SAG	Crude oil	62.3 ± 0.9	30.5 ± 1.6	92.8 ± 1.3	7.0 ± 3.0	0.28
E21-E22	Multi-cycle WAG[2]	Crude oil	66.8 ± 1.0	33.2 ± 1.8	100.0 ± 1.4	7.3 ± 1.6	0.25

$R_{f,WF}$ = recovery factor by waterflood; $R_{f,\,CO2foam}$ = recovery factor by CO_2 foam; $R_{f,tot}$ = total recovery factor
[1]Mean average last 0.5PVs at the end of the CO_2 foam flood. Uncertainties are given as standard deviation of the mean.
[2]Base case without surfactant.

Apparent viscosity trends during SAG EOR (green curves, Figure 3) were the same as for foam stability scans without oil present (Figure 1). The increase in apparent viscosity was more rapid for each SAG injection (green curves) than either of the co-injection experiments (orange curves), likely due a faster increase in CO_2 saturation. However, it is possible that a higher apparent viscosity could have been obtained by continuing the co-injection at 1.0 ft/day (Figure 3, solid orange curve). CO_2 foam apparent viscosity for multi-cycle SAG (Figure 3, dashed green curve) increased continuously for each cycle and reached an average value of 100.7 cP at the end of injection. In terms of mobility control, multi-cycle SAG was superior in creating conditions for high viscous displacement during EOR corefloods at FCM conditions (Figure 3).

The presence of the alkane *n*-Decane oil did not negatively influence foam generation or propagation for either injection strategy (Figure 3). *N*-Decane is a non-polar oil molecule with has no ability to alter wetting-state like that of heavy polar molecules (i.e. asphaltenes and resins). Aging carbonate rock in *n*-Decane does not alter wettability [33] and so stable foam was able to be created *in-situ*, even in the presence of oil.

Accelerated oil recovery rate was observed from the start of both co-injections (Figure 3, open and closed orange circles) and most of the oil was produced after 1.0 PV of CO_2 foam injected. For single-cycle and multi-cycle SAG (solid green circles), no oil was recovered during the initial surfactant slug before CO_2 was injected. Thus, diffusion dominated oil recovery above that of viscous displacement by foam at FCM conditions. CO_2 diffusion is a dominant recovery mechanism at core-scale with the potential to recover nearly 100% of the oil [34, 35]. At constant f_g = 0.70, CO_2 diffusion recovered the same amount of oil regardless of injection mode, on average 29.7 ± 2.2% OOIP. As observed in Figure 3, however, total recoveries did not reach the ultimate recovery potential of 100% OOIP. This is because stable foam lamellas can create barriers that hinder direct contact between the discontinuous CO_2 phase and unrecovered oil, negatively impacting oil recovery. Values of incremental oil recovery during CO_2 foam are listed in Table 2.

4.3.2 Multiple-Contact Miscible Conditions

Viscous forces are desirable for mobility control and fluid diversion during CO_2 foam EOR. Co-injection at 1.0 ft/day and multi-cycle SAG gave best results in terms of both apparent viscosity and EOR at FCM conditions. These injection strategies were therefore used to determine sensitivity in presence of multicomponent crude oil at MCM conditions. Base case experiments were also performed without surfactant solution as a reference. Results are presented in Figure 4.

The initial waterflood recovered on average 62.0 ± 2.8% OOIP (blue curves, Figure 4) with two-phase production, a characteristic of less water-wet conditions [36]. The measured apparent viscosities during CO_2 foam experiments demonstrated little to no *in-situ* foam generation. No increase in foam apparent viscosity was observed above that of the base cases without surfactant. Two possible explanations are offered for the absence foam generation:

- Oil composition is known to influence lamella creation and foam stability, and its presence can make some foams more unstable than others. Foam stability can be reduced with decreasing carbon content in crude oils. It has also been observed elsewhere that a lower number alkanes are more destabilizing than higher number alkanes, because their shorter hydrocarbon chains are more easily imbibed into foam plateau borders to solubilize with surfactants leading to oil spreading and foam instability [10, 37-39].
- Wettability alteration by crude oil towards less water-wet conditions can hinder foam generation and strength. It has previously been reported that foam cannot be generated at wettability conditions other than strongly water-wet due to the lack of water-wet snap-off sites [40, 16, 13]. The multi-component crude oil likely induced a shift towards oil-wet conditions, which caused the lamellas to detach from the pore walls, restricting foam generation.

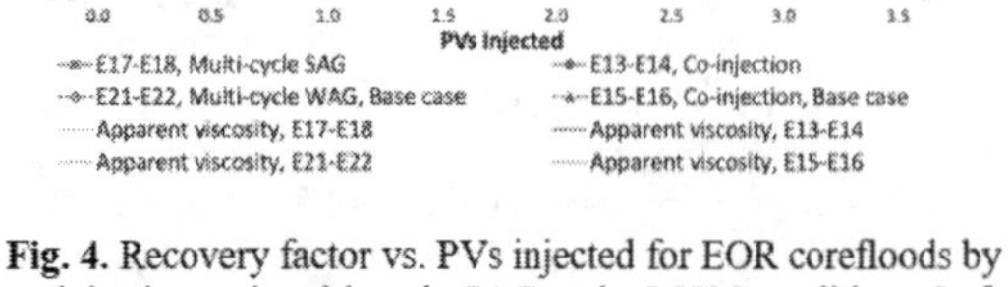

Fig. 4. Recovery factor vs. PVs injected for EOR corefloods by co-injection and multi-cycle SAG under MCM conditions. Left of the vertical dashed black line is waterflood (blue curves) and to the right is CO_2 foam (red curves). CO_2 foam injections are represented by the solid lined curves, while the base cases are dashed lines. The secondary y-axis shows apparent viscosity.

CO_2 foam oil recoveries were on average 30.6 ± 3.0% OOIP for all injection strategies with and without surfactant, at both FCM and MCM conditions. Therefore, incremental oil recoveries were also driven by CO_2 diffusion at MCM conditions (Figure 4). Bernard and Holm observed the same effect in their core-flood experiments with CO_2 foam [44]. This restricts evaluation of injection strategies in terms of production efficiency. Longer induction periods were observed, however, for continuous oil banks to develop under MCM conditions in Figure 4. Between 0.3 to 0.4 PVs of CO_2 foam were injected before oil production was observed (red curves). Total recoveries were higher at MCM conditions (93.2 ± 2.7% OOIP) compared to FCM conditions (71.2 ± 3.2% OOIP) as the waterfloods at MCM conditions left behind lower residual oil saturations (Table 2). Further sensitivity analysis on the effect of CO_2 fractions on foam behavior was considered using the history-matched core-scale simulation model.

4.4 Associated CO_2 Storage

A secondary objective for implementing CO_2 foam for mobility control in EOR is the potential for storing CO_2 as a part of carbon capture, utilization, and storage (CCUS). To calculate CO_2 stored, the volume of CO_2 produced was measured from volumetric conservation of injected and produced fluids, and subtracted from the volume of CO_2 injected. Calculations were made for the FCM and MCM experiments. Based upon mass balance, the value for CO_2 storage (given as a fraction of total PV) is equal to the saturation change of oil and water in the core after CO_2 foam.

Figure 5 shows CO_2 storage in fraction of PV for both co-injections (top) and multi-cycle SAGs (bottom). A direct correlation between the amounts of CO_2 stored and the water/oil saturation change was observed regardless of injection strategy. For every amount of fluid produced, the equivalent amount of CO_2 was stored. CO_2 storage potential was 9.0% greater for multi-cycle SAGs compared to co-injections at MCM. However, CO_2 storage potential was 17.1% greater at FCM conditions (for either injection strategy), compared to MCM, due to improved CO_2 foam displacement and increased CO_2 trapping by capillary forces in more water-wet core plugs.

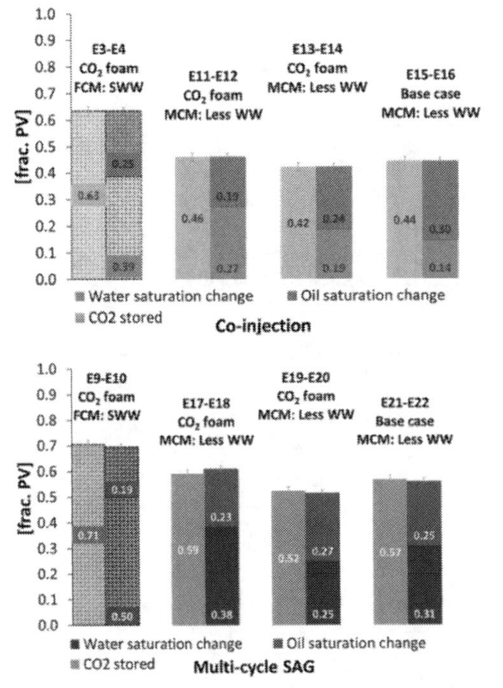

Fig. 5. CO_2 storage potential during CO_2 foam EOR corefloods at FMC and MCM conditions. Top: Co-injection CO_2 foam floods and base case (without surfactant). Bottom: Multi-cycle SAG CO_2 foam floods and base case. Orange bars represent CO_2 stored in fraction of total PV, and the blue and red bars are fractional change in water and oil saturation, respectively. Textured bars indicate FCM conditions, whereas solid bars are experiments at MCM conditions. Note wettability of cores at top of each bar.

4.5 Core-Scale Model Validation and Sensitivity Study

The core-scale simulation model utilized the foam stability measurements in Figure 2 to fit the empirical local equilibrium foam model by curve fitting regression [23, 30]. The value for the maximum gas mobility reduction, *fmmob*, however, was reduced to 41.5 in agreement with previous findings on field core material to reflect more realistic conditions for the field system. An acceptable history-match for experiment E13-E14 was obtained for both waterflood and co-injection. The waterflood match was achieved by tuning the oil relative permeability curve to match oil/water production rate and BHP. The co-injection was matched by tuning the oil and water relative permeabilities to match cumulative oil/water production and BHP. Figure 6 shows the history-match.

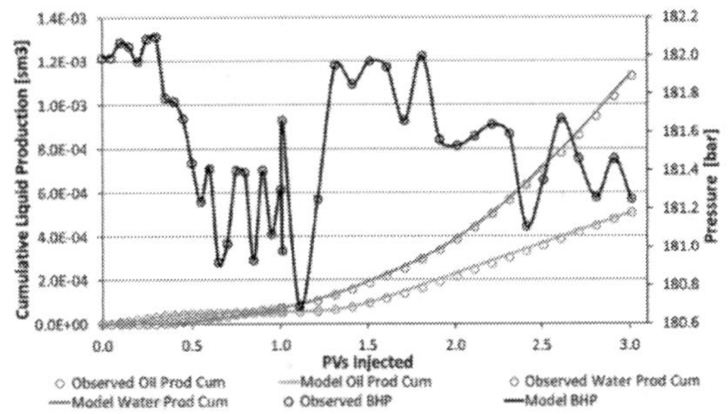

Fig. 6. Observed (open circles) and modeled (curves) cumulative oil/water production during waterflood and co-injection of the history-matched experiment. Production well BHP is shown on the secondary y-axis.

The validated model was first used to investigate the effect of grid resolution during co-injection. The generated case was identical to the base case history-match, but used the fine scale grid (5 times finer in the x, y, and z directions) to evaluate change in CO_2 mobility reduction and oil recovery. CO_2 mobility reduction is generally inferred from delayed CO_2 breakthrough and an increased response in injection pressure. However, simulation results showed limited effect of grid resolution on co-injection in terms of injection well pressures, CO_2 breakthrough, and cumulative recoveries. As limited variations were observed, with the fine grid model, further sensitivity studies utilized the coarser base grid.

4.5.1 Effect of Foam Quality

The effect of foam quality (CO_2 fraction) was studied using the base grid to determine its impacts on oil recovery and CO_2 mobility reduction. Cases injecting higher CO_2 fractions were set to assess model sensitivity on amounts of CO_2 injected and its influence on oil recovery. The base case history-matched model used the optimal CO_2 fraction of 0.70 measured in laboratory. Further sensitivity cases were set to CO_2 fractions of 0.80, 0.90, and 0.95. Figure 7 shows cumulative oil production (solid curves) for the base case (f_g=0.70) and CO_2 fraction sensitives. With increasing CO_2 fractions, the oil recovery rate accelerated but all cases recovered the same volume of cumulative oil. Hence, CO_2 miscibility dominated oil displacement as also observed in the CO_2 foam EOR corefloods experiments in Figure 3 and 4.

Injection pressures for the various CO_2 fractions are shown in Figure 7 (dashed curves). All injection pressures followed the same trend. Higher pressures, however, were observed as the fraction of CO_2 decreased and the fraction of surfactant solution increased, creating a higher apparent viscosity foam (blue and red dashed curves, Figure 7). This is consistent with trends observed in foam quality scans where the optimal CO_2 fraction and highest apparent viscosity was observed at 0.70 (cf. Figure 2, left, green and orange circles). At fractions above this, apparent viscosities declined.

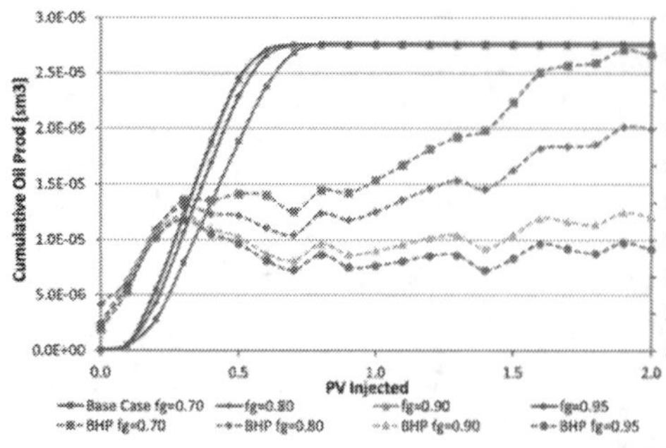

Fig. 7. Cumulative oil production versus time for the base case (blue solid curve), and three sensitivities with different gas (CO_2) fractions. Injection pressure versus time shown on the secondary y-axis for the base case (blue dashed curve), and three sensitivities with different gas (CO_2) fractions.

4.5.2 Effect of Surfactant

CO_2 diffusion and miscibility are evidently the dominant oil recovery mechanisms in both simulation and laboratory experiments. Therefore, a similar case to the history-matched experiment was set-up without surfactant to investigate pure CO_2 versus CO_2 foam displacement. The injection schedule was kept identical to the base case co-injection, except only water was injected as the aqueous phase. Figure 8 shows injection pressure (dashed curves) and cumulative CO_2 production (solid curves) during co-injection with surfactant (blue curves) and without surfactant (red curves).

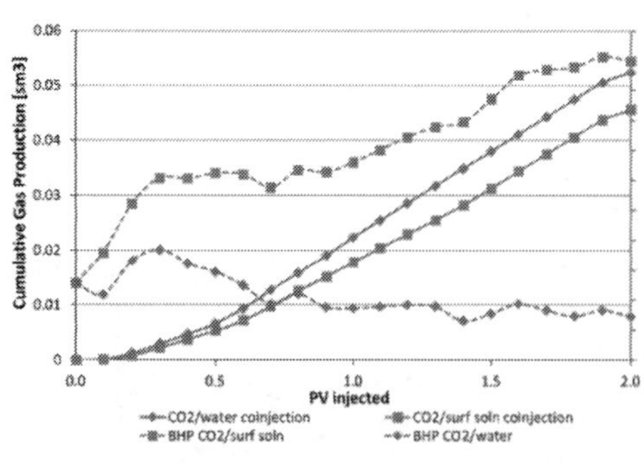

Fig. 8. Injection well pressure (dashed curves) and cumulative gas (CO_2) production (solid curves) for the base case with surfactant present (blue curve) and a case without surfactant present (red curves).

The injection well pressure was significantly lower for the case without surfactant compared to the base case with surfactant (Figure 8, red dashed curve). This indicated higher CO_2 mobility and increased CO_2 production in absence of foam (Figure 8, red solid curve). Analysis of liquid production showed the same cumulative volume of oil recovered in both cases (green curves, Figure 9), but additional water was produced during CO_2 foam injection (blue dashed curve). Hence, a larger storage potential for CO_2 was obtained with CO_2 foam displacement. Similar behavior was also observed in Figure 5, where CO_2 storage was higher for multi-cycle SAG because of increased water displacement during foam injection compared to CO_2 alone.

The similar volumes of oil produced with and without surfactant, demonstrated the dominance of CO_2 miscibility over viscous displacement by foam (green curves, Figure 9). The core-scale model is consistent with laboratory observations indicating that miscibility and diffusion are the governing displacement forces in small core-scale systems. This creates a challenge when upscaling core-scale foam behavior to the field-scale as reservoir heterogeneity and gravity effects will likely be more dominant in the field.

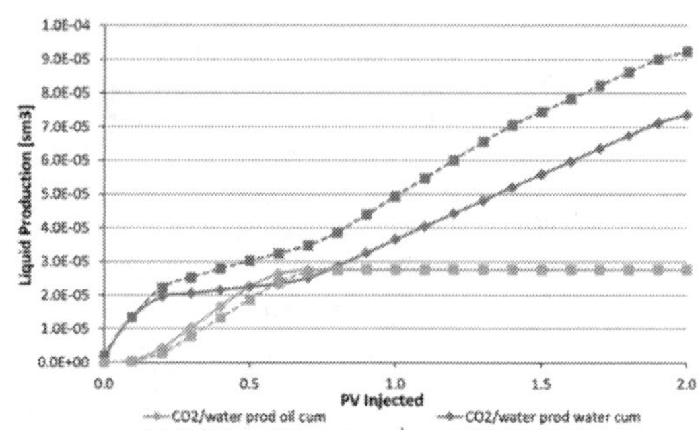

Fig. 9. Cumulative liquid production during CO_2/water co-injection (solid curves) and CO_2/surfactant solution (dashed curves). Green curves correspond to cumulative oil produced and blue to cumulative water.

5. Conclusions

This work investigated various injections strategies (co-injection and SAG) for CO_2 foam mobility control, EOR and CO_2 storage to assist in the design of a CO_2 foam field pilot. The sensitivity study involved both experimental laboratory work and numerical modeling. Supercritical CO_2 foam behavior in brine-saturated systems was investigated to generate foam *in-situ*, without the presence of oil. Additionally, CO_2 foam EOR corefloods were conducted in the presence of refined oil at first-contact miscible (FCM) conditions and in the presence of crude oil at multiple-contact miscible (MCM) conditions after waterflooding to investigate the impact of oil and miscibility on foam generation and stability. Key findings from this work are:

- Multi-cycle SAG resulted in the highest apparent viscosity foam of 120.2 cP during *in-situ* CO_2 foam stability scans compared to co-injection (56.0 cP) and single-cycle SAG (18.2 cP) without oil present. Multi-cycle SAG also achieved the highest apparent viscosity foam of 100.7 cP for the CO_2 foam EOR corefloods, with refined oil present, at FCM conditions.
- Incremental oil recoveries during tertiary CO_2 foam injections were on average 30.6% OOIP for all injection strategies, with and without surfactant, at both FCM and MCM conditions. At MCM conditions, CO_2 foam was not generated as a result of wettability

alteration by crude oil and foam destabilization in presence of crude oil.

- CO_2 diffusion and miscibility with oil were the dominant recovery mechanisms as observed in laboratory corefloods and numerical core-scale sensitivity studies on foam quality.
- A validated numerical core-scale model captured the observed foam behavior from laboratory corefloods. The foam model was not sensitive to grid resolution and corroborated laboratory observations of core-scale foam behavior. Further work is ongoing to upscale the laboratory observations to field-scale simulations.
- A direct correlation between the amounts of CO_2 stored and water/oil saturation change was observed in the laboratory corefloods. For every amount of fluid produced, the equivalent amount of CO_2 was stored regardless of injection strategy. CO_2 storage potential was 17.1% greater at FCM conditions, compared to MCM, due to increased displacement by CO_2 foam and CO_2 trapping by capillary forces in more water-wet core plugs. Core-scale simulations indicated higher CO_2 storage potential with CO_2 foam because of increased water displacement, compared to cases without foam.

Reservoir heterogeneity and gravity dominate displacement at the field-scale, therefore the high apparent viscosities and viscous displacement forces provided by multi-cycle SAG in the foam stability scans are favorable. The shear-thinning behavior of the foam system can also mitigate injectivity loss near the injection well, where flow rates are high. Additionally, a SAG injection scheme provides better injectivity control, when operating close to formation fracture pressure, due to the ability to switch to CO_2 injection for foam dry-out. Reported oil recoveries, from CO_2 foam EOR corefloods, cannot be upscaled to predict field performance as CO_2 diffusion will have less effect on displacement at the length scales existing in the field.

Acknowledgements

The authors wish to acknowledge the Norwegian Research Council CLIMIT program for financial support under grant number 249742 - *CO_2 Storage from Lab to On-Shore Field Pilots Using CO_2-Foam for Mobility Control in CCUS* and industry partners; Shell Global Solutions, TOTAL E&P USA, Inc., and Equinor ASA. The authors also thank the field operator for cooperation.

Nomenclature

API	American Petroleum Institute
BHP	bottom hole pressure
CCUS	carbon capture, utilization, and storage
EOR	enhanced oil recovery
FCM	first-contact miscible
f_g	gas fraction
fmmob	maximum gas mobility reduction
K	permeability
MCM	multiple-contact miscible
MMP	minimum miscibility pressure
MRF	mobility reduction factor (foam model)
OOIP	oil originally in place
PV	pore volume
PVT	pressure, volume, temperature
$R_{f,tot}$	total recovery factor
$R_{f,WF}$	recovery factor by waterflood
$R_{f,CO2\ foam}$	recovery factor by CO_2 foam
SAG	surfactant-alternating-gas
S_{wi}	irreducible water saturation
S_o	oil saturation
S_{or}	residual oil saturation
t	time
u_g	gas superficial velocity
u_{liq}	liquid superficial velocity
μ_{app}	apparent viscosity
∇p	pressure gradient
SWW	strongly water-wet
WW	water-wet

References

1. Bernard, G.G., Holm, L.W. and Harvey, C.P. (1980). Use of Surfactant to Reduce CO_2 Mobility in Oil Displacement. *SPE Journal* **20** (4): 281-292.
2. Latil, M. *Enhanced Oil Recovery* (1980) Paris: Editions Technip.
3. Bond, D.C. and Holbrook, O.C. Gas Drive Oil Recovery Process. (1958) United States Patent Office. 30 December. Patented number 2,866,507.
4. Nzekwu, B.I. and Bennion, D.W. (1987) Mobility Control in Dynamic Gravity Segregation Flow Systems. *Journal of Canadian Petroleum Technology* **26** (4): 80-87. JCPT87-04-08.
5. Rossen, W.R. (1996). Foams in Enhanced Oil Recovery. In Foams Theory, Measurements, and Applications. eds. Prud'homme, R.K. and Khan, S.A. **57**, ch. 11, 414-464, Marcel Dekker, Inc. New York
6. A.H. Falls, G.J. Hirasaki, T.W. Patzek, P.A. Gauglitz, D.D. Miller, and T. Ratulowski. (1988). *SPE Reserv Eng*. **3**: 844
7. Holm, L.W. (1968). The Mechanisms of Gas and Liquid Flow Through Porous Media in the Presence of Foam. *Transactions* **243**, 359-369.
8. Bernard, G.G., Holm, L.W. and Jacobs, W.L. (1965). Effect of Foam on Trapped Gas Saturation and on Permeability of Porous Media to Water. SPE Annual Fall Meeting, 3-6 October, Denver, Colorado, US. SPE-1204.
9. Emadi, A., Sohrabi, M., Farzaneh, S.A. and Ireland, S. (2013). Experimental Investigation of Liquid-CO2 and CO2-Emulsion Application for Enhanced Heavy Oil Recovery. SPE EAGE and SPE Europec, 10-13 June, London, United Kingdom. SPE-164798-MS.
10. Schramm, L.L. and Novosad. (1992). The destabilization of foams for improved oil recovery by crude oils: Effect of the nature of the oil. *Journal of Petroleum Science and Engineering* **7** (1-2), 77-90.
11. Prud'homme, R K. and Khan, S A. (1996). Foams, Theory, Measurements, and Applications. New York: Marcel Dekker Incorporated.
12. Farajzadeh, R., Andrianov, A., Krastev, R., Hirasaki, G.J. and Rossen, W.R. (2012). Foam-oil interactions

in porous media: Implications for foam assisted enhanced oil recovery. Advances in Colloid and Interface Science 183-184, 1-13.

13. Schramm, L.L. and Mannhardt, K. (1996). The effect of wettability on foam sensitivity to crude oil in porous media". *Journal of Petroleum and Engineering* **15** (1), 101-113.
14. Fredriksen, S., Alcorn, Z., Frøland, A. et al. (2019). Surfactant Prefloods During Carbon Dioxide Foam Injection for Integrated Enhanced Oil Recovery in Fractured Oil-Wet Carbonates. *SPE Journal*. SPE-190168-PA (in press; posted 5 April 2019). https://doi.org/10.2118/190168-PA.
15. Shan, D. and Rossen, W.R. (2002). Optimal Injection Strategies for Foam IOR. SPE/DOE Improved Oil Recovery Symposium, 13-17 April, Tulsa, Oklahoma. SPE-75180-MS.
16. Farajzadeh, R., Andrianov, A., Krastev, R., Hirasaki, G.J. and Rossen, W.R. (2012). Foam-oil interactions in porous media: Implications for foam assisted enhanced oil recovery. *Advances in Colloid and Interface Science* **183-184**, 1-13.
17. Jones, S.A., Laskaris, G., Vincent-Bonnieu, S., Farajzadeh, R. and Rossen, W.R. (2016). Surfactant Effect on Foam: From Core Flood Experiments To Implicit-Texture Foam Model Parameters. SPE Improved Oil Recovery Conference, 11-13 April, Tulsa, Oklahoma. SPE-179637-MS.
18. Groenenboom, J., Kechut, N.I., Mar-Or, A., Vincent-Bonnieu, S. (2017). Foam Assisted WAG: Injection Strategies to Optimize Performance. SPE/IATMI Asia Pacific Oil & Gas Exhibition, 17-19 October, Jakarta, Indonesia. SPE-186991-MS.
19. Hoefner, M.L. and Evans, E.M. (1995). CO_2 Foam: Results from Four Developmental Fields Trials. *SPE Reservoir Engineering* **10** (4), 273-281.
20. Chou, S.I., Vasicek, S.L., Pisio, D.L., Jasek, D.E., Goodgame, J.A. (1992). CO2 Foam Field Trial at North Ward Estes. 67th SPE Annual Technical Conference and Exhibition, October 4-7, Washington, D.C.
21. Harpole, K.J., Siemers, W.T. and Gerard, M.G. (1994). CO_2 Foam Field Verification Test at EVGSAU: Phase IIIC – Reservoir Characterization and Response to Foam Injection. SPE/DOE 9th Symposium on Improved Oil Recovery, 17-20 April, Tulsa, OK, USA.
22. Henry, R.L.; Fisher, D.R.; Pennel, S. P.; Honnert, M.A. (1996). Field Test of Foam to Reduce CO2 Cycling. SPE/DOE Tenth Symposium on Improved Oil Recovery, April 21–24, Tulsa, OK, US. SPE/DOE 35402
23. Alcorn, Z. P., Fredriksen, S. B., Sharma, M. et al. (2019). An Integrated Carbon-Dioxide-Foam Enhanced-Oil-Recovery Pilot Program With Combined Carbon Capture, Utilization, and Storage in an Onshore Texas Heterogeneous Carbonate Field. SPE Reservoir Evaluation & Engineering-Reservoir Engineering. SPE-190204-PA (in press; posted 27 March 2019). https://doi.org/10.2118/190204-PA.
24. Graue, A., Viksund, B G. and Baldwin, B A. (1999). Reproducible Wettability Alteration of Low-Permeable Outcrop Chalk. *SPE Reservoir Evaluation and Engineering* **2** (02), 134-140. SPE-55904-PA.
25. Steinsbø, M., Brattekås, B., Fernø, M.A., Ersland, G. and Graue, A. (2014). Supercritical CO_2 Injection for Enhanced Oil Recovery in Fracture Chalk. International Symposium of the Society of Core Analysts, 8-12 September, Avignon, France. SCA2014-092.
26. Jian, B., Puerto, M C., Wehowsky, A., Dong, P., Johnston, K P. and Hirasaki, G J. (2016). Static Adsorption of an Ethoxylated Nonionic Surfactant on Carbonate Minerals. *Langmuir* **32** (40), 10244-10252.
27. Steinsbø, M., Brattekås, B., Ersland, G., Bø, K., Opdal, I.E., Tunli, R., Graue, A. and Fernø, M.A. (2015). Foam as Mobility Control for Integrated CO2-EOR in Fractured Carbonates. 18th European Symposium on Improved Oil Recovery, 14-16 April, Dresden, Germany.
28. Hirasaki, G. J. and Lawson, J.B. (1985). Mechanisms of Foam Flow in Porous Media: Apparent Viscosity in Smooth Capillaries. *SPE Journal* **25** (2), 176-190.
29. Johnson, E.F., Bossler, D.P. and Naumann, V.O. (1959). Calculation of Relative Permeability From Displacement Experiments. *Trans. AIME* 1959, *216*, 370-372.
30. Sharma, M., Alcorn Z.P., Fredriksen, S.B., Fernø, M., Graue, A. (2017). Numerical Modeling Study for Designing CO2 Foam Filed Pilot. EAGE IOR Symposium, 24-27 April, Stavanger, Norway.
31. Rognmo, A.U., Fredriksen, S.B., Alcorn, Z.P., Sharma, M., Føyen, T., Eide, Ø., Graue, A. and Fernø, M. (2018). Pore-to-Core EOR Upscaling for CO2-foam for CCUS. SPE EUROPEC featured at the 80th EAGE Annual Conference and Exhibition, 11-14 June, Copenhagen, Denmark. SPE-190869-MS.
32. Eide, Ø., Haugen, Å., Svenningsen, S., Hoff, K., Ersland, G., Fernø, M.A. and Graue, A. (2012). Tertiary Liquid and Supercritical CO_2 injection in Chalk and Limestone at Strongly Water-Wet and Near Neutral-Wet Conditions. International Symposium of the Society of Core Analysts, 27-30 August, Aberdeen, Scotland.
33. Graue, A., Tonheim, E. and Baldwin, B. 1994. Control and Alteration of Wettability in Low-Permeable Chalk". Proceedings of the 3rd International Symposium on Evaluation of Reservoir Wettability and its Effect on Oil Recovery, 21-23 September, Laramie, WY.
34. Eide, Ø., Fernø, M.A., Alcorn, Z.P. and Graue, A. (2016). Visualization of Carbon Dioxide Enhanced Oil Recovery by Diffusion in Fractured Chalk. *SPE Journal* **21** (1), 112-120.
35. Fernø, M A., Steinsbø, M., Eide, Ø., Ahmed, A., Ahmed, K. and Graue, A. (2015). Parametric study of oil recovery during CO_2 injections in fractured chalk: Influence of fracture permeability, diffusion length and water saturation. *Journal of Nat Gas Sci and Eng* **27** (2), 1063-1073.
36. Jadahunandan, P.P. and Morrow, N.R. 1(995). Effect of Wettability on Waterflood Recovery for Crude-Oil/Brine/Rock Systems. *SPE Reservoir Engineering* **10** (1), 40-46.

37. Suffridge, F.E., Raterman, K.T. and Russell, G.C. (1989). Foam Performance Under Reservoir Conditions. 64th SPE ATCE, 8-11 October, San Antonio, Texas, US. SPE-19691-MS.
38. Vikingstad, A. K., Skauge, A., Høiland, H. and Aarra, M. (2005). Foam-oil Interactions Analysed by Static Foam Tests. *Colloids and Surfaces A: Physicochem. Eng. Aspects* **260** (2005), 189-198.
39. T.S. Kristiansen and T. Holt. (1992). Properties of Flowing Foam in Porous Media Containing Oil. Presented at the SPE/DOE Eighth Symposium on Enhanced Oil Recovery, Tulsa, OK, USA, April 22-24, 1992. SPE/DOE 24182.
40. Prud'homme, R K. and Khan, S A. (1996). *Foams, Theory, Measurements, and Applications*. New York: Marcel Dekker Incorporated.

https://doi.org/10.1051/e3sconf/202014603005

Two-Phase Fluid Flow Experiments Monitored by NMR

Thomas Hiller[1], *Gabriel* Hoder[2], *Alexandra* Amann-Hildenbrand[3], *Norbert* Klitzsch[3*], and *Norbert* Schleifer[4]

[1]Leibniz Institute for Applied Geophysics, Hannover, Germany
[2]SCK.CEN, Mol, Belgium
[3]RWTH Aachen University, Aachen, Germany
[4]Wintershall Holding GmbH, Barnstorf, Germany

Abstract. We present a newly developed high-pressure nuclear magnetic resonance (NMR) flow cell, which allows for the simultaneous determination of water saturation, effective gas permeability and NMR relaxation time distribution in two-phase fluid flow experiments. We introduce both the experimental setup and the experimental procedure on a tight Rotliegend sandstone sample. The initially fully water saturated sample is systematically drained by a stepwise increase of gas (Nitrogen) inlet pressure and the drainage process is continuously monitored by low field NMR relaxation measurements. After correction of the data for temperature fluctuations, the monitored changes in water saturation proved very accurate. The experimental procedure provides quantitative information about the total water saturation as well as about its distribution within the pore space at defined differential pressure conditions. Furthermore, the relationship between water saturation and relative (or effective) apparent permeability is directly determined.

1 Introduction

Petrophysical parameters play an important role in many geological applications and are subject of various research projects. For reservoir simulations, laboratory fluid flow experiments are crucial in order to determine parameters, which can be used to calculate fluid redistribution in the subsurface. Regularly, core analysis is done under ambient conditions on dry plugs or completely water saturated samples, i.e., single-phase fluid flow is measured in order to derive intrinsic permeability [1, 2]. However, when it comes to the characterization of low permeable material (tight sandstones, shales) below the mD-range, the experimental procedures need to be adapted for low flow rates and high fluid pressures. This is especially difficult for the determination of the effective permeability of a fluid in presence of another fluid, which can be up to three orders of magnitude lower than the intrinsic permeability [3]. In a two-phase fluid flow system, different factors have to be accounted for, which are often strongly coupled and interdependent like e.g. stress dependence, water saturation and capillary pressure. The latter is highly important, as gas flow through low permeable (partially) water saturated rocks is usually controlled by capillary pressure, i.e., as water is drained from the pores with increasing differential gas pressure [4].

In order to derive correlations between effective permeability, water saturation and capillary pressure, commonly, several experiments are carried out in different setups (core-flooding experiments, centrifuge, porous plate, mercury injection [5, 6, 7, 8, 9]. Generally, effective permeability experiments are conducted in different steps, each consisting of the pre-saturation and an effective permeability measurement. For each saturation level, the flow setup has to be dis- and reassembled [10, 11]. This procedure is very time consuming and bears the risk of sample damage caused by repetitive loading / unloading cycles. Additionally, changes in water saturation during gas flow experiments cannot be directly measured in such a conventional experimental flow cell. In this work, we present a newly developed NMR flow cell, which can be loaded up to a confining pressure of 30 MPa. The apparatus allows the determination of water saturation during ongoing fluid flow experiments. Here, we introduce the experimental procedure as well as its results on an initially fully water saturated sandstone sample. After sample installation, the drainage process was continuously monitored in terms of changing water saturation and effective gas permeability. For more experimental results, obtained with the introduced NMR flow cell, we refer to [12].

2 Theoretical Background

2.1 Single and Two-Phase Fluid Flow

In a single-phase system, permeability depends only on the rock's intrinsic properties, i.e. pore size distribution and tortuosity [9]. In laboratory experiments, permeability can be determined on cm-sized cylindrical plugs with either water or gas. When using water as the permeating liquid, Darcy's law for incompressible media is used [13]:

$$Q = -\frac{k_{\text{water}} A}{\eta} \frac{\Delta p}{\Delta x}. \tag{1}$$

* Corresponding author: NKlitzsch@eonerc.rwth-aachen.de

Parameters are the volume flow rate $Q\ [\mathrm{m^3/s}]$, the differential pressure Δp [Pa] across the sample, fluid viscosity η [Pa s], cross-sectional area $A\ [\mathrm{m^2}]$, sample length Δx [m] and intrinsic permeability $k_{\mathrm{water}}\ [\mathrm{m^2}]$. When using gas in the experiments, one has to account for gas compressibility (eq. 2) and slip flow (eq. 3). Assuming the validity of the ideal gas law, integration across the sample length and between the up- and downstream pressures, p_1 and p_2, yields Darcy's law for compressible media [9]:

$$Q = -\frac{k_{\mathrm{gas}} A}{\eta}\frac{(p_1^2 - p_2^2)}{2p_2\Delta x}. \qquad (2)$$

The coefficient $k_{\mathrm{gas}}\ [\mathrm{m^2}]$ is the apparent gas permeability, which is only valid for given fluid pressure conditions. It is well known, that gas permeability increases with decreasing mean fluid pressures (p_{mean}), which is due to the increasing molecule/pore wall interactions at low gas densities [14, 15, 16]. In single-phase flow systems, the so called "Klinkenberg" or "slip flow" correction is routinely done according to the following linear relationship with k_∞ being the intercept on the y-axis, thus at infinite high gas pressure:

$$k_{gas} = k_\infty\left(1 + \frac{b}{p_{mean}}\right) \qquad (3)$$

$$\text{with } b = \frac{8c\lambda p_{mean}}{d}. \qquad (4)$$

In eq. 4, the slip factor b [Pa] is directly related to d [m], the average or mean pore diameter of the porous medium and to parameters describing the property of the gas phase (λ [m], the mean free path length of the gas molecule, $c \approx 1$, the Adzumi constant and p_{mean}[Pa]). Accordingly, the smaller the pores, the larger the slip flow effect. In two-phase systems, this is extremely difficult to measure and so far only studied for pre-defined water saturations [10, 11, 17, 4]. This correction was not done here as it out of the scope of the present work.

In a two-phase fluid system, one distinguishes between the wetting and the non-wetting fluid. Both phases can be displaced by the other phase, which changes the saturation profile within the porous medium. Hereby, the terms drainage and imbibition are used, which refer to the displacement of the wetting phase by the non-wetting phase and vice versa. For most siliciclastic rocks, water is considered the wetting phase and gas the non-wetting phase. The mobility of both phases (effective permeability) is strongly coupled with fluid saturation and capillary pressure. Increasing capillary pressure results in decreasing water saturation (drainage process), which in turn results in more available fluid pathways for gas flow, thus an increased effective gas permeability. The smaller the pores, the larger is the capillary pressure, which has to be applied to drain water from the pore space. This relationship is described by the Washburn equation for the simplified model of cylindrical pores [18]:

$$P_c = \frac{2\gamma\cos\theta}{R}, \qquad (5)$$

with P_c the capillary pressure, γ [N/m] the interfacial tension between the wetting and non-wetting phase (here water and gas), θ [°] the surface contact angle of the wetting phase and R the equivalent capillary pore radius. In non-oil contaminated reservoir rocks, θ is usually assumed to be zero. In the present study, the pressures detected on the up- and downstream pressure side during the drainage experiment (p_1, p_2) correspond to the pressures of the gas and water phase, respectively. Hence, in the two-phase fluid flow experiments conducted in this study, the assumption is made that P_c equals the differential pressure, $\Delta p = p_1 - p_2$.

2.2 NMR T$_2$ relaxation measurements

In this work, the principles of nuclear magnetic resonance (NMR) relaxation measurements are utilized to support petrophysical flow experiments - especially monitoring the change of water saturation S_w, but also to qualitatively resolve the size of the corresponding water filled pores.

The NMR relaxation mechanism results from the interaction of a porous medium with its pore-filling fluid containing a detectable amount of hydrogen protons ^{1}H (here water). In most laboratory and well-logging applications, the magnetic moments (spins) of hydrogen protons are aligned with a strong static magnetic field $\boldsymbol{B}_0$ and therefore, yield a minute net magnetization. The characteristic precession frequency of the magnetization around the static magnetic field is called Larmor frequency $\omega_{\mathrm{L}} = \gamma_{\mathrm{H}}\boldsymbol{B}_0$ and depends solely on the strength of $\boldsymbol{B}_0$ and the gyromagnetic ratio γ_{H} of hydrogen. Larmor frequencies of commonly available applications can vary from approx. 0.2 MHz (4.7 mT) to 0.4 GHz (9.4 T) [19, 20]. A NMR relaxation measurement is started by applying an energizing electromagnetic pulse with the appropriate Larmor frequency (creating a secondary electromagnetic field $\mathbf{B}_1$) and thus tipping away all spins from their equilibrium state. After the pulse is switched off, the spins relax back into their equilibrium state. This relaxation process is measured and the resulting NMR signal is given by

$$E(t) = \sum_i \frac{V_i}{V_0}\exp\left(\frac{-t}{T_{2,i}}\right), \qquad (6)$$

with V_0 the total water filled pore volume, and V_i the volume of pore class i relaxing with the characteristic relaxation time $T_{2,i}$. The sum of individual amplitudes $\sum V_i/V_0 \to E(t=0) \to E_0$ is commonly referred to as initial amplitude E_0 and is directly proportional to the amount of excited hydrogen protons and therewith, a direct measure for water content, and hence, saturation $S_{\mathrm{w(NMR)}}$.

The NMR T_2 relaxation process itself is a superposition of three independent mechanisms [19, 21]: (i) the bulk relaxation T_b of the pore fluid, (ii) the surface relaxation T_s due to the interaction of pore fluid and rock matrix and (iii) diffusional relaxation T_d caused by spins diffusing through a non-uniform magnetic field. In this work, we assume no internal field gradients and therefore neglect diffusional relaxation. However, in the presence of minerals that exhibit high magnetic susceptibilities, diffusional relaxation must be accounted for [22, 23, 24]. Furthermore, we also assume the so-called fast diffusion regime where $\rho_s R/D \ll 1$, with surface relaxivity ρ_s, characteristic pore size R and self-diffusion coefficient of water D [25]. This assumption is reasonable considering the pore sizes of the tight sandstone samples used in this study. Therefore, in a water saturated porous media the effective relaxation time T_2 is given by

$$\frac{1}{T_2} = \frac{1}{T_{2,b}} + \frac{1}{T_{2,s}} + \frac{1}{T_{2,d}} \rightarrow \frac{1}{T_{2,b}} + \rho_s \frac{S}{V}. \quad (7)$$

By inspection of eq. 7 one can deduce the following straightforward relationship: the larger the pore is, the smaller is the surface-to-volume ratio S/V, and hence, the longer is the relaxation time T_2 and vice versa. The surface relaxivity ρ_s is a mineral parameter and relates surface inhomogeneities to accelerated relaxation and has dimensions of velocity m/s. Generally, ρ_s is assumed constant for a particular type of porous media and needs to be determined via calibration [26]. Depending on the knowledge of ρ_s and considering eqs. 6 and 7, it is possible to directly infer the pore size (or S/V of the pore) from NMR relaxometry data.

To find the individual amplitudes V_i/V_0 as a function of relaxation time $T_{2,i}$ (cf. eq. 6), the so-called relaxation time distribution (RTD), a linear system of equations has to be solved. Generally, and because this inverse problem is ill-posed [27], this is achieved by a regularized (smoothed) least-square minimization [28] of the form

$$\min\|\mathbf{Gm} - \mathbf{d}\|_2^2 + \lambda^2\|\mathbf{Lm}\|_2^2 \quad (8)$$

with $\mathbf{d} = E(t)$ the NMR data vector, $\mathbf{m} = [V_i/V_0]$ the model vector and the forward operator $\mathbf{G}$ (cf. eq. 6). Here, the smoothness constraint on $\mathbf{m}$ is applied by a first-order derivative matrix $\mathbf{L}$. The regularization parameter λ is found via the L-curve criterion and chosen such that the inversion misfit is in the order of the data noise while keeping a sufficiently smooth RTD [29, 28].

3 Experimental Methods & Procedures

3.1 NMR Flow Cell Setup

The experimental setup used in this study consists of multiple components, which allow for the combined measurement of fluid flow and NMR. An overview of the complete assembly is given in Fig. 1. The high-pressure flow cell, which is designed for samples of 30 mm in diameter and 15 – 60 mm in length, is placed vertically in the center of a Halbach magnet. The sample itself is placed between two NMR-inert PEEK (polyether ether ketone) pistons containing conduits for in- and outflow. Grooves on the pistons allow for the even distribution of the gas phase across the sample surface. A rubber sleeve encases the piston/sample arrangement. O-rings prevent influx of the confining pressure oil. The RF coil, which is a 13 mm long copper radio frequency coil (rf-coil), is located within the confining compartment surrounding the rubber sleeve and is centrally positioned around the sample plug. In order to apply a constant confining pressure, p_{conf}, (up to 30 MPa) the compartment is filled with a NMR-inert synthetic oil (Fluorinert^{TM} FC-40), regulated by a HPLC pump (Shimadzu LC-6A). Temperature fluctuations within the flow cell are monitored with a resistance thermometer PT 100 with an accuracy of 0.05 K. On the outside, a wooden cylinder encases the entire setup.

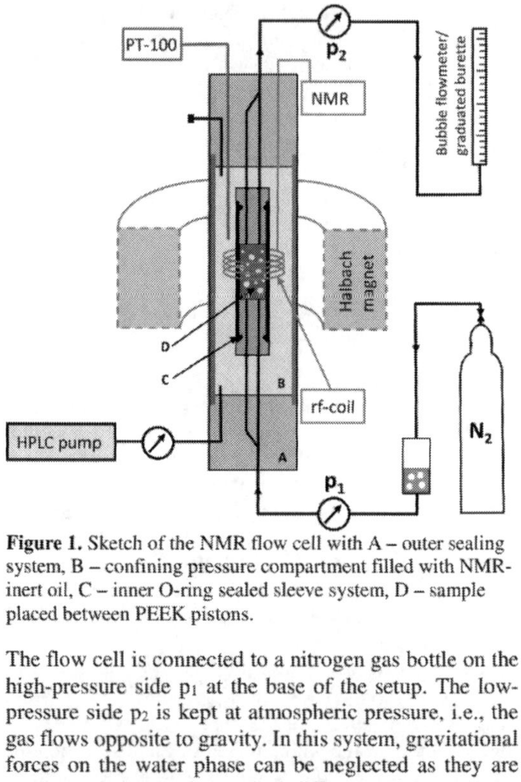

Figure 1. Sketch of the NMR flow cell with A – outer sealing system, B – confining pressure compartment filled with NMR-inert oil, C – inner O-ring sealed sleeve system, D – sample placed between PEEK pistons.

The flow cell is connected to a nitrogen gas bottle on the high-pressure side p_1 at the base of the setup. The low-pressure side p_2 is kept at atmospheric pressure, i.e., the gas flows opposite to gravity. In this system, gravitational forces on the water phase can be neglected as they are much smaller than the applied differential gas pressures of at least 100 kPa. A water reservoir moisturizes the gas stream in order to prevent drying of the sample. The downstream capillary is connected to either a bubble flow meter for gas flux measurements or a graduated pipette to determine single-phase water flux (top of the setup). Additionally, two pressure transducers (Keller, 0.05% FSO) at the inlet/outlet of the flow cell continuously monitor the pressure on both sides.

3.2 NMR Data Processing

All NMR relaxation measurements presented in this work are conducted with a low field Halbach NMR setup working at a frequency of 4 MHz [30, 31] and using the

Carr-Purcell-Meiboom-Gill (CPMG) pulse sequence [32, 33]. The shortest possible echo time in our setup is $TE = 320\ \mu s$ and, depending on the saturation and relaxation behavior of the corresponding sample, up to 2500 echoes were recorded, yielding signal lengths up to 0.8 s. Initially, each CPMG echo train was averaged until a signal-to-noise ratio of $S/N \sim 200$ was reached. Over the course of the drainage experiment and due to the decreasing water saturation (and hence, decreasing signal amplitude), this constraint could not be maintained while keeping an equal NMR measurement interval of 30 minutes.

According to Curie's law the magnetization is negatively correlated with temperature, i.e., for water, a temperature decrease of 1 K yields a magnetization increase of ~0.4 % and vice versa. However, in preliminary experiments we found, that even small temperature fluctuations have a much stronger effect on the initial amplitude of the NMR signal as expected. For testing purposes, we installed a 100 % water filled dummy sample (PVC container filled with degassed tap water) in the flow cell and the confining compartment was filled with the NMR-inert oil – the confining pressure was allowed to be atmospheric. Like in a drainage experiment, the temperatures inside the flow cell T_{cell} and the laboratory T_{room} as well as the initial NMR amplitudes E_0 were continuously monitored over a period of 30 hours. From Fig. 2 a strong positive correlation between the NMR amplitude (NMR_{raw}) and the temperature measured within the flow cell (T_{cell}) is evident ($R^2 = 0.9990$, cf. inset in Fig. 2). While the temperature inside the flow cell T_{cell} (solid black line) varied by approximately ± 0.1 K, the initial amplitude varied by about ± 10 [a. u.] or 4 %. This value is two orders of magnitude higher than the theoretical increase. As the NMR amplitude correlates linearly with the temperature fluctuation (see inset of Fig. 2), the latter can be used to correct the NMR amplitudes. The temperature-corrected signal (NMR_{corr}) fluctuates less than 0.2 % around its mean value. For this reason, we conduct this pre-test always on the fully water saturated samples before starting the actual drainage experiment. We found, that a measurement time of approx. 48 hours yielded stable correlations between NMR amplitude and temperature fluctuations.

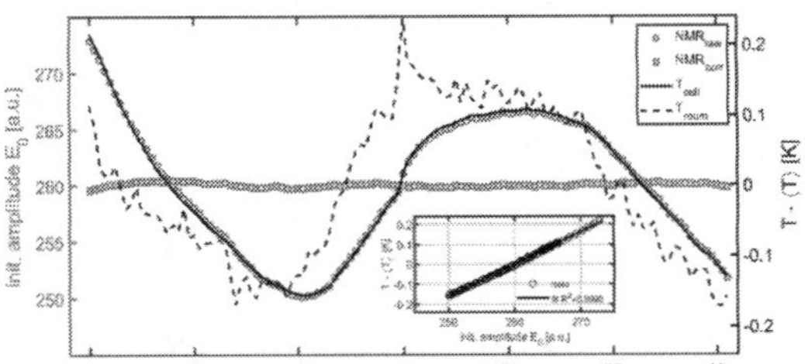

Figure 2. Raw (red circles) and corrected (red squares) initial NMR amplitudes E_0 together with the flow cell (black solid) and room temperature (black dashed) variations for a water sample; <T> denotes the mean temperature of the flow cell or the room, respectively. The inset shows the correlation between the raw amplitude and cell temperature fluctuation.

The effect of temperature on bulk relaxation is well understood and can easily be approximated [19]. In the case of our test measurement with 100 % water, the measured bulk relaxation time follows the theoretical relation. Unlike modern NMR devices, our setup is not temperature regulated. We hypothesize that the temperature dependence of the NMR amplitude observed in our measurements is device or setup dependent, e.g., temperature effects on the electronic components of the resonant circuit. With the procedure described above, we are able to account for the temperature fluctuations and derive reliable water saturation information from our experiment.

3.3 Single-Phase Flow – Intrinsic Gas Permeability

To qualitatively and quantitatively compare single-phase and two-phase flow results, we use for both types of measurements the NMR flow cell under equal confining pressure conditions. To determine the intrinsic permeability k_∞, the dry sample is installed into the flow cell and the capillaries on the high-pressure side are connected to a nitrogen gas source. After application of the confining pressure, the gas permeability experiments are conducted at different mean fluid pressures p_{mean}. Here, steady state experiments are conducted at different upstream pressures p_1, while keeping the downstream pressure p_2 at ambient conditions. The volume flow Q, is measured on the outflow side with the attached bubble flow meter until steady state flow conditions are established. Apparent and intrinsic permeability values are determined from eqs. 2 and 3.

3.4 Two-Phase Flow – Effective Gas Permeability

Prior to all drainage experiments, air is removed from the sample in a vacuum desiccator. Thereafter, we saturated the sample with brine (10.6 g/L $MgSO_4$ solution) and installed it into the flow cell. There, a confining pressure of 15 MPa is established. To obtain a sufficient amount of data for the temperature correction described above, repetitive NMR measurements (30 min interval) are performed on the fully water saturated and pressurized sample for a time span of at least 48 hours.

Drainage experiments are performed according to the step-wise procedure usually applied for gas breakthrough determination of sealing lithologies [34]. The drainage experiment is conducted by increasing the gas pressure on the high-pressure side in different intervals ($p_1 = 0.3 - 2.5$ MPa), while the outflow side is constantly held at atmospheric pressure ($p_2 =$ const.). At low differential pressures, only water is displaced from the sample but no gas flows through it. After exceeding a threshold pressure, which equals the capillary pressure of the smallest pore along the percolation path, the gas phase breaks through and a saturation gradient establishes along the sample, i.e., along the pressure gradient. The whole drainage experiment is monitored by NMR measurements (30 min interval), as well as temperature and pressure measurements (30 s interval). Because of the saturation gradient and the sensitive range of the NMR coil of about

 https://doi.org/10.1051/e3sconf/202014603005

2 cm, we measure an average water saturation over this range. First gas flow being detected on the low-pressure side indicates that the capillary breakthrough pressure has been overcome. At this point, sufficient water has been displaced from the pore space and at least one gas conducting percolation path has been established. The number of percolation pathways increases with increasing gas pressure. Moreover, more pores are desaturated at the high-pressure side than at the low-pressure side, i.e., there is a saturation gradient, which changes with increasing gas pressure. Regularly, the gas outflow rate is detected with a bubble flow meter. After gas outflow and NMR signal (saturation) stabilize, the next pressure difference is applied. From the bubble flow meter and from the NMR data at stationary conditions, we calculate effective permeability of the gas phase, water saturation and relaxation time distribution, respectively. They in turn can be related to the applied pressure difference.

4 Results & Discussion

4.1 Sample Characterization & Intrinsic Properties

The test plug (R1), used in this study, was provided by the Wintershall Holding GmbH and is a tight reservoir sandstone from a depth of about 4000 m. It is a coarse-grained sandstone from the Rotliegend formation, containing illite as pore filling mineral. Some petro-physical properties of sample R1 are provided in Tab. 1.

Table 1. Petrophysical properties (logarithmic mean NMR relaxation time T_{lgm}, NMR porosity Φ_{NMR}, Helium porosity Φ_{He}, Archimedes porosity Φ_{Arch}, intrinsic gas permeability k_∞, and Klinkenberg slip factor b) of the sample used in this study.

T_{lgm} [ms]	Φ_{NMR} | Φ_{He} | Φ_{Arch} [%]	k_∞ [m^2]	*b* [MPa]
4.8	5.8 | 6.7 | 6.9	4 x 10^{-17}	0.20

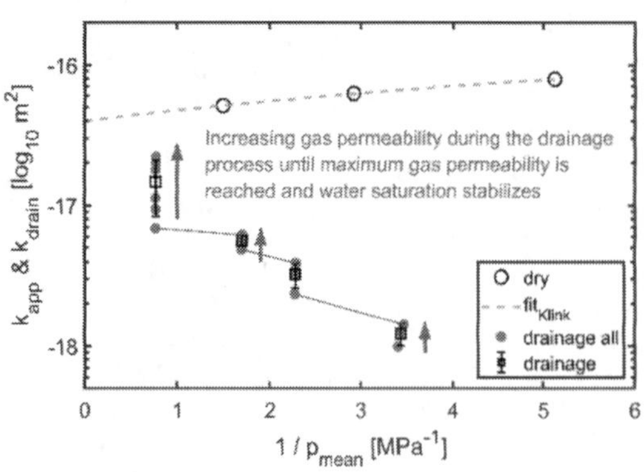

Figure 3. Apparent gas permeability for the dry R1 sample (circles) and effective gas permeability during drainage (squares) as a function of reciprocal mean pressure p_{mean}. The dashed line is a linear fit to the dry data yielding an intrinsic gas permeability of $k_\infty = 4\ 10^{-17}\mathrm{m}^2$.

Porosity was determined via Archimedes' principle. The intrinsic (Klinkenberg corrected) permeability is $4\ 10^{-17}\mathrm{m}^2$ and the slip factor is 0.2 MPa (cf. Tab. 1). In Fig. 3, the apparent permeability k_{app} as a function of reciprocal mean pressure p_{mean} is shown (black circles). The Klinkenberg fit to derive the intrinsic permeability k_∞ is depicted with the gray dashed line.

In Fig. 4 a typical NMR measurement is shown. The NMR transient (solid gray line) and the corresponding multi-exponential fit (solid black line) are depicted in Fig. 4a. The signal consists of 750 echoes with an inter-echo time of 320 µs. The signal-to-noise ratio (SNR) of this particular measurement is 180. Figure 4b shows the corresponding inverted relaxation time distribution (RTD). The RTD was derived as described above using a smoothed least squares fit with 20 relaxation times per decade. The regularization parameter was chosen as such, that the fitting error (rms = 0.07) is equal to the noise level of the data (dashed black line in Fig. 4a). The corresponding solid gray line in Fig. 4b represents the cumulative relaxation time distribution. The vertical black dashed line in Fig. 4b indicates the characteristic relaxation time of the RTD often referred to as logarithmic mean relaxation time T_{lgm}.

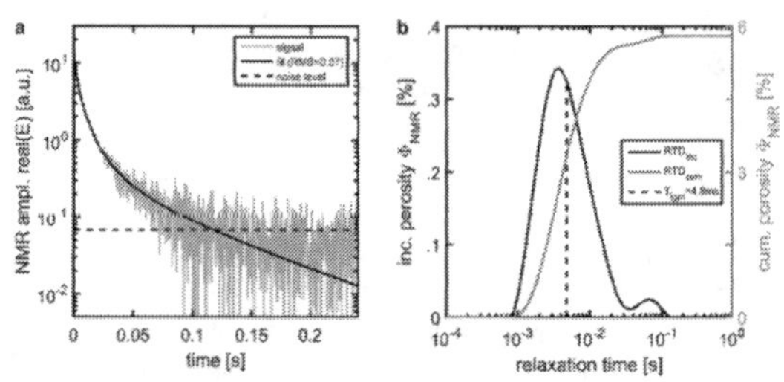

Figure 4. (a) T2 NMR signal (gray solid) and corresponding fit (black solid) for the full saturated sample R1; **(b)** corresponding incremental (black solid) and cumulative (gray solid) relaxation time distribution.

The RTD for sample R1 is characterized by one pronounced peak at about 4 ms and a minor peak at about 70 ms. Following the standard classification of RTDs [19] more than 95% of the signal originates from pores with relaxation times smaller than 33ms and hence, classifies as bulk volume irreducible (BVI) and clay bound water (CBW). This already indicates the tight character of the sample.

A pre-characterization of the intrinsic properties before conducting the actual drainage experiment is generally advisable. Thereafter, the experimental protocol can be adjusted accordingly, i.e., with respect to the choice of initial pressure difference or the amount of pressure steps. If, for instance, samples exhibit a narrow pore size distribution, only a few (densely spaced) pressure steps can be applied before the irreducible water saturation is reached.

4.2 Drainage Experiment

The drainage experiment for sample R1 was conducted at 15 MPa confining pressure. The differential pressure ranged from 0.2 to 2.4 MPa and the sample was drained to a $S_{w,irr}$ of 0.13. The temperature over the course of the

 https://doi.org/10.1051/e3sconf/202014603005

entire experiment was (19.97 ± 0.21 °C).

During the drainage experiment, we applied five distinct pressure steps (cf. Fig. 5a). At the first differential pressure of 0.2 MPa, no significant saturation decrease and no gas flow could be measured over the course of 24 h. Effective gas permeabilities were measured for the first time once breakthrough occurred and gas flow was established. This happened after approx. 50 h at a differential pressure of 0.4 MPa. In the following, differential pressure was increased approx. every 24 h depending on the equilibration of water saturation and effective gas permeability.

The temperature fluctuation inside the flow cell is depicted in Fig. 5b. The fluctuations show amplitudes of ± 0.5 K over the course of the experiment and decrease towards the end of the experiment. Fig. 5c shows the raw and temperature-corrected NMR amplitudes in gray and black, respectively. There are several events (Fig. 5b, c) where the change of temperature clearly influences the NMR amplitudes. For instance, between 24 h and 72 h the temperature varied between -0.5 K and 0.5 K yielding a saturation fluctuation of ± 10 %. Figure 5c visualizes the impact of the aforementioned temperature correction. Without correction, water saturation (raw in Fig. 5) increases after the onset of gas flow ($\geq$ 48 h), which is clearly unphysical considering the given experimental setup, but slightly decreases after temperature correction (corr. in Fig. 5). Effective gas permeabilities increased by more than an order of magnitude during the entire measuring cycle from $9.9\ 10^{-19}\ \mathrm{m}^2$ to $2.2\ 10^{-17}\ \mathrm{m}^2$ at differential pressures of 0.4 MPa and 2.4 MPa, respectively.

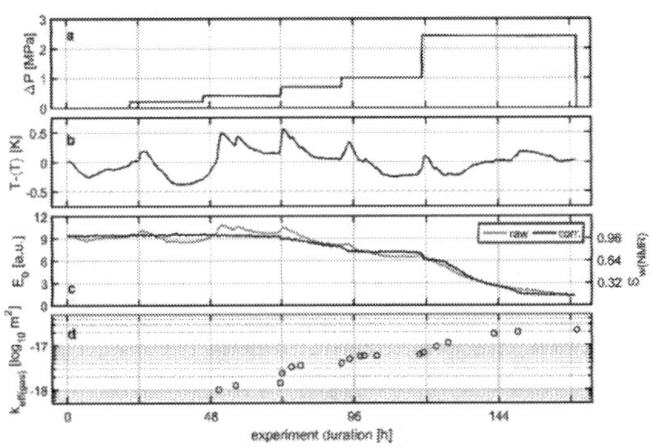

Figure 5. Parameters recorded during the drainage experiment: **(a)** differential pressure; **(b)** temperature fluctuation inside flow cell; **(c)** raw (gray) and temperature-corrected (black) initial NMR amplitude or water saturation; **(d)** effective gas permeability.

The increase of effective gas permeability with increasing differential pressure clearly occurs due to the drainage of successively smaller pores. This process can be visualized by contiguously plotting the RTD in a carpet plot, see panel **I** in Fig. 6. Even though we think that the drainage of successively smaller pores dominates the observed RTD changes, the shift towards shorter relaxation times with decreasing saturation might also be caused by the associated changes in the surface-to-volume ratio of drained pores. For example, water trapped in the corners of a desaturated angular pore contributes to the RTD with considerably smaller relaxation times than the originally saturated pore [35]. Panel **I** in Fig. 6 shows the relative amplitudes of the RTD (incremental porosity cf. Fig. 4b) ranging from 0.005 % (white) to 0.35 % (black) over the entire duration of the experiment. For better visualization, all amplitudes smaller than 0.005 % are not shown. The orange points in panel **I** indicate the logarithmic mean relaxation time T_{lgm} for every individual relaxation time distribution. It shows the change of the (logarithmic) mean size of the water saturated pores over the course of the experiment. For reference, the corresponding full saturation relaxation time distribution is shown in panel **II**. In panel **III** the temperature-corrected saturation $S_{\mathrm{w(NMR)}}$ is plotted, which is given by the ratio of the initial NMR amplitudes $E_0(t > 0)$ to the first initial NMR amplitude at full saturation $E_0(t = 0)$.

During the first 48 h of the drainage experiment, the saturation stays constant and all relaxation time distributions are very similar (cf. Fig. 4b). Like for the saturation, there is no visible change of relaxation times at the first pressure step of 0.2 MPa (after 22 h), i.e., the pressure is not high enough to establish a percolation pathway in the pore network. When increasing the differential pressure to 0.4 MPa (between $48 - 72$ h) gas breakthrough is observed and saturation, as well as T_{lgm}, decrease only slightly. At 0.7 MPa differential pressure ($72 - 90$ h), saturation and T_{lgm} decrease significantly mainly due to the drainage of larger pores (> 20 ms). After a first strong decrease in saturation and T_{lgm} at 1 MPa differential pressure the decrease levels off and all values stay constant for approximately 24 h. The final increase to 2.4 MPa differential pressure yields a rather strong drainage effect and a final water saturation of $S_{\mathrm{w(NMR)}} = 0.13$. Here, the strong drainage relates to pores having relaxation times between $1 - 10$ ms. We cross-checked the final saturation by weighing the sample after the experiment and it was found to be $S_{\mathrm{w(weigh)}} = 0.15$, so comparable to the value derived from the NMR measurement.

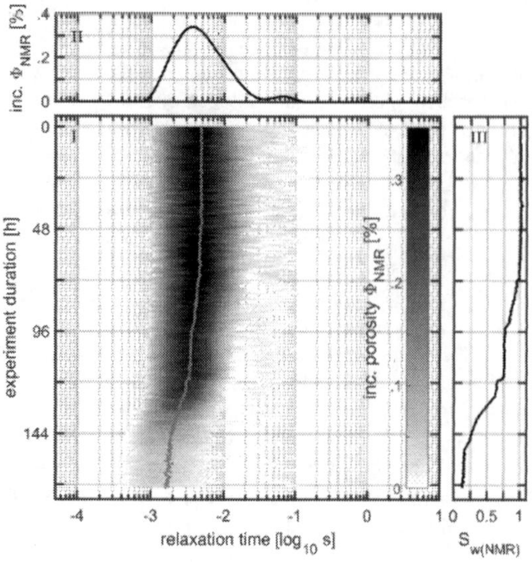

Figure 6. panel **(I)** RTDs over experiment duration; panel **(II)** full saturation RTD; panel **(III)** water saturation over experiment duration; orange points in panel (I) refer to the logarithmic mean relaxation time T_{lgm}.

4.3 History Matching

The simultaneous recording of average water saturation (water production) and pressure gradient throughout a drainage experiment allows to history match recorded data by a two-phase flow simulator. A 1D model is used for the simulation assuming a homogeneous rock in all three dimensions. Software packages available are similar to reservoir black-oil simulators [36] and based on immiscible displacement of the fluid phases [37, 38]. They fit experimental data at distinct time steps, e.g., pressure gradient and water production, by adjusting capillary pressure and relative permeability as these curves are related to pressure and saturation. Analytical functions as Corey (power law), LET, Log (beta) etc. or measured data can be selected. In this study the software Cydar© [39] was used to history match the data. In order to constrain the number of solutions, the measured relative gas permeability curve was fitted using Corey exponents and considered invariable throughout the history match. The history match was run by simulating the NMR water saturations and pressure recordings and adjusting the analytical curves of the relative permeability of water and the drainage capillary pressure. Figure 7 shows a comparison of the measured and simulated water production, the measured and simulated pressure gradients and the relative permeability curves of gas and water, as well as the drainage capillary pressure data. The capillary pressure function includes an entry pressure (P_e) where gas starts to enter the pore space and displaces the water phase. The value of this entry pressure is consistent with the experimental differential pressure where a decrease of the NMR water saturation was observed. The analytical functions simulating the experimental data in 1D are according to [40, 41].

$$S_{wn} = (S_w - S_{wirr})/(1 - S_{wirr}), \tag{9}$$

$$P_c = P_e - \alpha \log S_{wn}, \tag{10}$$

$$k_{rw} = k_{rw0}(S_{wn})^{Nw}, \tag{11}$$

$$k_{rg} = k_{rg0}(1 - S_{wn})^{Ng}, \tag{12}$$

with S_w the measured water saturation, S_{wirr} the residual water saturation, P_e the pore entry pressure, α as fitting parameter, k_{rw} and k_{rg} the relative permeabilities of water and gas respectively, and Nw and Ng the Corey fitting parameters.

Figure 7a+b show the history match results for water saturation and differential pressure from the drainage experiment of sample R1. In Fig. 7c, the measured relative permeability (circles) is plotted. The fitted relative permeability for gas (dashed lines) and water (solid lines) are derived from history matching the monitored water saturation, pressure gradient and effective permeabilities. For the presented data, the derived Corey parameters yield a robust estimation of relative gas permeability over the entire water saturation range. Additionally, capillary pressure data can also be obtained from the history match, as shown in Fig. 7d.

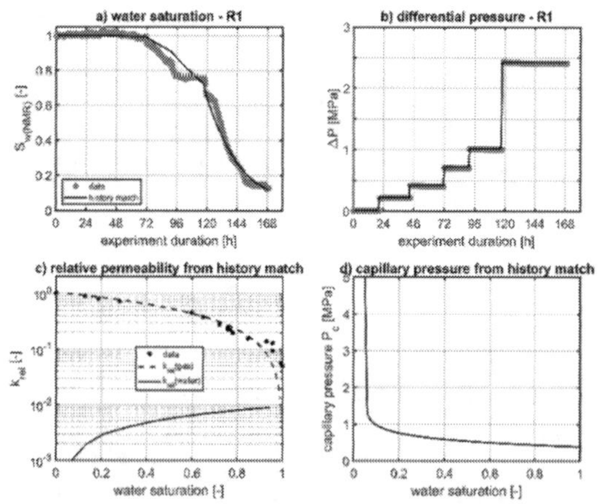

Figure 7. **(a)** History matched water saturation and **(b)** differential pressure for sample R1; **(c)** estimated relative permeabilities and **(d)** capillary pressure derived from history matching the monitored data for all four samples. The circles in panel a, b and c show the corresponding experimental data.

5 Conclusion

With the combined NMR and fluid flow measurements, we overcome the lengthy experimental series of different experiments that are generally used to characterize fluid flow properties of rocks (repetitive pre-saturation tests and permeability experiments). The test measurements proved crucial and resulted in significant improvements for the workflow and thus data quality.

The combination of NMR relaxation measurements and two-phase fluid flow experiments allows for the continuous allocation of water saturation $S_{w(\text{NMR})}$ to differential pressure Δp (respectively capillary pressure

P_c). Moreover, it provides information about the changing water distribution within the pore space, i.e., about the size of the water filled pores, over the course of the experiment. Furthermore, with the our NMR flow cell setup, we directly measure the relationship between water saturation and effective gas permeability $k_{\mathrm{eff(gas)}}$.

After accounting for temperature fluctuations, the monitored changes in water saturation proved very accurate. However, for future experiments, it may be advantageous to place the whole setup in a temperature-controlled container.

The data recorded with the NMR flow cell, can be utilized for the parametrization of the relations between effective permeability, water saturation and capillary pressure by an inverse simulation of the experiment, i.e., by a history match. Hence, the results can readily enter a reservoir simulation.

The authors gratefully acknowledge Wintershall Holding GmbH for funding the iLoPS research project, within this study was conducted. We also like to thank Bernhard Krooss, Jonas Kaiser and Lena Kämpfner for their valuable input and help with the experimental setup.

References

1 E.J. Peters, *Advanced Petrophysics: Geology, Porosity, Absolute Permeability, Heterogeneity and Geostatistics*, **1**, Live Oak Book Company (2012)

2 M. Miller, B. Lieber, G. Piekenbrock and T. McGinness, "Low Permeability Gas Reservoirs How Low Can You Go?" SPWLA Middle East Regional Symposium (2007)

3 K.W. Shanley, R.M. Cluff and J.W. Robinson, "Factors controlling prolific gas production from low-permeability sandstone reservoirs: Implications for resource assessment, prospect development, and risk analysis," AAPG bulletin, **88**, 1083-1121 (2004)

4 A. Amann-Hildenbrand, J.P. Dietrichs and B.M. Krooss, "Effective gas permeability of Tight Gas Sandstones as a function of capillary pressure–a non-steady-state approach," Geofluids, **16**, 367-383 (2016)

5 W.R. Purcell, "Capillary Pressures - Their Measurement Using Mercury and the Calculation of Permeability Therefrom," Trans., AIME **186**, 39-48 (1949)

6 J.S. Osoba, J.G. Richardson, J.K. Kerver, J.A. Hafford and P.M. Blair, "Laboratory Measurements of Relative Permeability," Trans., AIME **192**, 47-56 (1951)

7 M.M. Honarpour, F.Koederitz and A. Herbert, *Relative permeability of petroleum reservoirs*, CRC Press Inc, Boca Raton, FL (1986)

8 C.L. Vavra, J.G. Kaldi and R.M. Sneider, "Geological applications of capillary pressure: a review (1)," AAPG Bulletin, **76**, 840-850 (1992)

9 D. Tiab and E.C. Donaldson, *Petrophysics: Theory and Practice of Measuring Reservoir Rock and Fluid Transport Properties*, 4th ed., Waltham (USA): Gulf Professional Publishing (2015)

10 R.K. Estes and P.F. Fulton, "Gas slippage in permeability measurements," J. Pet. Tech, **8**, 69-73 (1956)

11 J.A. Rushing, K.E. Newsham and V.a.K.C. Fraassen, "Measurement of the Two-Phase Gas Slippage Phenomenon and its Effect on Effective Gas Permeability in Tight Gas Sands," SPE Annual Technical Conference and Exhibition (2003)

12 A. Amann-Hildenbrand, M. Shabani, T. Hiller, N. Klitzsch, N. Schleifer and B. M. Krooss, " Gas Slippage in Partially Saturated Tight Rocks and During Drainage," *SCA2019, #79* (2019)

13 H.P.G. Darcy, *Les Fontaines publiques de la ville de Dijon*, Victor Dalamont (1856)

14 A. Kundt, E. Warburg, "Über Reibung und Wärmeleitung verdünnter Gase," Annalen der Physik, **232**, 177-211 (1875)

15 M. Knudsen, "Die Gesetze der Molekularströmung und der inneren Reibungsströmung der Gase durch Röhren," Annalen der Physik, **333**, 75-130 (1909)

16 L.J. Klinkenberg, "The permeability of porous media to liquids and gases," Drilling and production practice (1941)

17 L. Qingjie, L. Baohua, L. Xianbing, Y. Shouguo, "The effect of water saturation on gas slip factor by pore scale network modeling," SCA 2002 Symposium (2002)

18 E.W. Washburn, "The dynamics of capillary flow," Phys. Rev., **17**(3),. 273-283 (1921)

19 G.R. Coates, L. Xiao, M.G. Prammer, *NMR logging: principles and applications* (1999)

20 A.A. Behroozmand, K. Keating and E. Auken, "A review of the principles and applications of the NMR technique for near-surface characterization," Surveys in Geophysics, **36**, 27-85 (2015)

21 K.J. Dunn, D.J. Bergman and G.A. LaTorraca, *Nuclear magnetic resonance: petrophysical and logging applications* (2002)

22 K. Keating and R. Knight, "A laboratory study of the effect of magnetite on NMR relaxation rates," J. Appl. Geophys., **66**, 188-196 (2008)

23 K. Keating and R. Knight, "A laboratory study of the effect of Fe (II)-bearing minerals on nuclear magnetic resonance (NMR) relaxation measurements," Geophysics, **75**, F71-F82 (2010)

24 J. Mitchell, T.C. Chandrasekera, M.L. Johns, L.F. Gladden and E.J. Fordham, "Nuclear magnetic resonance relaxation and diffusion in the presence of internal gradients: The effect of magnetic field strength," Phys. Rev. E, **81**(2), 026101 (2010)

25 K.R. Brownstein and C.E. Tarr, "Importance of classical diffusion in NMR studies of water in biological cells," Phys. Rev. A, **19**(6), 2446-2453 (1979)

26 W.E. Kenyon, "Petrophysical principles of applications of NMR logging," The Log Analyst, **38**(2), 21-43 (1997)

27 J. Hadamard, *Lectures on Cauchy's problem in linear partial differential equations*, Yale University Press, New York (1923)

28 R.C. Aster, B. Borchers, C.H. Thurber, *Parameter Estimation and Inverse Problems*, 2nd ed. (2013)

29 P.C. Hansen, *Rank-Deficient and Discrete Ill-Posed Problems: Numerical Aspects of Linear Inversion*, SIAM Philadelphia, (1998)

30 K. Halbach, "Design of permanent multipole magnets with oriented rare earth cobalt material," Nucl. Instrum. Methods, **169**, 1-10 (1980)

31 S. Anferova, V. Anferov, J. Arnold, E. Talnishnikh, M.A. Voda, K. Kupferschläger, P. Blümler, C. Clauser and B. Blümich, "Improved Halbach sensor for NMR scanning of drill cores," Mag. Res. Imag., **25**, 474-480 (2007)

32 H.Y. Carr and E.M. Purcell, Phys. Rev., **94**(3), 630-638 (1954)

33 S. Meiboom and D. Gill, "Effects of diffusion on free precession in nuclear magnetic resonance experiments," Rev Sci. Instrum., **29**(8), 688-691 (1958)

34 P. Egermann, J.M. Lombard and P. Bretonnier, "A fast and accurate method to measure threshold capillary pressure of caprocks under representative conditions," *International Symposium of the Society of Core Analysts* (2006)

35 O. Mohnke, R. Jorand, C. Nordlund and N. Klitzsch, "Understanding NMR relaxometry of partially water-saturated rocks," Hydrol. Earth Syst. Sci., 19, 2763-2773 (2015)

36 K. Aziz and A. Settari, *Petroleum reservoir simulation* (1979)

37 E.F. Johnson, D.P. Bossler and V.O. Naumann, "Calculation of relative permeability from displacement experiments," Pet. Trans., AIME, **216**, 370-372 (1959)

38 S.C. Jones and W.O. Roszelle, "Graphical techniques for Determining Relative Permeability from Displacement Experiments," J. Pet. Tech., **30**, 807-817 (1978)

39 R. Lenormand and G. Lenormand, *Cydar© User Manual*, Cydarex© Company Brochure (2016)

40 A.T. Corey, "The interrelation between gas and oil relative permeabilities," Producers monthly, **19**, 38-41 (1954)

41 R.G. Bentsen and J. Anli, "Using parameter estimation techniques to convert centrifuge data into a capillary-pressure curve," Soc. Pet. Eng. J., **17**, 57-64 (1977)

High-Resolution Inline Density Measurements: Insight on Multiphase Flow and Transport Phenomena in Porous Media

Jelayne Falat[1], *Adam* Fehr[1], *Ali* Telmadarreie[1], and *Steven* Bryant [1, 2]

[1]Univeristy of Calgary, Chemical and Petroleum Engineering Department, 2500 University Dr. NW T2N 1N4
[2]Canada Excellence Research Chair in Materials Engineering for Unconventional Oil Reservoirs

Abstract. Understanding fluid flow in porous media is essential with complex and multiphase fluid flow. We demonstrate that high-resolution in-line density measurements are a valuable tool in this regard. An in-line densitometer is used in fluid flow in porous media applications to quantify fluid production and obtain quantitative and qualitative information such as breakthrough times, emulsion/foam generation, and steam condensation.

In order to determine the potential applications for in-line densitometry for fluid flow in porous media, a series of sand pack floods were performed with a densitometer placed at the outlet of a sand pack. All fluids passed through the measurement cell at experiential temperatures and pressures. An algorithm was developed and applied to the density data to provide a quantitative determination of oil and water production. The second series of tests were performed at high temperature and pressure, with a densitometer placed at the inlet and outlet of a sand pack, for steam applications. In both series of experiments, data acquisition was collected at 1 hertz and the analyzed density data was compared to results from the conventional effluent analysis, including Dean-Stark, toluene separations, magnetic susceptibility measurement, and flash calculations where applicable.

The high-resolution monitoring of effluent from a flow experiment through porous media in a system with two phases of known densities enables two-phase production to be accurately quantified in the case of both light and heavy oil. The frequency of measurements results in a high-resolution history of breakthrough times and fluid behavior. In the case of monitoring steam injection processes, reliable laboratory tests show that in-line density measurements enable the determination of steam quality at the inlet and outlet of a sand pack and qualitative determination of steam condensation monitoring

The use of in-line densitometry provides insight on the monitoring of complex fluid flow in porous media, which typical bulk effluent analysis is not able to do. The ability to measure produced fluids at high resolution and extreme temperatures reduces mass balance error associated with the effluent collection and broadens our understanding of complex fluid flow in porous media.

Introduction

Traditional core flood systems utilize two/three-phase separators or fraction collectors downstream of the core to collect effluent. While phase separators prove reliable in conventional applications, they are limited in their ability to handle highly viscous samples and complex fluids like emulsions and nanofluids. Fraction collectors may cause an error in mass balance, due to misalignment of tubes, rate of sample entering tubes, movement of the collector's arm between tubes, etc. Additionally, fraction collectors result in low temporal resolution, as all effluent is mixed in the tube, resulting in an average composition over a given tube volume. Once the effluent is collected, analyzing it can be both time-consuming and challenging [1]. The dean-stark analysis is accurate; however, the temporal resolution is very low and, in many cases, only a single end-point oil production value is obtained. Methods such as NMR, solvent extraction, Karl-Fischer, and the like are costly, time-consuming, and limited to the resolution of the effluent collection method. In-line densitometry enables real-time quantification of produced water and oil and gives a valuable qualitative understanding of phase behavior and fluid flow in porous media.

A high-temperature and high-pressure density meter can be integrated into a core flood set-up to measure effluent density and distinguish between crude oil and injection fluid. The method discussed involves weighing the produced oil to determine the weight of the residual oil remaining. The density measurements are used along with effluent analysis to facilitate laboratory studies of enhanced oil recovery.

Olsen et al. (2017) described a method for using a densitometer for quantifying oil production in two-phase core flood experiments and looked at the understanding obtained by in-line density data for dynamic events such as water breakthrough, gas breakthrough, and connate water production. They conducted immiscible displacements at constant temperature and pressure conditions using North Sea light crude oil and were able to determine that the densitometer produced comparable data to that from an acoustic separator [2].

In this paper, we look at the potential applications and the challenges of utilizing in-line densitometry in core flooding. We demonstrate the benefit of using a high temporal resolution effluent analysis technique as a tool for understanding phase behavior and fluid flow in porous media. We look at the challenges of quantifying heavy oil and assess the use of in-line densitometry in emulsion, foam, and nanofluid core flood experiments. Additionally, we look at the applications of in-line densitometry for steam flooding and the potential understanding gained from being able to determine dynamic water saturation histories during hot gas flooding, condensation monitoring for steam flooding, and real-time volumetric steam quality data. Conventionally, acquiring such information is challenging and may require complex dynamic imaging techniques (i.e. CT, MRI) which are costly. This paper introduces easy and continuous methods using high-resolution densitometry for measuring and analyzing the data that cannot be obtained with conventional core flood configurations.

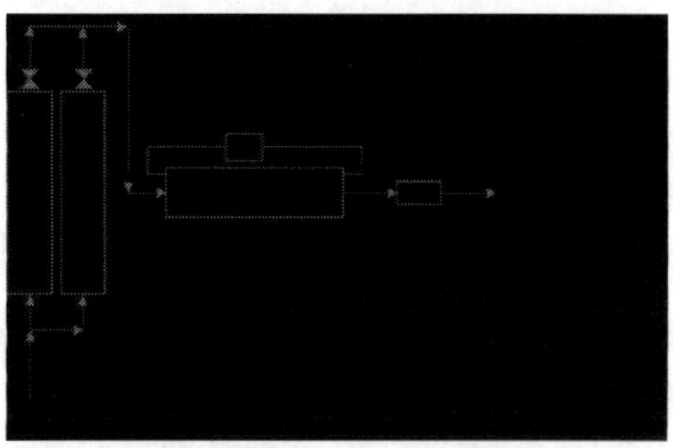

Figure 1: Simple schematic of multiphase core flood apparatus with the densitometer (DMA) located at the outlet of the porous media.

Methodology

The Anton Paar density meter (DMA HPM) operates on the principle of the oscillating U-tube, which has been used for over 50 years. The U-tube is excited by applying an electrical voltage to a Piezo element, causing it to oscillate at a characteristic frequency [3]. When fluid enters the measurement cell, the oscillation frequency is changed and is accurately detected by a second Piezo element. Each fluid has a characteristic frequency which correlates to its density, as per the formula:

$$\rho = A\tau^2 - B$$

Where p is the density of the fluid, A and B are instrument constants, and τ is the oscillation period. The instrument constants A and B are determined in an instrument adjustment, where two reference standards, such as dry air and pure degassed water, are measured in order to determine the relationship between density and the period of oscillation. Using this relationship, unknown sample densities can be determined using their oscillation. The measuring cell is made from Hastelloy C-276, which can withstand high mechanical stress and is highly chemical resistant. The stated cell volume is 2 mL; however, the sensitive measuring volume is 0.35 ± 0.03 mL [2].

Multi-phase core floods were performed using a densitometer placed at the outlet of the sand pack. Density was measured at a frequency of 1 hertz and all measurements were made at the same temperature and pressure as the sand pack by using a circulating water bath and a back-pressure regulator (BPR). The density meter was placed so that all fluid entering the sand pack passed through the measuring cell, before reaching the BPR and fraction collector. Flow rates used were < 0.35 ml/sec, in order to ensure that all fluid passing through the measuring cell was measured. If the flow rate exceeds the limit of the sensitive measuring volume, the accuracy of the analysis is affected.

Steam floods were performed in such a way that additives could be mixed with superheated steam resulting in wet steam laden with chemical additives that are present in the liquid phase of the injected mixture. These mixtures produced an increase in pressure drop relative to the back pressure of the apparatus which was then normalized against the pressure drops of steam of similar total heat values. Table 1 and Table 2 show the details of steam/steam foam experiments.

The dynamic saturation test was performed with nitrogen and water co-injected at a 4:1 gas to liquid ratio at varied flow rates to evaluate the equilibrium water saturation for a given flow rate at high-pressure high-temperature conditions. The back pressure of the system is adjusted in such a way that compressibility of flow is negligible and therefore unaccounted for.

Table 1: The sand pack dimensions, geometry, permeability, porosity, composition, and pressure tap locations for steam foam experiments.

Dimensions	40.84cm X 2.244cm
Geometry	Cylindrical Unconsolidated
Permeability	2.27 Darcy
Porosity	40.14%
Composition of Media	Silica Sand 140-270
Tap Locations	Absolute Pressure at 0, 10cm, dP at 10, 20, 30, 40 cm
Operating Pressure	200 psig
Operating Temperature	200 °C

Table 2: The sand pack dimensions, geometry, permeability, porosity, composition, and pressure tap locations for the dynamic fluid saturation experiment.

Dimensions	30.65cm X 1.5748cm
Geometry	Cylindrical Unconsolidated
Permeability	35.52 Darcy
Porosity	35.87 %
Composition of Media	Silica Sand 50-70
Tap Locations	Inlet, Outlet
Operating Pressure	750 psig
Operating Temperature	100 °C

Example Calculations

Olsen et al. (2017) determined a method for using in-line densitometry for quantification of produced water and oil in a two-phase system using the following equation [2]:

$$CumProdWater\,(0,N) = \sum_{0}^{N} \frac{MeasDens\,(i) - LowerEnvelope(i)}{Upper\ Envelope\,(i) - LowerEnvelope(i)} * [InjVol(i) - InjVol\,(i-1)] \quad (1)$$

Where the Lower Envelope is the density of the lightest phase, and the Upper Envelope is the density of the heaviest phase, oil and water, respectively.

The produced oil fraction can then be determined by subtracting the produced water from the total injection volume

$$CumPr(0,N) = CumInjFluid(0,N) - CumProdWater(0,N) \quad (2)$$

It is noted that the upper and lower envelopes must be carefully determined in order to ensure the accuracy of the analysis.

This equation can also be applied to miscible displacements, where the densities of the two pure phases are known.

In the case of highly viscous oil, the density measurement is dampened by a coating of oil on the walls of the measurement cell. This, in turn, lowers the apparent density. In this case, a moving upper envelope must be used. This is determined by applying a macro to the data set, which detects the change in the upper-density value:

$$Moving\ Upper\ Envelope = IF\left(\frac{Measured\ Density\,(i) - Measured\ Density\,(i-1)}{CumProdVol\,(i) - CumProdVol(i-1)} > 0.001, Upper\ Envelope\,(i-1), Measured\ Density\ Value\right) \quad (3)$$

Liquid volumetric fraction calculation:

$$LVF = \frac{\rho Measured - \rho Gas}{\rho Water - \rho Gas} \quad (4)$$

Therefore, the vapor volumetric fraction can be assumed to be

$$VVF = 1 - LVF \quad (5)$$

The produced water/liquid can be calculated when the system pressure is significantly higher than the phase envelope pressure (boiling pressure) of the liquid phase. This is executed in discrete steps with each interval being each individual density measurement.

$$Produced\ Liquid\ per\ Data\ Interval = LVF * \frac{Total\ Volume\ Produced}{Unit\ Time} * Unit\ Time \quad (6)$$

Using the above equation, the cumulative produced liquid can be calculated as follows:

$$Cumulative\ Produced\ Liquid = \sum_0^N Produced\ liquid\ per\ Data\ Interval \quad (7)$$

Then using that information we can calculate the liquid saturation within the porous media as follows:

$$S_w = \frac{Cumulative\ Produced\ Liquid - Cumluative\ Injected\ Liquid}{Pore\ Volume} \quad (8)$$

For steam vapor, volumetric qualities are calculated in the same way as Equation (4)

The mobility reduction of steam foams is expressed as a mobility reduction factor.

$$Residual\ Resistance\ Factor\ (RRF) = \frac{Pressure\ Drop\ of\ Steam\ with\ Additive}{Equiblibirum\ Pressure\ Drop\ of\ Saturated\ Steam} \quad (9)$$

At steady state the equation for volumetric flow rate

$$Volumetric\ Flow\ Rate = \frac{Mass}{Unit\ Time} \times \frac{Unit\ Volume}{Mass} \quad (10)$$

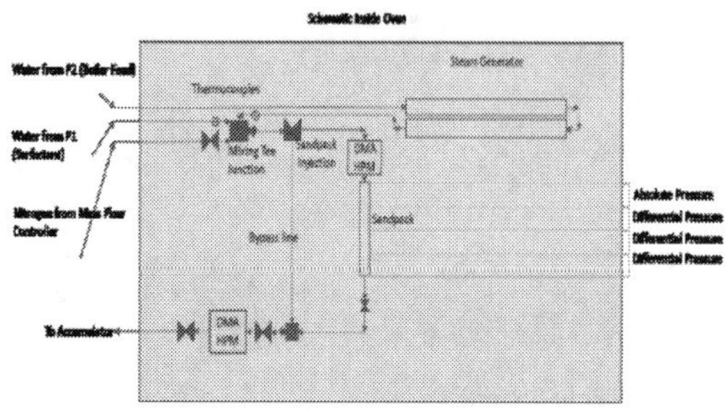

Figure 2: Schematic of the oven internals during steam flood/ hot gas flood experiments, the hot gas flood and the 170 °C steam flood only had one differential pressure measurement. The density measurement cell was inside the oven at 0.494 cubic centimeters of dead volume downstream the outlet of the porous media for all high-temperature experiments discussed in this paper.

RESULTS AND DISCUSSIONS

Here we summarized the capability of a high-resolution inline densitometer for flow experiments in porous media considering various conditions including high temperature, high pressure, variable oil viscosity, and density differentials.

1. OIL RECOVERY MONITORING AND BREAKTHROUGH TIME

The objective of this section is to show how we can precisely monitor the fluid production namely oil recovery in dynamic flow experiments.

The densitometer can be used to accurately determine initial oil in place (OOIP). Additionally, mass balance errors associated with the effluent collection and analysis techniques are eliminated by providing in-line results. Figure 3 to Figure 5 show the densitometer results for conventional displacing experiments with Dodecane as the oil phase.

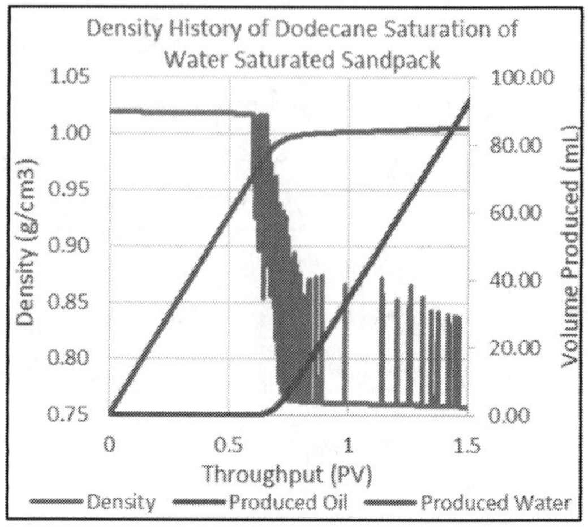

Figure 3: Oil displacing water in water-saturated sand pack. Density monitoring for determination of original oil in place.

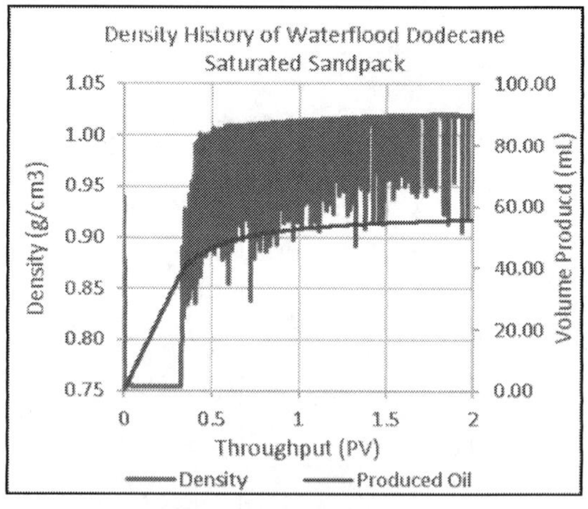

Figure 4: Density history of waterflood through Dodecane saturated sand pack.

As shown in Figure 4, the high-resolution monitoring of effluent from a flow experiment through porous media in a system with two phases of known densities enable oil production to be accurately quantified. The frequency of measurements results in a high-resolution history of breakthrough times and fluid behavior.

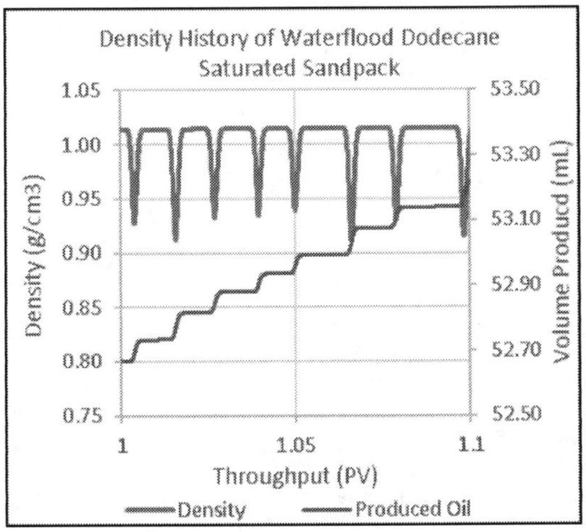

Figure 5: A closer look at density history of waterflood through Dodecane saturated sand pack

A closer look at the density history enables us to identify that the oil is being produced in "trains" (Figure 5) rather than as an emulsion, which is similar to the observation made by Olsen et al. [2]. This is because the density levels off at the density of the water phase, based on the calibration, before dipping down towards the density of the oil. If the oil was being produced as an emulsion, there would be smaller oscillations in density, at a density representative of the quality of the emulsion.

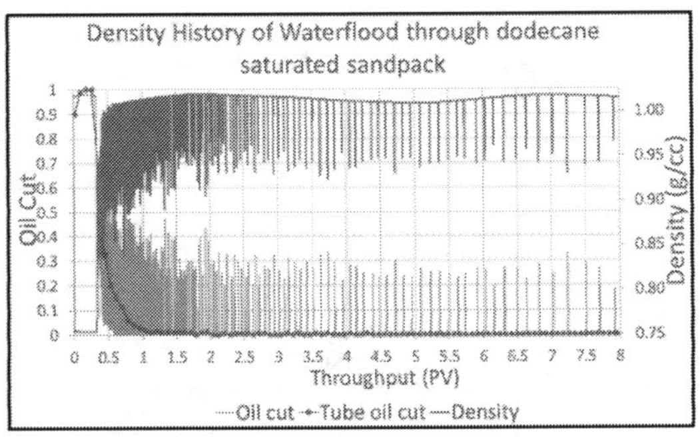

Figure 6: Comparison of density calculation to the manual determination of oil production.

Here we demonstrate that the densitometer provides reliable results when compared to effluent fraction collection and analysis. The mass balance from the effluent analysis was 99.6% and the mass balance from calculated throughput analysis of the densitometer was 100.2%. The mass balance can be improved by utilizing a circulating water bath in the measurement cell of the densitometer, which was considered in the next section. As shown in Figure 6, effluent analysis cannot accurately capture the oil production performance, especially after 2 PV. On the other hand, the oil cut from density calculation shows that the oil is still producing even after 8 PV of the injection. The change in amplitude of the density curve provides us information about the volume of oil production. The change is the frequency of the density will tell us how often the oil is producing. For example, in Figure 6, the frequency of density after 5 PV is lower than that of the early time stage (1-2 PV) supporting the fact that the oil production rate decreases after 5 PV.

Heavy oil recovery calculation; potential and challenges

As shown above, using the density meter for oil saturation allows for a quick determination of original oil in place. This is more crucial when working with heavy crude oil. The effluent analysis is difficult with heavy oil, as it coats the collection vessel and makes visual determination inaccurate. Solvent separations are also time-consuming.

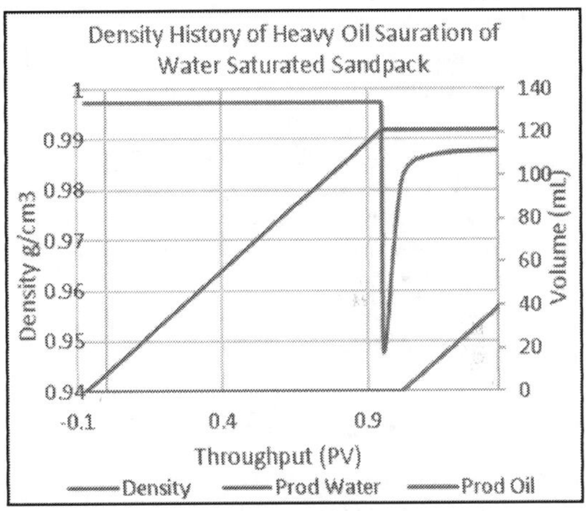

Figure 7: Density history of heavy oil saturation in water saturated sand pack showing breakthrough time.

To verify the effectiveness of oil saturation with the density meter the effluent was passed through the measurement cell and collected into a separatory funnel. The OOIP was 111 mL from analysis of the separatory funnel and 113 mL by the densitometer. The percent relative difference between the two methods was 1.79%.

https://doi.org/10.1051/e3sconf/202014601005

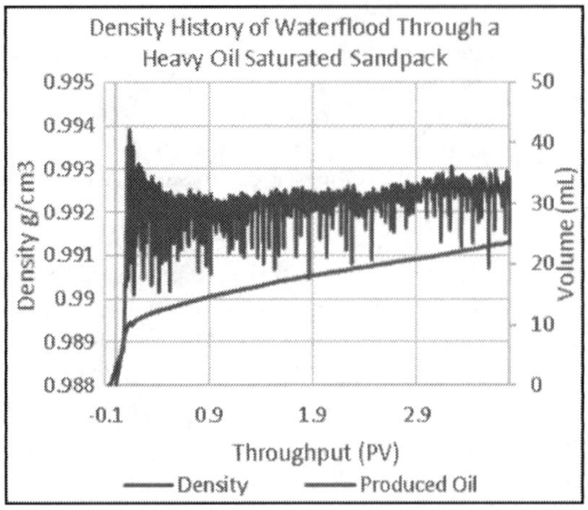

Figure 8: Density history of waterflood through heavy oil saturated sand pack

The quantification of heavy oil through the densitometer is challenging, due to its tendency to coat the walls of the measurement cell. However, if the upper envelope (UE) is adjusted accordingly (See equation 3- Moving upper envelope), it is possible to obtain accurate quantitation of heavy oil production. In order to verify the densitometer results, the effluent was collected into a separatory funnel and analyzed. According to the manual method, 23.6 mL of oil was produced. According to the density meter results, 24.8 mL of oil was produced. The methods had a percent relative difference of 5.1%. It should be mentioned that several core flood tests with heavy oil were performed and the average percent relative difference was less than 10%. The density meter eliminates challenging effluent analysis, including the need for time-consuming Dean-Stark analysis.

2. Nanoparticle transport in porous media

There is a linear relationship between the density and concentration of nanofluid, enabling the densitometer to be used as an inline analyzer for quantification of nanofluid in a two-phase system (Figure 9). Errors could arise from adsorption of a stabilizing agent to the porous media and fines migration that can be potentially detected by high-resolution density measurement.

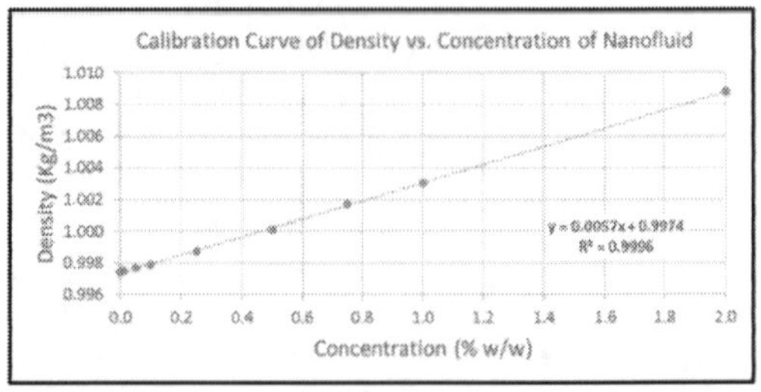

Figure 9: The density of silica nanoparticle at different concentrations. High-resolution density measurement allows accurate detection of nanoparticle with a relatively small increment of nanoparticle concentration.

Following is an example of densitometer capability to detect nanoparticle transport in porous media. The results were also correlated with the magnetic susceptibility measurement of the effluent samples. A suspension of iron oxide nanoparticles was injected in a water-saturated porous media (properties of porous media is mentioned in Table 2) at room condition. Thereafter, the nanofluid displaced with post flush of water. The iron oxide nanoparticles have magnetic properties so they can be detected by magnetic susceptibility measurement. The magnetic susceptibility is normalized based on the magnetic susceptibility of porous media without iron oxide nanoparticles (Figure 10b) any value above zero shows the presence of iron oxide nanoparticles. More detail on such experiments and magnetic susceptibility measurement can be found elsewhere [4].

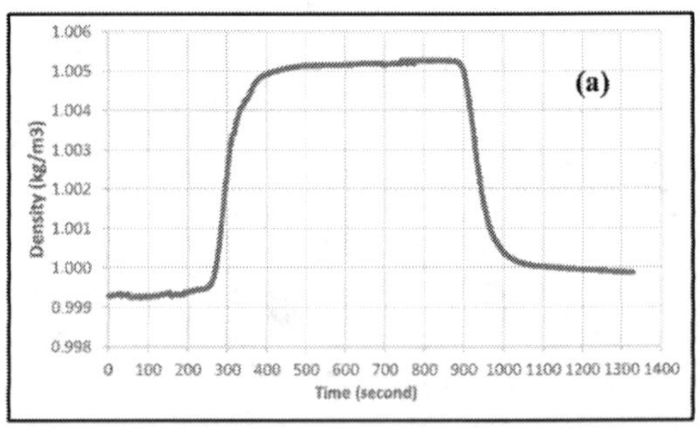

The normalized magnetic susceptibility profile shows that the particle breakthrough occurred around 250-300 minutes. However, the density profile can accurately show the breakthrough time (260 seconds) since it is a continuous measurement with 1 Hz frequency.

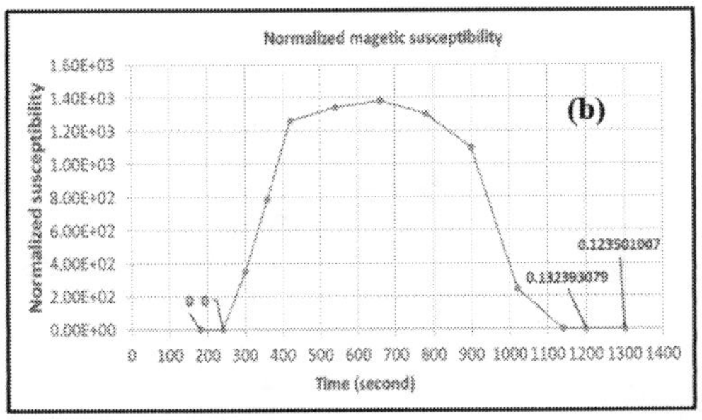

Figure 10: Density profile (a) and magnetic susceptibility profile (b) of iron oxide nanoparticle showing the breakthrough time and nanoparticle transport behavior. Post flush of water started at 600 seconds.

Interestingly the density profile at 1200-1300 seconds is higher than the beginning (water density) suggesting the nanoparticles still present in the effluent sample. The dispersed iron oxide nanoparticles in water create a very dark color liquid which is easily detectable with the naked eye even in low concentration (e.g. 0.05 wt %). The effluent sample at 1200 and 1300 seconds in Figure 10b was clear by naked eyes, but the susceptibility measurement suggests that there is still nanoparticle in porous media supporting the density results.

3. Monitoring saturated steam and hot gas multiphase flows through porous media

Steam/saturated steam flow in porous media presents an especially complex multiphase flow scenario where densitometry can provide some qualitative and semi-quantitative data. High-resolution densitometry enables us to monitor –but not quantify- steam condensate production of the porous media, and we observe distinct variances during foam production as it compares to wet steam flow. Volumetric steam qualities can also be calculated from the density histories, and consequently, mass qualities are also able to be calculated when assuming the system is at steady state.

Steam flood (baseline) is from 4000 seconds to 6850 seconds as shown in Figure 11. This flood had a flash calculated steam quality of injection of 0.4-0.6, however, the flash calculation does not account for the heat losses associated with the steam moving through the stainless-steel tubing while being injected into the porous media. Densitometry enables the analyst to have experimental data to determine what the true steam qualities of injections are. In this scenario, the mass flow rate of the steam was 10 g/min and the thermodynamic flash calculations provide a mass quality result of 50% +/10% while the density meter results are less than 10% mass quality for both the inlet and the outlet. The density meters placed at the inlet and outlet also enables us to observe that in this experiment, the outlet mass quality is higher than at the inlet due to flashing off liquid as the absolute pressure decreases during propagation through the porous media.

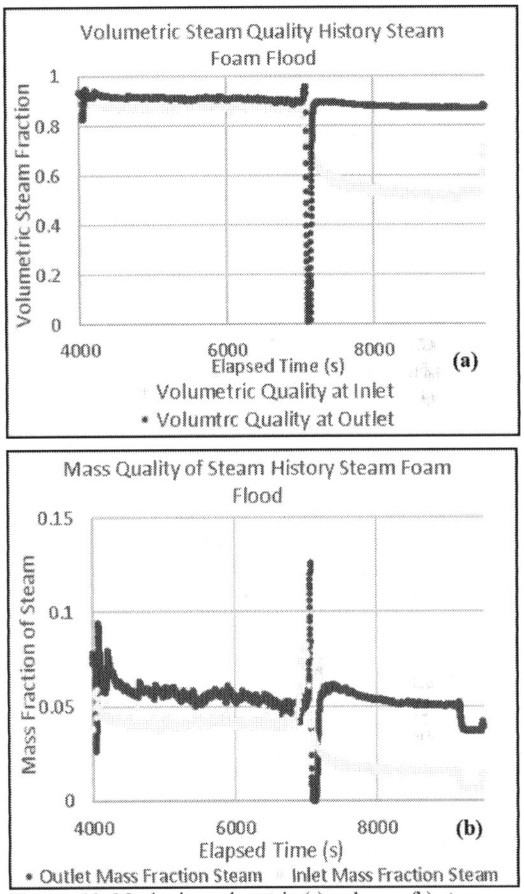

Figures 11: Monitoring volumetric (a) and mass (b) steam quality (inlet/outlet) with the help of inline densitometer. An ambient oven temperature of 170 °C was used.

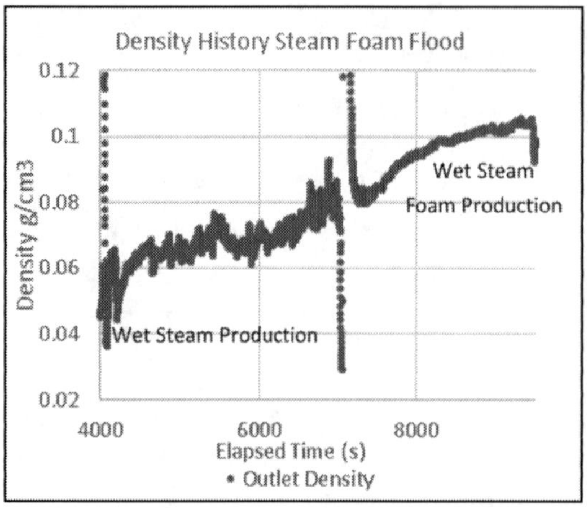

Figure 12: Foam/no-foam boundary identification for steam foams at the outlet of porous media. An ambient oven temperature of 170 °C was used.

Figure 12 shows the steam (4000 seconds to 6850 seconds) and steam foam flood experiment. The steam foam flow has a distinctive characteristic compared to the pure steam flow at the same steam quality. Steam foam production has a distinct lack of random fluctuations that are observed in the saturated steam production.

Steam condensation can be monitored with the density profile (Figure 13). This will enable us to evaluate the breakthrough times of steam and find out when vapor began being produced at the outlet.

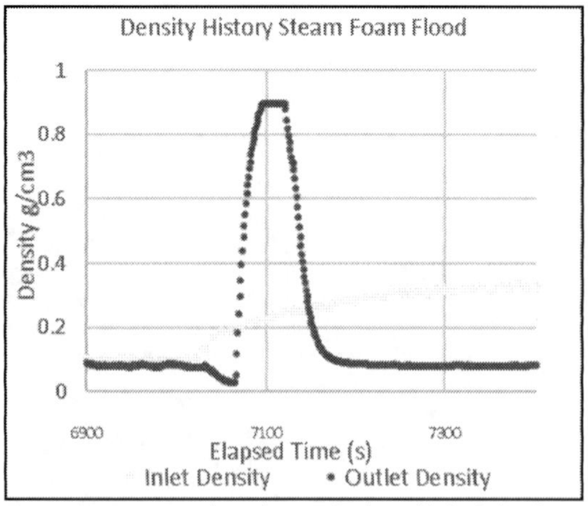

Figure 13: Steam condensation monitoring with the help of inline densitometer.

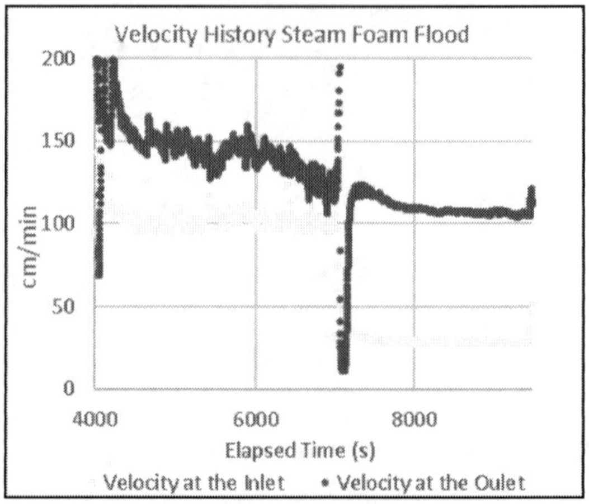

Figure 14: The true velocities of steam moving into the porous media. Steam flood (baseline) is from 4000 seconds to 6850 seconds. An ambient oven temperature of 170 °C was used.

The true velocities of steam cannot accurately be calculated with flash calculations and heat loss modeling. Figure 14 demonstrates the capability of calculating volumetric flow rates during steam flooding. This is not possible to calculate accurately without the use of densitometry due to the complex nature of the heat transfer that is involved with steam floods such as oven convention speeds, insulation, etc. Calculating the velocity based on flash calculations results in a falsely high velocity. The calculations all assume the system in is a steady state.

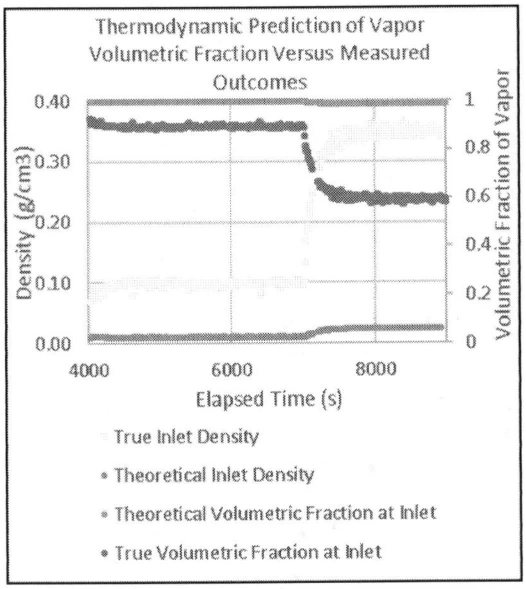

Figure 15: A comparison of the theoretical calculated density and the volumetric fraction of the steam being injected into the porous media, highlighting the comparison between thermodynamic estimations of steam volumetric quality and measured steam volumetric quality using the density meter.

The true velocities of the being injected into the porous media can readily be calculated using the stream density and the mass flow rate which provides us with accurate flow velocities of steam. Using the density meter has enabled more accurate predictions of velocities than with flash calculations alone due to real-time measurements of density, or specific volume of the injected phase as highlighted in Figure 15. The instrument also provides a without a doubt assurance of what phase of the state the injected fluid is in which aides greatly in steam foam floods since occasionally rapid changes in pressure cause condensate production. In some cases, it's helpful to troubleshoot whether the performance of a chemical is simply due to the lack of the presence of a steam phase (instrument limitation) or to distinguish that a chemical simply does not perform in the presence of the condensable gas phase.

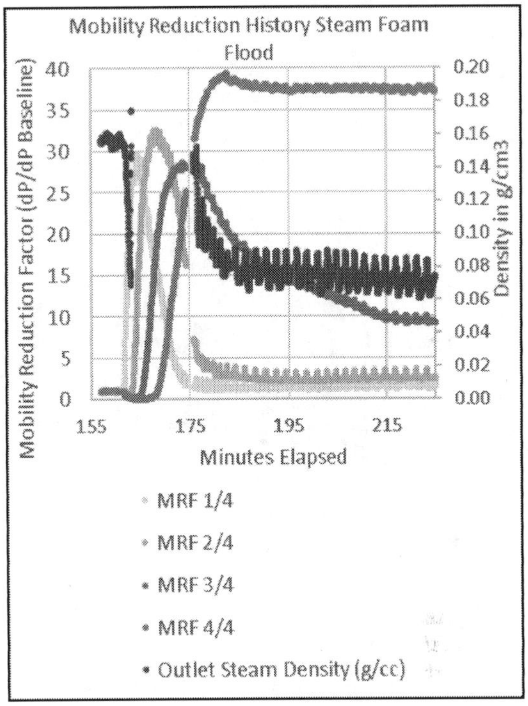

Figure 16: Mobility reduction factor and density profile during steam foam injection. The baseline was performed from 130 minutes until 165 minutes, which is only partially visible in this plot. This steam flood was performed at 200 °C with the outlet at the saturation pressure regulated with a back-pressure regulator.

As shown in Figure 16, initially we see the equilibrium steam density at baseline. The mixing water is then forced to stop mixing with the superheated steam prior to injection and is substituted by a specially designed foaming solution. Immediately a reduction in density occurs simultaneously to the increase of differential pressure in the first quarter of the sand pack. As resistance to flow increases the bubble point temperature of the steam also increases, this means that heat must be delivered to the porous media in the form of latent heat which generates condensate. We also observe the true breakthrough of the steam phase, which normally would be assumed to be at the maximum differential pressure of the last section of the sand pack, however, we can observe that steam actually begins to forms at the maximum differential of the 3rd quarter of the sand pack. This is caused by the flashing of the condensate and should not be considered the actual breakthrough time of the injected steam front.

Hot gas injection flow monitoring

The ability to dynamically measure liquid saturation allows us to analyze the equilibrium water saturations during multiphase flow conditions as shown in Figure 17.

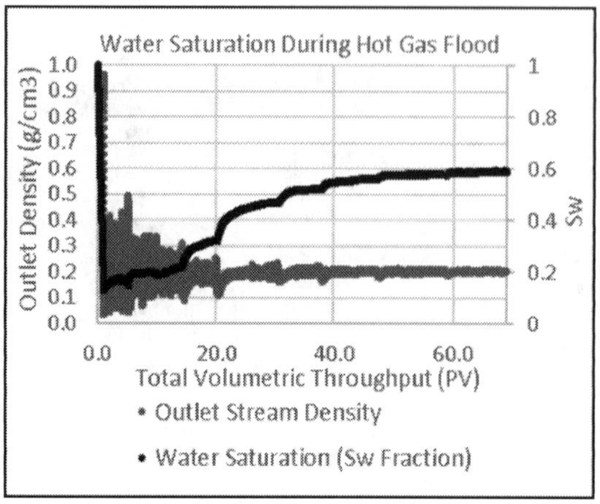

Figure 17: A plot demonstrating the use of densitometry to calculate dynamic fluid saturation. This test was performed at 100 °C and 750 psig back pressure. Total Volumetric flow rates were 0.5, 1.0, 1.5, 3.0, 6.0, 10.0, 15.0, 20.0 28.0 mL/min respectively

Quality Control of Dynamic Fluid Saturation

It should be noted that beyond ~21cc/min does not provide the density meter with enough time to measure "every molecule" that passes through it, the residence time within the measurement cell is not long enough however the saturation values still aligned quite closely with manual mass balance performed by weighing the isolated sand pack before and after the flood. Manual mass balance results in a water saturation of 0.497 in comparison to the 0.587. The relative difference between the two methods is 16.6%. According to manual mass balance, 10.66 mL of water was left within the sand pack, and according to the densitometer 12.56 mL of water are left inside the sand pack, this error could be due to the fact that once velocities above ~21cc/min were used at the last two velocity intervals the materials flowing through the densitometer did not have enough residence time to be accounted for in the cell. The error is magnified in the S_w calculations because of the reduced pore volume, a larger pore volume would reduce the impact of these uncertainties. The total volume (of water) injected was 281.67cc, the total volume produced was 281.08cc. A mass balance deficit of -0.2% is obtained. The densitometer has proven to be an effective tool for mass for two-phase flow based on the data obtained.

Summary

Understanding fluid flow in porous media is challenging. Laboratory experiments at real reservoir condition are key to achieve this goal. However, the nature of experiments (pressure, temperature, fluid viscosity, multiphase flow, etc.) makes the data analysis and interpretation challenging and sometimes impossible. This study introduces the high-resolution inline density measurement at experimental condition (i.e. high pressure, high temperature) as a unique tool for understanding fluid flow in porous media. Following are some of the unique achievement of data analysis utilizing densitometer (DM):

- Besides oil recovery calculation in conventional light oil experiments, DM can significantly reduce the time, chemical usage for oil recovery calculation in tests including heavy viscous oil without compromising the accuracy.
- Valuable information in high-temperature experiments including monitoring steam condensation, steam/steam foam propagation and breakthrough time, water saturation changes, and steam front velocity calculation.
- Understanding complex fluids flow by identifying foam/no foam boundary at reservoir condition without the need for visualization.
- Monitoring and detecting nanoparticles or fine transport in porous media.

References

[1] A. Baldygin, D. S. Nobes, and S.K. Mitra. New Laboratory Core Flooding Experimental System. Industrial & Engineering Chemistry Research 2014 53 (34), 13497-13505 (2014).

[2] D. Olsen. Using a Densitometer for Quantitative Determinations of Fluid Density and Fluid Volume in Core Flooding Experiments at Reservoir Conditions. The International Symposium of the Society of Core Analysts, Vienna, Austria, 27 August – 1 September (2017).

[3] Instruction Manual, DMA HPM, Document no. C34IB06A.fm, Anton Paar GmbH, Graz, Austria (2005).

[4] A., Donath. Magnetic Imaging of Nanoparticles Flow through Porous Media. Master thesis, University of Calgary (2017).

The digital rock analysis of biogenically induced reservoir heterogeneities in Cretaceous reservoirs of Saudi Arabia

Ivan Deshenenkov[1,*], and *Camilo Polo*[2]

[1]Geological Operations Department, Saudi Aramco, Dhahran, Saudi Arabia
[2]Northern Area Reservoir Characterization Department, Saudi Aramco, Dhahran, Saudi Arabia

Abstract. The characterization of the fluid-flow properties in biogenically altered formations is a key for successful exploration campaigns. This study assessed reservoir quality and evaluated permeability of the biogenically-modified Cretaceous carbonate section in Saudi Arabia. When dealing with these bioturbated carbonates, characterization of sedimentary heterogeneities is often overprinted by the complex spatial geometries of burrows. The high-resolution 3D X-ray microscopy imaging and analysis of these bioturbated sections lead to a better understanding of the interconnectivity between permeable burrows and tight matrix. The analysis deploys multiscale imaging from whole core to sub-micron scale. The coarse imaging at 20-50 microns resolution helped to identify bioturbated sections to select samples for higher resolution tomography. The 3D sample tomograms were segmented to define burrows and matrix distributions, where samples were extracted for thin-sections, scanning electronic microscopy (SEM) and mercury injection capillary pressure (MICP) analyses to refine the pore sizes and rock types. The analysis showed an intricate, highly connected, mixed horizontal and inclined burrow system dominated by *Thalassinoides*. Intergranular porosity, associated with the fill of *Thalassinoides,* constitutes a mechanism for permeability enhancement in a tight matrix. Increased permeability is associated with higher dolomite content that might be used as a sweet spot identifier from wireline logs.

Introduction

Textural and mineralogical modifications of the original sedimentary fabric by organisms at the time of sediment deposition are known as bioturbation. Biogenically driven chemical and physical alteration of the primary sedimentary fabric due to burrowing infauna in carbonate formations can result in fabric-selective dolomitization [1]. Recent studies have demonstrated that in tight carbonate formations this type of modification can lead to the enhancement of reservoir permeability [1]. Enhancement associated with bioturbation can be expressed as dual-permeability flow-media and shares some conceptual similarities to fracture systems. Sediment modifications in the primary sedimentary fabric include burrowing, churning and continuous reworking of particles. One of the fundamental input parameters into reservoir models is permeability. Commonly, biogenically induced dual-permeability flow media exhibit poor reservoir characteristics owing that only trace fossils (i.e., the permeable conduits) are able to transmit fluids, with little to none interaction with the tighter matrix [1]. This requires detailed understanding of the geological heterogeneities to characterize their impact on petrophysical and reservoir engineering studies.

This paper documents the influences of bioturbation on reservoir quality from the Cretaceous Limestone (CL), Saudi Arabia. Permeability distributions were investigated using digital and conventional techniques to refine different rock types. We present a case study of biogenic enhancement in tight facies of a carbonate formation and provide a conceptual framework for burrow-associated permeability improvement analyzed through three-dimensional (3D) X-ray computed tomography (CT) imaging acquired at multiple scales for digital rock characterization supported by mercury injection capillary pressure (MICP) experiments and high-resolution scanning electron microscopy (SEM) imaging.

The objective of the work was to integrate the CT images, thin-section photomicrographs, MICP and SEM data to simulate permeability in high-resolution 3D images and upscale it to the plug and whole-core levels to define if permeability is linked to specific macroscale geological features of the rock.

This may only be examined through multi-scale imaging and modeling. Whole core sample analyses was essential in calculation of the reservoirs heterogeneity to assess permeability anisotropy. The comparison of permeability at different scales is

* Corresponding author: Ivan.Deshenenkov@aramco.com

necessary for upscaling laboratory-measured properties to grid cells of the full-field geological models.

1 Geological context

The CL is a tight formation (<1mD) in Saudi Arabia. To understand the causes of flow restriction, this study captures systematically different pore types, their relationship, distribution, connectivity and their impact on reservoir fluid flow behavior. It is observed that pores are not related to any depositional surface and are rather formed due to mesogenetic corrosion of highly micritized, tight carbonate rock bodies. Primary pores are almost completely destroyed during the process of shallow burial diagenesis. Separated larger pores are both fabric as well as non-fabric selective type. They are the main contributors to pore volume within an overall pervasive (micritic) matrix pore dominated system of wackestone and packstone.

1.1 Lithology

Sedimentological analysis based on core samples reveals that burrowed, tight facies range from mudstones to mud-dominated packstones and wackestones with abundant skeletal fragments and pellets.

1.2 Depositional environments

Overall, sediment deposition took place on a low-gradient carbonate platform along the eastern border of the Arabian Shield. The preponderance of fine-grained lithology along with intense bioturbation including well-developed *Thalassinoides* suggests deposition in sheltered lagoonal settings in the inner part of the platform during the Lower Cretaceous.

Several porosity generation and destruction events are recorded in these rocks. A eugenic phase of porosity enhancement took place initially with the dissolution of aragonite shells followed by compaction related porosity reduction. During prolonged subsidence of this passive margin sequence, a good part of the pores were occluded due to calcite cementation.

Pressure solutions are also recorded within argillaceous mudstone, which dissolved carbonate minerals locally and deposited them in the surrounding pores, causing further reduction in porosity. Mesogenic porosity enhancement is found as an important event for this formation. It is presumed that a phase of organic acid generation and carbonate mineral dissolution took place during the onset of hydrocarbon expulsion from deeper source beds. This process was very important for the CL as most of the porosities are developed through secondary pores.

2 Methods

2.1 Conventional core observations

Core descriptions, thin-section petrography and SEM assessments allowed the identification of an intricate, highly connected, mixed horizontal and inclined burrow system dominated by *Thalassinoides*. Selective dolomitization occurs in burrow fills and is commonly surrounded by a non-dolomitized lime mud matrix. This selective burrow dolomitization results in intergranular porosity and constitutes a mechanism for vertical permeability enhancement in an otherwise low-permeability matrix. Preferential vertical burrowing direction suggests that dwellings of bioturbating infauna constitute selective fluid-flow networks. Core pictures displaying oil-stained *Thalassinoides* burrows show that oil staining is concentrated preferentially in the fill of burrows (fig. 1, a). Thin-section photomicrographs provide information about rock texture and porosity types in carbonates that are essential for rock-type classification. Photomicrographs (fig. 1, b, c) illustrate dolomitized burrows, muddy matrix and intergranular porosity concentrated in the burrow fill due to preferential dolomitization associated to *Thalassinoides*. Burrows are mainly vertical with a minor horizontal component, and may span a depth of up to one meter below the sea floor, with typical diameters between five and ten millimeters.

Bioclastic material is partially leached and cemented by drusy calcite. This section of the sample shows good porosity with common fine moldic pores and micropores. A calcite cemented fracture separates the dolomitized and non-dolomitized parts of the sample. Microporosity is commonly observed in the bioclastic wackestone.

Fig. 1. a) Core picture and b, c) thin section photomicrography illustrating the type of biogenic enhancement of the formation.

2.2 Scanning electron microscopy

Samples containing burrows and matrix were selected from each whole core sample. Continuous magnification of SEM observations together with energy dispersive spectroscopy (EDS) show that most of the pores preserved in CL is moldic. Images (fig. 2, a-d) highlight the dominance of solution pores within the formation intervals. Dolomitization of grains selectively increases intercrystalline porosity locally, but this phenomenon is less common. Reservoir quality as a whole is poor due to poor connectivity of secondary dissolution or intercrystalline pores. Burrows have large dolomite grains. Matrix has tight micritic calcite as well as a few large dolomite grains.

The main goal of the SEM-EDS measurements was to establish the presence of dolomite grains in the micritic system. Fig. 2 (e, f) shows SEM-EDS points. The results demonstrate that grains are mainly calcite and dolomite.

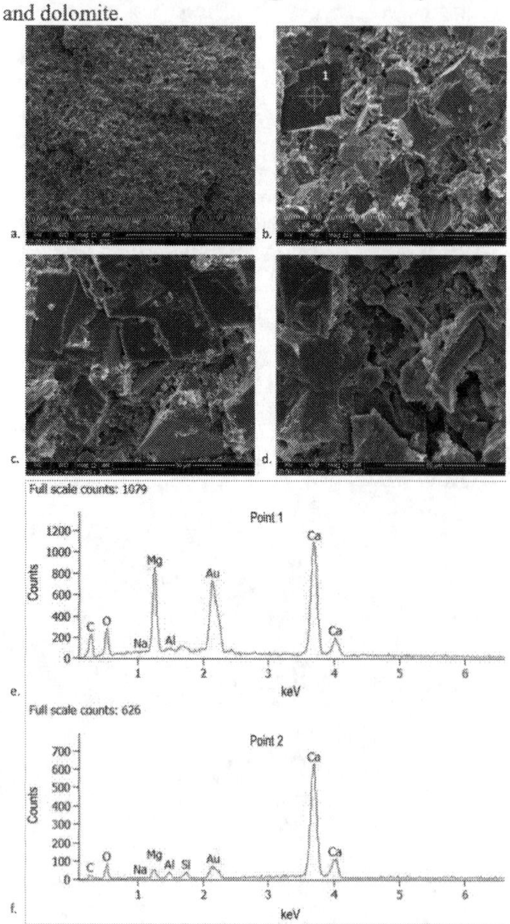

Fig. 2. a-d. Continuous magnification (150x-2600x) of the CL samples with SEM imaging combined with e. dolomite EDS spectrum (point 1 in b.), f) Calcite EDS spectrum (point 2 in image b.), Au peaks (gold) in the spectra are due to the gold sample coating.

3 Digital rock analysis of the CL bioturbated samples

The digital rock analysis workflow includes the whole core samples computed tomography analysis, burrow segmentation, identification of burrow-modified regions, and the selection of smaller samples for higher resolution work. The smaller samples were scanned at medium resolution (4 μm) to identify the matrix to burrow interface, and at high-resolution (0.5 μm) to study morphology and properties of burrows and matrix. Based on initial image analysis results, appropriate volumes were taken for MICP. Once MICP results were available, sub-samples that cover the entire porosity spectrum in the rock were selected and extracted. The next phase of the workflow involved the determination of burrow and matrix properties through high-resolution multi-scale CT imaging. In addition, a combination of high-resolution CT and SEM-EDS data were used to refine the mineralogy of the bioturbated zones. Image processing and various analyses lead to computations of porosity and permeability of the burrowed and matrix sections of the rocks. The final stage of the workflow is to integrate the properties of burrows and matrix to produce up-scaled properties. These results can also be integrated to the whole core and wireline log scale; however, this is beyond the scope of this paper.

3.1. Whole core samples imaging and analysis

The whole core samples were selected based on the sedimentological descriptions of 200 ft of the whole core and 1,000 ft of well logging data. Sampling locations were identified to represent variations in lithofacies, texture and porosity type. In total, thirty 25 cm long whole core samples were taken for low-resolution (25 μm) scanning. Examples of the whole core samples scans are illustrated in fig. 3.

The grayscale image of the whole core samples (fig. 3) shows that it is composed of two distinct lithology regions: matrix and longitudal circular shaped *Thalassinoides* burrows.

Based on the low-resolution scanning three type of sub-samples have been selected for the following studies: advanced high-resolution computed tomography (16 samples), MICP measurements (8 samples) and conventional core analysis to define total porosity and absolute permeability (37 samples). Visual examination of all samples was carried out to ensure the plugs were physically suitable for the measurements and free of visible fractures or anomalies.

Sub-samples were cut from representative regions in the core that preserved changes in the geological features seen in the whole core CT images to provide integrated characterization among available data. This is an effective way of understanding rock structure at multiple scales, which would lead to proper evaluation

 https://doi.org/10.1051/e3sconf/202014601004

of the impact of rock quality and geological heterogeneity on the permeability of the reservoir rocks.

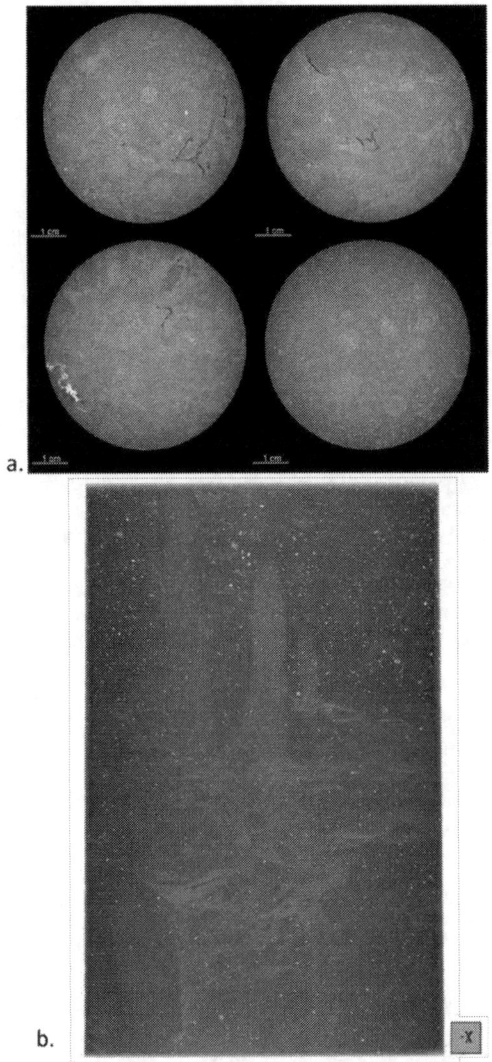

Fig. 3. Whole core samples tomography: a) Z-slices and b) 3D render illustrating borrows through the whole core sample.

3.2 Mercury injection capillary pressure data analysis

The laboratory-measured MICP samples are carefully cut from the whole core to represent each phases (burrow and matrix) in the sample (usually 1 inch diameter plug).

The MICP data indicate the prevalence of micropores in matrix (fig. 4, a) and mesopores in bioturbated regions (fig. 4, b) of the formation. MICP data show a narrow, unimodal distribution of the pore-throat radii, highlighting the development of mesopores in burrows, with the moderate entry pressure and gentle saturation gradient indicating fluid flow via mesopores. MICP data observations on the matrix sections indicate that pore-throat size distribution is bimodal, representing two types of micropores: 0.03 μm and 0.3 μm in diameter.

The difference in the pore-throat diameters is interpreted as the cause of higher permeability in the burrowed sections.

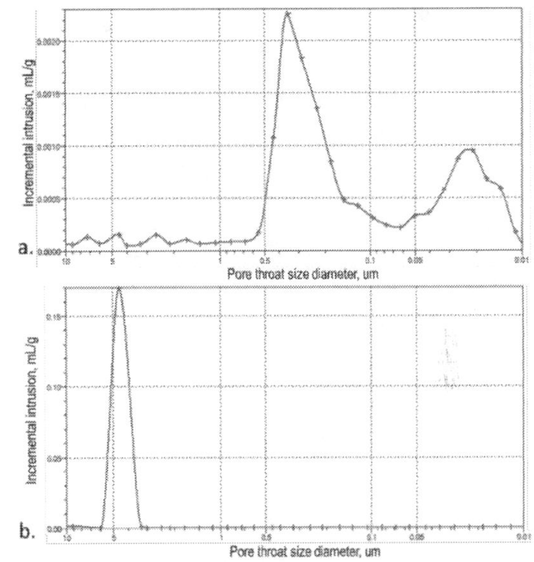

Fig. 4. Pore-throat size diameters distribution in a) matrix and b) burrow sections of the rock.

Absolute permeability variation associated with difference in the pore-throat size diameters has been investigated with conventional core analysis (CCAL) of the plugs.

3.3 Conventional core analysis of plugs

Plug samples (1.5 inch diameter) were taken from each lithology to represent the statistical property variations along the entire core length. The samples were cleaned in hot solvent extractors and were measured for helium porosity and gas permeability in standard equipment at ambient conditions using a calibrated steady-state permeameter with nitrogen gas as flowing medium.

CCAL demonstrates clear separation between bioturbated and matrix intervals on porosity-permeability graph (fig. 5). Matrix is tighter and exhibits a smaller range of porosities in comparison to the burrows porosity range. The porosity in this burrowed phase is the main contributor to the flow. On average, total porosity of these tight bioturbated plugs exceeds matrix porosity by 5-7% and permeability difference consists 1-2.5 orders of magnitude.

CCAL results underscore the importance of the bioturbated sections analyses; however, it does not provide information on permeability (k) anisotropy in vertical (k_v) and horizontal (k_h) directions. Another important task was to assess the impact of the burrow to matrix volume relationship with bulk permeability of the sample. High-resolution digital rock analysis was utilized to answer these questions.

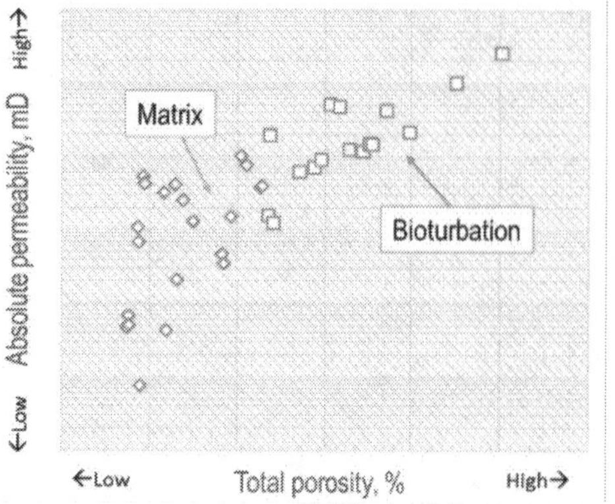

Fig. 5. Porosity-permeability cross-plot derived from CCAL measurements of the full (1.5 inch) plugs with embedded classification of bioturbated (orange) and matrix (blue) samples.

3.4 High-resolution digital rock analysis

Multi-scale digital rock analysis is conceptually well-suited for bioturbated rocks analysis and modeling [2] as it allows varying resolution and provides a systematic procedure for coarsening and refinement. Darcy's model is to approximate pressure and fluxes on a coarse grid in large-scale discontinuities, whereas fine-scale effects are captured through basis functions computed numerically by solving local Stokes-Brinkman flow problems on the underlying fine-scale geocellular grid. The Stokes-Brinkman equations give a unified approach to simulating free-flow and porous regions using a single system of equations, they avoid explicit interface modeling, and reduce to Darcy or Stokes flow in certain parameter limits [3]. The modeling demonstrates how fine-scale flow in bioturbated pore networks can be represented within a coarse-scale Darcy-flow model in fractures of the burrow to matrix interface and multi-scale elements. The flow computed solving the Stokes-Brinkman equations in burrow and matrix. This method is a promising path toward direct simulation of highly detailed rock models in bioturbated formations. The high-resolution CT imaging was conducted to build pore structures for the flow analysis.

3.4.1 Imaging and segmentation of full-plugs

To perform high-resolution CT imaging, 1.5 inch samples that include a combination of burrow and matrix were extracted and scanned at 4 µm resolution. Orthogonal slices of burrow and matrix at this resolution are shown in fig. 6. It illustrates that small pores are not present at this resolution (fig. 7) within the matrix. The throat sizes distribution derived from MICP data (see fig. 4) shows that for determining the flow properties of these burrow-modified cores, the required resolution is at least 5 µm for burrow-through-flow and below 1 µm for the matrix flow unit. The higher resolutions scans are typically acquired to understand the fabric of the rock, and to define large-scale geologic objects in the samples. Therefore, a range of resolutions is needed to fully describe the plugs.

Fig. 6. Example of the burrowed structure, full-plug scale, 4 µm resolution: 3D render and three orthogonal projections.

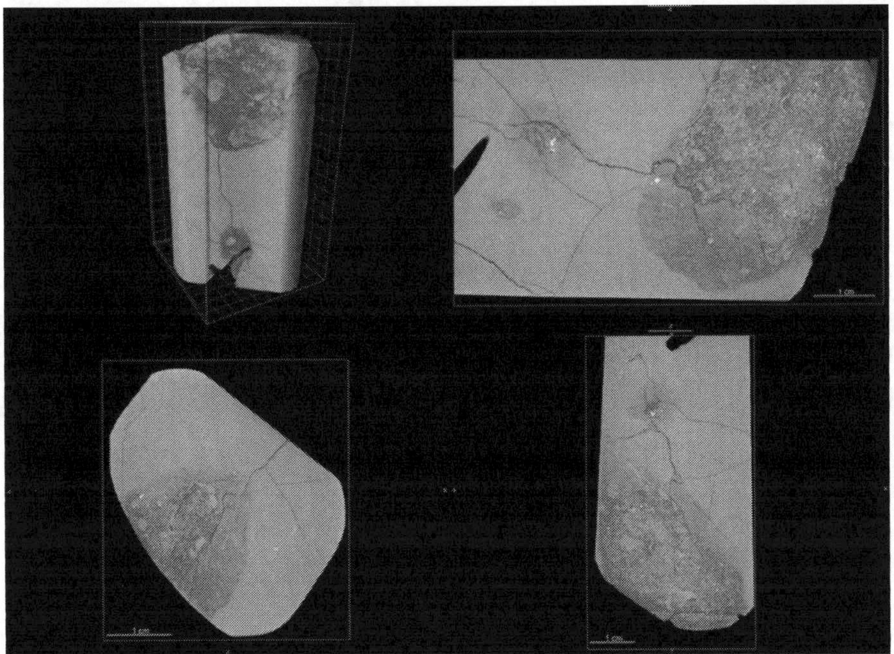

Fig. 7. Example of the porosity segmentation in the burrowed structure, full-plug scale, 4 μm resolution: 3D render and three orthogonal projections.

3.4.2 Imaging and segmentation of sub-plugs

The analysis of the burrow flow unit requires drilling of sub-plugs in the matrix to burrow interface locations and scanning them at 1 μm resolution; smaller separate burrow volumes were analyzed at maximum 0.5 μm that revealed a connected pore system in the burrow phase. Figure 8 shows a cross section CT slice for the burrow to matrix flow unit. The digital rock analysis confirmed the SEM data. The large dolomitic grains can be seen predominantly in the burrow phase, and the space between them is filled with micritic calcite (fig. 9). Based on the MICP data, the matrix flow unit would require imagining by focused ion beam SEM beyond the resolution of CT images. Therefore, permeability of matrix for pore-scale modeling was assigned based on the minimum CCAL of samples with matrix only. The modeling results will represent the pessimistic estimate of the whole system permeability.

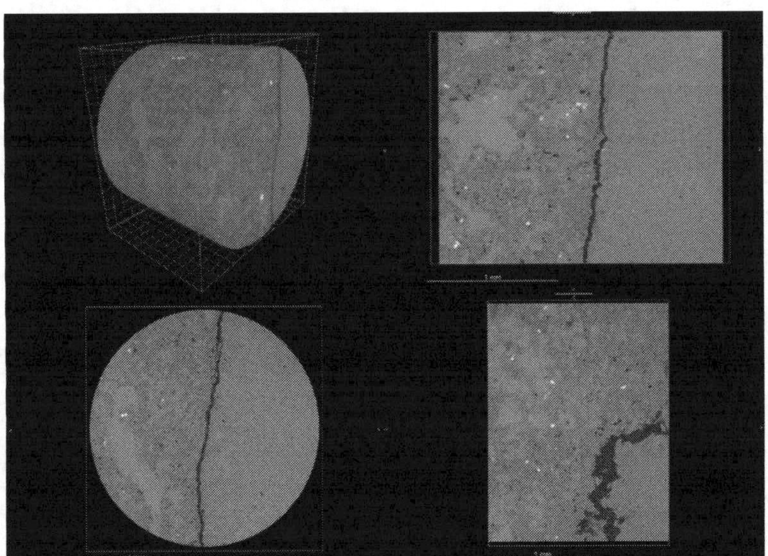

Fig. 8. Example of the high-resolution tomography in the matrix to burrow interface, sub-plug scale, 1 μm resolution: 3D render and three orthogonal projections.

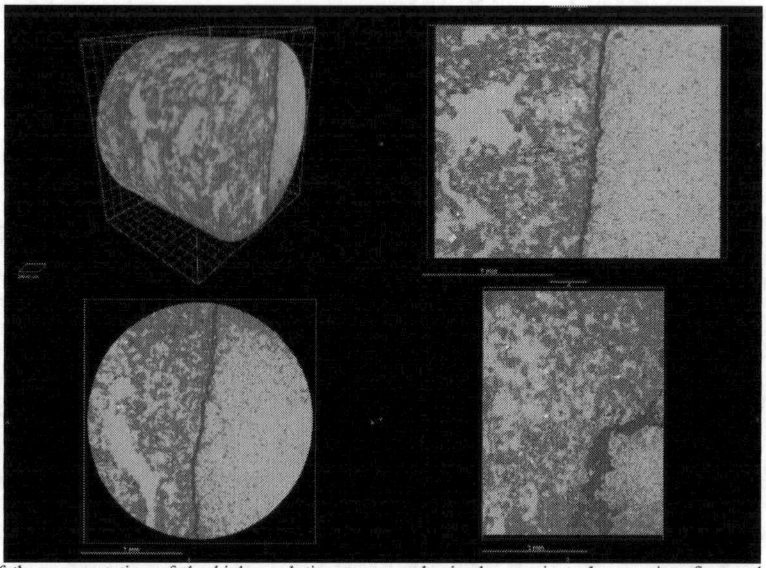

Fig. 9. Example of the segmentation of the high-resolution tomography in the matrix to burrow interface, sub-plug scale, 1 µm resolution: 3D render and three orthogonal projections. Color key: red – porosity, green – dolomite, cyan – calcite.

3.4.3 Numerical modeling of the flow properties

Numerical modeling of permeability in burrows was based on the highest magnification tomography at 0.5 µm resolution. 3D images were segmented to allow porosity and permeability computation at subsample scales. Two main phases have been identified: pores and grains.

Incompressible flow in a porous rock matrix typically obeys Darcy's law and is described by a first-order elliptic system in which Darcy's law was combined with a mass-conservation equation to relate the pressure and the total (interstitial) velocity. Incompressible flow in open domains, on the other hand, obeys the Stokes equations [3, 4]. The Stokes-Brinkman equations combine Darcy and Stokes models into a single equation. This model gives a unified approach to model flow in both the fractures and the intergranular porous subdomains using a single system of equations. In the free-flow (or fluid) domain, it is assumed that permeability tends to infinity and sets the effective viscosity equal to the fluid viscosity, otherwise it transforms into the coupled Darcy-Stokes equations, which reintroduces the requirement for interface conditions and computational intractability.

The original structure, porosity segmentation, velocity field and illuminated streamlines are illustrated in fig. 10. The porosities of burrows are in the range of 14 to 19 pu, while the permeability is in the range of 5.25 to 2.55 mD for the studied samples that compares well with literature sources [5].

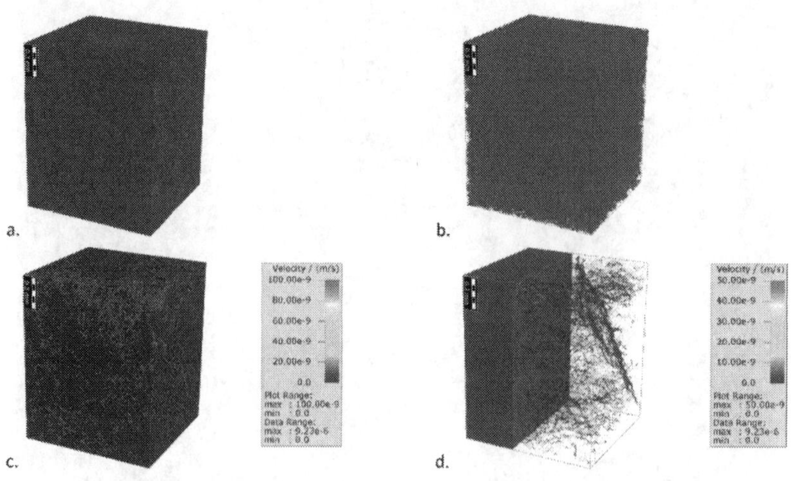

Fig. 10. Flow modeling in the burrow phase: a) structure, b) porosity, c) velocity field and d) structure and illuminated streamlines.

3.4.4 Modeling of the fluid flow through composite rocks

Flow-based upscaling procedure based on Stokes-Brinkman equation computes effective permeability on coarser grids. To build the coarser grid, it is required to estimate permeability of the unresolved porous materials within the grid. Three sub-samples were selected to characterize the burrow and matrix system under sub-micron CT imaging (fig. 11): a) burrow phase, b) burrow to matrix interface and c) matrix.

As discussed before, permeability of 0.01 mD was pessimistically assigned to matrix based on CCAL results of the typical matrix samples. Burrow permeability was estimated with Stokes-Brinkman equation in 2.4.3.

The extracted properties of burrow and matrix then were combined into the larger scale network to evaluate permeability and interconnectivity. Two composite models were tested: 1) direct communication between burrows and matrix, and 2) inter-burrow flow through the fracture interface. In both cases, the fluid-flow was modeled with Stokes-Brinkman numerical solver under no-slip boundary conditions.

Unresolved porous burrow and tight matrix fabrics have been propagated in 3D volumes through utilization of the machine learning algorithm based on the Extra Trees engine [6] complemented with resolved porosity. This classifier implements a meta estimator that fits a number of randomized decision trees on sub-samples of the burrow and matrix grayscale dataset and uses averaging to improve the predictive accuracy and control over-fitting.

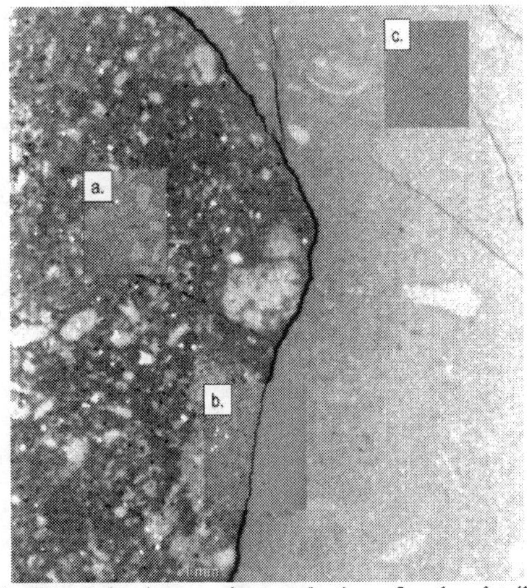

Fig. 10. Sub-micron volume selections for the detailed analysis: a) burrow, b) burrow to matrix interface and c) matrix.

Examples of the original grayscale image and classification results are shown on fig. 11 and fig. 12.

The results of the modeling illustrate permeability enhancement and porosity-permeability relations in the bioturbated system in three orthogonal directions. The interaction of the burrow system with the fracture network contributes to the bulk enhancement of flow between dead-end burrows and improves permeability of the system by 1-2 orders of magnitude in comparison with the case of direct communication between burrow and matrix (fig. 13).

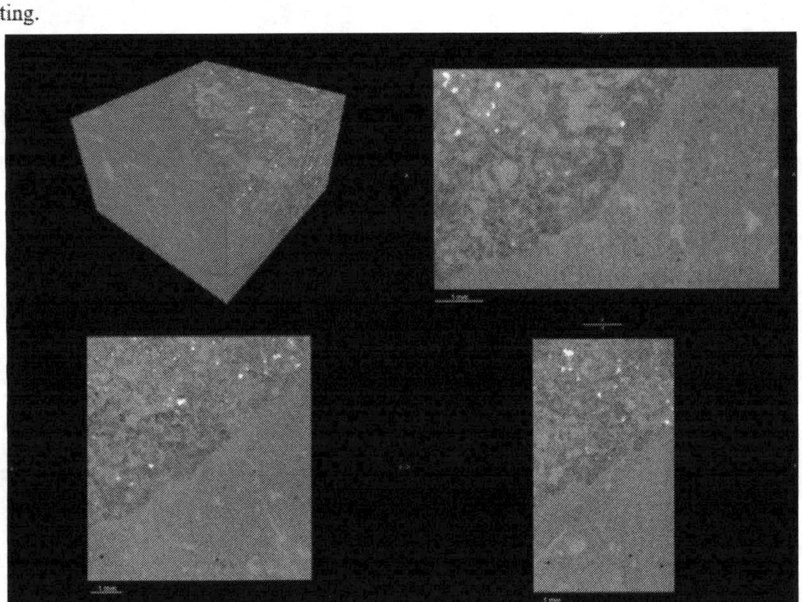

Fig. 11. Example of the grayscale volume used for the machine learning classification, 0.5 μm resolution: 3D render and three orthogonal projections.

https://doi.org/10.1051/e3sconf/202014601004

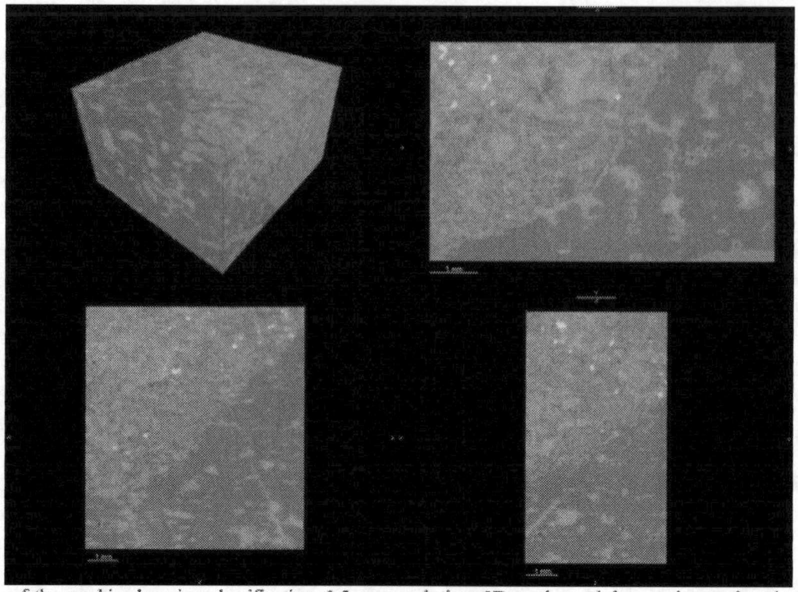

Fig. 12. Example of the machine learning classification, 0.5 μm resolution: 3D render and three orthogonal projections. Color key: green – high-density minerals, purple – the matrix phase and brown – the burrow phase.

Fig. 13. Velocity field in composite models for both scenarios. Warmer colors show better permeability areas.

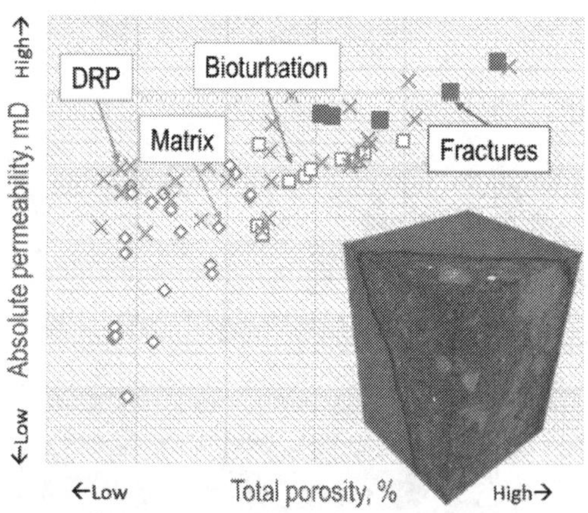

Fig. 14. Porosity-permeability cross-plot derived from CCAL measurements of the full (1.5 inch) plugs compared to digital rock analysis data (gray crosses). Embedded classification: fractured (red), bioturbated (orange) and matrix (blue) samples.

CCAL confirms enhancement of the system permeability by the fractures in the burrow to matrix interface. Highest permeability CCAL samples have been scanned and demonstrate various degrees of the fracturing. The porosity-permeability graph in fig. 14 shows five fractured samples together with the 3D render of one of them.

Comparison of the CCAL data against digital rock analysis results (see fig. 14) shows general match between two datasets with commonly underestimated porosity of matrix samples due to limitations of the scanning resolution not showing the microporosity smaller than 0.5 μm.

3.4.5 Relation between burrow volume fraction and permeability

The developed digital rock model was used to simulate the permeability ratios K_v/K_h as a function of burrow and dolomite volume fractions shown on fig. 15. The volume fractions of burrows vary from 0.2 to 0.75 and volume fractions of dolomite vary from 0.15 to 0.65 that is comparable with [7]. Increase of both burrow and dolomite fractions content is leading to higher K_v/K_h ratios.

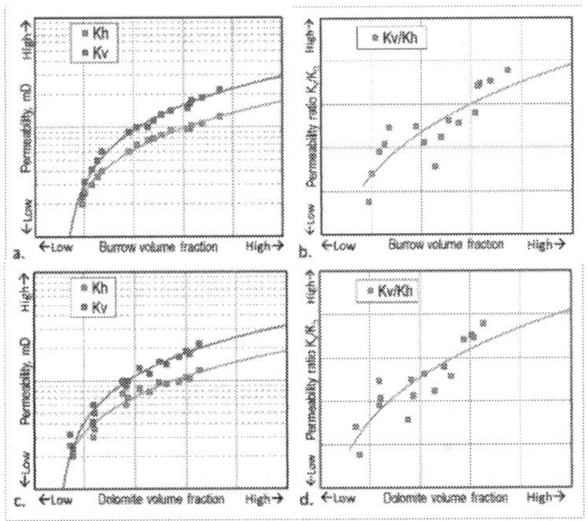

Fig. 15. Permeability as a function of a) burrow and c) dolomite volume fractions and permeability ratio K_v/K_h as a function of b) burrow and d) dolomite volume fractions.

These relations (fig. 15) can be used for the permeability anisotropy prediction based on dolomite content derived from core or well log data analysis to improve permeability simulation in the grid of geological models.

Conclusions

Multiscale digital rock analysis was completed on whole core samples, full plugs and high-resolution sub-plugs to investigate the geological and petrophysical properties of the bioturbated CL at varying resolution from 25 µm to 0.5 µm. Integrated analysis of the CT images, thin-sections, CCAL and MICP data resolved permeability of burrow and matrix as well as permeability ratio in vertical and horizontal directions.

In the case study presented herein, high burrow connectivity suggests that a continuous flow network is likely to be established. The analysis shows, this is primarily influenced by the: 1) concentration and preferential orientation of permeable trace fossils; 2) degree of permeability contrast between burrows and surrounding matrix; and 3) three-dimensional connectivity of the *Thalassinoides* network system. Here, *Thalassinoides* occupy a significant volume of the rock compared to the fractures and this fact enhances the permeability and transmissivity of otherwise low-permeability matrix.

Detailed thin-section descriptions and MICP experiments improved the understanding of the geological variations in the 3D rock volume. This analysis has led to an improved geological interpretation of the diagenetically altered carbonate rocks. The overall analysis offered improved understanding of the heterogeneity in the formation and its impact on multi-scale permeability measurements.

There is at least one order of magnitude difference in permeability of burrow and matrix. Permeability of matrix appears isotropic. Permeability of the burrow fraction is one to three orders of magnitude larger than the matrix permeability; the results show enhanced burrow permeability with higher dolomite fractions. The flow between dead-end burrows occurs through the matrix or fracture interface as presented by the composite models. The K_v/K_h ratio as a function of burrow and dolomite volume fractions are identified. The ratio of K_v/K_h is always above one for bioturbated samples.

Increased permeability associated with the higher dolomite content resulting from intensely bioturbated intervals might be used as sweet spots identifier from wireline logs. Recognition of bioturbated intervals and the role they play in reservoir petrophysics may aid in both the optimization of exploration and field development campaigns in similar biogenically modified intervals in the study area.

The authors acknowledge the management of the Saudi Arabian Oil Company (Saudi Aramco) for permission to publish this paper. The authors thank Jose Estevez Gonzalez and Keith MacPherson (Saudi Aramco) for their review and comments on the paper. The authors appreciate Aqeel Furaish, Hussain Alhilal and Ali Zahrani (Saudi Aramco) for their assistance with data acquisition.

References

1. M.K. Gingras, G. Baniak, J. Gordon, J. Hovikoski, K.O. Konhauser, A. La Croix, R. Lemiski, C. Mendoza, S.G. Pemberton, C. Polo, J.P. Zonneveld. Chapter 27: Porosity and Permeability in Bioturbated Sediments. *Developments in Sedimentology,* **vol.64**, pp.837-868. Amsterdam, Elsevier (2012).

2. M. Dernaika, M. Al Mansoori, M. Singh, T. Al Dayyani, Z. Kalam, R. Bhakta, S. Koronfol, Y.N. Uddin. Digital and Conventional Techniques to Study Permeability Heterogeneity in Complex Carbonate Rocks. "Petrophysics", **vol.59(3)**, pp.373-396 (2018).

3. P. Popov, L. Bi, Y. Efendiev, R.E. Ewing, G. Qin, J. Li, Y. Ren. Multiphysics and Multiscale Methods for Modeling Fluid Flow Through Naturally Fractured Vuggy Carbonate Reservoirs. *SPE 105378*. SPE Middle East Oil and Gas Show and Conference, Bahrain (2007).

4. H.C. Brinkman. A Calculation of the Viscous Force Exerted by a Flowing Fluid on a Dense Swarm of Particles. *Flow, Turbulence and Combustion*, **vol.1(1)**, p.27 (1949).

5. M.K. Gingras, S.G. Pemberton, M. Smith. Bioturbation: Reworking of Sediments for Better or Worse. "Oilfield Review", **vol.26(4)**, pp.46-58 (2015).

6. P. Geurts, D. Ernst, L. Wehenkel. Extremely Randomized Trees. "Machine Learning", **vol.63(1)**, pp.3-42 (2006).

7. Y.N. Uddin, E.A. Clerke, B. Bammel, S. Koronfol, A. Grader. Quantitative Study of Formation Vertical Permeability in Middle East *Thalassinoides* Burrowed Carbonate Mudstone. *SPE-186001-MS*. SPE Reservoir Characterization and Simulation Conference and Exhibition, Abu Dhabi, UAE (2017).

 https://doi.org/10.1051/e3sconf/202014603007

Novel technique to measure mutual bulk fluid diffusion using NMR 1-D gradient

Son Dang[*], *Carl Sondergeld*, and *Chandra Rai*

Mewbourne School of Petroleum and Geological Engineering, University of Oklahoma, USA

Abstract. Many modelling and theoretical studies have shown that diffusion can be a significant transport mechanism in low-permeability porous media. Understanding the process allows engineers to better predict reservoir performance during both primary production and enhanced recovery in unconventional reservoirs. Direct measurement of effective diffusion in tight rocks is difficult, due to small pore volumes and the lack of techniques to actually monitor the process. Conventional diffusion measurements generally require fluid sampling, which induces a pressure transient which changes the mass transfer mechanism. Previously, we introduced a novel technique to measure tortuosity in nano-porous media by simultaneously monitoring methane versus nitrogen concentrations at high pressure using transmission Infrared Spectroscopy (IR). To complete the estimation of effective diffusion, bulk fluid diffusion coefficient also needs to be measured. In this study, we demonstrate the usage of Nuclear Magnetic Resonance (NMR) 1-D imaging to examine the dynamic change of Hydrogen Index (HI) across the interface between two bulk fluids. The experiment was conducted between a crude oil sample and methane; fluid samples were pressurized within an NMR transparent ZrO_2 pressure cell which operates at pressures up to 10,000 psi. The Hydrogen Index (HI) profile was continuously measured and recorded for 7 days. The results provided oil the swelling factor and the concentration profile as a function of both time and distance. These data then were fitted with Maxwell-Stefan equation to precisely back calculate the diffusion coefficient between oil and gas samples at high pressure. Accurate estimation of tortuosity and fluid diffusion is critical for the gas injection strategy in a shale formation. Greater tortuosity and smaller fluid diffusion rate lead to longer injection and production times for desirable economic recovery.

1 Introduction

Recent studies, including simulation works (Li et al., 2018[1]), experimentation (Li et al., 2019[2]; Dang et al., 2019[3]), and production modelling (Cronin et al., 2018[4]), suggest that matrix diffusion is a major mass transport mechanism, along with advection. Advection of a flow in porous media is governed by fluid properties (such as viscosity, density, and compressibility) and matrix permeability. Diffusion is governed by fluid diffusivity (either free diffusivity or multi-component diffusivity) and porous media' tortuosity. The Sherwood number, commonly cited in surface science and catalyst studies, is used a dimensionless factor accounting to the relative contribution between diffusion and advection in overall mass transport (Coutelieris et al., 2002[5]). With high porosity media, such as conventional rocks, the impact of advection overpowers the role of diffusion. While in tight rocks, with matrix permeability in the order of nano-Darcy, the impact of diffusion is not negligible.

With the development of unconventional shale gas and oil, the need to reevaluate these transport mechanisms, especially in nano-porous media, is essential. This does not only apply for primary production, but also for Enhanced Oil Recovery (EOR) processes. The understanding of behaviors of light gas molecules diffusing into rock matrix with included reservoir fluids, is important to optimize the efficiency of gas injection. Tortuosity measurement in nano-porous media is limited. Recently, Fleury and Brosse, 2017[6] and Dang et al., 2018[7] proposed different techniques to estimate tight matrix's tortuosity. The results suggested for rock samples with porosity, less than 7%, that tortuosity can vary from 4 to 25, a considerable range. In contrast, tortuosity values for conventional rocks, with porosity greater than 10%, are around 2 (Iversen and Jorgensen, 1993[8])

Besides matrix tortuosity, bulk fluid diffusivity also controls effective diffusion. The molecular diffusivity in the gas phase can be computed with Sigmund, 1976[9] correlation. In term of experimentation, this parameter can be estimated via several methods, including monitoring pressure profile while the oil phase contacts the gas phase inside a closed cell (Guo et al., 2009[10]). The drawback of this technique is that pressure is not maintained constant throughout the measurement. Another technique is monitoring the oil phase swelling while injecting gases at a constant pressure (Jamialahmadi et al., 2016[11]). The swelling data is acquired by tracking the change in elevation of oil-gas interface. However, with the pressure above first contact minimum miscibility pressure (MMP), the interface becomes so vague to defy precise monitoring. In this study, we propose a new method using NMR 1-D gradient to monitor the dynamic change between oil-gas interfaces, from which bulk fluid diffusivity can be determined.

[*] Corresponding author: dangthaison@ou.edu

2 Experimental setup and fluid samples

For pressurization experiments, we used a Daedalus® cell, made of NMR transparent ZrO_2; the cell can be operated up to 10,000 psi internal pressure. The oil phase was injected into the pressure cell via a downstream port, while the gas phase would be later injected into the cell through the upstream port. The cell was then positioned inside the NMR spectrometer, in which the oil-gas interface would be aligned in the middle of the gradient scanning window. From its inlet, the cell was connected to a syringe pump system, which was used to compress gases from supply cylinders, then inject gasses into the test cell at a test pressure. The pressure was maintained constant throughout the diffusion process. **Fig. 1** illustrates major components of the experimental setup.

NMR gradient profiles were acquired using Oxford 2 MHz GeoSpec™ spectrometers, and Green Imaging acquisition and processing software. The magnet temperature was set at 35°C throughout the experiments. The gradient scanning window was set at 7 cm, using DNK sequence. A new scan was repeated every 1 hour. Dynamic change of hydrogen index (HI) profiles across the oil-gas interface during the diffusion process, allows us to extract bulk diffusion parameters. In this study, the fluid samples included a dead oil from Meramec formation, and the injection gas is methane. **Fig. 2** represents a HI profile at the beginning of a 7-days experiment with methane injection pressure of 6000 psi, which is above minimum miscibility pressure (MMP).

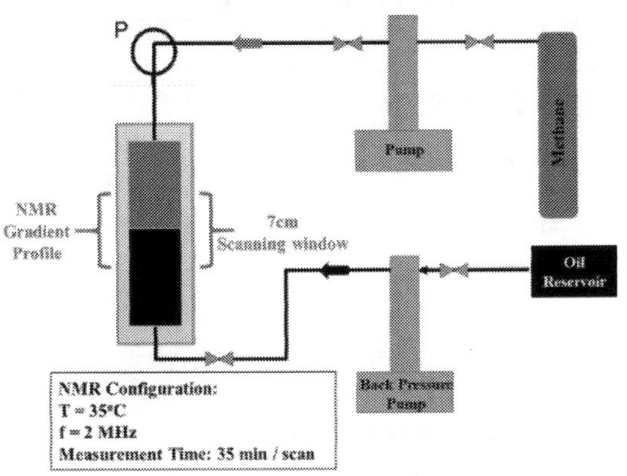

Fig. 1. Experimental setup, including NMR transparent ZrO_2 pressure cell, placed within 2 MHz NMR spectrometer. The oil phase was injected from a downstream port; the gas phase was injected through an upstream port, with the pressure controlled by a computer controlled Teledyne ISCO syringe pump system. The cell was positioned inside the spectrometer, in which the oil-gas interface was located in the middle of the gradient window.

3 Experimental results

With the contrast in HI between the gas and the oil phases, we can monitor the change in elevation of the interface (**Fig. 2**). During the diffusion process, in which pressure is maintained constant, methane molecules would diffuse into the oil phase at certain rate. This phenomenon dynamically changes the HI profile of the oil phase as a function of time. **Fig. 3** shows HI profile of the oil phase from the beginning of the experiment up to 7 days. HI at any point within the oil phase decreases over time, but the reduction rate is different depend of the relative position, with respect to the oil-gas interface. **Fig. 4** shows the effect of methane diffusion on HI trends at four different positions, note the marker colors correspond with the position, labeled in **Fig. 3**.

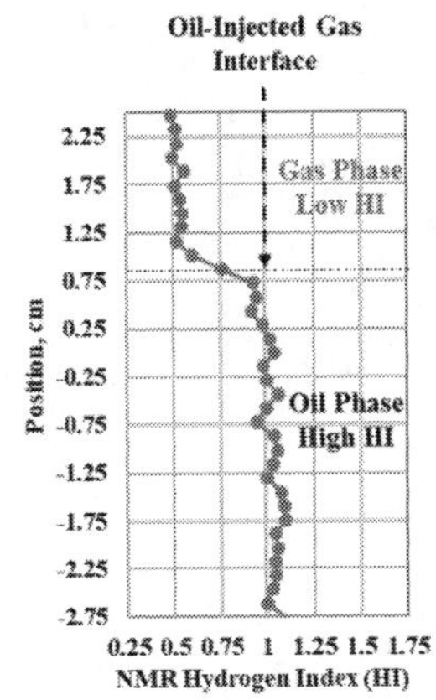

Fig. 2. An HI profile across the oil-gas interface. The position values are based on the relative position of the test cell in the NMR scanning window. The contrast in HI between the gas (low value) and the oil (high value) phases, allows to determine the dynamic position of the oil-gas interface during the experiment.

As expected, at the positions closest to the interface, HI decreased at the fastest rate, and quickly approached a constant value, HI_{final}. Moving further away from the interface, HI decreased with slower slope, but eventually reached the same HI_{final}. HI_{final} is the result of the mixing between original oil's molecules and injection gas' molecules (methane in this case) at a particular concentration. From the experimental results, this specific concentration can be regarded as the maximum concentration of methane that can be diffused into the oil body; this parameter is a function of pressure and temperature. Using each of these HI trends, the methane diffusion coefficient can be calculated. However, with the fluctuation in the data, due to NMR signal-to-noise, the integral of HI profile is used to better estimate the diffusivity.

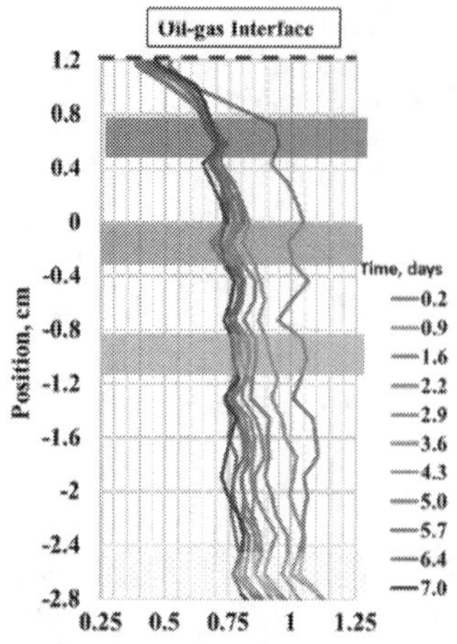

Fig. 3. HI profile within the oil phase as a function of time. The decreasing of the HI value within the oil phase is due to the methane diffusion. HI temporal profiles at different positions (color labels) within the oil phase are plotted in **Fig. 4**.

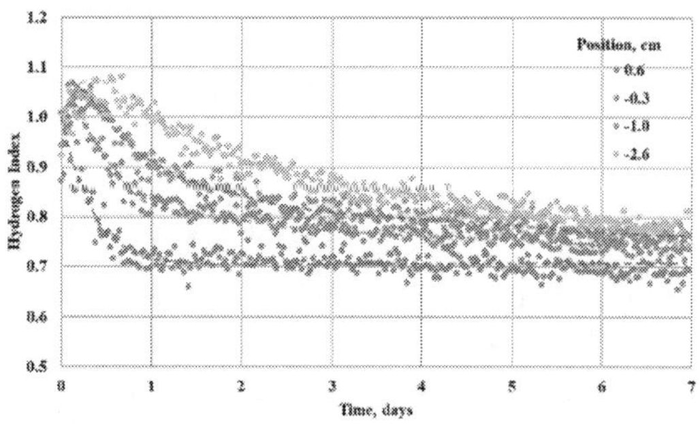

Fig. 4. HI profiles at different positions within the oil phase. The marker colors correspond to highlighted depths in **Fig. 3**. At the position closest to the oil-gas interface, HI decreases with the faster rate, and eventually approach a constant value, HI_{final}. HI_{final} corresponds to the maximum concentration of diffused methane into the oil phase.

At a particular position and time, the HI value is the molar averaging between the HI value of the original oil and the HI value of methane at 6000 psi (**Eq. 1**). With the known HI value of the original oil is 1, and known HI value of methane at 6000 psi is 0.525 (both values can be extracted from the HI profile at the beginning, i.e. at t=0), relative methane concentration can be calculated, then plotted as a function of time. **Fig. 5** shows the integral of HI profile (red) and calculated methane concentration in the oil phase (blue.)

$$HI_{oil\ phase}(t) = (1-\alpha)HI_{original\ oil} + \alpha HI_{methane\ @\ 6000psi}$$

$$\alpha(t) = \frac{c_{methane}(t)}{c_{methane}(t) + c_{oil}(t)} \tag{1}$$

$$HI_{original\ oil} = 1$$

$$HI_{methane\ @\ 6000psi} = 0.525$$

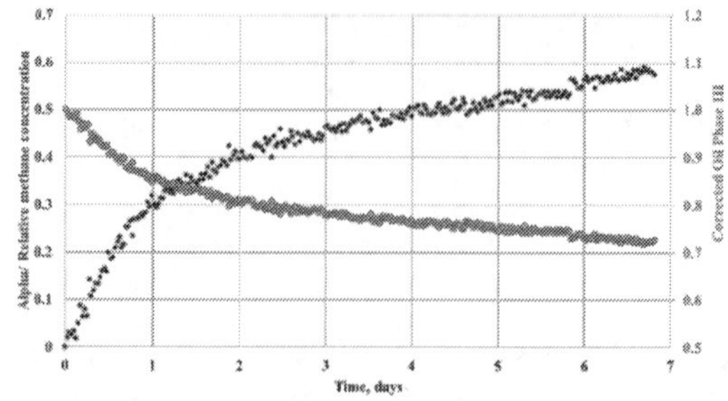

Fig. 5. The integral of HI respective to position, plotted as a function of time (red). Calculated relative methane concentration is also plotted (blue). The data is used to estimate methane diffusion coefficient.

4 Discussions

Fig. 4 show HI trends at different positions within the oil phase. Depending on the relative position to the interface, the HI reduction rate would be different; however, they all approach a similar value of HI_{final}. In the other words, for a specific reservoir fluids and injection gasses, at a particular P-T condition, there is a maximum concentration of gas that can be introduced into the oil phase (Whitman, 1923[12]). While modelling the dual diffusion of injection gases into reservoir fluids and vice versa, this maximum concentration should be considered as the boundary condition; and now, we can estimate it from laboratory measurements.

Using Fick's second law (**Eq. 2**), diffusion coefficient can be estimated from the relative methane concentration profile. Note instead of using methane profile at a single position within the oil phase, we used the integral to reduce the impact of data oscillations. C_s is methane concentration at the oil-gas interface. This is usually derived from the late-time diffusion data; however, with this study, C_s can be directly calculated from HI_{final}.

$$\frac{C(x,t)}{C_s} = 1 - \text{erf}(z)$$

$$z = \frac{x}{2\sqrt{DT}} \tag{2}$$

Fig. 6 shows the fitting results of different diffusion coefficients using the relative methane concentration profile. The whole profile throughout 7 days can be fitted with diffusion coefficients ranging from 6.5×10^{-10} to 8.5×10^{-10} m^2/s; however, it is clear that the diffusion rate decreasing as a function of time. This is considered as experimental artifact. While Fick's Law was solved for infinite boundary condition, our test cell has limited volume. As soon as the first gas molecule travelling toward the oil phase, and approaches the end of the cell, the diffusion rate would be reduced.

By reviewing literature on diffusion measurements (Renner, 1986[13]; Grogan et al., 1988[14]; Jamialahmadi et al., 2006[11]), bulk diffusion rate can be varied within 2 orders of magnitude (10^{-10} – 10^{-8} m^2/s). Translating this to field EOR applications, to

 https://doi.org/10.1051/e3sconf/202014603007

efficiently inject a same reservoir volume, the injection time can be also varied within 2 orders of magnitude.

5 Conclusions

Knowing the importance of diffusion as one of major drive mechanisms in tight rocks, our studies focus on defining and estimating key parameters, allows engineers to model the process. They include porous matrix tortuosity and bulk fluid diffusivity. For example, in gas injection EOR, the combination of these two factors governs how fast injection gas molecules travel into the porous matrix to interact with a reservoir fluids' mobility, hence injection and production strategy can be optimized. In this paper, a new method to estimate fluid diffusion coefficient between methane and a crude sample was presented using NMR 1-D gradient. The technique directly captures the dynamic change of methane concentration within the oil body, reflected through the change in the HI value.

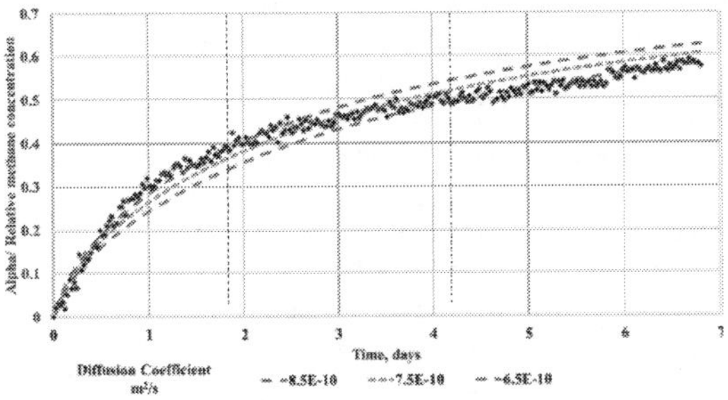

Fig. 6. The fitting of diffusion coefficient using the relative methane concentration profile. The diffusion rate can be precisely fitted within 25% of error, despite the experimental artifact, which makes diffusion rate decreases when gas molecules approach the end of the test cell.

This work was funded by the University of Oklahoma Unconventional Shale Consortium. We have benefitted from valuable conversations with Jairo Corredor, John Mansoori, Tom Smagala and Richard Newhart. The laboratory support of Gary Stowe, Micaela Langenvin and Jeremy Jernigen is greatly appreciated. We thank the members of Integrated Core Characterization Center (IC^3) for helping us during experimentation.

References

1. L. Li, Y. Su, J. J. Sheng, "Investigation of gas penetration depth during gas huff-n-puff EOR process in unconventional oil reservoirs", Society of Petroleum Engineers, 2018
2. L. Li, Y. Su, J. J. Sheng, Y. Hao, W. Wang, Y. Ly, Q. Zhao, H. Wang, "Experimental and numerical study on CO_2 sweep volume during CO_2 huff-n-puff enhanced oil recovery process in shale oil reservoirs", Energy Fuels, 2019
3. S. T. Dang, C. H. Sondergeld, C. S. Rai, "Effects of temperature and gas pressurization on the interpretation of nuclear magnetic resonance (NMR) hydrocarbon measurements in organic rich shales", URTeC, 2019
4. M. Cronin, H. Emami-Meybodi, R. T. Johhns, "Diffusion-dominated proxy model for solvent injection in ultra-tight oil reservoirs", Society of Petroleum Engineers, 2018
5. F. A. Coutelieris, M. E. Kainourgiakis, A. K. Stubos, "The effects of Peclet on the Sherwood number in high porosity granular media", Studies in Surface Science and Catalysis, **Vol 144**, p. 753-760, 2002
6. M. Fleury, E. Brosse, "Transport in tight rocks", AGU Hydrobiogeochemical Properties of Caprock, June, 2017
7. S. T. Dang, C. H. Sondergeld, C. S. Rai, A, O. Tinni, N. Drenzek, 2018, "A first step in evaluating the role of diffusion in EOR in tight shale formations", SCA, 2018
8. N. Iversen, B. B. Jorgensen, "Diffusion coefficients of sulfate and methane in marine sediments: influence of porosity", Geochimica et Cosmochimica Acta **57**, 571–578, 1993
9. P. M. Sigmund, "Prediction of molecular diffusion at reservoir conditions, Part 1-measurement and prediction of binary dense gas diffusion coefficients", J. Can. Pet. Technol., **15**, pp. 48-57, 1976
10. P. Gou, Z. Wang, P. Shen, J. Du, "Molecular diffusion coefficients of the multicomponent gas-crude oil systems under high temperature and pressure", Ind. Eng. Chem. Res, **48**, p. 9023-9027, 2009
11. M. Jamialahmadi, M. Emadi, H. Muller-Steinhagen, "Diffusion coefficients of methane in liquid hydrocarbons at high pressure and temperature", Journal of Petroleum Science and Engineering. **Vol 52**, 1-2, p. 47-60, 2006
12. W. G. Whitman, "The two-film theory of absorption", Chem. Metall. Eng., **29**, pp. 147-152, 1923
13. T. A. Renner, "Measurement and correlation of diffusion coefficients for CO_2 and rich-gas applications", Society of Petroleum Engineers, 1988
14. A. T. Grogan, V. W. Pinczewski, G. J. Ruskauff, F. M. Orr, "Diffusion of CO_2 at reservoir conditions: models and measurements", Society of Petroleum Engineers, 1988

https://doi.org/10.1051/e3sconf/202014602003

A Surface Complexation Model of Alkaline-SmartWater Electrokinetic Interactions in Carbonates

Moataz Abu-Al-Saud*, Amani Al-Ghamdi, Subhash Ayirala, and Mohammed Al-Otaibi

Saudi Aramco, EXPEC Advanced Research Center, Dhahran, Saudi Arabia

Abstract. Understanding the effect of injection water chemistry is becoming crucial, as it has been recently shown to have a major impact on oil recovery processes in carbonate formations. Various studies have concluded that surface charge alteration is the primary mechanism behind the observed change of wettability towards water-wet due to SmartWater injection in carbonates. Therefore, understanding the surface charges at brine/calcite and brine/crude oil interfaces becomes essential to optimize the injection water compositions for enhanced oil recovery (EOR) in carbonate formations. In this work, the physicochemical interactions of different brine recipes with and without alkali in carbonates are evaluated using Surface Complexation Model (SCM). First, the zeta-potential of brine/calcite and brine/crude oil interfaces are determined for Smart Water, NaCl, and Na_2SO_4 brines at fixed salinity. The high salinity seawater is also included to provide the baseline for comparison. Then, two types of Alkali (NaOH and Na_2CO_3) are added at 0.1 wt% concentration to the different brine recipes to verify their effects on the computed zeta-potential values in the SCM framework. The SCM results are compared with experimental data of zeta-potentials obtained with calcite in brine and crude oil in brine suspensions using the same brines and the two alkali concentrations. The SCM results follow the same trends observed in experimental data to reasonably match the zeta-potential values at the calcite/brine interface. Generally, the addition of alkaline drives the zeta-potentials towards more negative values. This trend towards negative zeta-potential is confirmed for the Smart Water recipe with the impact being more pronounced for Na_2CO_3 due to the presence of divalent anion carbonate $(CO_3)^{-2}$. Some discrepancy in the zeta-potential magnitude between the SCM results and experiments is observed at the brine/crude oil interface with the addition of alkali. This discrepancy can be attributed to neglecting the reaction of carboxylic acid groups in the crude oil with strong alkali as NaOH and Na_2CO_3. The novelty of this work is that it clearly validates the SCM results with experimental zeta-potential data to determine the physicochemical interaction of alkaline chemicals with SmartWater in carbonates. These modeling results provide new insights on defining optimal SmartWater compositions to synergize with alkaline chemicals to further improve oil recovery in carbonate reservoirs.

1. Introduction

Water ionic composition in the waterflooding process plays a crucial role in oil recovery for carbonate formations [1]. The chemistry of injected water strongly affects the reservoir wettability, which has been proven and observed in numerous laboratory and field studies for both carbonates and sandstones [2-4]. Modifying the water chemistry to alter the rock wettability is known as SmartWater or low salinity [5]. However, the root causes of this wettability alteration effect, which takes place at the pore-level, remain poorly understood especially for carbonates [6]. This lack of fundamental understanding of root causes of wettability alteration has resulted in conflicting studies, where some studies have observed an increase in oil recovery while other cases have not shown an increment in oil recovery [7]. Various pore-scale mechanisms have been proposed to delineate the wettability alteration process in carbonates. Some of these plausible mechanisms include electric double layer [4], in-situ soap generation (saponification effect) [8], and multi-ion exchange [9]. The surface charges of carbonate/brine and crude oil/brine are altered in such pore-scale processes, which affects the zeta-potential measurements used in understanding the rock wettability [10,11].

Recently, the synergy between tailored water salinity and enhanced oil recovery (EOR) has gained a lot of attraction [12]. For polymer EOR, it has been shown that the use of tailored salinity (SmartWater) reduces the polymer consumption used to increase the injected water viscosity [13]. For surfactant-based EOR, it has been observed that decreasing the water salinity increases the effect of surfactant on crude oil/brine through efficiently reducing the interfacial tension [14]. It is well-known that surfactants are adsorbed on carbonate rock surfaces. Therefore, a large amount of chemical surfactant is consumed before the chemical reaches the crude oil/brine

* Corresponding author: moataz.abualsaud@aramco.com

interface. To minimize surfactant adsorption on carbonate rock surfaces, alkaline is added to the injected water [15]. The alkaline chemical reacts with both crude oil/brine and calcite/brine interfaces. The reaction of alkali and naphthenic acid in crude oil generates in-situ soap that locally reduces the interfacial tension of the crude oil/brine interface. For the calcite surface, the adsorption of alkali changes the electric charges balance on the calcite/brine interface which alters the carbonate wettability towards more water-wet state [15]. Therefore, understanding the electrokinetics of brine with alkaline chemicals in carbonates is crucial to define the optimal water compositions for improving oil recovery.

There are several electrokinetic studies involving zeta-potentials that have reported the synergy effect of water salinity, crude oil, and alkaline in carbonates for crude oi/brine [17, 18] and calcite/brine interfaces [15, 16]. To the best of our knowledge, the role of individual ions in different brine compositions in the presence of alkaline chemicals for both the calcite/brine and crude oil/brine interfaces has not been previously studied using a surface chemistry model and lab measurements. In this work, we study the electrokinetics of with various brine recipes interacting with crude oil and calcite with and without alkali chemicals using a surface complexation model (SCM). We validate the modeling results with zeta-potentials of calcite/brine and crude oil/brine interfaces measured in the lab.

The outline of this work is as follows; a brief summary of the experimental setup for zeta-potential measurement is presented in section 2, followed by the surface complexation model (SCM). Then, the results of the SCM and lab measurements are shown in section 3. The conclusion and summary are included in section 4.

2. Materials and Methods

2.1. Experiment

Rock sample:
Pure calcium carbonate is used to represent the calcite rock sample. The purity of the calcite disk sample is measured using x-ray diffraction (XRD), which is composed of 99 wt. % of $CaCO_3$ as shown in **Figure 1**. The calcite purity confirms that there are no mineral impurities interfering with the ion adsorption on the calcite surface. The calcite disk is manually grinded using a granite mill for 30 minutes at atmospheric conditions far from contaminants to avoid surface impurities.

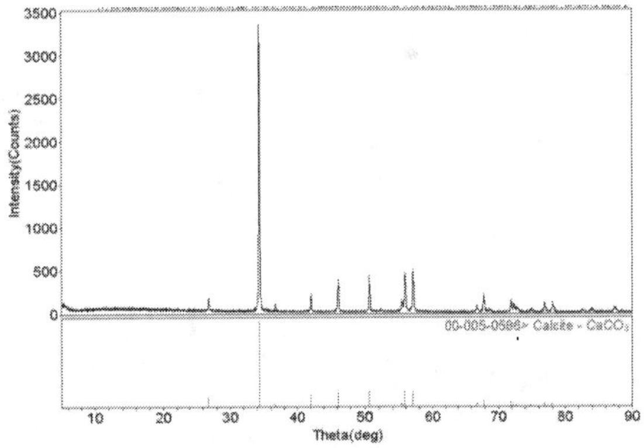

Figure 1: X-ray diffraction (XRD) diffractograms of calcite disk with reference patterns of pure calcite compound. The x-ray wavelength is 1.54 Å.

Brine and Crude Oil Properties:
Synthetic brine ionic compositions are prepared by adding different salts to deionized water. The ionic composition of the considered brine samples is listed in **Table 1**. The crude oil properties are shown in **Table 2**.

Alkaline chemicals:
Two types of Alkali (NaOH and Na_2CO_3) are added at 0.1 wt% concentration (1000 ppm) relevant to each brine recipe. Typically, alkali chemicals increase the brine pH level, which causes calcium and magnesium ions to precipitate as $Ca(OH)_2$ and $Mg(OH)_2$ respectively. Therefore, the addition of alkali to seawater is not considered to avoid precipitation, as seawater contains a high ionic concentration of hardness ions such as Mg^{2+} and Ca^{2+}. **Table 3** lists the equilibrium pH level of different brines including alkali chemicals. The brine pH is measured after reaching equilibrium with alkaline and calcite suspensions without manually adjusting the pH level.

Zeta-potential measurement:
The measurement of zeta-potential is conducted using phase analysis light scattering technique (zeta PALS) applied to calcite in brine and crude oil emulsion in brine samples. 0.2 g of grinded pure calcite particles were thoroughly mixed with 50 cm^3 of different brines for a minimum of 48 hours to ensure suspension and equilibrium are achieved. The ratio of solid/liquid and liquid/liquid is fixed throughout the experiment to guarantee consistency in the reported values. The brine pH level is not manually adjusted. The oil droplets and calcite particles size distribution in different brines is determined using optical microscopy, similar to the approach in [19]. The average calcite particle size distribution is estimated to be around 10 μm, while the average oil droplet size is between 14-35μm. Sonification is used to avoid calcite particle aggregation or crude oil coalescence in brine suspensions. The zeta-potential values are determined using Smoluchowski approximation based on the electrophoretic mobility of brine suspensions. Each zeta-potential measurement is

 https://doi.org/10.1051/e3sconf/202014602003

repeated three times to ensure consistent results. The measurement variation for each reported zeta-potential value is within 3 mV. Additional details on the sample preparation and experimental zeta-potential measurement procedures can be found in [19].

Table 1. The composition of different synthetic brines used in this study.

	Brine samples (concentration mg/L)			
	Seawater	SmartWater	NaCl	Na_2SO_4
Ions				
Na^+	18,300	1824	2266	1865
Cl^-	32,200	3220	3495	-
Ca^{2+}	650	65	-	-
Mg^{2+}	2,110	211	-	-
SO_4^{-2}	4,290	429	-	3896
HCO3-	120	-	-	-
Total dissolved Solids, ppm	57,670	5,761	5,761	5,761
pH	7.45	7.4	6.3	6.11

Table 2. Crude Oil Properties

API	27.1
Acid Number mg KOH/g	0.47
Base Number mg KOH/g	0.04
Saturates (%)	50.6
Asphaltenes (%)	1.6
Resins (%)	20.7
Aromatics (%)	27.1

Table 3. Equilibrium pH values of calcite suspension in different electrolytes containing alkali.

Type of electrolyte	pH $CaCO_3$	pH $CaCO_3$ (NaOH 0.1 wt.%)	pH $CaCO_3$ (Na_2CO_3 0.1 wt.%)
Na_2SO_4	9.85	12	10
NaCl	9.4	11.33	10
SmartWater	8.6	11.7	10
Seawater	7.45	-	-

2.2 Surface Complexation Model

The surface complexation model (SCM) describes the equilibrium state of ion adsorption based on specified surface reactions. The surface reactions of ions give rise to surface electric charges. For calcite/brine/crude-oil system, the adsorption of ions on crude-oil/brine and calcite/brine interfaces determine the surface charges and the corresponding zeta-potentials. The SCM has been employed to gain insight on the effect of electrokinetics on wettability in the context of brine chemistry in carbonates [20,21,24,25]. Brady *et al.* [21] used SCM based on surface reactions proposed in [22,23] to predict zeta-potentials for both rock/brine and brine/crude-oil interfaces in sandstone and carbonate rocks. However, the SCM has not been validated with experimental zeta-potential measurements. Mahani *et al.* [24] studied the electrokinetics of carbonate-based rocks with different water salinities using SCM. The SCM results were qualitatively validated with different carbonate/brine zeta-potential measurements. Song *et al.* [25] applied SCM and reported quantitative agreement with experimental zeta-potential measurements of synthetic calcite and multiple brine recipes. The SCM surface reactions are based on the model proposed in [26], which includes different SCM reactions compared to the models in [21,24]. The SCM work of Song *et al.* [25] has been recently extended to include surface reactions of organic and inorganic impurities occurring in natural carbonates [27]. In this work, we use SCM with surface reactions similar to the approach in [25] to predict zeta-potentials for pure calcite and different brine recipes with and without alkali chemicals. In addition, we determine brine/crude oil zeta-potentials by modeling SCM reactions at the brine/crude oil interface. **Tables 4 and 5** list the surface reactions and the corresponding equilibrium constants for both calcite and crude oil surfaces. The non-integer surface charge values of $>CaOH^{-0.75}$ and $>CO_3H^{+0.75}$ hydrated calcite sites is due to the structure of calcite crystal. The ionic bonding between Ca and O atoms from $(CO_3)^{-2}$ gives an effective charge of +1/3 for Ca atoms [26]. In this model, a charge of +1/4 is assumed for Ca atoms, following the approach in [25,26]. The crude oil surface reactions are similar to the models used in [20,21,28]. The SCM equations are solved using PHREEQC software [29], whereby the double layer model is specified. In the model, the concentration of adsorbed surfaces complexes (adsorbed ions) determines the total surface charge as follows:

$$\sigma = \frac{F}{SA}\Sigma z_i c_i, \qquad (1)$$

where σ is the surface charge density (C/m^2), F is the Faraday constant (96493.5 C/mol), S is the surface material mass (g), A is the specific surface area (m^2/g), z_i is the ionic electric charge, and c_i is the adsorbed ion concentration (mol). The surface charge and surface-potential are related through the Gouy-Chapman model [29]:

$$\sigma = (8000\epsilon_0\epsilon_w RTI)^{1/2}\sinh\left(\frac{\nu F\Psi}{2RT}\right), \qquad (2)$$

where ϵ_0 is the vacuum permittivity $\left(\frac{C^2}{mJ}\right)$, ϵ_0 is the water relative permittivity, Ψ is the surface-potential (V), R is the gas constant $\left(\frac{J}{mol\,K}\right)$, T is the temperature (K), I is the brine ionic strength (mol/l), and ν is the electrolyte ionic charge which is assumed to be unity in PHREEQC [29]. The bulk concentration of ions interacts with the adsorbed ions at the surface due to coulombic forces. Hence, the apparent equilibrium constants are considered to include the effect of bulk concentration of the ions. The apparent and intrinsic equilibrium constants (listed in Tables 4 and 5) are described through the Boltzmann distribution [29]:

$$\mathrm{K_{app}} = \mathrm{K_{int}} \exp\left(\frac{Z_c F \Psi.}{RT}\right), \qquad (3)$$

where Z_c is the net change of the surface charge at the surface due to surface reaction. The zeta-potential ζ can be approximated from the surface-potential based on the linearized Debye-Huckel theory [25], which is valid for $|\Psi| \leq 25$ mV [30]:

$$\zeta = \psi \exp(-\kappa d_s), \qquad (4)$$

where κ is the inverse Debye length-scale, and d_s is the slipping plane distance from the outer Helmholtz plane. For brines with ionic-strength of 0.1 mol/l (the considered brine recipes except for sea water), the slipping distance d_s is 0.33 nm [25, 26], while the Debye length κ^{-1} is 0.97 nm. For the seawater case (ionic strength of 1.1 mol/l), d_s is 0.1 nm ($d_s = 0.1/c^{0.5}$ [26]), while κ_{sw}^{-1} is 0.29 nm. The site density for calcite surface is 4.95 sites/nm^2 [26, 27], while the crude oil surface has a site density of 0.47 sites/nm^2 [28]. The calcite specific surface area is 1 m^2/g [26], while the crude oil specific area is 0.5 m^2/g [28]. Additional details of the above SCM equations are elaborated in [29,31].

Table 4. Surface Complexation Reactions and Parameters for the Calcite Surface.

Calcite Surface Reaction	Equilibrium Constant ($\log_{10} K_{int}$)
$> CaOH^{-0.75} + H^+ \leftrightarrow\ > CaOH_2^{+0.25}$	0.4
$> CO_3H^{+0.75} + OH^- \leftrightarrow\ > CO_3^{-0.25} + H_2O$	0.5
$> CaOH^{-0.75} + Ca^{2+} \leftrightarrow\ > CaOH..Ca^{+1.25}$	1.53
$> CaOH^{-0.75} + Mg^{2+} \leftrightarrow\ > CaOH..Mg^{+1.25}$	1.15
$> CO_3H^{+0.75} + SO_4^{2-} \leftrightarrow\ > CO_3H..SO_4^{-1.25}$	1.5
$> CO_3H^{+0.75} + CO_3^{2-} \leftrightarrow\ > CO_3H..CO_3^{-1.25}$	2.23
$> CO_3H^{+0.75} + HCO_3^- \leftrightarrow\ > CO_3H..HCO_3^{-0.25}$	0.09
$> CaOH^{-0.75} + Na^+ \leftrightarrow\ > CaOH..Na^{+0.25}$	0.22
$> CO_3H^{+0.75} + Cl^- \leftrightarrow\ > CO_3H..Cl^{-0.25}$	0.65

Table 5. Surface Complexation Reactions and Parameters for the Crude Oil Surface.

Crude Oil Surface Reaction	Equilibrium Constant ($\log_{10} K_{int}$)
$-COOH \leftrightarrow -COO^- + H^+$	6
$-N + H^+ \leftrightarrow -NH^+$	4
$-COOH + Ca^{+2} \leftrightarrow -COOCa^+ + H^+$	-4
$-COOH + Mg^{+2} \leftrightarrow -COOMg^+ + H^+$	-4.3
$-COOH + Na^+ \leftrightarrow -COONa + H^+$	-4

3. Results and Discussion

Figures 2 to 7 compare the SCM and experimental measurements of the zeta-potential values for different brine recipes with and without alkali. First, the zeta-potentials for brine/calcite interface are analyzed, followed by the brine/crude oil interface. The intrinsic equilibrium constants are varied in the SCM to match the measured zeta-potentials. The number of fitting parameters is equal to the number of surface reaction equations (nine equations for calcite/brine interface, and five equations for the crude oil/brine interface). The site density and specific surface area of the considered surfaces are fixed (see **Section 2.2**). For the brine/calcite interface (**Figures 2-4**), the SCM results follow the trends observed in experimental data, and quantitatively agree with the lab zeta-potential measurements especially for NaCl, SmartWater, and Na_2SO_4 brines. For the seawater case without alkali, the SCM underestimates the zeta-potential as can been seen in **Figure 2**. When NaOH alkaline is added to the brine recipes, the change in zeta-potential values of calcite/brine interface is almost negligible and in agreement with the experimental zeta-potential measurements as illustrated in **Figure 3** (within 2.5 mV difference). The addition of $(OH)^-$ ions increases the negative charge, and decreases the adsorption of H^+ protons while adding Na^+ ions increase the charge positivity resulting in an insignificant total change in the surface charge. For the Na_2CO_3 alkaline, the SCM calcite/brine zeta-potentials decrease for the considered brine recipes as displayed in **Figure 4**. This decrease in zeta-potential values is mainly due to the presence of divalent anion carbonate $(CO_3)^{-2}$, which forms a surface complex on the calcite surface with a -1.25 charge as illustrated in reaction 6 in **Table 3**. Calcite precipitation/dissolution has not been considered in SCM, which is likely to contribute to the slight discrepancy observed in the SmartWater recipe case in **Figure 4**. The intrinsic equilibrium constants (fitting parameters) in **Table 4** are in agreement with the work of Song et al. [25] except for the Cl^- ion which has the largest discrepancy. The SCM confirms that the total surface charge and zeta-potential of calcite/brine interface strongly depend on the divalent ions (Ca^{2+}, Mg^{2+}, CO_3^{2-}, and SO_4^{2-}) due to large intrinsic equilibrium constants compared to the other ions. The SCM predicts that the most effective ion in pushing the calcite surface charge to a negative value is the divalent anion carbonate $(CO_3)^{-2}$, which has the largest equilibrium constant.

For the crude oil/brine interface cases, the SCM zeta-potential results qualitatively match the experimental observations. The predicted zeta-potentials are in general negatively charged as shown in **Figure 5**, which agree with the experimental results. The equilibrium constants in **Table 5** agree with the values reported in the literature [21,28] (the discrepancy in the equilibrium constants are within a value of one for each reaction). At pH > 7 (greater than the isoelectric point for crude oil/brine interface with a similar acid number [32]), the carboxylic acids in crude oil are not completely protonated (reaction 1 in **Table 5**). Also, the protonation of nitrogen bases (reaction 2 in **Table 5**) is not sufficient to switch the crude oil surface to a positive charge. When alkali is added to the brine recipe (**Figures 6 and 7**), the variation in SCM zeta-potentials is less than 2 mV. The zeta-potentials slightly increase in the positive side, especially for NaOH alkaline. This slight increase is not expected as the protonation activity decreases with the increase of the brine pH level.

However, the reactions in SCM predict that the adsorption of Mg^{2+} and Ca^{2+} (reactions 3 and 4) slightly compensates for the decrease in H^{+}, which results in a small increase in the zeta-potential. Based on the model predictions of calcite/brine and brine/crude oil zeta-potentials, the Na_2SO_4 brine with Na_2CO_3 alkaline provides the largest interface negative charges (**Figures 4 and 7**). The modeling results support the conclusion drawn from the experiments that the Na_2SO_4 brine with Na_2CO_3 alkaline is the preferable brine recipe to synergize with anionic surfactants due to its large negative zeta-potential. This large negative zeta-potential increases the water-wetness towards carbonate reservoir [16] and reduces the anionic surfactant retention [15]. Further improvement and refinement in the SCM will be considered as this is a starting point for ongoing research in electrokinetic modeling. Such improvements include using an SCM triple-layer model, non-linear diffuse double layer model to infer zeta-potential from the surface-potential, and adding dissolution/percipation reactions for the calcite/brine interface. Also, including the crude oil chemistry (acid/base numbers) to be part of the surface site density [33] of nitrogen base and carboxylate groups will improve the SCM for the crude oil/brine interface.

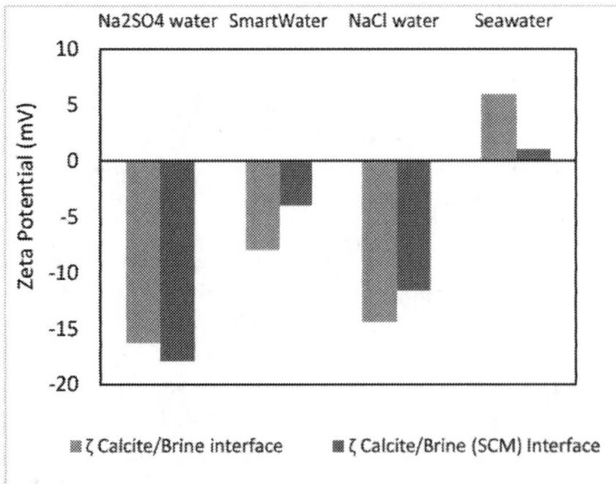

Figure 2: Experimental and SCM ζ- potential values at the calcite/brine interface

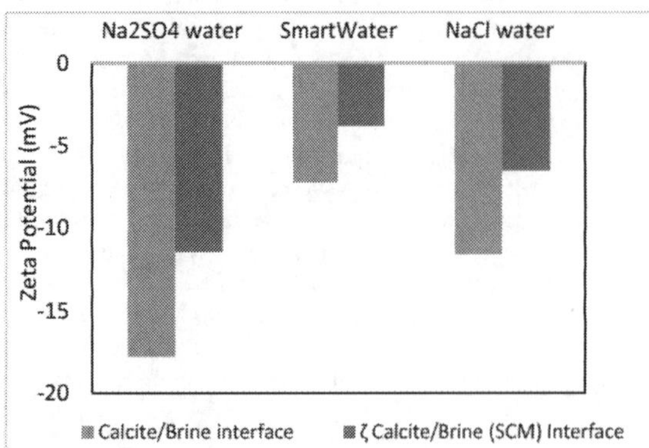

Figure 3: Experimental and SCM ζ- potential values at the calcite/brine interface. Brine is containing 0.1 wt.% NaOH.

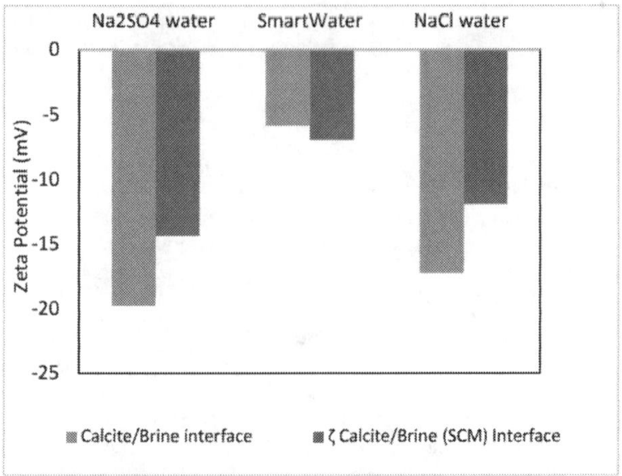

Figure 4: Experimental and SCM ζ- potential values at the calcite/brine interface. Brine is containing 0.1 wt.% Na_2CO_3.

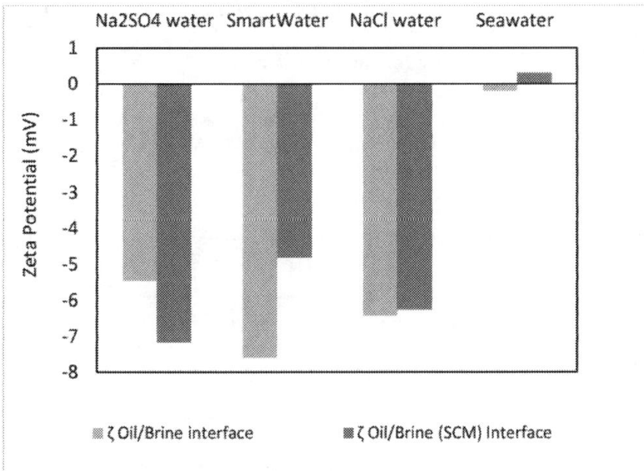

Figure 5: Experimental and SCM ζ- potential values at oil/brine interface.

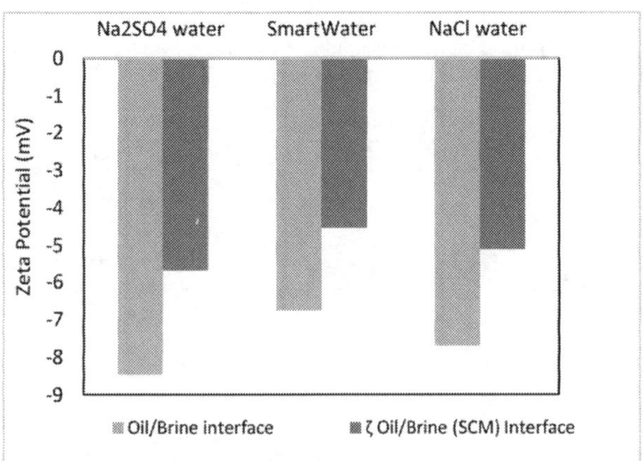

Figure 6: Experimental and SCM ζ- potential values at oil/brine interface. Brine is containing 0.1 wt.% NaOH.

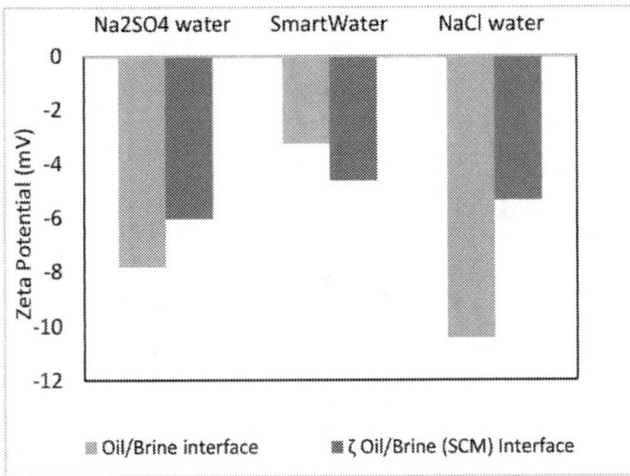

Figure 7: Experimental and SCM ζ- potential values at oil/brine interface. Brine is containing 0.1 wt.% Na_2CO_3.

4. Conclusions

This work explores the electrokinetic interactions of different brine recipes with and without alkali in carbonates using Surface Complexation Model (SCM). Various SCM calculations of brine/calcite and crude oil/brine zeta-potentials are conducted to evaluate the synergy between different brine recipes and alkali chemicals. The proposed SCM predicts zeta-potential results that are consistent with the experimental measurements for choosing the Na_2SO_4 brine with Na_2CO_3 alkaline to be the most suitable alkaline-based recipe due to its large negative zeta-potentials. In alkaline/surfactant EOR process, the negative electrokinetic calcite charge repels the anionic surfactant causing a reduction in surfactant retention. The modeling of zeta-potential results quantitatively agree with the experiments for the calcite/brine interface. For the crude oil/brine interface, the SCM zeta-potential results reasonably match the experimental measurements. Further correlation of the carboxylic acid group and nitrogen base crude oil active surface sites with the acid/base number is required [33] to quantitatively predict accurate zeta-potentials and gain additional insights on the electrokinetics of crude oil surfaces.

Acknowledgements

The authors thank Hussain Saleem and Dong kyu Cha, Saudi Aramco for their help with zeta-potential measurements.

References

1. T. Austad, S. Strand, E. Høgnesen, P. Zhang. "Seawater as IOR fluid in fractured chalk". *In SPE international symposium on oilfield chemistry* 2005 Jan 1. Society of Petroleum Engineers.
2. A. Yousef, S. Al-Saleh, A. Al-Kaabi, M. Al-Jawfi. "Laboratory investigation of the impact of injection-water salinity and ionic content on oil recovery from carbonate reservoirs". *SPE Reservoir Evaluation & Engineering*. 2011 Oct 1;14(05):578-93.
3. A. Yousef, J. Liu, G. Blanchard, S. Al-Saleh, T. Al-Zahrani, R. Al-Zahrani, H. Al-Tammar, N. Al-Mulhim. "Smart waterflooding: industry". *In SPE Annual Technical Conference and Exhibition* 2012 Jan 1. Society of Petroleum Engineers.
4. D. Ligthelm, J. Gronsveld, J. Hofman, N. Brussee, F. Marcelis, H. Van der Linde. "Novel Waterflooding Strategy by Manipulation Of Injection Brine Composition". *In EUROPEC/EAGE conference and exhibition* 2009 Jan 1. Society of Petroleum Engineers.
5. N. Morrow, J. Buckley. "Improved oil recovery by low-salinity waterflooding". *Journal of Petroleum Technology*. 2011 May 1;63(05):106-12.
6. H. Mahani, A. Keya, S. Berg, W. Bartels, R. Nasralla, W. Rossen. "Insights into the mechanism of wettability alteration by low-salinity flooding (LSF) in carbonates". *Energy & Fuels*. 2015 Mar 3;29(3):1352-67.
7. M. Jackson, D. Al-Mahrouqi, J. Vinogradov. "Zeta-potential in oil-water-carbonate systems and its impact on oil recovery during controlled salinity water-flooding". *Scientific reports*. 2016 Nov 23;6:37363.
8. P. McGuire, J. Chatham, F. Paskvan, D. Sommer, F. Carini. "Low salinity oil recovery: An exciting new EOR opportunity for Alaska's North Slope". *In SPE western regional meeting* 2005 Jan 1. Society of Petroleum Engineers.
9. A. Lager, K. Webb, C. Black, M. Singleton, K. Sorbie. "Low salinity oil recovery-an experimental investigation-1". *Petrophysics*. 2008 Feb 1;49(01).
10. D. Al Mahrouqi, J. Vinogradov, M. Jackson. "Zeta-potential of artificial and natural calcite in aqueous solution". *Advances in colloid and interface science*. 2017 Feb 1;240:60-76.
11. M. Jackson, J. Vinogradov, G. Hamon, M. Chamerois. "Evidence, mechanisms and improved understanding of controlled salinity waterflooding part 1: Sandstones". *Fuel*. 2016 Dec 1;185:772-93.
12. S. Ayirala, A. Yousef. "A state-of-the-art review to develop injection-water-chemistry requirement guidelines for IOR/EOR projects". *SPE Production & Operations*. 2015 Feb 1;30(01):26-42.
13. S. Ayirala, E. Uehara-Nagamine, A. Matzakos, R. Chin, P. Doe, P. Van den Hoek. "A designer water process for offshore low salinity and polymer flooding applications". *In SPE Improved Oil Recovery Symposium* 2010 Jan 1. Society of Petroleum Engineers.
14. W. Alameri, T. Teklu, R. Graves, H. Kazemi, A. AlSumaiti. "Low-salinity water-alternate-surfactant in Low-permeability Carbonate Reservoirs". *In IOR 2015-18th European Symposium on Improved Oil Recovery* 2015 Apr 14.
15. G. Hirasaki, C. Miller, M. Puerto. "Recent advances in surfactant EOR". *SPE Journal*. 2011 Dec 1;16(04):889-907.

16. G. Hirasaki G, D. Zhang. "Surface chemistry of oil recovery from fractured, oil-wet, carbonate formations". *Spe Journal*. 2004 Jun 1;9(02):151-62.
17. Q. Liu, M. Dong, X. Yue, J. Hou. "Synergy of alkali and surfactant in emulsification of heavy oil in brine". *Colloids and Surfaces A: Physicochemical and Engineering Aspects*. 2006 Feb 1;273(1-3):219-28.
18. K. Chan, D. Shah. (1981) "The Physico-Chemical Conditions Necessary to Produce Ultralow Interfacial Tension at the Oil/Brine Interface". In: Shah D.O. (eds) *Surface Phenomena in Enhanced Oil Recovery*. Springer, Boston, MA
19. M. Alotaibi, A. Yousef. "The role of individual and combined ions in waterflooding carbonate reservoirs: electrokinetic study". *SPE Reservoir Evaluation & Engineering*. 2017 Feb 1;20(01):77-86.
20. Q. Xie, A. Sari, W. Pu, Y. Chen, P. Brady, N. Al Maskari, A. Saeedi. "pH effect on wettability of oil/brine/carbonate system: Implications for low salinity water flooding". *Journal of Petroleum Science and Engineering*. 2018 Sep 1;168:419-25.
21. P. Brady, J. Krumhansl, P. Mariner. "Surface complexation modeling for improved oil recovery". *In SPE improved oil recovery symposium* 2012 Jan 1. Society of Petroleum Engineers.
22. P. Van Cappellen, L. Charlet, W. Stumm, P. Wersin. "A surface complexation model of the carbonate mineral-aqueous solution interface". *Geochimica et Cosmochimica Acta*. 1993 Aug 1;57(15):3505-18.
23. O. Pokrovsky, J. Schott, F. Thomas. "Dolomite surface speciation and reactivity in aquatic systems". *Geochimica et Cosmochimica Acta*. 1999 Oct 1;63(19-20):3133-43.
24. H. Mahani, A. Keya, S. Berg, R. Nasralla. "Electrokinetics of carbonate/brine interface in low-salinity waterflooding: effect of brine salinity, composition, rock type, and pH on ζ-potential and a surface-complexation model". *SPE Journal*. 2017 Feb 1;22(01):53-68.
25. J. Song, Y. Zeng, L. Wang, X. Duan, M. Puerto, W. Chapman, S. Biswal, G. Hirasaki. "Surface complexation modeling of calcite zeta-potential measurements in brines with mixed potential determining ions (Ca2+, CO32−, Mg2+, SO42−) for characterizing carbonate wettability". *Journal of colloid and interface science*. 2017 Nov 15;506:169-79.
26. F. Heberling, T. Trainor, J. Lützenkirchen, P. Eng, M. Denecke, D. Bosbach. "Structure and reactivity of the calcite–water interface". *Journal of colloid and interface science*. 2011 Feb 15;354(2):843-57.
27. J. Song, S. Rezaee, L. Zhang, Z. Zhang, M. Puerto, O. Wani, F. Vargas, S. Alhassan, S. Biswal, G. Hirasaki. "Characterizing the Influence of Organic Carboxylic Acids and Inorganic Silica Impurities on the Surface Charge of Natural Carbonates Using an Extended Surface Complexation Model". *Energy & Fuels*. 2019 Jan 11;33(2):957-67.
28. T. Takeya, M. Shimokawara, Y. Elakneswaran, T. Nawa, S. Takahashi. "Predicting the electrokinetic properties of the crude oil/brine interface for enhanced oil recovery in low salinity water flooding". *Fuel*. 2019 Jan 1;235:822-31.
29. D. Parkhurst, C. Appelo. "Description of input and examples for PHREEQC version 3: a computer program for speciation, batch-reaction, one-dimensional transport, and inverse geochemical calculations". *US Geological Survey*; 2013.
30. J. Israelachvili. "Intermolecular and surface forces". *Academic press*; 2011 Jul 22.
31. D. Dzombak, F. Morel. "Surface complexation modeling: hydrous ferric oxide". *John Wiley & Sons*; 1990 Mar 16.
32. J. Buckley, K. Takamura, N. Morrow. "Influence of electrical surface charges on the wetting properties of crude oils". *SPE Reservoir Engineering*.1989. 4(03):332-340.
33. M. Bonto, A. Eftekhari, H. Nick. "An overview of the oil-brine interfacial behavior and a new surface complexation model". *Scientific reports*. 2019 Apr 15;9(1):6072.

E3S Web of Conferences **146**, 03008 (2020)
SCA 2019
https://doi.org/10.1051/e3sconf/202014603008

Effects of gas pressurization on the interpretation of NMR hydrocarbon measurements in organic rich shales

*Son Dang**, *Carl Sondergeld*, and *Chandra Rai*

Mewbourne School of Petroleum and Geological Engineering, University of Oklahoma, USA

Abstract. The estimation of total hydrocarbons (HCs) in place is one of the most important economic challenges in unconventional resource plays. Nuclear magnetic resonance (NMR) has proven to be a valuable tool in directly quantifying both hydrocarbons and brines in the laboratory and the field. Some major applications of NMR interpretation include pore body size distributions, wettability, fluid types, and fluid properties. However, for tight formations, the effects of the factors on NMR relaxation data are intertwined. One purpose of this study is to review the interpretation of NMR response of HCs in a tight rock matrix through illustrated examples.

When comparing NMR data between downhole wireline and laboratory measurement, three important elements need to be considered: 1) temperature differences, 2) system response differences, and 3) pressure (mainly due to the lost gasses.) The effect of temperature on HCs would be presented with experimental results for bulk fluids. Whereas, the effect of pressure is investigated by injecting gas back into rock matrix saturated with original fluids. The experiments were performed within an NMR transparent Daedalus ZrO_2 pressure cell, which operates at pressures up to 10,000 psi.

The results show that, at ambient temperature and pressure, NMR responds to a fraction of HCs, which is volatile enough to be observed as an NMR relaxation sequence. The invisible fraction of HCs to NMR sequence at ambient condition can be up to 20% of the total extractable HCs. Molecular relaxation is impacted by fluid viscosity, pore size, and surface affinity. In other words, the fluid with higher viscosity (either due to temperature or gas loss), presenting in smaller pore, or highly affected by the pore surface, will relax faster, and would be partially invisible to NMR, especially in the field. This is critical to the interpretation of NMR response for liquid rich source rocks, in which all of the above molecular relaxing restrictions can be found. Thus, engineers can underestimate movable HCs by using routine core analysis data.

1 Introduction

Nuclear Magnetic Resonance (NMR) has been proven to be a useful tool to evaluate formation characteristics in both the laboratory and the field. NMR responses are induced during the relaxation of nuclear spins, for a particular pulse sequence. At a specific magnetic-field strength, scanning frequency is tuned to resonate with the Lamour's frequency of hydrogen found in reservoir fluids, such as brine, oils, gases, and bitumen (Brown 1961[1]; Bryan et al., 2002[2]; Hirasaki et al., 2003[3]). Common NMR parameters used to interpret formation properties, include T_1 as the longitudinal relaxation time, and T_2 as the transverse relaxation time. Combination of T_1 and T_2 relaxation data can provide important information about formation and fluid properties. For conventional reservoirs, NMR has been used to estimate saturated porosity, and pore size distribution (Keynon et al., 1986[4]), from which permeability can be calculated (Coates et al., 1991[5]; Kenyon et al., 1995[6][7]; Straley et al, 1994[8].)

Recently developed applications of NMR, focusing on unconventional tight rocks, include the partitioning of pore surface affinity (oil-wet versus water-wet in shales) (Odusina et al., 2011[9]; Valori and Nicot, 2019[10]) and the characterization of *in situ* fluids. However, the interpretation is not necessarily straightforward due to the coexistence of multiples fluids within a complicated pore structure (inorganic pores versus organic pores) (Sinha et al., 2017[11]). **Fig. 1** presents an example in which the same HCs yield different NMR responses, due to its containment within different pore types. **Fig. 1-a** is the T_1-T_2 map of a preserved sample with dominant inorganic pores (TOC < 0.5w %); whereas **Fig. 1-b**is the T_1-T_2 map of an adjacent preserved sample (0.73ft away) with dominant organic pores (TOC = 5w %.)

(a)
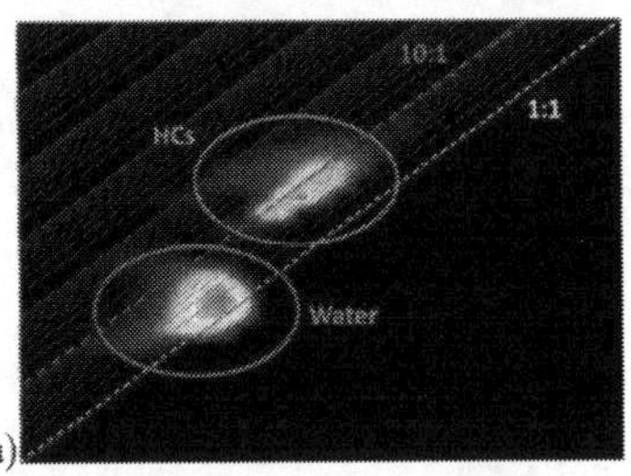

* Corresponding author: dangthaison@ou.edu

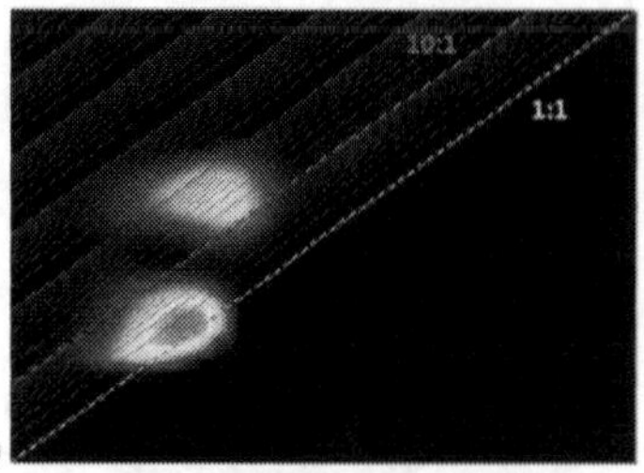

(b)

Fig. 1. T_1-T_2 maps of two preserved tight samples with T_2 is on the x-axes, and T_1 is on the y-axes. Brine NMR response is close to 1:1 line (yellow dash line). Sample (a) with dominant inorganic pores, HCs NMR response is between 1:1 line and 10:1 line (red dash line). Sample (b) with dominant organic pores, HCs NMR response is above 10:1 line, suggesting stronger affinity between HCs and pore surface.

It is essential to verify log-derived parameters using laboratory measurements. Besov et al., 2017[12] demonstrated how the differences in NMR configuration (such as scanning frequency, echo spacing, signal to noise ratio, and magnetic field gradient) can make the comparison between laboratory and field data complicated. Another major concern for almost all laboratory versus log comparison studies is the core condition, even for preserved samples (Blount et al., 2018[13]). In this study, we will review the understanding of NMR response to HCs in bulk fluids as a function of temperature (Dang et al., 2019[14]), as well as the new insights on how gas pressurization/ depressurization influences NMR response on organic rich tight rocks.

2 Experimental instrument and samples

NMR distribution, including T_2 relaxation and T_1-T_2 maps, were acquired at the frequency of 2 MHz, using Oxford GeoSpec™ spectrometers, and Green Imaging acquisition and processing software. The magnet temperature was set at 31°C throughout the experiments. The optimal echo spacing, of 114 μs, was chosen to capture fast relaxation components in the shale samples (including fluids in small pores and heavy HCs components), while preventing the interference of the fluorine signal from internal machine parts.

For pressurization experiments, the samples were placed inside a Daedalus® cell, made of NMR transparent ZrO_2; the cell can be operated up to 10,000 psi internal pressure. The cell was positioned inside the NMR spectrometer, in which samples were aligned with the uniform section of the permanent magnetic field. From its inlet, the cell was connected to a Teledyne ISCO syringe pump system, which was used to compress gases from supply cylinders, and then inject gasses into the test cell at a designed pressure. To understand the dynamic response of only the HCs components in rock samples during gas pressurization and depressurization, NMR transparent gasses were chosen. Injected gas candidates include CO_2, N_2, or CD_4 (deuterated methane). **Fig. 2** illustrates major components of the experimental setup. Note there was no confinement applied on rock samples, gasses were injected all around samples.

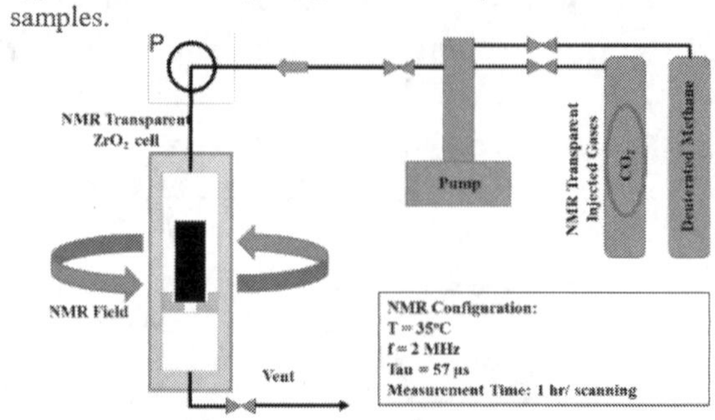

Fig. 2. General experimental setup, including NMR transparent ZrO_2 pressure cell, placed within 2 MHz NMR spectrometer. The cell was connected to a pump system to its inlet, and to a vent line from its outlet. NMR transparent injected gasses were chosen, including CO_2, N_2, or deuterated methane.

Four organic rich tight samples were selected for this study from various shale plays; two of them were preserved samples. Routine petrophysical characterizations were performed and presented in **Table. 1**, including total porosity (the sum of high-pressure gas-filled porosity and NMR saturated porosity), mineralogy measured by transmission Fourier Transform Infrared (FTIR), and TOC measured with a LECO® instrument.

Sample ID	Porosity	TOC	Mineralogy, w%				Sample Condition
	p.u	w%	Total Clays	Total Carbonates	Quartz + Feldspars	Others	
A	11.2	7.2	43	0	42	15	Non-preserved
B	5.8	5	33	18	48	1	Preserved
C-1	5.1	4.9	16	62	13	9	Preserved
C-2	8.5	7.2	32	45	15	8	Non-preserved

Table 1. Petrophysical characterization of four samples. These samples are from three different tight formations.

3 Measureable fractions of HCs under NMR spectroscopy

For T_2 relaxation acquisition, if a magnetization vector relaxes faster than the setup echo spacing, protons inducing those magnetization vectors cannot be observed with the NMR spectrometer. Fast relaxing components in organic tight rocks include: 1) fluid bound in small pores, 2) highly viscous fluid components (bitumen or heavy alkanes), 3) fluids with strong affinities to pore surfaces and 4) existence of paramagnetic minerals (pyrite or siderite). The existence or coexistence of these factors make the interpretation of NMR response of HCs in shale samples complicated.

Fig.1 illustrates the impact of surface affinity or pore type on NMR response. With the same HCs, sample (a) with more inorganic pores (or weaker surface affinity), HCs signal was observed in slower relaxation region; whereas, sample (b) with more organic pore type (or stronger surface affinity), HCs signal was observed in a faster relaxation region.

Dang et al., 2019[14] synthesized a set of six oil samples from the distillation of a same produced crude under inert condition. The crude was heated to elevated temperatures to generate oil samples with successively increasing mean molecular weights. Since NMR measurements were performed at 31°C, the results suggested that NMR response for these oil samples correlated well with < C17 fraction. **Fig. 3** shows good agreement between NMR response and Rock-Eval® S_1 intensity (Dang et al., 2019[14]). S_1 is associated with the measured fraction of total HCs with vaporization temperature equal or less than 300°C.

Previous studies (Chakravarty et al., 2018[15]; Hirasaki et al., 2003[3]) on NMR properties of petroleum reservoir fluids also suggests that besides tool configuration and echo times, NMR response is sensitive to the mobile fraction of HCs, which is clearly a function of fluid composition and temperature.

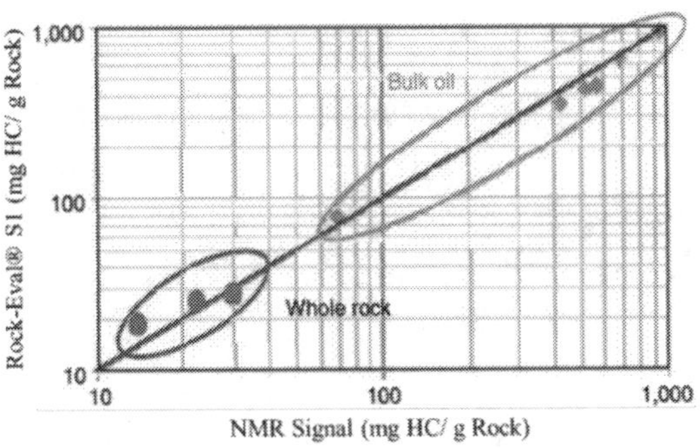

Fig. 3. Comparison between NMR response for HCs and Rock-Eval® S_1 (Dang et al., 2019.) S_1 intensity represents the fraction of HCs components vaporized equal to or less than 300°C, or roughly <C17 fraction. Blue data points are from bulk oil samples; whereas red data points are from rock samples.

4 Impact of gas pressurization or depressurization on NMR response

As previously mentioned, the mobile fraction of HCs is governed by overall fluid composition and temperature. Since NMR response is sensitive to HCs mobility, it is critical to understand the role of gas loss while rock samples are brought up to surface conditions. Gas loss changes overall HCs composition, make the comparison between NMR benchtop measurement and downhole wireline parameters inexact. On the other hands, NMR sensitivity to fluid mobility makes it a useful tool to evaluate HCs remobilization for gas injection EOR studies.

To focus on the dynamic change of *in situ* HCs within rock matrix, experiments were performed with NMR transparent gasses. **Fig. 4** shows NMR T_2 spectra of sample C-1 while CO_2 was injected at different pressures. As injection pressure increased, T_2 spectra were observed to shift to the slower relaxation region (1-10ms), and signal intensity in the fast relaxation region (0.1-1ms) also decreases. These changes suggested viscosity of the original fluid was reduced, or we remobilization of HCs.

Along with the acquisition of T_2 relaxation, we also obtained T_1-T_2 maps, which help to separate brine from the HCs signals. HCs volume in rock matrix could be calculated by subtracting brine volume from total fluid volume. **Fig. 5** shows the normalization of HCs volume in rock matrix as a function of injection pressure. In addition to the shift toward slower T_2 relaxation, HCs volume detected by NMR response also increases, respectively to injection pressure. This result confirmed the existence of a fraction of HCs, which was invisible for NMR under routine laboratory condition. For some samples, the HCs volume can increase by 25vol% from initial condition to 5000 psi pressurized condition. **Fig. 6** illustrates the remobilization of HCs during gas injection, which reduces fluid viscosity and bring out a fraction of HCs originally undetectable for NMR.

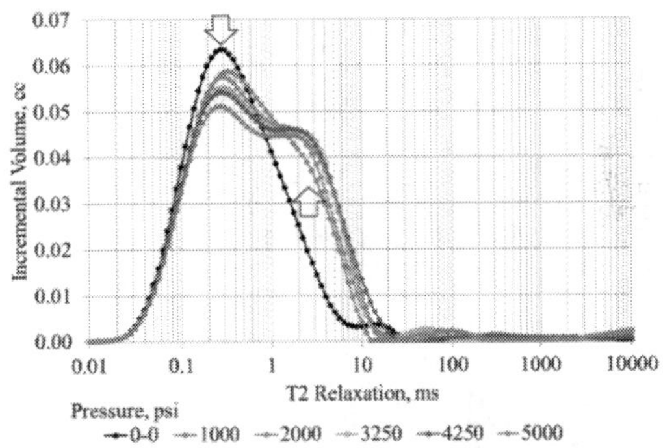

Fig. 4. T_2 spectra of sample C-1 with the injection of CO_2. Black spectrum is NMR response at original condition. T_2 were observed to shift to slower relaxation region (1-10ms), and signal intensities in the fast relaxation region (0.1-1ms) decreases. These changes suggest the reduction of fluid viscosity.

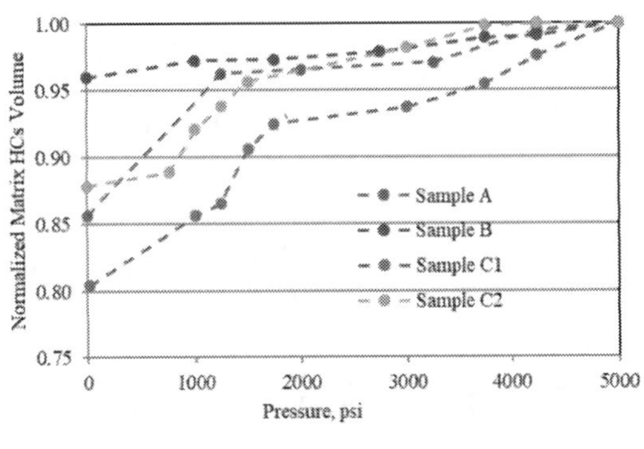

Fig. 5. Normalized HCs volume as a function of injection pressure. HCs volume detected by NMR increases with injection pressure. This suggests the existence of a fraction of HCs, originally invisible for NMR under routine laboratory condition.

Fig. 7 shows the hysteresis of normalized HCs volume as a function of pore pressure between pressurization and depressurization. While pressurization data confirms the remobilization of HCs, depressurization data suggests gas trap phenomenon inside the HCs phase during depressurization. Translating to field applications, huff-n-puff EOR can be effective by increasing remaining HCs' mobility. Due to gas trapping mechanism, engineers can plan for shorter subsequent injection cycles.

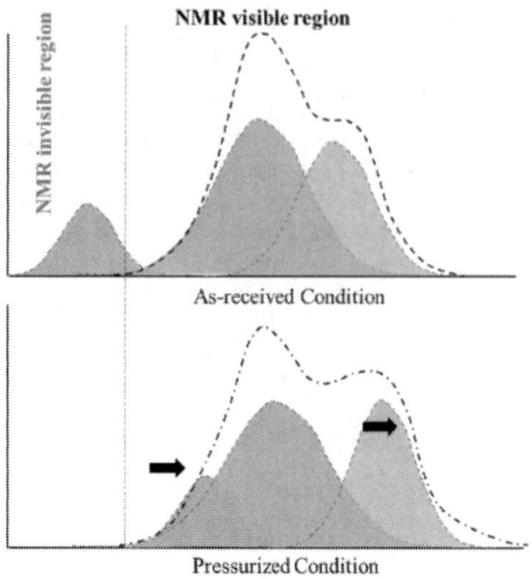

Fig. 6. Scheme for the remobilization of HCs during gas pressurization experiments. Blue peak corresponds to brine, green peak corresponds to light/ originally mobile HCs fraction, and gray peak corresponds to heavy HCs components. Gas pressurization reduces fluid oil viscosity, brings HCs signals toward slower relaxation regions. This makes more HCs detectable by NMR, comparing to the original condition without gas injection.

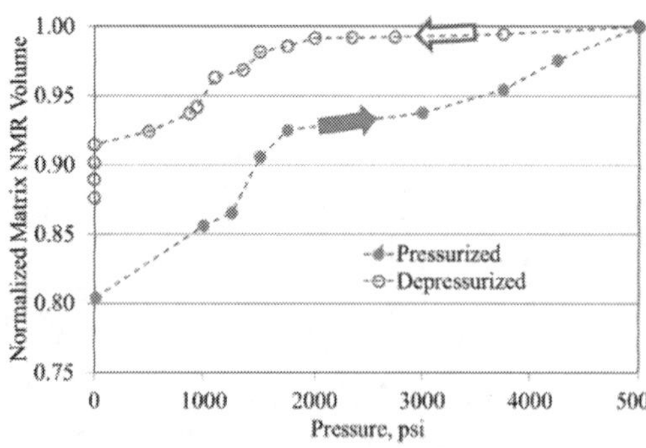

Fig. 7. Normalized HCs volume detectable under NMR as a function of gas pressure during pressurization and depressurization. The hysteresis between two data sets, suggest gas trapping during depressurization.

5 Application to study huff-n-puff in shales

The dynamic interaction between injection gas and reservoir fluid is not only a function of pressure, but also a function of injection/production strategy. With the introduction of gas into rock matrix, mobility of reservoir fluid would be enhanced, **Fig. 4**. Under a certain mass transport mechanism (convective flow or diffusion or both), additional volumes of HCs can be recovered after primary production. **Fig. 8** shows T_2 relaxation response of a shale sample throughout a huff-n-puff experiment with CO_2 as the injected gas. The experiment included four cycles with subsequent increasing of injection pressure (3750 psi, 4250 psi, 5000 psi, 5000 psi). During each cycle, gas injection was conducted for 1 day at a constant pressure, followed by 1-day production phase, in which pressure was reduced to blowdown under a controlled program. T_2 spectra were divided into three regions, namely non-movable fluids in matrix, movable fluid in matrix, and expelled fluids, respective to different relaxation regions. During the experiment, the signal of remaining non-movable fluid in matrix did not change; whereas, the signal of movable fluids in matrix reduced, but was volumetric balanced to the increasing of expelled fluid fractions (**Fig. 9-a and 9-b**).

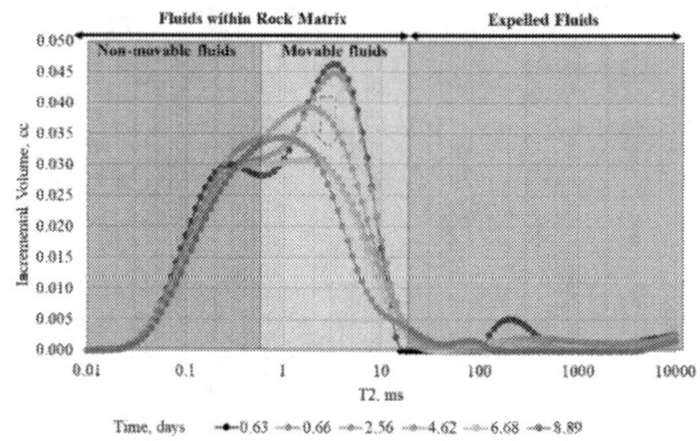

Fig. 8. T_2 relaxation response of a shale sample throughout a huff-n-puff experiment with CO_2 as the injected gas. T_2 spectra were divided into three regions, namely non-movable fluids in matrix (gray box, 0.1-1ms), movable fluid in matrix (green box, 1-10ms), and expelled fluids (red box, >10ms).

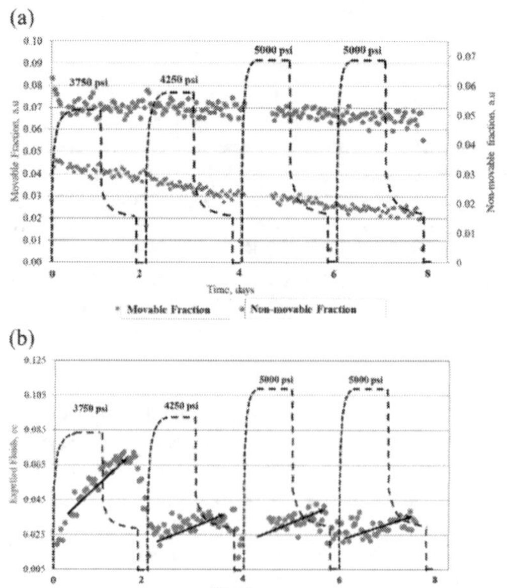

Fig. 9. (a) Concentration profiles of non-movable fluid in matrix (gray) and movable fluid in matrix (green). (b) Concentration profile of expelled fluids (red). While the magnitude of non-movable fluid in matrix did not change during the experiment, movable fluid fraction reduced, but volumetric balanced to expelled/ recovered fluid concentration.

6 Conclusions

In this study, we investigate the effect of gas pressurization by injecting gas into a rock matrix saturated with original fluids. The experiments were performed within a NMR transparent ZrO_2 pressure cell, placed inside a 2MHz NMR spectrometer. At a particular temperature and pressure, NMR responds to a fraction of HCs which mobile enough to be observed as an NMR relaxation sequence (roughly <C17). The invisible fraction of HCs to NMR sequence at ambient condition can be as much as 20% of the total extractable HCs. Molecular relaxation is impacted by fluid viscosity, pore size, and surface affinity. In other words, the HCs fraction with higher viscosity (either due to temperature or gas loss), presenting in smaller pore, or highly affected by the pore surface, will relax faster, and would be partially invisible to NMR. This is critical to the interpretation of NMR response for liquid rich source rocks, in which all above molecular relaxing restrictions can be found. The comparison between downhole wireline parameters and benchtop NMR measures can be incompatible. Engineers can also underestimate movable HCs by using routine core analysis data.

We also demonstrate the use of NMR spectroscopy to evaluate the efficiency of huff-n-puff EOR processes in tight rocks. This technique allows the quantification of HC recovery without compromising pressure conditions. Relaxation data also provides important knowledge on both fluid remobilization mechanism and mass transportation mechanism, during huff-n-puff.

This work was funded by the University of Oklahoma Unconventional Shale Consortium. We thank the members of Integrated Core Characterization Center (IC[3]) for helping us during experimentation.

References

1. R. J. S. Brown, "Proton relaxation in oils", Nature 198: 387-388, 1961
2. J. Bryan, F. Manalo, Y. Wen, "Advances in heavy oil and water property measurements using low field nuclear magnetic resonance", SPE International Thermal Operations and Heavy Oil Symposium and International Horizontal Well Technology Conference, Calgary, 4-7 November, 2002, SPE-78970-MS
3. G. J. Hirasaki, S. W. Lo, Y. Zhang, "NMR properties of petroleum reservoir fluids", Magnetic Resonance Imaging, **Vol. 21**, Issues 3-4, p. 269-277, 2003
4. W. W. Kenyon, P. I. Day, C. Straley, J. F. Willemsen, "A three-part study of NMR longitudinal relaxation properties of water-saturated sandstones", SPE Formation evaluation, 622-636, 1986, SPE 15643
5. G. R. Coates, R. C. A. Peveraro, A. Harwick, et al., "The magnetic resonance imaging log characterized by comparison with petrophysical properties and laboratory core data", SPE ATCE, Dallas, 6-9 October, 1991, SPE-22723-MS.
6. B. R. Kenyon, B. R., C. Straley, C., G. Gubelin, G., and C, Morris, "Nuclear magnetic resonance imaging-technology for the 21st century", Oil Field Review, 19-33, 1995
7. W.E. Kenyon, H. H. Takezaki, C. C. Straley, P. N. Sen, P. N., M. Herron, A. Matteson, M. J. Petricola, "A laboratory study of nuclear magnetic resonance relaxation and its relation to depositional texture and petrophysical properties", Carbonate Thamama group, Mubarraz field, Abu Dhai, 1995, SPE 29886.
8. C. D. Straley, C. D. Rossini, H. Vinegar, P. Tutunjian, C. Morriss, "Core analysis by low field NMR" , 1994, SCA-9404
9. E. Odusina, C. H. Sondergeld, C. S. Rai, "An NMR study on shale wettability", Canadian Unconventional Resources Conferences, Calgary, 15-17 November, 2012, SPE-147371-MS.
10. A. Valori, B. Nicot, "A review of 60 years of NMR wettability", Society of Petrophysicists and Well-Log Analysts, 2019
11. A. Sinha, S. Dang, C. Sondergeld, C. Rai, "Impact of solvent extraction on surface area measurements in organic-rich shales using nitrogen adsorption", Unconventional Resources Technology Conference, 2017, URTEC-2017-2668849
12. A. Besov, A. Tinni, C. Sondergeld, C. Rai, W. Paul, D. Ebnother, T. Smagala, "Application of laboratory and field NMR to characterize the Tuscaloosa marine shale", Society of Petrophysicists and Well-Log Analysts, 2017
13. A. Blount, A. McMullen, M. Durand, T. Croft, B. Driskill, "Maintaining and reconstructing in-situ saturations: a comparison between whole core, sidewall core, and pressurized sidewall core in the Permian basin", Society of Petrophysicists and Well-Log Analysts, 2018
14. S. T. Dang, C. H. Sondergeld, C. S. Rai, "Interpretation of nuclear-magnetic-resonance response to hydrocarbons: application to miscible enhanced-oil-recovery experiments in shales", Society of Petroleum Engineers, 2019
15. A. Chakravarty, A. Tinni, C. S. Rai, C. H. Sondergeld, "NMR considerations in shales at elevated temperature", Unconventional Resources Technology Conference, 2018, URTEC-2018-2902883